P9-DCS-153

Marshall Brain's

How STUFF Works

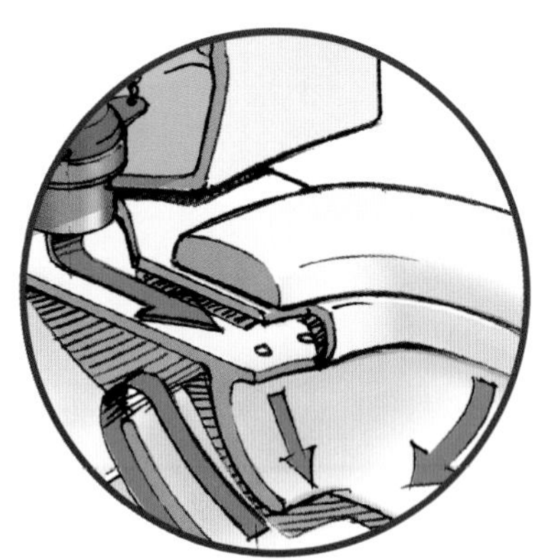
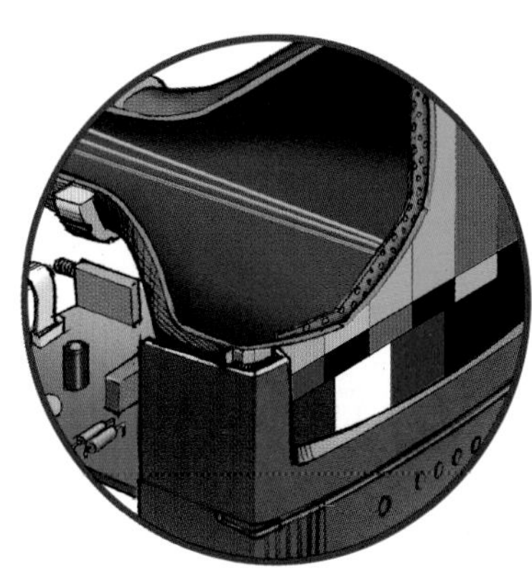
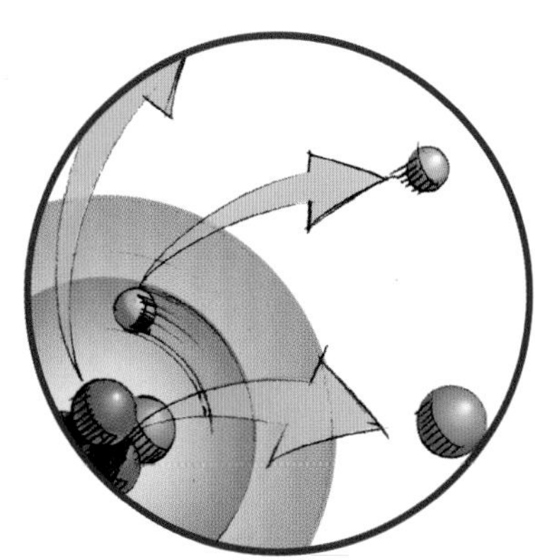
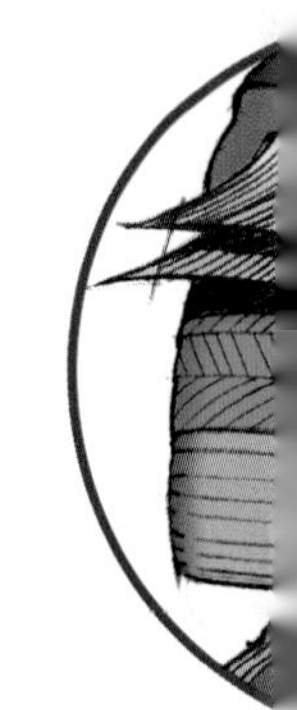

Marshall Brain's

How STUFF Works

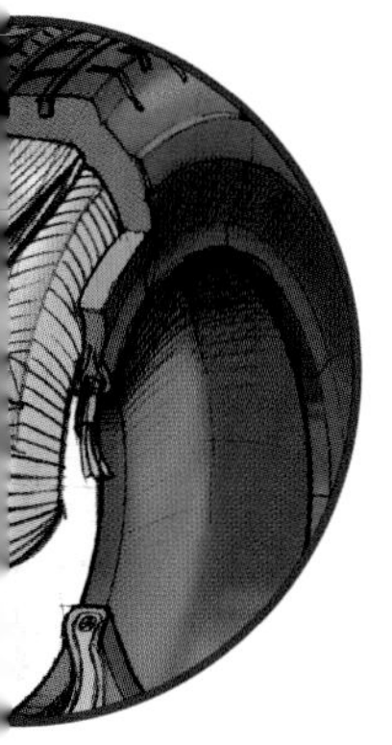

Marshall Brain

and the Staff at HowStuffWorks.com

Hungry Minds™

New York, NY • Cleveland, OH • Indianapolis, IN

Hungry Minds™

Published by
Hungry Minds, Inc.
909 Third Avenue
New York, NY 10022
www.hungryminds.com

For general information on Hungry Minds' products and services please contact our Customer Care Department within the U.S. at 800-762-2974, outside the U.S. at 317-572-3993, or fax 317-572-4002.

For sales inquiries and reseller information, including discounts, premium and bulk quantity sales, and foreign-language translations, please contact our Customer Care Department at 800-434-3422, fax 317-572-4002, or write to
Hungry Minds, Inc.,
Attn: Customer Care Department,
10475 Crosspoint Boulevard, Indianapolis, IN 46256.
Library of Congress Cataloging-in-Publication Data available from publisher
ISBN: 0-7645-6518-4

How Stuff Works, Inc.
EDITOR-IN-CHIEF: Marshall Brain
ART DIRECTOR: Rick Barnes
EDITORIAL DIRECTOR: Katherine Fordham Neer
CONTRIBUTING WRITERS: Marshall Brain, Kevin Bonsor, Craig Freudenrich, PH.D, Tom Harris, Katherine Fordham Neer, Karim Nice, Jeff Tyson, Brian Adkins, Gerald Gurevich, and Lucas Hoffman, MD.,
COPY EDITORS: Julia Layton, Melissa Russell-Ausley
RESEARCHER: Gary Brown

Hungry Minds, Inc.
VICE PRESIDENT AND PUBLISHER: Kathy Nebenhaus
EDITORIAL DIRECTOR: Cindy Kitchel
CREATIVE DIRECTOR: Michele Laseau
COPY EDITOR: Kitty Jarrett
SPECIAL EDITORIAL ASSISTANCE: S. Kristi Hart
MANUFACTURING BUYER: Kevin Watt

Design and Layout
ILLUSTRATIONS: Charles Floyd, Studio Alchemy
PHOTOS: Rick Barnes and Roxanne Reid
U.S Navy photo of attack submarine courtesy of Electric Boat Corporation by Jim Brennan.
Helicopter photo courtesy of U.S. Navy; photo taken by Staff Sgt. Eric C.Tausch.
LAYOUT DESIGN: Sheilah Barrett, Sheilah Barrett Design
LAYOUT: Cynthia Anderson, Studio Alchemy
COVER AND INTERIOR DESIGN: Michele Laseau and Rick Barnes

For additional engaging content visit **www.howstuffworks.com**

Manufactured in the United States of America.
10-9-8-7-6-5-4-3-2-1

Preface

Hello!

Imagine standing in a room full of strangers, and compare that to standing in a room full of close friends. It's a completely different feeling. Learning about all the different technologies in our world changes your relationship with them, too—the devices and systems become transparent, accessible, and friendly. You look at them in a whole new way. You see how they all interrelate with each other and with you, and it is fascinating. It is empowering to master their secrets.

Any man-made object that you see today represents some sort of technology. That's pretty obvious when you hold a cell phone in your hand. But it is also true when you hold a loaf of bread. Bread is actually a biochemical technology—a very old one—that makes wheat taste better. And it is true of a guitar, which is a technology for creating music from the vibrations of strings.

Today we are all surrounded by, and dependent on, an incredible collection of technologies that affect every part of our lives. Just take a tour of your home—there is the refrigerator, the microwave oven, the telephone, the toaster, the stereo system, the television, the VCR, the computer, the air conditioner, the cell phone, the car in the garage—the list is nearly endless and includes thousands of objects. It also includes big things like electrical power and cable TV, things we take completely for granted until something like an earthquake or a hurricane knocks them out. And it includes little mundane items like paper clips and masking tape.

Technology is the story of human beings making our lives better. It is the story of literally millions of people cooperating together to build things that improve the world we live in. It is also a story of astounding creativity and invention. We tend to think of books and sculpture and painting as creative, but once you understand the incredible ingenuity that goes into something like the Global Positioning System or a cell phone, you realize that they are works of art that we happen to carry with us and use constantly. The fact that they can be made so reliable and so inexpensive that millions of people can carry them in their pockets is a startling testimony to the power of the human mind and spirit.

In this book, I would like to take you behind the scenes and show you how the modern world works. It is an amazing story, and one that will change the way you look at everything around you.

Sincerely,

Contents

We pick up and read books like this one all the time. They are made up of words and pictures printed on paper. The final product looks fairly simple, but have you ever wondered how it all gets put together? It turns out that creating a book like this involves an unbelievable amount of work performed by dozens of people!

The first step is coming up with the words and the art. This book contains more than 135 different articles written by nine different people. Once written, the articles have to be edited so that they all have the same voice and reading level, and then they have to be trimmed or expanded to fit into the space available. The "fitting" part is especially interesting—it's never a problem when you publish on the Web, but it is extremely important when printing on paper. Craig, Jeff, Karim, Kevin, Tom, and I did most of the writing. Katherine and I did the editing and the fitting.

Each article in this book also has art, in the form of 3-D drawings, diagrams, and photos, as well as attachments like sidebars and lists. Charles did an amazing job on the illustrations and Rick did the photographs. Each illustration is hand drawn, scanned, colored, and proofed. For photos we purchased and disassembled objects, photographing each one from many different angles. Katherine herded the sidebars together.

With words, illustrations, photos, and sidebars assembled, the book can start to come together. To do this, three different things need to happen:

- The cover
- The layout
- Copyediting and proofing

The cover of any book normally takes on a life of its own, and this cover was no different. A wide variety of people (perhaps 50) had opinions on it and gave input and Michele pulled all of those ideas together to create the final cover that you see today. I bet that if you were to add it all up, something like 500 or 1,000 hours of time goes into a cover like this. The cover is normally the first thing that people start working on and the last thing that is finished.

In layout, you create a "template" that will act as a standard format for holding all the words and pictures. Then you lay the pictures, sidebars, titles, page numbers, and everything else into that template. During the layout process, you wrap all of the words around everything else on the page. Most pages in this book have pieces of art and other features in different positions, and the words flow around them. Layout is a fascinating, detailed process. If you have ever seen a dry-stacked stone wall in New England, where every stone fits together perfectly to make a wall that needs no mortar or cement, then you understand something about the layout process. Sheilah and Michele created the template, and then Cynthia fit all of the pieces into it.

Copyediting and proofing is a two- or three-phase process where you go over the text cleaning up mistakes in grammar, spelling, and style. Cindy, Kitty, and Katherine did all of this work. It takes a certain stamina to go over hundreds of pages of text with a fine-tooth comb like that.

Now you have something that is very close to a book. By adding the table of contents, the index, and the front and back matter (all the "stuff" that you find at the beginning and end of the book), you get to the point where the book is finished and it can go to press. A press run is its own remarkable process that could fill an entire book itself!

Meanwhile, there has been another process going on that gets the book into the bookstore. This process is led by Kathy and implemented by Michele, Cindy, and the sales force. Michele and Cindy created a "blad" (basic layout and design)—a 10-page mini-book that acts as a sample. It contains three short articles, a sample cover, and different pieces describing the book. The sales force uses the blad to talk with buyers at bookstores, discount stores, warehouses, and so on to generate interest in the book and orders. As these orders are fulfilled, you are able to find the book on a shelf in a store, and from there you buy it and take it home.

What's incredible is that all of these different threads are running together. The words, art, sidebars, proofing, cover, layout, and sales processes are all taking place at the same time, with everyone coordinating and working together to get everything done on time. At any given point, perhaps two dozen people are all doing something on the book.

If you have ever seen a movie being produced, you will recognize the same sort of process. In a big movie, hundreds of people are working together to get a project done in a short amount of time. Large engineering projects are the same way. For example, in a huge project like the Apollo mission to the moon, thousands of people work together under very tight deadlines. You see big and small projects like this all around you every day. Everything from the newspaper to the water coming out of the tap to bread on the shelf in the grocery store involves groups of people working together.

And that is one of the most amazing things about being human—the ability that people have to work together with a common goal. It allows us to accomplish extraordinary things in a short amount of time, and it is the underpinning of our society as a whole.

How We Come up with Ideas

We publish two or three articles every day on our Web site, and that means that today we have thousands of topics. Picking the new topics that we cover each day is something we think about all the time. We have a conference room in the office filled with Post-It notes that chronicle all the articles scheduled for the next 4 or 5 weeks, all of the ideas people have come up with, and all of the partial articles that we started but stalled out on. There are ideas everywhere!

We get our ideas from lots of different places. People send us hundreds of emails every day. We have a big suggestion area on the site as well. We look at logs of the site's search engine to see what's popular. We also watch the news and try to match topics with current news events. In addition, the authors develop pet projects based on their personal interests. Ideas are pitched to the group, who then gives each idea a thumbs-up or a thumbs-down.

We take a team approach to article creation. We write, illustrate, edit, format, and publish the article as a team. The average article takes about 3 weeks from start to finish, although we've been known to produce articles in less than 24 hours when we need to.

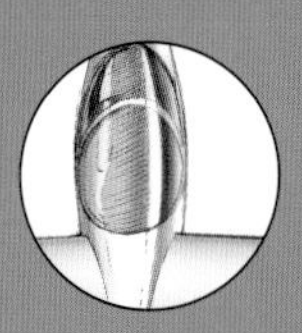

chapter one

IN THE AIR

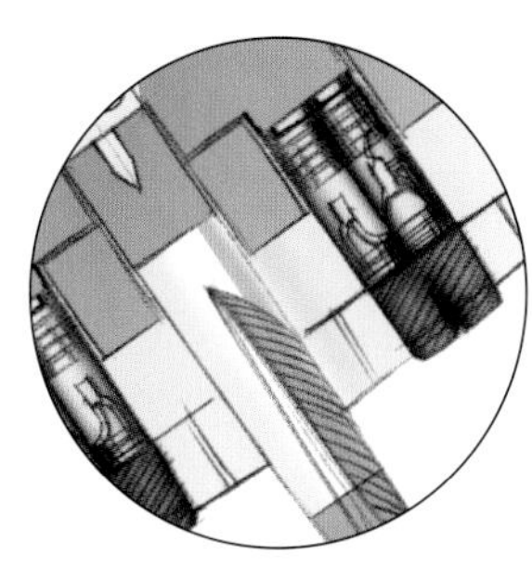

How **AIRPLANES** Work

If you have ever flown on a 747, you know that it is a gigantic, extraordinary machine. A 747 can carry 500 or 600 people. It weighs up to 870,000 pounds (395,000 kg) at takeoff. Yet it rolls down the runway and, as though by magic, lifts itself into the air and can fly up to 7,000 nautical miles without stopping. It is absolutely incredible when you think about it! That ability to rise into the air is the magic behind an airplane.

HSW Web Links

www.howstuffworks.com

How Blimps Work
How CargoLifter's Airship Will Work
How the Concorde Works
How Gliders Work
How Hot Air Balloons Work
How Space Shuttles Work

Before you can understand how airplanes work, it is important to look at the four basic aerodynamic forces: drag, thrust, weight and lift. When any object moves through a fluid like air or water, it has to push some of the fluid out of the way. The energy it takes to clear a path through the fluid creates drag. If you stick your hand out of the window of your car while driving down the road, you can feel the drag that your hand creates.

You see a great example of drag reduction when you watch downhill skiers in the Olympics. You'll notice that, whenever they get the chance, they squeeze down into a tight crouch. By making themselves "smaller," skiers decrease the drag they create, and that allows them to move faster down the hill.

A passenger jet always retracts its landing gear into the body of the airplane after takeoff to reduce drag. Just like the downhill skier, the pilot wants to make the aircraft as small as possible as it moves through the air. The amount of drag produced by the landing gear of a jet is huge.

Thrust is the force that overcomes drag and keeps the airplane moving through the air. Airplanes create thrust using propellers, jet engines, or rockets. A propeller acts like a very powerful version of a household fan, pulling air past the blades.

Gliders are airplanes that have no way to produce thrust. Instead, a glider converts the energy of its altitude into forward motion.

Weight is one is the easiest aerodynamic forces to explain because we all experience weight every time we pick something up. Every object on earth has weight (including air). It's amazing that a typical 747 can weigh up to 870,000 pounds (that's 435 tons!) and still manage to get off the runway.

Lift is the aerodynamic force that overcomes weight and holds an airplane in the air. On airplanes, the wings create most of the lift required to keep the plane aloft.

In order for an airplane to fly straight and level, the following relationships must be true:

Thrust = Drag
Lift = Weight

If, for any reason, the amount of drag becomes larger than the amount of thrust, the plane will slow down. If the thrust is increased so that it is greater than the drag, the plane will speed up. Similarly, if the amount of lift drops below the weight of the airplane, the plane will descend. By increasing the lift, the pilot can make the airplane climb.

Wings and Lift

The wing is the most important part of an airplane: It's what gets the airplane in the air, and it is the most amazing part of the airplane. It took inventors and engineers hundreds of years to understand how to create such a surface to generate lift.

Two parts of the wing are important to creating lift: the shape of the wing's cross-section and the wing's angle of attack. The wing's cross-section determines the wing's performance. Some wings perform well at low speeds, and some perform better at high speeds. Some wings perform well at supersonic speeds, and others do not. The shape of the wing is very important because it controls how air molecules flow over the wing as it slices through the air.

Most people have seen the familiar airfoil shape. It is important to understand that the airfoil shape is not required to create lift. Stunt planes, which fly upside-down, often have a symmetric cross-section rather than an airfoil. And even airplanes with a normal airfoil shape can fly upside–down. The key is the angle of attack.

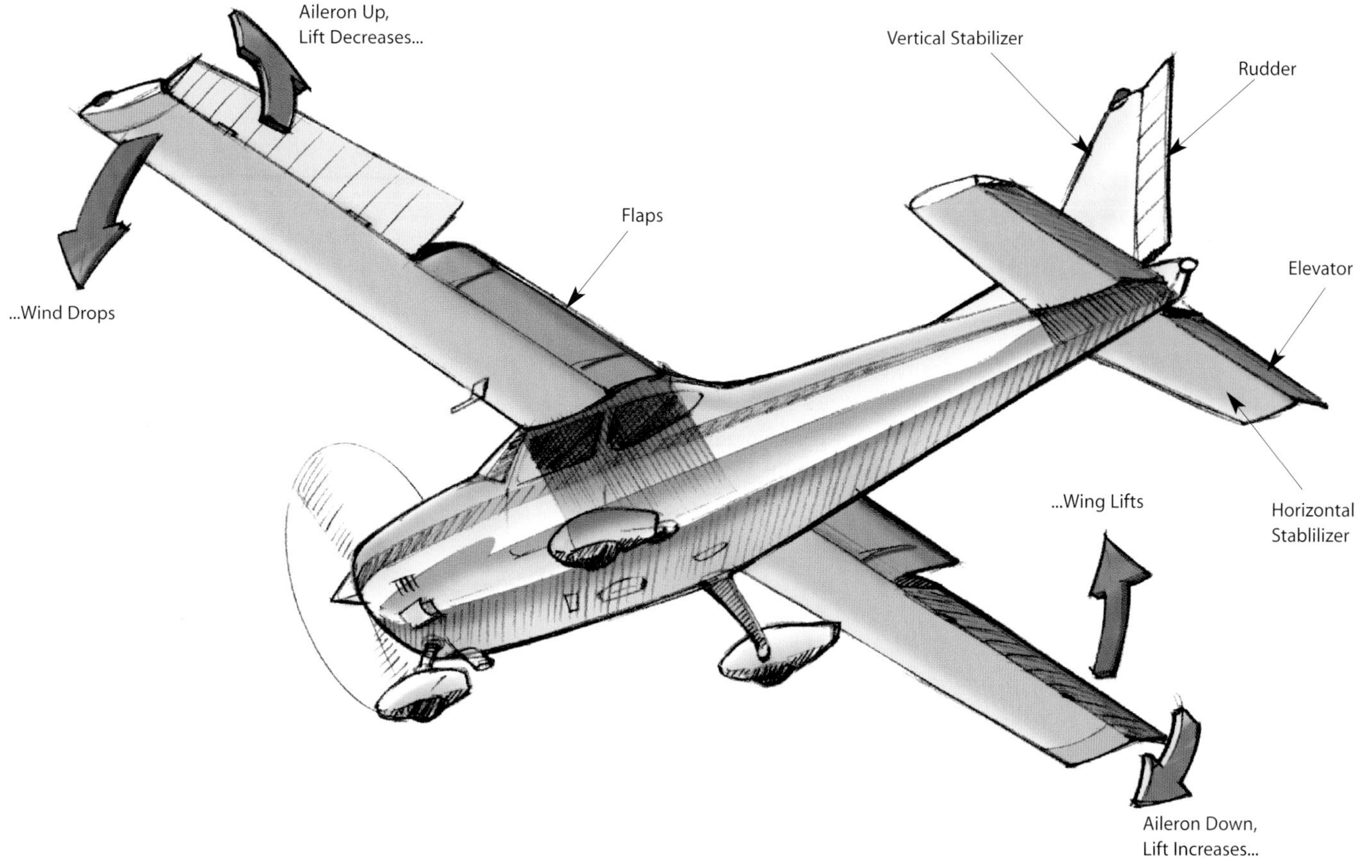

If you have gone down the road in a car with your hand out the window, you know how angle of attack works. As you change the angle of your hand in the wind, you can create an upforce or a downforce. You can imagine that if you attached a large piece of cardboard to your hand and angled it, you could potentially generate enough force to take your arm off. Wings use this same effect when they create lift.

The advantage the airfoil shape (or a symmetric wing shape) has over a piece of cardboard is that it lets the wing slice through the air, generating lift without creating a lot of drag.

The shape of the wing is designed to deflect air downward. In deflecting the air downward, areas of low and high pressure form around the wing. Both the top and bottom surface of the wing work to direct the air downward. When the top of the wing pulls air downward, it creates areas of lower pressure above the wing, and when the bottom of the wing pushes air downward, it creates areas of higher pressure below the wing. The net result is lift on the wing.

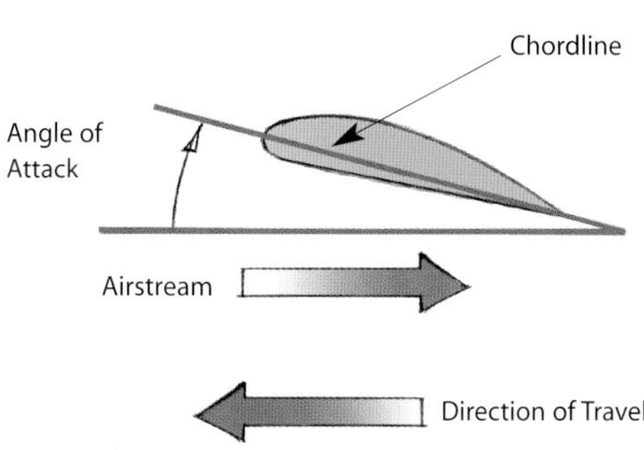

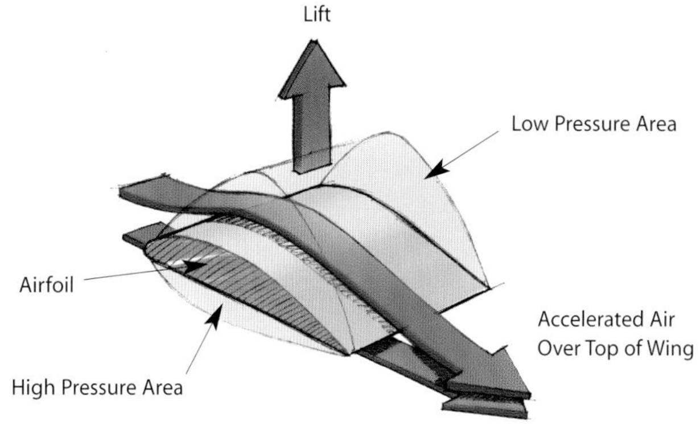

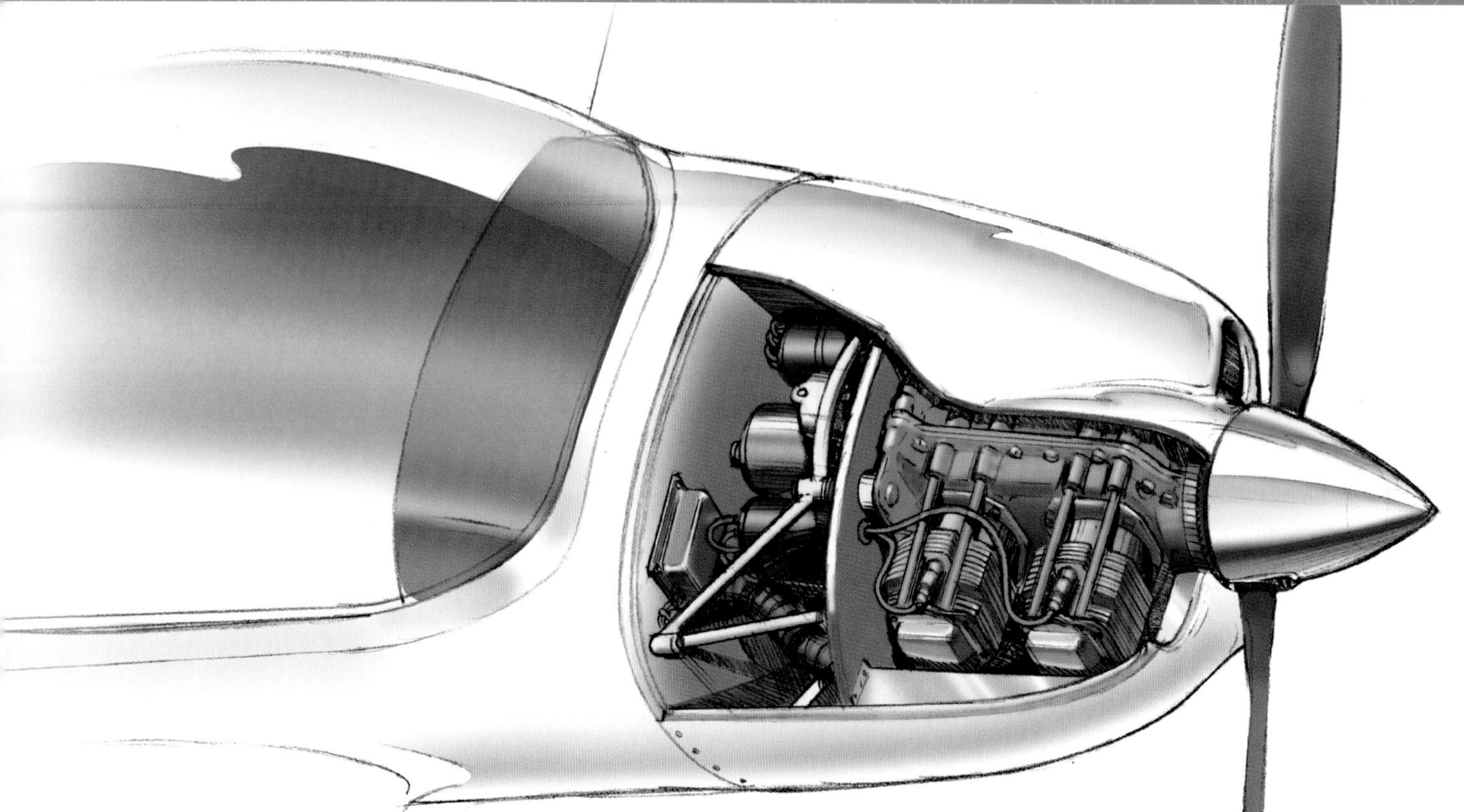

In practice, calculating the pressure distribution around a wing and the exact amount of lift it creates is very difficult. Sophisticated fluid dynamic computer models have made calculating lift easier, but many lift calculations are still based on experimental data that engineers have gathered by conducting extensive wind tunnel tests on thousands of airfoil shapes. The data allows engineers to predictably calculate the amount of lift and drag that airfoils can develop in various flight situations.

The lift coefficient of an airfoil is a number that relates its lift-producing capability to air speed, air density, wing area, and angle of attack. The lift coefficient of a given airfoil depends on the angle of attack.

Here is the standard equation for calculating lift by using a lift coefficient:

$$L = C_l \times \frac{1}{2} \times p \times V^2 \times A$$

The variables in this equation have the following meanings:

- L = Lift
- C_l = Coefficient of lift (which is obtained from NACA charts)
- p = Air density
- V = Velocity
- A = Wing area

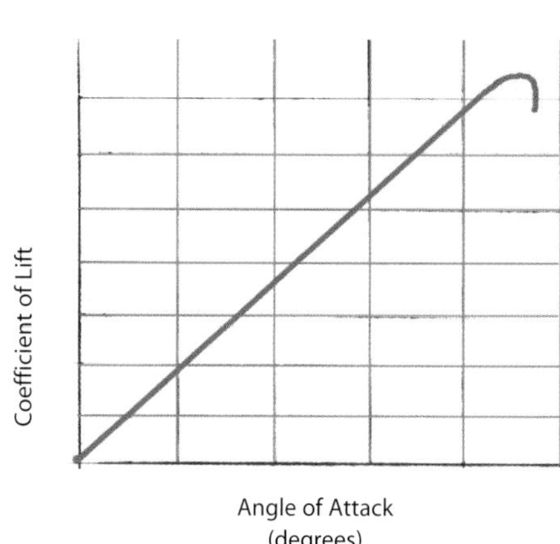

When you want to calculate the lift for a wing, you look up a chart for the airfoil shape you are interested in. The chart should show a graph of the lift coefficient versus angle of attack. After you find the angle of attack of the wing, you can use the chart to determine the lift coefficient.

When you know the lift coefficient, you can plug in the air density, wing area, and velocity to calculate the lift. One thing you'll notice about the lift calculation equation is that the lift is highly dependent on velocity. A wing going twice as fast as another can generate four times as much lift.

Flaps

In general, the wings on most planes provide an appropriate amount of lift (along with minimal drag) while the plane is operating in cruising mode. For example, in cruising mode a Boeing 747-400 travels approximately 560 mph (901 kph). However, when it is taking off or landing, its speed is less than 200 mph (322 kph). This dramatic change in the wing's working conditions means that different airfoil shapes work better during different parts of a flight.

Airplane wings therefore have movable sections called flaps. During takeoff and landing, the flaps extend rearward and downward from the trailing edge of the wings. The flaps alter the shape of the wing, allowing the wing to turn more air and thus create more lift. When the flaps are extended, the drag on the wings increases, so the flaps are put away during the rest of the flight.

The Propeller and the Engine

Probably the most important parts of an airplane, after the wing, are the propeller and the engine. The propeller (or, on jet aircraft, the jets) provides the thrust that moves the plane forward.

A propeller is really just a special, spinning wing. If you look at a cross-section of a propeller, you will find that a propeller has an airfoil shape and an angle of attack that changes along the length of the propeller. The angle is greater toward the center of the propeller because the speed of the propeller is slower close to the hub. Many large propeller aircraft have more elaborate three-blade or four-blade propellers with adjustable pitch mechanisms. These mechanisms allow the pilot to adjust the propeller's angle of attack, depending on air speed and altitude.

Controlling Airplane Direction

The tail of the airplane has two small wings, called the horizontal and vertical stabilizers, which the pilot uses to control the direction of the plane. Both are symmetrical airfoils, and both have large flaps on them that the pilot controls with a control stick to change their lift characteristics.

With the horizontal tail wing, the pilot can change the plane's angle of attack and therefore control whether the plane goes up or down. With the vertical tail wing, the pilot can turn the plane left or right.

The pilot controls flaps by using cables in small planes and by using hydraulic or electric motors in large planes.

Airplanes are one of the most amazing things that you see on a daily basis. It's truly fascinating how so many mechanical parts combine with the laws of aerodynamics to form a machine that is able to fly.

747-400 Facts

Length:	232 feet, or 71 m
Height:	63 feet, or 19 m
Wingspan:	211 feet, or 63 m
Wing area:	5,650 square feet, or 525 m^2
Empty weight:	538,000 pounds, or 244,033 kg
Max. takeoff weight:	870,000 pounds, or 394,625 kg
Max. landing weight:	630,000 pounds, or 285,763 kg (explains why planes may need to dump fuel for emergency landings)
Engines:	Four turbofan engines, 57,000 pounds of thrust each
People capacity:	Up to 660 people and their luggage
Fuel capacity:	Up to 57,000 gallons, or 216 kl
Max. range:	7,200 nautical miles
Cruising speed:	490 knots
Takeoff distance:	10,500 feet, or 3,200 m

How **HELICOPTERS** Work

Helicopters are the most versatile flying machines in existence today. A helicopter gives the pilot access to 3-D space in a way that no airplane can. If you have ever flown in a helicopter, you know that its abilities are exhilarating! Its incredible flexibility means that a helicopter can fly almost anywhere.

HSW Web Links

www.howstuffworks.com

How Airplanes Work
How Blimps Work
How Hot Air Balloons Work
How Hoverboards Will Work
How Rocket Engines Work

Helicopters have a number of unique abilities that airplanes do not have. The signature of a helicopter is its ability to hover over a point on the ground. While hovering, a helicopter can also spin 360 degrees so that the pilot can look in any direction.

Another unique feature of a helicopter is its ability to fly backward and sideways. A helicopter can fly sideways down a road, for example, while a photographer films a scene through the side window. A helicopter that is flying in any direction can also stop quickly in midair and begin hovering. All these maneuvers are impossible in an airplane because an airplane must fly forward at all times in order to develop lift from its wings.

The Parts of the Helicopter

Imagine that you would like to create a machine that can simply fly straight up. Don't even worry about getting back down for the moment—up is all that matters. If you are going to provide the upward force with a wing, then the wing has to be moving in order to create lift. (See "How Airplanes Work," page 2, for details.)

Using rotary motion is the easiest way to keep a wing in continuous motion. So you can mount two or more wings on a central shaft and spin the shaft, much like the blades on a ceiling fan. The rotating wings of a helicopter are shaped just like the airfoils of an airplane wing, but generally the wings on a helicopter's rotor are narrow and thin because they are spinning so fast. The helicopter's rotating wing assembly is normally called the main rotor. If you give the main rotor wings a slight angle of attack on the shaft and spin the shaft, the wings will start to develop lift.

In order to spin the shaft with enough force to lift a human being and the vehicle, you need an engine of some sort. Reciprocating gasoline engines and gas turbine engines are the most common types of engines in helicopters. The engine's driveshaft connects through a transmission to the main rotor shaft.

This arrangement works great until the moment the vehicle leaves the ground. At that instant there is nothing to keep the body of the helicopter from spinning just like the main rotor does. So, in the absence of anything to stop it, the body will spin in an opposite direction to the main rotor. To keep the body from spinning, you need to apply a force to it.

The normal way to provide a force to the body of the helicopter is to attach another set of small rotating wings to a long boom. These wings are known as the tail rotor. The tail rotor produces thrust just like an airplane's propeller does. By producing thrust in a sideways direction, counteracting the engine's desire to spin the body, the tail rotor keeps the body of the helicopter from spinning. Normally, the tail rotor is driven by a long drive shaft that runs from the main rotor's transmission back through the tail boom to a small transmission at the tail rotor.

Going Up and Down

In order to be able to control a helicopter, both the main rotor and the tail rotor need to be adjustable. The pilot wants to be able to control the amount of thrust from the main rotor so the helicopter can move up and down. The pilot also wants to be able to control the amount of thrust from the tail

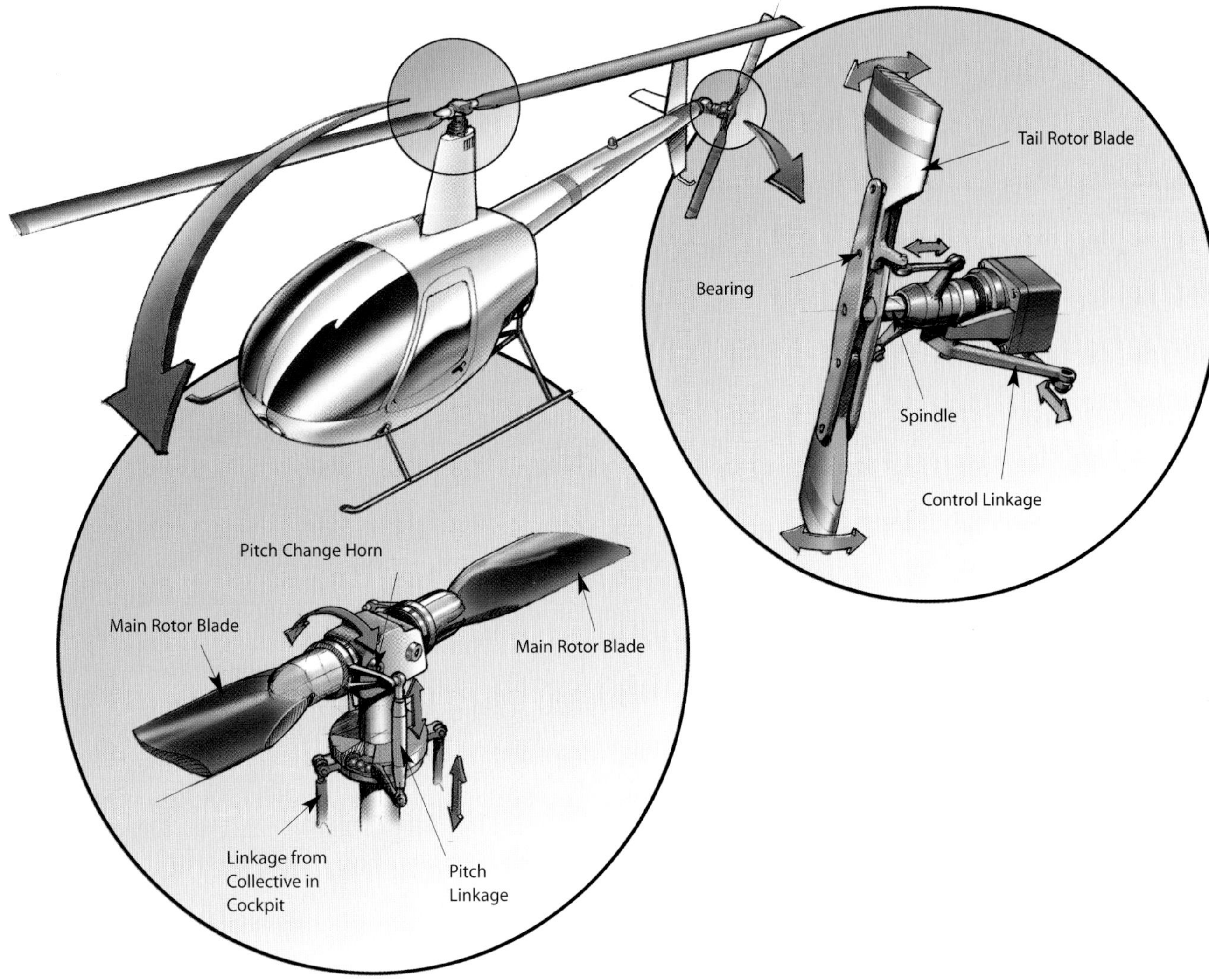

rotor so that the helicopter can rotate on its axis.

You want the ability to change the angle of attack on the tail rotor wings so that you can control the amount of sideways thrust it produces. A helicopter pilot therefore has two foot pedals that directly control the angle of attack of the tail rotor blades.

You can use the same arrangement to control the angle of attack of the wings in the main rotor. This control is called the collective, and it is a lever that the pilot adjusts with one hand. The collective also has a twist-grip on it that controls the speed of the engine. Using the collective control and the throttle, the pilot can make the helicopter rise or descend, or the pilot can perfectly balance the main rotor's thrust so that the helicopter hovers in midair.

Going Forward, Backward, and Sideways

So far you have designed a helicopter that can go up and down, and this is relatively easy. Now you need to add the ability to go forward, backward, and sideways. This is the part of a helicopter that is extraordinary.

To make a helicopter fly in a certain direction, you need to adjust the wings on the main rotor so that they produce more thrust on one side of the rotor than on the other. Because the wings are rotating, you have to adjust the angle of each wing constantly, depending on which side of the rotor it is on.

The main rotor hub where the rotor's drive shaft and blades connect has to be extremely strong as well as highly adjustable. The swash plate assembly is the component that makes this possible.

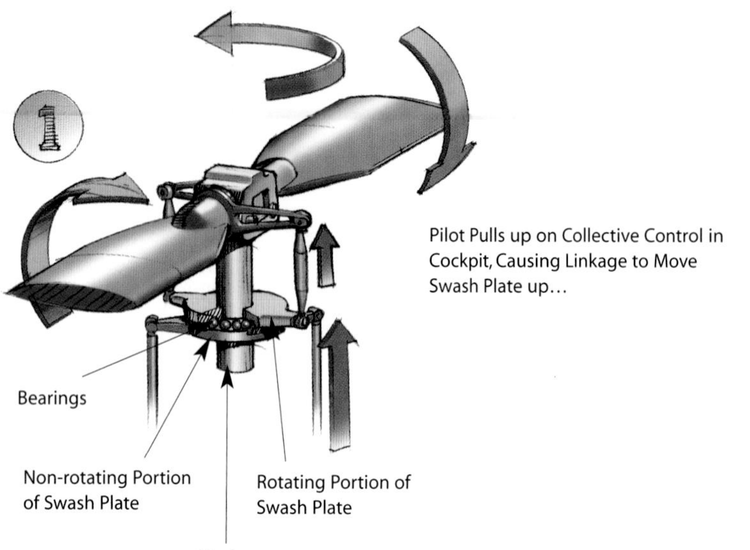

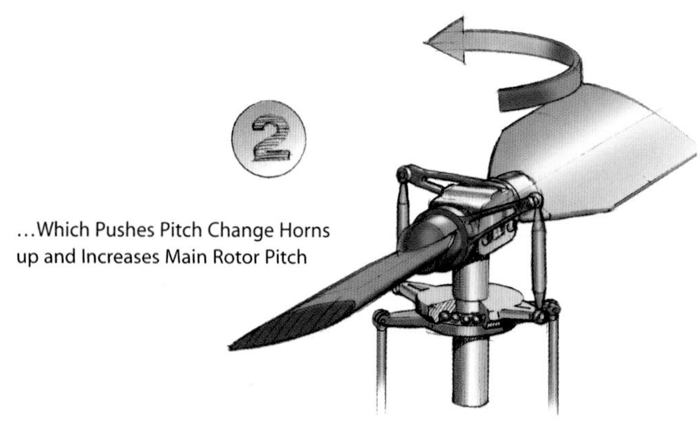

The swash plate assembly changes the angle of attack of the main rotor's wings as the wings revolve. A steep angle of attack provides more lift than a shallow angle of attack. The cyclic control tilts the swash plate assembly so that the angle of attack on one side of the helicopter is greater than it is on the other. On one side of the helicopter the angle of attack, and therefore the lift, is greater.

The pilot uses two controls to adjust the swash plate:

- **Collective control**—The collective control raises the entire swash plate assembly as a unit. This has the effect of changing the pitch of both blades simultaneously and controls the total thrust of the main rotor.
- **Cyclic control**—The cyclic control pushes one side of the swash plate assembly up or down. This has the effect of changing the pitch of the blades unevenly, depending on where they are in the rotation. The result is that the rotor's wings have a greater angle of attack (and therefore more lift) on one side of the helicopter and a lesser angle of attack (and less lift) on the opposite side. The unbalanced lift causes the helicopter to tip and move forward, backward, and sideways.

The swash plate assembly consists of two plates, called the fixed and the rotating swash plates separated by bearings:

- The rotating swash plate rotates with the drive shaft and the rotor's blades. The rotating swash plate connects to the rotor's wings with control rods that allow it to change the angle of attack of the wings. As you angle the rotating swash plate in different directions, it changes the direction of the main rotor's thrust.
- The fixed and rotating swash plates are connected with a set of bearings between the two plates. These bearings allow the rotating swash plate to spin on top of the fixed swash plate. The pilot controls the angle of the fixed swash plate, and therefore the rotating swash plate, with a control called the cyclic.

Hovering in a helicopter requires experience and skill. The pilot adjusts the cyclic to maintain the helicopter's position over a point on the ground. The pilot adjusts the collective to maintain a fixed altitude; this is especially important when the helicopter is close to the ground. The pilot adjusts the foot pedals to maintain the direction in which the helicopter is pointing. As you can imagine, windy conditions can make hovering a real challenge.

Like a ballerina at a ballet or an organist at a pipe organ, it truly takes both feet and both hands in a balancing act to fly a helicopter!

How **BLIMPS** Work

You have probably seen blimps providing TV coverage at sporting events. Blimps are a common sight at football games, golf tournaments, and auto races. Basically, blimps (a.k.a., airships) are gigantic helium-filled balloons that can move through the air under their own power. Unlike airplanes and helicopters, blimps can stay aloft for days!

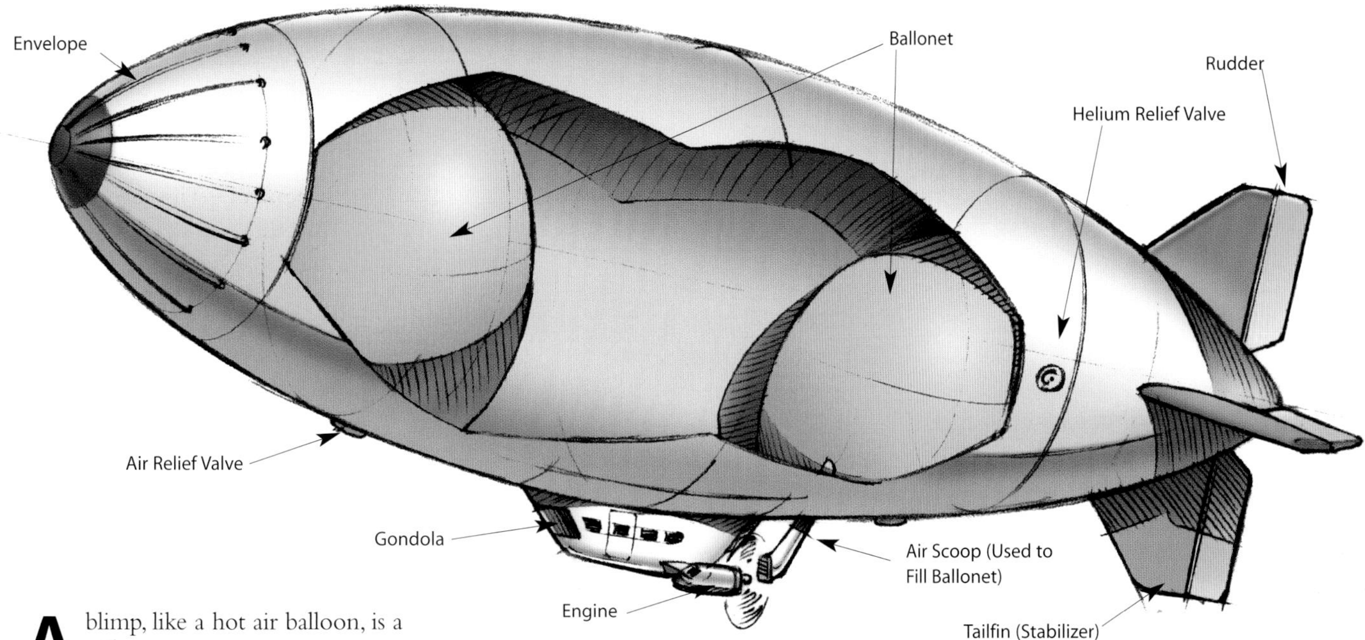

A blimp, like a hot air balloon, is a lighter-than-air (LTA) flying machine. (See "How Hot Air Balloons Work," page 14.) Unlike a hot air balloon, a blimp has shape and structure. A pilot uses engines and rudders to fly and maneuver a blimp a lot like it's an immense, slow-moving helicopter.

The Envelope

The envelope is by far the largest part of any blimp. It is an immense bag that holds the helium gas, and it can hold from 70,000 to 20,000,000 cubic feet (1,900 to 550,000 m^3) of helium, depending on the size of the blimp and its load capacity. One cubic foot of helium can lift 0.064 pounds (about 1 kg/m^3), so a small blimp that holds 100,000 cubic feet of helium can lift about 6,400 pounds, including the weight of the blimp itself. A typical blimp that you see at a football game holds about 200,000 cubic feet of helium, is about 200 feet long, and weighs about 13,000 pounds.

The Ballonets

One important part of a blimp that you can't see is the ballonets. Inside the envelope are one or two smaller balloons, filled with air rather than helium. Because air is heavier than helium, the ballonets are deflated or inflated with air to make the blimp lighter or heavier. They also keep the balloon level in flight—a process called "trimming."

The most astonishing thing about blimps is that a gigantic object can hang motionless in midair. It can do that because of a very small difference in the weight of helium and nitrogen atoms.

Did You Know?

Blimp pilots undergo a comprehensive training program and are Federal Aviation Administration certified for lighter-than-air craft.

A typical blimp at a sporting event has a ground crew that follows the blimp wherever it goes, with the following support vehicles:

- **Bus**—The Administrative office
- **Tractor-trailer**—The electrical/mechanical shop
- **Van**—The command car and utility vehicle

Airships

There are three main types of airships:

Rigid—Usually long (more than 360 ft, or 120 m) and cigar shaped, with an internal metal frame and gas-filled bags. Example: Hindenburg

Semi-rigid—Pressurized gas balloon (envelope) attached to a lower metal keel. Examples: Norge, Italia

Nonrigid—Large gas-filled envelopes. Examples: Goodyear, MetLife, Fuji

All airships have solid gondolas, engine-powered propellers, and solid tailfins.

How **CRUISE MISSILES** Work

A cruise missile is basically a small, pilotless airplane. A cruise missile has an 8.5-foot (2.61 m) wingspan, is powered by a small turbofan engine, and can fly 500 to 1,000 miles (800 to 1,600 km), depending on the configuration. A cruise missile's job is to deliver a 1,000-pound (450-kg) high-explosive bomb to a precise location—the target. The missile is destroyed when the bomb explodes. Cruise missiles cost between $500,000 and $1,000,000 each, so they're a pretty expensive way to deliver a 1,000-pound package!

HSW Web Links

www.howstuffworks.com

How Airplanes Work
How Gas Turbine Engines Work
How Helicopters Work
How Helium Balloons Work
How Rocket Engines Work

There are a number of variations in cruise missile design, and there are a variety of ways cruise missiles can be launched: from submarines, destroyers, or aircraft. When you hear about hundreds of cruise missiles being fired at targets such as Iraq, they are almost always Tomahawk cruise missiles launched from destroyers.

Dimensions and Specifications

A cruise missile is 20 feet (6.25 m) long and 21 inches (0.52 m) in diameter. At launch, it includes a 550-pound (250-kg) solid rocket booster and weighs 3,200 pounds (1,450 kg); it also has a fuel load of 800 to 1,000 pounds (363 to 450 kg), or approximately 150 gallons (600 l). The booster falls away after it has burned its fuel. Then the wings, tail fins, and air inlet unfold, and the turbofan engine takes over. This engine weighs just 145 pounds (65 kg) and produces 600 pounds (273 kg) of thrust. A cruise missile burns RJ4 fuel and has a cruising speed of 550 mph (880 kph).

Guidance

The hallmark of a cruise missile is its incredible accuracy. A common statement used when talking about cruise missiles is, "It can fly 1,000 miles and hit a target the size of a single-car garage." A cruise missile is also very effective at evading detection by the enemy because it flies very low to the ground, out of the view of most radar systems.

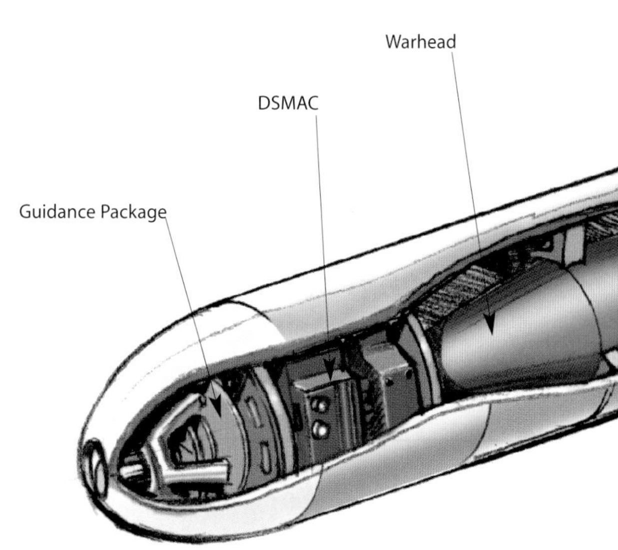

Four different systems help guide a cruise missile accurately to its target:

- **IGS**—Inertial Guidance System
- **Tercom**—TERrain COntour Matching
- **GPS**—Global Positioning System
- **DSMAC**—Digital Scene Matching Area Correlation

The IGS is a standard acceleration-based system that can roughly keep track of where the missile is, based on the accelerations it detects in the missile's motion. An IGS uses gyroscopes (to create a solid inertial frame of reference) and accelerometers (to detect acceleration in all three dimensions). Information from the accelerometers helps the IGS computer keep track of its position in space.

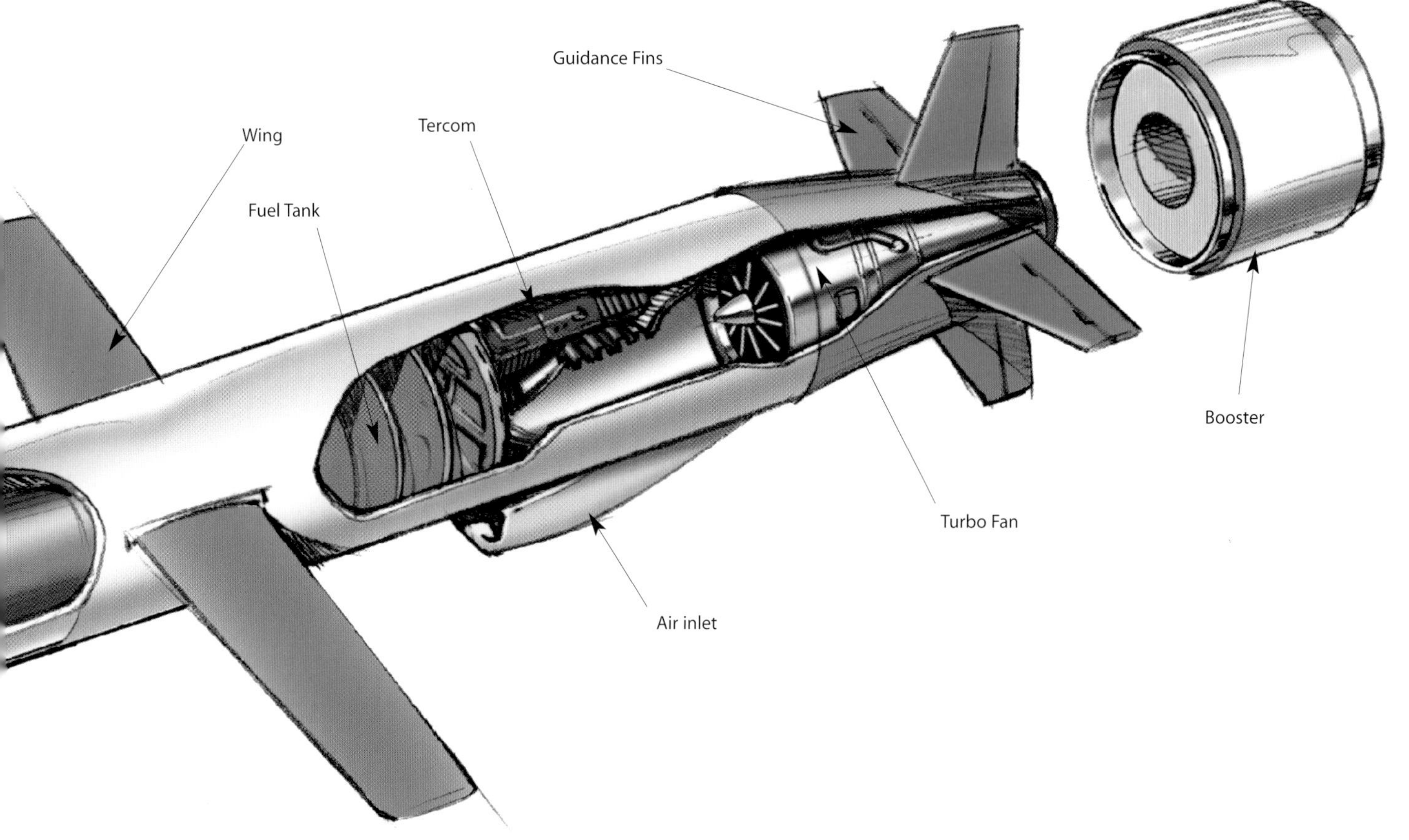

Tercom uses an on-board 3-D database of the terrain the missile will be flying over. The Tercom system "sees" the terrain it is flying over by using its radar system and matching this to the 3-D map stored in memory. The Tercom system gives a cruise missile its ability to hug the ground during flight.

The GPS uses the military's network of GPS satellites and an onboard GPS receiver to detect position with very high accuracy. Because the missile knows its exact position from the GPS and also knows the exact terrain from the Tercom system, it can fly very close to the ground.

When it is close to the target, the cruise missile switches to a "terminal guidance system" to choose the point of impact. The point of impact could be pre-programmed by its GPS coordinates or the Tercom map. The DSMAC system uses a camera and an image correlator to find the target, and it is especially useful if the target is moving. For example, a cruise missile can hone in on a moving truck or a specific building.

A cruise missile can also be equipped with thermal imaging or illumination sensors. Thermal sensors can "see" heat in, for example, a factory or an engine. Illumination sensors can detect an infrared laser. A spotter located well away from the target can aim a laser at the target, and the missile can hone in on the laser light.

A cruise missile is a lot like a small flying robot. It is able to guide itself for hundreds of miles before finally arriving at an exact location.

How GLIDERS Work

In its simplest form, a glider is an airplane without a motor. That sounds simple, but it turns out that the lack of the motor is complicated in a way. From paper airplanes to a space shuttle during reentry, there are many types of gliders. But an official glider—also known as a sailplane and designed from the ground up without an engine—is a remarkable and graceful machine. Gliding is often described as the closest that humans can get to soaring like a bird.

Three basic forces act on a glider: lift, gravity, and drag. The wings on a glider have to produce enough lift to balance the weight of the glider. The wings make more lift the faster the glider goes.

But the wings and the body of the glider also produce more drag the faster the glider goes. Because there is no engine on the glider to produce thrust, the only way to make the glider go faster is to angle the glider more sharply downward, trading altitude for speed—and generating more lift. So altitude is the source of energy that powers a glider.

The way you measure the performance of a glider is by its glide ratio. This ratio tells you how much horizontal distance a glider can travel compared to the altitude it has to drop. Modern gliders can have glide ratios better than 60:1. This means that they can glide for 60 miles (96 km) if they start at an altitude of 1 mile (1.6 km). In comparison, a commercial jetliner might have a glide ratio of around 10:1.

Drag reduction is the most important way to increase the glide ratio. Reducing drag means that the glider does not need to use up as much altitude to maintain the speed that produces enough lift to keep the glider in the air. There are three big reasons gliders have much less drag than conventional planes:

- Gliders are as small and light as possible. They have a tiny frontal area, usually just tall and wide enough for one person to squeeze into the cockpit.
- The outer skin of the glider is as smooth as possible. A glider is built from one smooth piece of fiberglass or carbon composite, so there are no rivets or seams in the material to cause drag.
- The wings have a high aspect ratio, which means they are very long compared to their width. In other words, glider wings are very long and thin. This makes the wings efficient: They produce little drag for the amount of lift they generate.

Getting Up

Without an engine, a glider's big problem is getting up to altitude. The most common launching method is an aero-tow. A conventional-powered airplane tows the glider up into the sky, using a long rope. The glider pilot controls a quick-release mechanism located in the glider's nose and releases the rope at the desired altitude. Right after release, the glider and the tow plane turn in opposite directions, and the glider begins its unpowered flight. The tow plane is then free to return to the airport and set up for another tow.

Staying Up

Gliding is a sport, and one of the most interesting and challenging parts of it is keeping the glider in the air. If the glide ratio were the only thing involved in staying airborne, gliders couldn't stay in the air nearly as long as they do. An experienced pilot can soar for the entire day if the weather is right. The key to staying in the air for long periods of time is to get some help from nature, in the form of updrafts, whenever possible. Glider pilots use two types of naturally occurring updrafts to regain lost altitude: thermals and ridge lift.

Thermals are columns of rising air created when the sun heats the earth's surface. As the air near the ground heats up, it expands and rises. Pilots keep an eye out for terrain that absorbs the sun's heat more rapidly than surrounding areas. These areas—areas like big asphalt parking lots, dark plowed fields, and rocky terrain—are great places to find thermal columns. Pilots also keep a lookout for newly forming cumulus clouds, which can be a sign of thermal activity, and soaring birds, which also use thermals. When a pilot finds a thermal, he or she turns back and circles in the column until reaching the desired altitude.

Ridge lift is created by winds blowing against mountains, hills, or other ridges. As the air reaches the mountain, it redirects upward and forms a band of lift along the side of slope. Ridge lift usually reaches no higher than a few hundred feet above the ridge that creates it. What ridge lift lacks in height, however, it makes up for in length. Pilots have been known to fly for 1,000 miles (1,600 km) along mountain chains by using ridge lift!

The Impact of Weight on Speed

A heavier glider will fly faster than a light glider. The actual glide ratio does not change— a heavier glider will sink faster but it does it at a higher airspeed. In order to add weight and fly faster, some gliders have ballast tanks that can hold up to 500 pounds of water. Ballast tanks are common for cross-country flying, as well as for performing acrobatic maneuvers such as loops.

The extra weight can reduce climb rates in a lifting environment (such as a thermal) and, possibly, shorten flight duration if suitable lift cannot be located. To prevent this, the pilot can dump the ballast at any time using dump valves.

Getting Down

Landing a glider is much like landing a conventional plane, except that a glider usually has only one small wheel, located directly under the pilot.

The wings on gliders are very strong, and the tips are reinforced to prevent damage in case they scrape along the ground during a landing.

How **HOT AIR BALLOONS** Work

The basic idea behind a hot air balloon is simple. It is a machine designed to float in midair. In the same way that you can perfectly balance the ballast in a submarine so that it can float in the middle of the ocean, a hot air balloon floats in an ocean of air. Many people describe flying in a huge hot air balloon as one of the most serene, enjoyable activities they've ever experienced. If you simply want to enjoy the experience of flying, there's nothing quite like it. There is no engine noise, there is no wind, and you are standing in a basket completely out in the open! If you actually need to get somewhere and get there quickly, a hot air balloon is probably not the right vehicle. You can't really steer it, and it only travels as fast as the wind blows. But for a Sunday afternoon adventure, it is perfect!

HSW Web Links

www.howstuffworks.com

How Airplanes Work
How Blimps Work
How CargoLifter's Airship Will Work
How Helicopters Work
How Helium Balloons Work

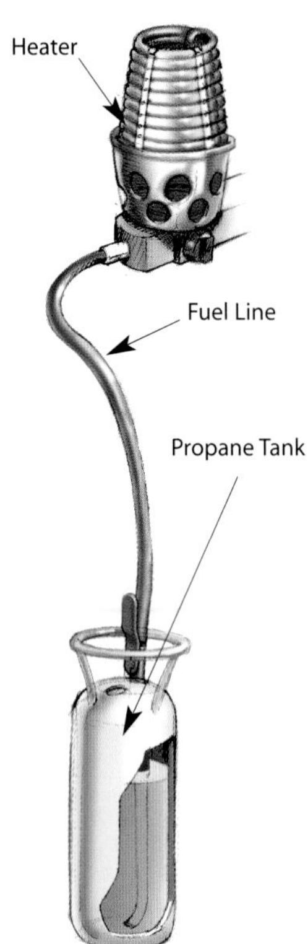

Here's how propane gas flows from the tank to the heater so that the balloon can float.

So why can a hot air balloon float? In order for something to float in a fluid, it has to displace some of the fluid. For example, when a 1,000-pound (454-kg) boat floats in water, it is displacing 1,000 pounds (454 kg) of water. Because water weighs about 62 pounds per cubic foot (1,000 kg/m^3), only about 16 cubic feet (0.45 m^3) of water have to be displaced for a 1,000-pound (454-kg) boat to float.

Air is much lighter than water. In fact, air is so light that we tend to think of it as being weightless. But it actually does have weight. A cubic foot of the earth's air weighs about 28 g (about 1 oz). In order for a hot air balloon to float, it has to displace enough air to equal the weight of the balloon itself, plus the basket and the passengers. At 28 g per cubic foot, that's a lot of air, and this is why hot air balloons have to be so huge.

Air Temperature

Hot air is lighter than cool air, and how much lighter it is depends on the temperature difference. When you heat air, it gets lighter because it expands. Heat causes the molecules in the air to speed up, which means they collide with surfaces more often and with greater force. It takes fewer hot air particles to maintain the same pressure as more cool air particles. A hot air balloon rises because it is filled with hot air and that is less dense than the surrounding colder, denser air.

As it turns out, you can only heat the air inside the balloon by about 100°F (38°C). Any hotter, and you have to worry about the balloon melting. At this temperature, the hot air is about 25% lighter than the air around it. In other words, 1 cubic foot (.028 m^3) of hot air weighs about 7g less than 1 cubic foot (.028 m^3) of air at room temperature.

Parts of a Hot Air Balloon

A hot air balloon needs three parts:

- **An envelope** (or a balloon) of lightweight material—To hold the hot air
- **A heater**—To keep the air in the envelope hot
- **A basket or platform**—To hold passengers

Modern hot air balloons heat the air by burning propane. The propane is stored in lightweight cylinders in the balloon basket. The cylinders are connected to the burner above the basket with flexible hoses.

Burning propane heats the air in the envelope. In most modern hot air balloons, the envelope is made of coated nylon, reinforced with sewn-in webbing. Nylon works very well in balloons because it is lightweight, it is fairly sturdy, and it has a high melting temperature. The skirt at the base of the envelope is coated with fire-resistant material to keep the flame from burning the balloon.

The hot air won't escape from the hole at the bottom of the envelope because buoyancy always keeps it moving up. The cold air outside gradually cools the air inside, however, so the pilot has to periodically fire the burners to stay aloft. If the pilot continually fires the burners, the balloon will continue to rise.

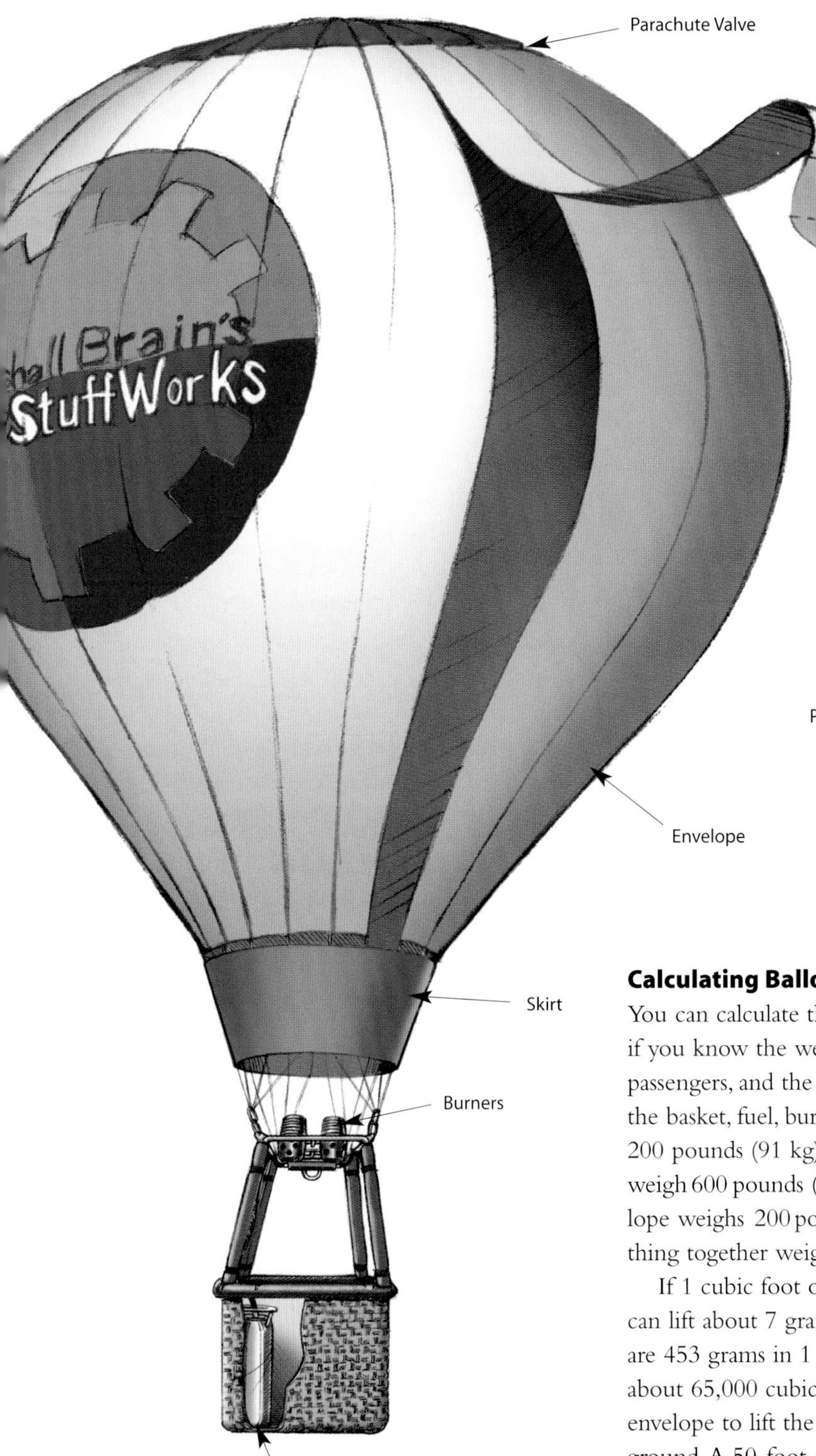

Calculating Balloon Size

You can calculate the size of a balloon easily if you know the weight of the basket, the passengers, and the envelope. Let's say that the basket, fuel, burners, and so on weigh 200 pounds (91 kg) total, the passengers weigh 600 pounds (272 kg), and the envelope weighs 200 pounds (91 kg). So everything together weighs 1,000 pounds (454 kg).

If 1 cubic foot of hot air in the envelope can lift about 7 grams (0.015 lb) and there are 453 grams in 1 pound, then it takes about 65,000 cubic feet of hot air in the envelope to lift the 1,000 pounds off the ground. A 50-foot-diameter sphere will hold about 65,000 cubic feet (1,841 m^3) of air. This means the envelope needs to be about as tall as a five-story building—and it needs to be that wide, too!

Hot air balloons gave human beings their first taste of flight in 1783. Today they still provide people with an unbelievable flying experience that is like no other.

Try This...

You can get a sense of how much air contracts and expands as its temperature changes by performing a simple experiment. Start with two plastic zipper-seal bags (1- gallon size) and blow them up. You can do this by zipping the bag closed, then unzipping a small hole at one end of the zipper. Blow each bag up like a balloon and seal it while holding pressure on the last breath. You want these bags to be full — you want the plastic on both inflated bags to be tense.

Now let the bags sit on the counter for a couple of minutes and cool off. You pumped 98.6°F (36° C) air into them, and you want the temperature to drop to room temperature. The bags will probably become a little less tense in the process of cooling (makes sense...), so add one more puff of air to make them tense again.

Now stick one of the bags into your freezer for about 3 minutes, while leaving the other one on the counter. When you take the bag in the freezer out it will have deflated by about 10% to 15%. It has deflated because cooler air is denser than warmer air. Compare the cold bag to the bag on the counter:

- The cold bag will not be tense at all. Then a funny thing will happen as the cold bag warms up…
- It will get tense again and return to its original size!

You can clearly see that warmer air takes up more space than cooler air. Therefore warmer air is lighter than cooler air, and that is what makes a hot air balloon float.

How THE CONCORDE Works

Imagine leaving London at 8:30 a.m. and arriving in New York at 7:30 a.m. the same day. Is this a form of time travel? No, it is an amazing piece of technology called the Concorde. This supersonic plane flies faster than sound and twice as high as most commercial jets. It punches through the sound barrier with a streamlined design, high-power jet engines complete with afterburners, and other high-tech features.

HSW Web Links

www.howstuffworks.com

- How Airplanes Work
- How Becoming an Airline Pilot Works
- How Gas Turbine Engines (and Jet Engines) Work
- How Space Planes Will Work
- How Space Shuttles Work

As any aircraft approaches the speed of sound (1,100 ft/s, or 343 m/s), the air pressure builds up in front of the aircraft and forms a "wall" of air. To punch through that wall, supersonic planes must be streamlined. The Concorde looks very different from normal commercial jets—it looks much more streamlined.

Special Features

It has four important special features:

- A needle-like fuselage
- Swept-back, delta-shaped wings
- A movable nose
- A vertical tail but no horizontal tail

If you compare the Concorde to a Boeing 747, the differences really stand out. The body of the Concorde is about the same length as that of a 747, but it is much narrower. The Concorde's body is less than 10 feet (3 m) wide. The long, narrow shape of the Concorde reduces the drag on the plane as it moves through the air.

The wing of the Concorde is thin, swept back, and triangle shaped. A 747's wings are big rectangles that are attached to the plane at an angle. The Concorde's wing has what is called a delta-wing design, and it has two big advantages for supersonic flight:

- It reduces drag by being thin and swept back (at a 55-degree angle to the fuselage).
- It provides enough stability in flight that no horizontal stabilizers are needed on the tail, and this reduces drag even more.

At subsonic speeds, delta wings can produce enough lift to keep the plane in the air if the angle of attack increases. This is why the Concorde takes off and lands at such an incredible angle.

The Concorde has a long, needle-shaped nose. The nose helps the aircraft penetrate the air, and it can be tilted down by 13 degrees on takeoff and landing so that the pilots can see the runway despite the huge angle of attack.

With its special fuselage, delta wings, high-tech nose, and tail, the Concorde's design enables it to fly at supersonic speeds and take off and land from regular airports.

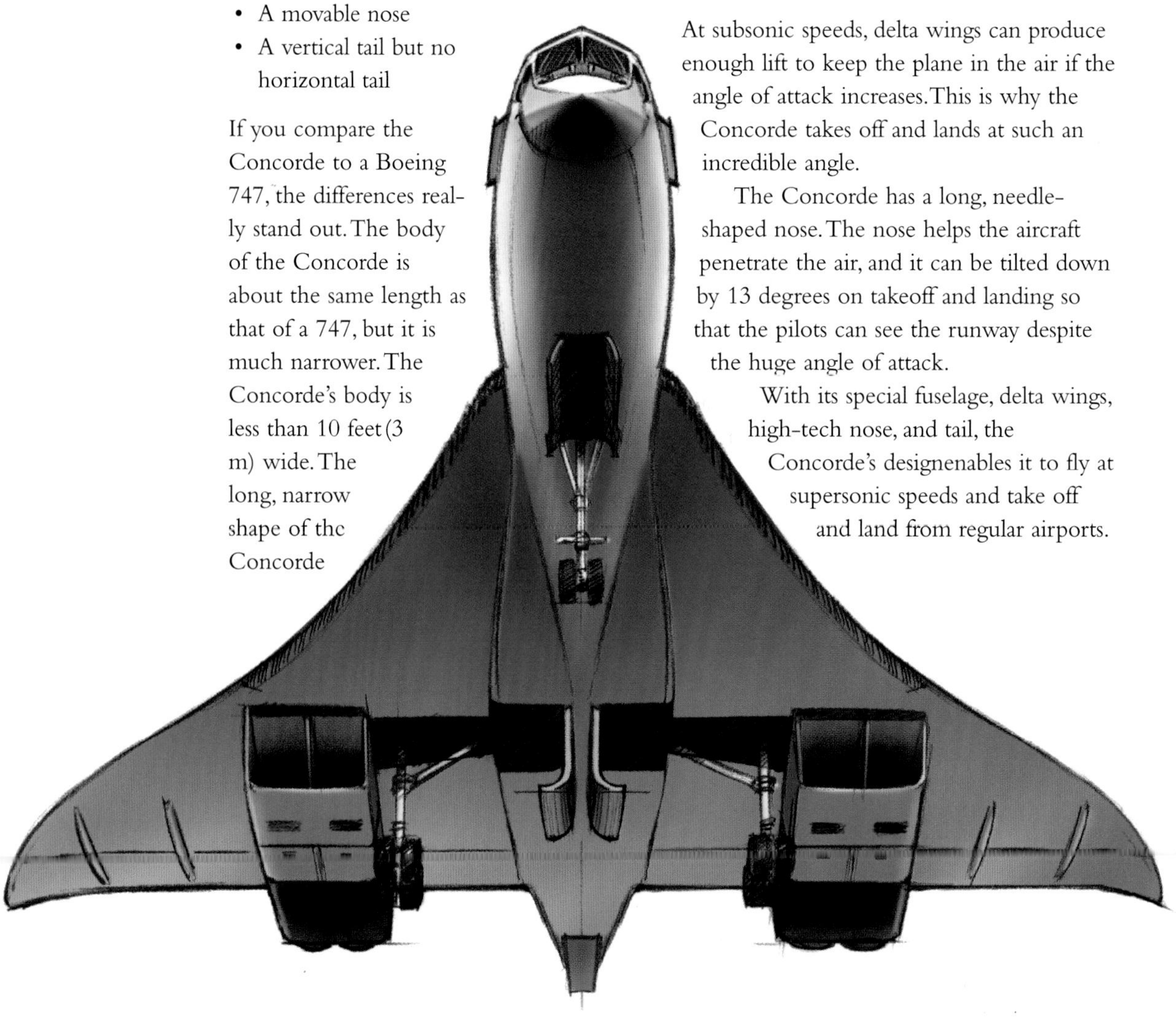

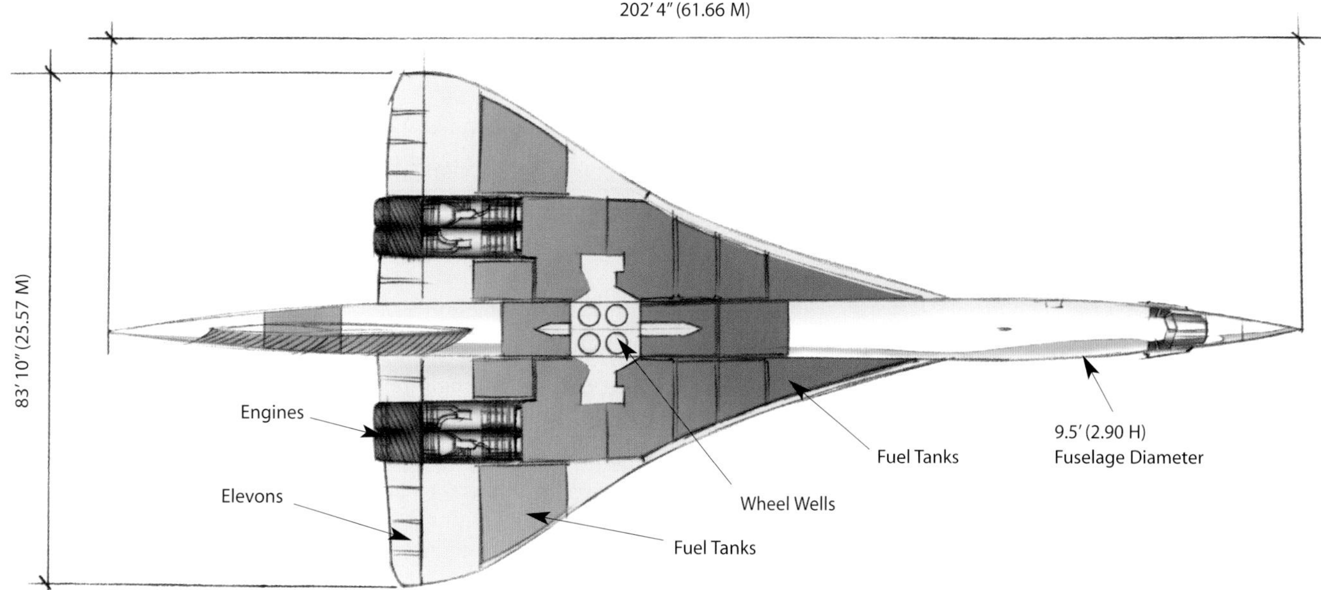

Engines

The engines on the Concorde are huge, and they provide the thrust needed for punching through the air at such incredible speeds. The Concorde has four Rolls Royce/Snecma Olympus 593 turbo jet engines. Each engine generates 18.7 tons (180 kilonewtons) of thrust, and the four engines combined burn 6,771 gallons (25,629 l) of fuel per hour!

Fuel Tanks

The Concorde has 17 fuel tanks located all over the plane that can hold a total of 31,569 gallons (119,500 l) of kerosene fuel. There are five tanks in each wing and four more in the fuselage. The Concorde also has three auxiliary, or trim, fuel tanks: two in front and one in the tail.

The Concorde uses fuel not just to keep the engines running but as a functional part of the airframe and for aerodynamic stability. During different parts of the flight, the pilots move fuel forward and backward in the trim tanks to change the center of gravity on the plane.

Heat Effects

The temperature of the Concorde's skin varies from 261°F (127°C) at the nose to 196°F (91°C) at the rear of the plane. All this heat has a surprising side effect: The Concorde actually grows by 7 inches (17.8 cm) in flight! The Concorde is made of a special aluminum alloy that is lightweight and more heat tolerant than titanium.

Future Supersonic Transports

In addition to the Concorde, other supersonic planes are currently being designed. Former president Ronald Reagan called for a program to develop a hyperspace transport, or national aerospace plane, capable of going from New York to Tokyo in two hours. Such planes will have to enter outerspace in a suborbital flight. They will have to have both air-breathing and rocket engines to achieve the appropriate speeds and, much like space shuttles, they will have to have systems to handle the intense heat of reentry.

Boeing 747 vs. Concorde

		Boeing 747	Concorde
Length	ft.	231	202
	m	70.5	61.7
Width	ft	20	9.5
	m	6.1	2.7
Altitude	ft	35,000	60,000
	m	10,675	18,300
Speed	mp/h	560	1,350
	km/h	901	2,172

How **SPACE SHUTTLES** Work

Basically, a space shuttle is a rocket that can land like an airplane. The big advantage of a space shuttle is that several different parts of it are reusable, and that lowers the cost of getting satellites and equipment into orbit. A typical shuttle mission lasts 7 to 8 days, but can be expanded to 14 days when necessary. On any mission there are four big phases: getting into orbit, living in space and doing the work of the mission, reentering the atmosphere, and landing.

HSW Web Links

www.howstuffworks.com

- How Air-Breathing Rockets Will Work
- How Light Propulsion Will Work
- How *Mars Odyssey* Will Work
- How Space Stations Work
- How Space Suits Work

Space shuttles are some of the biggest vehicles people have created. At launch a space shuttle is more than 150 feet (30 m) tall and weighs in at 4.5 million pounds (2.05 million kg). Ninety percent of a space shuttle's weight (almost 4 million lb, or 1.8 million kg) is fuel because it takes a huge amount of energy to accelerate the shuttle from 0 to 17,000 mph (28,000 kph) and to lift it from sea level into orbit 115 to 400 miles (185 to 643 km) above the earth.

Getting into Orbit

A space shuttle needs three different pieces in order to get into orbit:

- Two solid rocket boosters (SRBs)
- Three main engines of the orbiter
- An external fuel tank (ET)

The SRBs are the two big rockets that you see on either side of the external tank at launch. They are much more powerful than the three main engines and provide a good part of the thrust needed to get the shuttle up to speed. They also support the entire weight of the space shuttle when it is sitting on the launch pad.

Each SRB is a gigantic aluminum tube filled with solid rocket fuel. (See "How Rocket Engines Work," page 46, for details.) The SRBs ignite as the very last step of the launch sequence, because, like any other solid-fuel rocket, they cannot be shut down once they are lit.

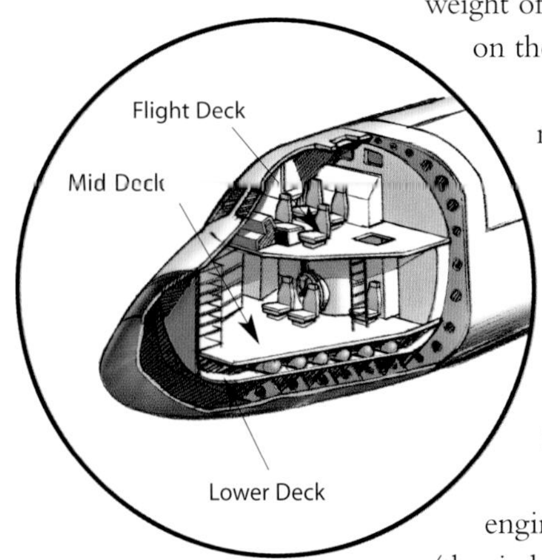

The orbiter has three main engines, which are located in the aft (that is, back) fuselage (in the body of the spacecraft). The main engines use about 526,000 gallons (2 million l) of fuel (a mixture of liquid hydrogen and liquid oxygen) from the ET and provide the remainder of the thrust (29%) to lift the shuttle off the pad and into orbit. Together they hold about 2.5 million pounds (1.1 million kg) of fuel, and they burn it all in about 2 minutes!

The three main engines burn liquid hydrogen and liquid oxygen stored in the external tank. Together these engines produce more than 1 million pounds (more than 4.4 million newtons) of thrust. They have two big advantages over the SRBs:

- They are adjustable, so the shuttle's computers can change the amount of thrust they produce during the flight.
- They are movable and can angle their exhaust nozzles up to 10.5 degrees. This gives the control system a way to make adjustments during flight to keep the shuttle exactly on track.

The main engines use up all the fuel in the external tank in about 8 minutes. The external tank breaks away, and the main engines don't do anything for the rest of the mission.

To move the shuttle around after the launch, there is a whole separate system called the orbital maneuvering systems (OMS), located in pods on the back end of the orbiter. These engines place the shuttle into final orbit, change the shuttle's position from one orbit to another, and slow down the shuttle for reentry.

Being in Orbit

When people are in orbit on a shuttle mission, they live inside the shuttle orbiter for up to 2 weeks. It is an absolutely amazing machine, one of the most complex vehicles ever created. It has to protect the crew from all the hazards of space, including the vacuum forces, intense heat and cold, solar radiation, micrometeorites, and the

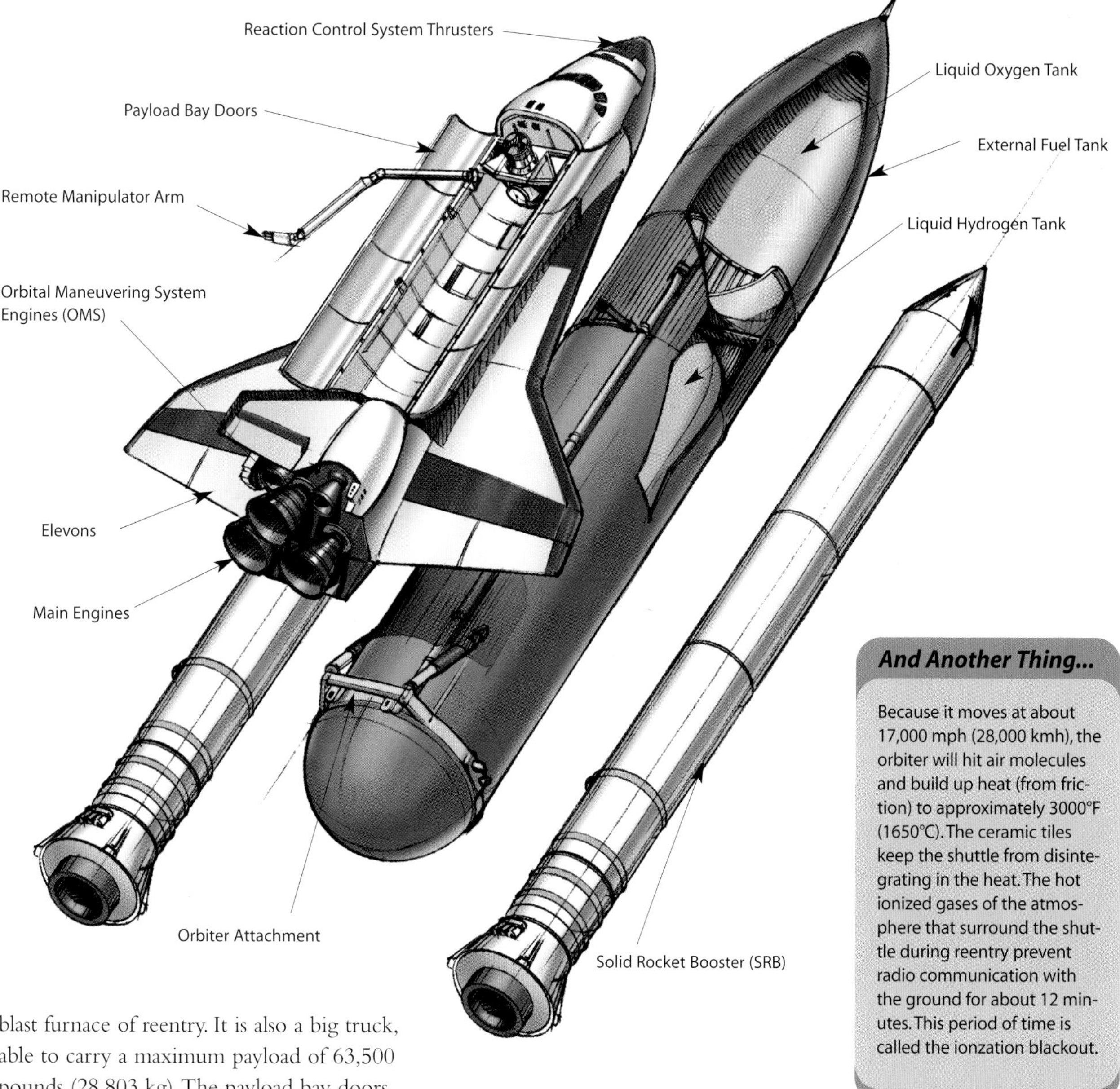

blast furnace of reentry. It is also a big truck, able to carry a maximum payload of 63,500 pounds (28,803 kg). The payload bay doors are 60 feet (18.3 m) long and 22.7 feet (6.9 m), wide and they enclose a diameter of 15 feet (4.6 m).

If you were on a spacewalk looking at the orbiter from outside, you would see several big pieces. The cargo bay is by far the biggest part of the shuttle, and it is always visible because the cargo bay doors are always open in orbit. Lining the inside of the cargo bay doors are huge thermal radiators that dump the shuttle's excess heat into space. Without the doors open, the shuttle would overheat.

Also inside the cargo bay is the remote manipulator arm. Astronauts use it to move large pieces of equipment in and out of the cargo bay. It is also a movable platform for spacewalking astronauts.

And Another Thing...

Because it moves at about 17,000 mph (28,000 kmh), the orbiter will hit air molecules and build up heat (from friction) to approximately 3000°F (1650°C). The ceramic tiles keep the shuttle from disintegrating in the heat. The hot ionized gases of the atmosphere that surround the shuttle during reentry prevent radio communication with the ground for about 12 minutes. This period of time is called the ionzation blackout.

Space Shuttle Data

		SRB	ET	Orbiter
Length	ft	150	158	122
	m	46	48	37
Diameter	ft	12	27.6	17
	m	3.7	8.4	5.4
Weight	lb	192,000	78,100	165,000
	kg	87,090	35,425	75,000
Full	lb	1,300,000	1,600,000	2,285,000
	kg	589,670	719,000	103,803
Thrust	lb	2,650,000		1,125,000
	N	11,700,000		4,800,000

Did You Know?

A full shuttle countdown takes days and involves conducting thousands of tests, fueling the external tank, and getting the crew on board, among many other things. The final launch preparations are going on through T -31 seconds, and this is where all the action is. If you were riding in the shuttle, here's what would happen:

- **T–31 s:** The on-board computers take over the launch sequence.
- **T–6.6 s:** The shuttle's main engines ignite one at a time (0.12 s apart). The engines build up to more than 90% of their maximum thrust.
- **T–3 s:** Shuttle main engines are angled into liftoff position.
- **T–0 s:** The SRBs are ignited and the shuttle lifts off the pad.
- **T+20 s:** The shuttle rolls to the right (180-degree roll, 78-degree pitch).
- **T+60 s:** The shuttle main engines are at maximum throttle.
- **T+2 min:** The SRBs separate from the orbiter and fuel tank at an altitude of 28 miles (45 km). They have parachutes and are recovered from the ocean for reuse. The main engines continue firing.
- **T+7.7 min:** The main engines are throttled down to keep acceleration below 3Gs so that the shuttle does not break apart.
- **T+8.5 min:** The main engines shut down.
- **T+9 min:** The external tank separates from the orbiter. The tank will burn up upon reentry.
- **T+10.5 min:** The OMS engines fire to place the shuttle in a low orbit.
- **T +45 min:** The OMS engines fire again to place the shuttle in a higher, circular orbit (about 250 miles, or 400 km, above earth).

In front of the cargo bay is the crew compartment, which contains quite a bit of the support equipment that keeps the shuttle alive—things like the computers, fuel cells, and life-support systems. The crew compartment has 2,325 cubic feet (66 m^3) of space with the airlock inside, or 2,625 cubic feet (nearly 100 m^3) with the airlock outside.

A shuttle has three decks:

- **Flight deck** (also known as the cockpit)— This uppermost deck contains all the controls and warning systems for the space shuttle.
- **Mid-deck**—This is the living quarters, including the galley, sleeping bunks, and toilet.
- **Lower deck** (equipment bay)— This deck contains life-support equipment, electrical systems, and other equipment.

The other big parts of the shuttle that you cannot miss are the wings and tail, which make it look like an airplane. These parts let the shuttle glide to a landing on a huge runway in Florida.

Covering the entire shuttle are heat-resistant tiles. They are made of silica and are fantastic insulators. You may have seen photographs showing a person holding a silica tile in bare hands while heating it with a blowtorch. These tiles insulate so well that part of the tile can be red hot while another part is cool to the touch.

Several different systems in the crew compartment provide the shuttle with all the comforts of home:

- **The life-support system**—The life-support systems provide air for the astronauts to breathe. Liquid oxygen and nitrogen mix from pressurized tanks to provide an earth-like atmosphere. Scrubbers and filters remove carbon dioxide, particles, and odor.
- **The electrical system**—Three fuel cells provide electrical power to the shuttle's power grid, which distributes power to everything that needs it in all areas of the ship. The fuel cells use hydrogen and oxygen to produce electricity, and as a byproduct they create pure water that the crew drinks.
- **The cooling system**—An incredibly involved cooling system takes heat from the shuttle and moves it to the thermal radiators so that the ship doesn't overheat. Because space is a vacuum, a space shuttle is like a gigantic thermos. Using infrared radiation from the thermal radiators is the only way to get rid of the heat.
- **Computer systems**—Five on-board computers handle all the flight systems, especially during launch, maneuvering, and landing. Four of the computers vote on every decision to make, so that if one of the computers fails, the other three can override it. The fifth computer is a backup in case all four of the main computers have a catastrophic problem.

Future Shuttles

NASA is currently exploring the idea of a one-piece, reusable shuttle with the X-33 and VentureStar designs. In addition, private entrepreneurs are competing for the $10 million X Prize for the first team to build and launch a fully reusable rocket that can boost three humans into a suborbital flight (60 miles, or 100 km, high) on two consecutive flights within 2 weeks. The technologies developed in this competition may lead to the development of commercial reusable spacecraft.

chapter two

POWER UP!

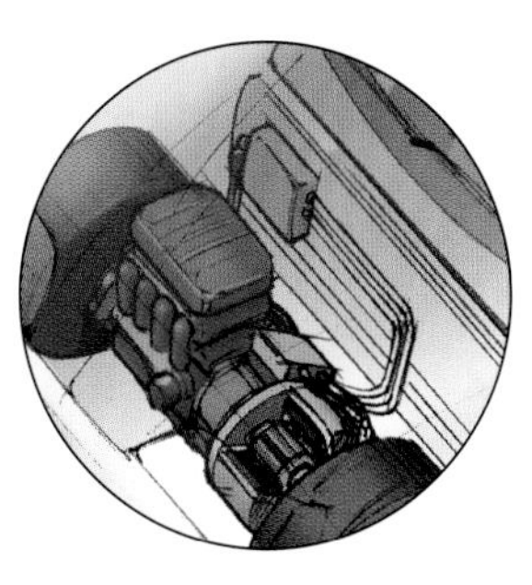

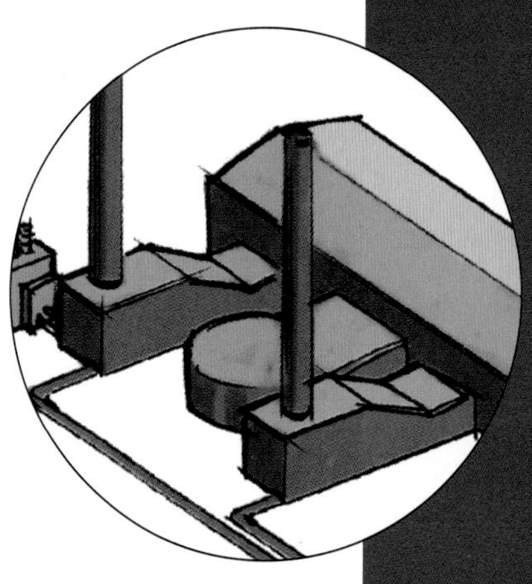

How **FOUR-STROKE GASOLINE ENGINES** Work

A car is one of the most complicated objects a person sees during a normal day. A car has thousands of parts, all of them functioning together day in and day out. The engine is a big part of the whole package. A modern engine is a computer-controlled marvel, but at its core it is incredibly simple.

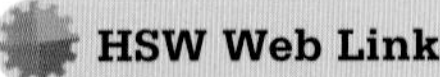

www.howstuffworks.com

How Automobile Ignition Systems Work
How Caterpillar Backhoe Loaders Work
How Fuel Injection Systems Work
How Mufflers Work
How NASCAR Safety Works

A car engine has one goal in mind: to convert gasoline and air into motion. The engine does this by burning the gasoline in thousands of tiny explosions, a process called internal combustion.

For an example of internal combustion, think of the Revolutionary War cannons you've probably seen in movies. The soldiers loaded a cannon with gunpowder and a cannon ball, and then they lit it. That is internal combustion, but what does it have to do with engines?

As another example, say that you took a big piece of plastic sewer pipe, maybe 3 inches (7.6 cm) in diameter and 3 feet (0.91 m) long, and you put a cap on one end of it. Then say that you sprayed a little WD-40 into the pipe or put in a tiny drop of gasoline. Then say that you stuffed a potato down the pipe. (It is not recommended that you do this, but say you did.) What you have here is a device commonly known as a potato cannon. When you introduce a spark, you can ignite the fuel. What is interesting, and the reason we are talking about this device, is that a potato cannon can launch a potato about 500 feet (152 m) through the air!

The potato cannon uses the basic principle behind any internal combustion engine: If you put a tiny amount of high-energy fuel (such as gasoline) in a small, enclosed space and ignite it, an incredible amount of energy is released, in the form of expanding gas. You can use that energy to propel a potato 500 feet (152 m). In this case, the energy translates into potato motion. You can also use it for more productive purposes. For example, if you create a cycle that allows you to set off explosions like this hundreds of times per minute, and then you harness that energy in a useful way, you have the core of a car engine.

The Four-Stroke Combustion Cycle

Almost all cars currently use what is called a four-stroke combustion cycle to convert gasoline into motion. The four-stroke approach is also known as the Otto cycle, in honor of Nikolaus Otto, who invented it in 1867. The four strokes are:

- The intake stroke
- The compression stroke
- The combustion stroke
- The exhaust stroke

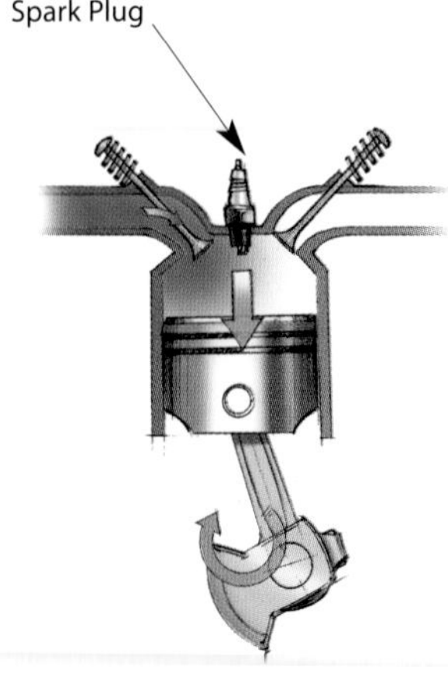

Intake

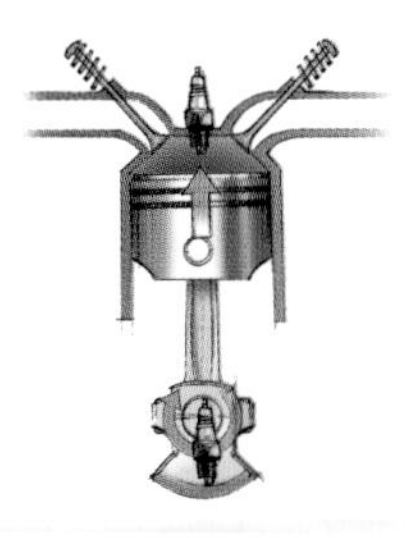

Compression

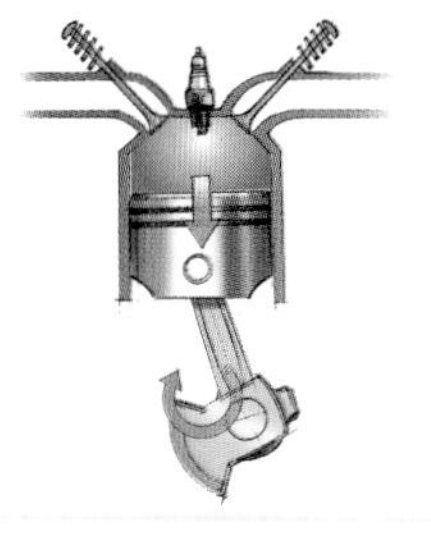

Combustion

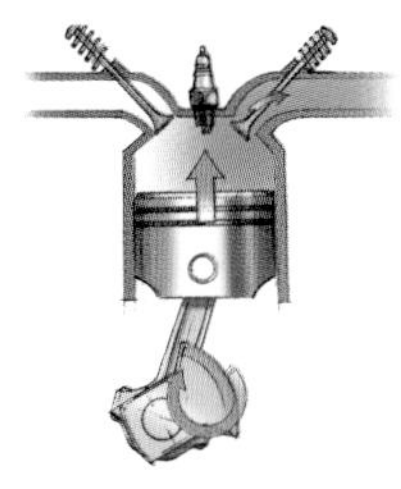

Exhaust

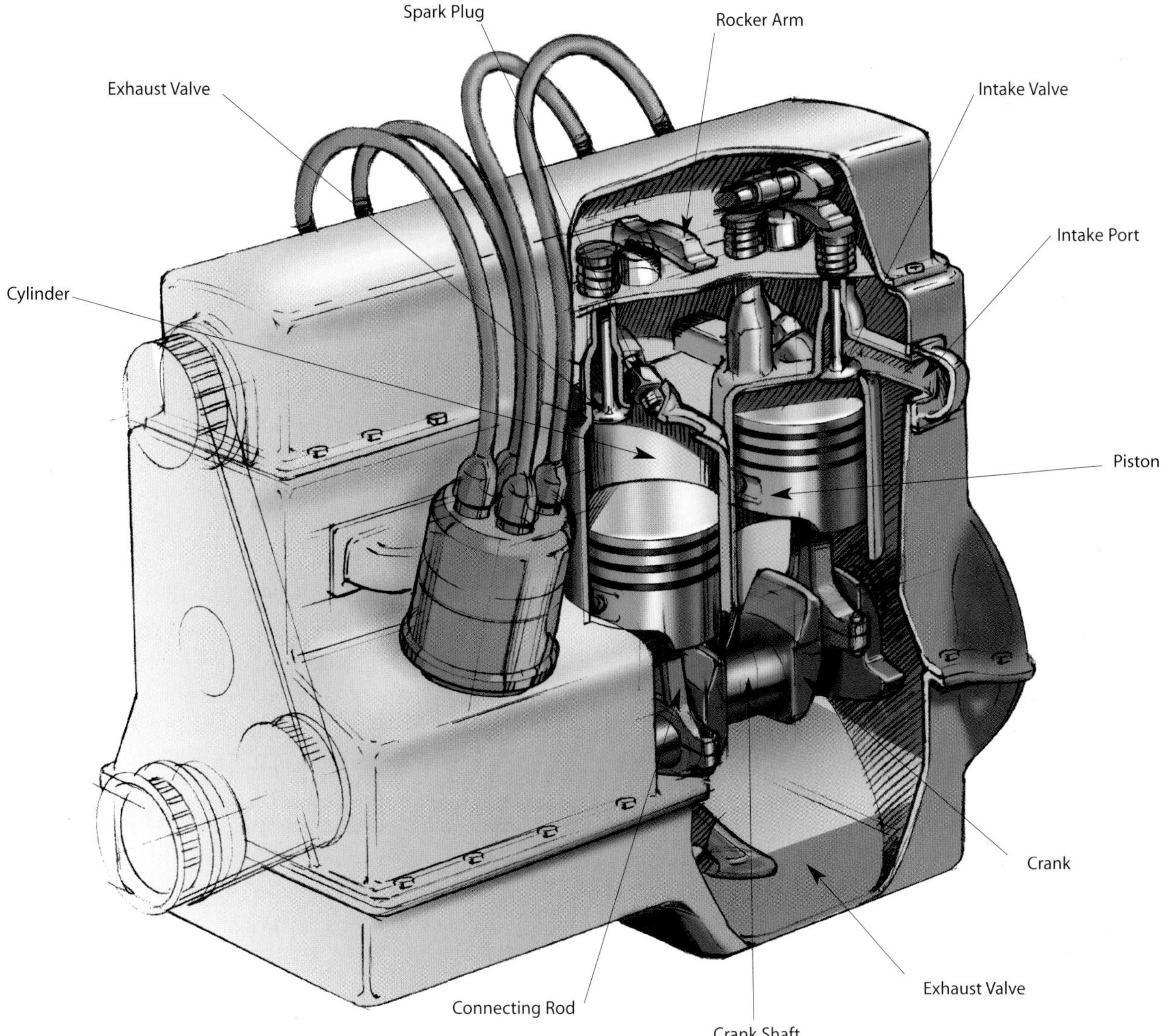

In the intake stroke, the piston starts at the top, the intake valve opens, and the piston moves down to let the engine take in a cylinder full of air and gasoline. Only a tiny drop of gasoline needs to be mixed into the air for this to work.

Next, the piston moves back up to compress this fuel/air mixture. Compression makes the explosion more powerful. When the piston nears the top of its stroke, the spark plug emits a spark to ignite the gasoline. The gasoline charge in the cylinder explodes, driving the piston down.

When the piston hits the bottom of its stroke, the exhaust valve opens and the exhaust leaves the cylinder to go out the tail pipe. Now the engine is ready for the next cycle, so it takes in another charge of air and gas, starting the cycle over.

The motion that comes out of a piston (or a potato cannon) is linear. In an engine, the crankshaft converts the linear motion into rotational motion. The rotational motion is needed because the goal is to rotate a car's wheels.

Internal vs. External Combustion Engines

There are different kinds of internal combustion engines (for example, piston engines and gas turbine engines). The gas turbine engine has interesting advantages and disadvantages over the piston engine, but its main disadvantage right now is that it has an extremely high manufacturing cost and is therefore much more expensive than the piston engine used in cars today.

Did You Know?

The main reason large, 4.0-liter engines have eight half-liter cylinders rather than one big 4-liter cylinder is smoothness. A V-8 engine is much smoother because it has eight evenly spaced explosions instead of one big explosion. Another reason is starting torque. When you start a V-8 engine, you are only driving two cylinders (one liter) through their compression strokes, but with one big cylinder you would have to compress four liters instead.

And Another Thing...

Why is nearly every car, motorcycle, moped, and even lawn mower powered by gasoline? Gasoline has an extremely high energy density, it is inexpensive compared to its alternatives, and it is easy and relatively safe to move around. For example, it takes about 1,000 pounds (454 g) of lead-acid batteries to store the same amount of energy that 7 pounds (1 gallon, or 3.8 l) of gas contains. It takes several hours to recharge the batteries, but it takes about 15 seconds to pump 1 gallon of gas. That is why you do not see very many electric cars right now: Gas is a lot easier.

There are not just internal combustion engines, but external combustion engines as well. Steam engines in old-fashioned trains and steamboats is a good example of an external combustion engine. The fuel (coal, wood, oil, or whatever) in a steam engine burns outside the engine to create steam, and the steam creates motion inside the engine.

Parts of a Car Engine

A car engine has the following parts:

- **Cylinder**—The core of the engine is the cylinder. The piston moves up and down inside the cylinder. Most cars have more than one cylinder (four, six, and eight cylinders are common). In a multi-cylinder engine, the cylinders are usually arranged in one of three ways: inline, V, or flat (also known as horizontally opposed, or boxer). Different configurations of cylinders have different smoothness, manufacturing-cost, and shape characteristics that make them suitable to specific vehicles.
- **Spark plug**—The spark plug supplies the spark that ignites the air/fuel mixture so that combustion can occur. The spark must happen at just the right moment for things to work properly.
- **Valves**—The intake and exhaust valves open at the proper time to let in air and fuel and to let out exhaust. Note that both the intake and exhaust valves are closed during compression and combustion so that the combustion chamber is sealed.
- **Piston**—A piston is a cylindrical piece of metal that moves up and down inside the cylinder.
- **Piston rings**—Piston rings provide a sliding seal between the outer edge of the piston and the inner edge of the cylinder. The rings serve two purposes:
  - They prevent the fuel/air mixture and exhaust in the combustion chamber from leaking into the sump during compression and combustion.
  - They keep oil in the sump from leaking into the combustion area, where it would be burned and lost. Many cars that are said to be "burning oil" are burning it because the rings no longer seal properly.
- **Connecting rod**—The connecting rod connects the piston to the crankshaft. It can rotate at both ends so that its angle can change as the piston moves and the crankshaft rotates.
- **Crankshaft**—The crankshaft turns the pistons' up-and-down motion into circular motion, just like a crank on a jack-in-the-box does.
- **Sump**—The sump surrounds the crankshaft. It contains some oil, which collects in the bottom of the sump.

Displacement

The combustion chamber is the area where compression and combustion take place. As the piston moves up and down, the size of the combustion chamber changes. It has a maximum volume as well as a minimum volume. The difference between the maximum and minimum is called the displacement and is measured in liters or CCs (cubic centimeters, where 1,000 CCs equals 1 l).

If you have a four-cylinder engine and each cylinder displaces 0.5 l, then the entire engine is a 2 l engine. If each cylinder displaces 0.5 l and there are six cylinders arranged in a V configuration, you have a 3 l V-6. Generally, the displacement tells you something about how much power an engine can produce. A cylinder that displaces 0.5 l can hold half as much fuel/air mixture as a cylinder that displaces 1 l. Therefore, you would expect about twice as much power from the larger cylinder (if everything else is equal). So a 2 l engine is roughly half as powerful as a 4 l engine.

It is amazing that something this simple plays such a huge role in our society. In the U.S., nearly every adult owns a car and uses it daily for everything from getting to work to picking up the groceries. Without the internal combustion engine, society as we know it would come to a near standstill.

How **HYBRID CARS** Work

Hybrid vehicles are all around us. Most of the locomotives we see pulling trains are diesel-electric hybrids. Some cities, such as Seattle, have diesel-electric buses that can draw electric power from overhead wires or run on diesel when they are away from the wires. Giant mining trucks are often diesel-electric hybrids. Submarines are also hybrid vehicles—some are nuclear-electric and some are diesel-electric hybrids. Any vehicle that combines two or more sources of power that can directly or indirectly provide propulsion power is a hybrid.

A car normally has an engine that is much larger than it needs. For example, when a car is cruising down the highway, it might need at most 20 horsepower to overcome wind resistance and rolling resistance in the tires. If the car has a 200-horsepower engine, then the other 180 horsepower is unused, and it is taking up space and adding lots of weight to the car. The reason you need 200 horsepower is for the rare moments when you want to "floor it" and get maximum acceleration. Ninety-nine percent of the time, all that extra power is deadweight.

A hybrid car tries to combine the best of both worlds. Gasoline engines have a range of speeds where they operate best and another range where they are inefficient. A hybrid car adds a second source of power—normally some sort of electric motor—to handle the parts of the drive that would waste a lot of gas.

Power sources in hybrid cars can be combined in different ways to create different types of hybrids. Two of the most common types are series hybrid cars and parallel hybrid cars.

Series Hybrid Cars

In a series hybrid car, the gasoline engine turns a generator, and the generator can either charge the batteries or power an electric motor that drives the transmission. The gasoline engine never directly powers the vehicle, so the car can run in its optimal power range all the time.

Parallel Hybrid Cars

Another type of hybrid, a parallel hybrid car, has a gasoline engine hooked to the transmission as usual. But it also has a set of batteries that supply power to an electric motor that is also connected to the transmission. The engine and the electric motor can turn the transmission at the same time.

The engine in a parallel hybrid car is powerful enough to move the car along on the freeway. But when it needs to get the car moving in a hurry, it needs help. That help comes from the electric motor and battery, which step in and provides the extra power. The electric motor can:

- Assist the gasoline engine, providing extra power while the car is accelerating or climbing a hill
- Provide some regenerative braking to capture energy during braking
- Recharge the batteries
- Start the engine, which eliminates the need for a starter

In some configurations, the electric motor might also power the car during stop-and-go driving, when a gasoline engine has its worst efficiency.

Hybrid cars provide better gas mileage and have lower emissions than gasoline engine–powered cars by combining two or more technologies. The idea is to get the best performance from each of the different sources of power.

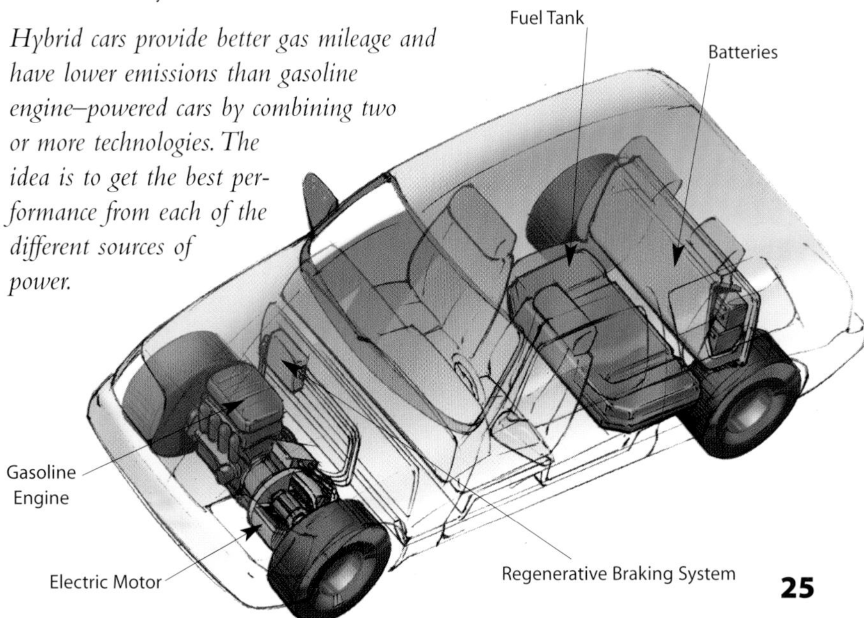

HSW Web Links

www.howstuffworks.com

- How Car Engines Work
- How Clutches Work
- How Diesel Engines Work
- How Horsepower Works
- How Manual Transmissions Work

How BATTERIES Work

If you look at any battery, you'll see that it has two terminals. One terminal is marked + (positive), and the other is marked – (negative). In an AA-, a C-, or a D-cell battery, the ends of the battery are the terminals. In a large car battery, two heavy lead posts act as the terminals.

HSW Web Links

www.howstuffworks.com

How Electric Motors Work
How Hybrid Cars Work
How Laptop Computers Work
How Power Paper Will Work
How Wires, Fuses, and Connectors Work

Electrons collect on the negative terminal of the battery. Inside the battery, a chemical reaction produces the electrons, and the speed of electron production (the battery's internal resistance) controls how many electrons can flow between the terminals. Electrons must flow through a wire, from the negative to the positive terminal, for the chemical reaction to take place. That is why a battery can sit on a shelf for a year and still have plenty of power: Unless electrons are flowing from the negative to the positive terminal, the chemical reaction does not take place. When you connect a wire, the reaction starts.

Battery Chemistry

The first battery was created by Alessandro Volta in 1800. To create his battery, Volta made a stack of alternating layers of zinc, blotting paper soaked in saltwater, and silver. This arrangement came to be known as a voltaic pile. If you attach a wire to the top and bottom of the pile, you can measure a voltage and a current. The pile can be stacked as high as you like, and each layer increases the voltage by a fixed amount. Each layer creates about 1 volt, so a 10–layer stack will have 10 volts.

Electrochemical Reactions

Probably the simplest battery you can create is a zinc/carbon battery, and by understanding the chemical reaction going on inside this battery, you can understand how batteries work in general.

Imagine that you have a jar of sulfuric acid (H_2SO_4). If you stick a zinc rod in it, the acid will immediately start to eat at the zinc. You will see hydrogen gas bubbles form on the zinc, and the rod and acid will start to heat up. Here's what is happening:

1) The acid molecules break up into two ions—two H^+ ions and one SO_4^{2-} ion.
2) The zinc atoms on the surface of the zinc rod lose two electrons ($2e^-$) to become Zn^{2+} ions.
3) The Zn^{2+} ions combine with SO_4^{2-} ions to create $ZnSO_4$, which dissolves in the acid.
4) The electrons from the zinc ions combine with hydrogen ions in the acid to create H_2 molecules (that is, hydrogen gas). You see the hydrogen gas as bubbles forming on the zinc rod.

If you now stick a carbon rod in the acid, the acid does nothing to it. But if you connect a wire between the zinc rod and the carbon rod, two things change:

- The electrons flow through the wire and combine with hydrogen on the carbon rod, so hydrogen gas begins bubbling off the carbon rod.
- There is less heat.

You can power a light bulb or similar load by using the electrons flowing through the wire, and you can measure the voltage and current in the wire. Some of the heat energy is turned into electron motion.

The electrons go to the trouble to move to the carbon rod because they find it easier to combine with hydrogen there. There is a characteristic voltage in the cell of 0.76 volts. Eventually the zinc rod dissolves completely or the hydrogen ions in the acid are used up and the battery dies.

In any battery—from car batteries to D-cell flashlight batteries—the same sort of electrochemical reaction occurs: Electrons move from one pole to another to create energy. The metals and electrolytes in the battery control the voltage, and each combination creates a characteristic voltage.

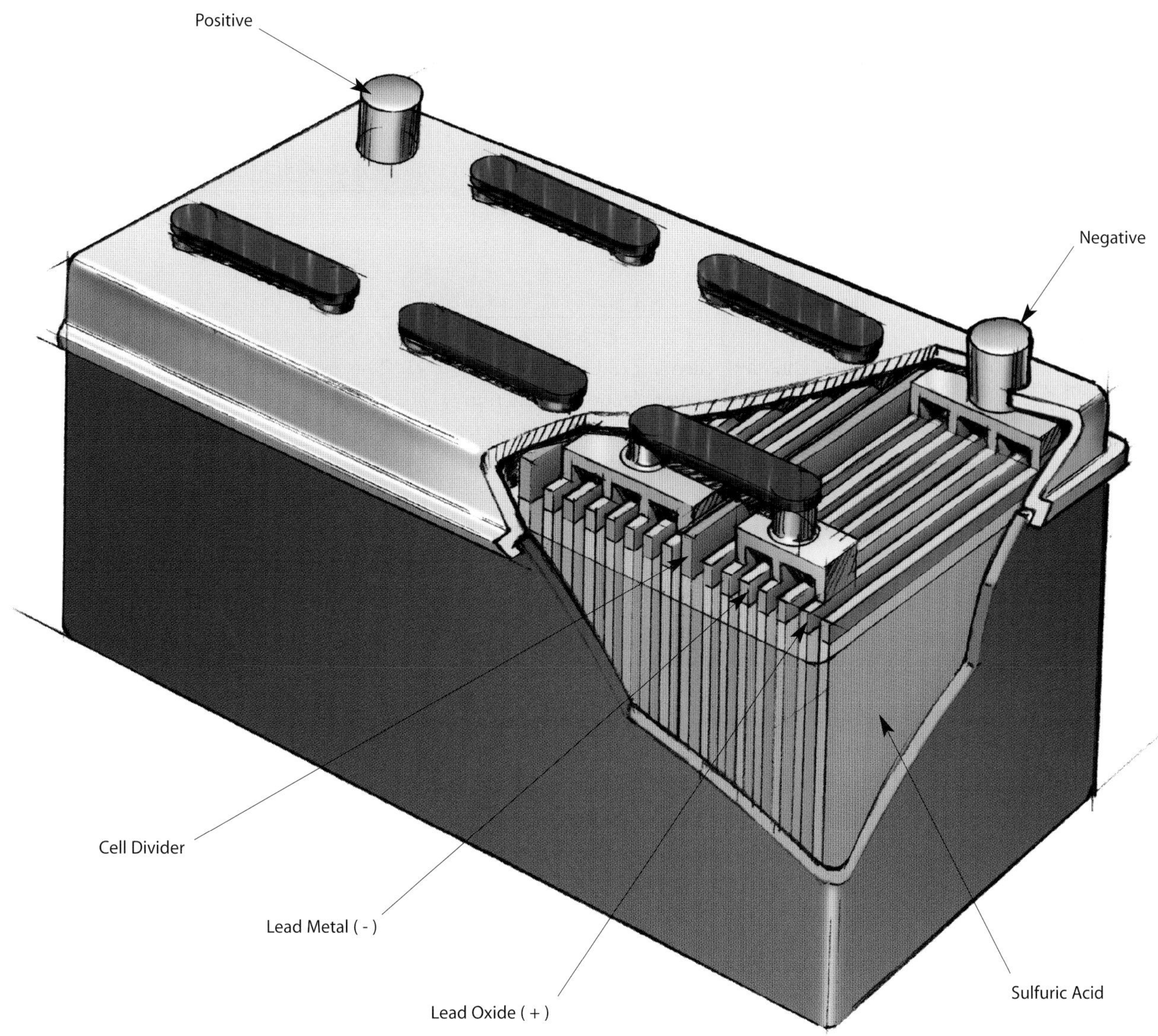

Battery Arrangements

In almost any device that uses batteries, you do not use just one battery (or cell) at a time. You normally group them together serially to form higher voltages or in parallel to form higher currents. In a serial arrangement, the voltages add. In a parallel arrangement the currents add.

Normally when you buy batteries, the package tells the batteries' ratings for voltage and current. For example, a digital camera might use four nickel-cadmium batteries that are rated at 1.25 volts and 650 milliamp-hours per cell. The milliamp-hour rating means that the cell can produce 650 milliamps for 1 hour. To a certain extent, you can scale milliamp-hours linearly, so this battery could produce 325 milliamps for 2 hours or 1,300 milliamps for a half hour. However, this scaling is not completely linear; for example, all batteries have a maximum current they can produce. Using the amp-hour rating, you can estimate how long the battery will last under a given load.

If you arrange four of these batteries in a serial arrangement, you can get 5 volts (1.25 X 4) at 650 milliamp-hours. If you arrange them in parallel you get 1.25 volts at 2,600 (650 X 4) milliamp-hours.

How DIESEL ENGINES Work

Rudolf Diesel developed the idea for the diesel engine and obtained the German patent for it in 1892. His goal was to create an engine with high efficiency. Gasoline engines had been invented in 1876 and, especially at that time, were not very efficient. Diesel's idea was to burn fuel at a much higher compression ratio.

HSW Web Links

www.howstuffworks.com

How Car Engines Work
How Caterpillar Backhoe Loaders Work
How Diesel Two-Stroke Engines Work
How Oil Refining Works
How Two-Stroke Engines Work

Did You Know?

If you have ever compared diesel fuel and gasoline, you know that they are different from each other. They certainly smell different. Diesel fuel is heavier and oilier, and it evaporates much more slowly than gasoline; its boiling point is actually higher than the boiling point of water. Diesel fuel is sometimes referred to as diesel oil because it is so oily.

Diesel fuel evaporates more slowly than gasoline because it is heavier. It contains more carbon atoms in longer chains than does gasoline; gasoline is typically C_9H_{20}, and diesel fuel is typically $C_{14}H_{30}$. It requires less refining to create diesel fuel, which is why it is generally less expensive than gasoline.

Diesel fuel has a higher energy density than gasoline. On average, a gallon of diesel fuel contains approximately 155,000,000 joules (147,000 BTUs), whereas a gallon of gasoline contains 132,000,000 joules (125,000 BTUs). This, combined with the greater efficiency of diesel engines, explains why diesel engines get better mileage than similar gasoline engines.

A normal gasoline engine takes in a mixture of gasoline and air, compresses it, and ignites the mixture with a spark. The problem with this approach is that it limits the amount of compression. If you compress gasoline vapor too much, it ignites on its own. You hear that as a knock in the engine.

Compression Ratio

A diesel engine takes in air, compresses it, and then injects fuel into the compressed air. The heat of the compressed air lights the fuel spontaneously. A diesel engine doesn't need a spark plug.

Because there is no fuel in the mix during the compression process, a diesel engine can use a much higher compression ratio than a gasoline engine. A gasoline engine compresses at a ratio of 8:1 to 12:1, whereas a diesel engine compresses at a ratio of 14:1 to as high as 25:1. The higher compression ratio means better efficiency.

Direct Fuel Injection

Diesel engines use direct fuel injection, which means the diesel fuel is injected directly into the cylinder. Gasoline engines generally use either carburetion, where the air and fuel are mixed long before they enter the cylinder, or port fuel injection, in which the fuel is injected just prior to the intake valve, outside the cylinder.

The fuel injector on a diesel engine is its most complex component and has been the subject of a great deal of experimentation. In any particular engine it may be located in a variety of places. Some diesel engines use special induction valves, precombustion chambers, or other devices to swirl the air in the combustion chamber or otherwise improve the ignition and combustion process.

Some diesel engines contain a glow plug of some sort. When a diesel engine is cold, the compression process may not raise the air to a high enough temperature to ignite the fuel. The glow plug is an electrically heated wire (think of the hot wires you see in a toaster) that helps ignite the fuel when the engine is cold so that the engine can start.

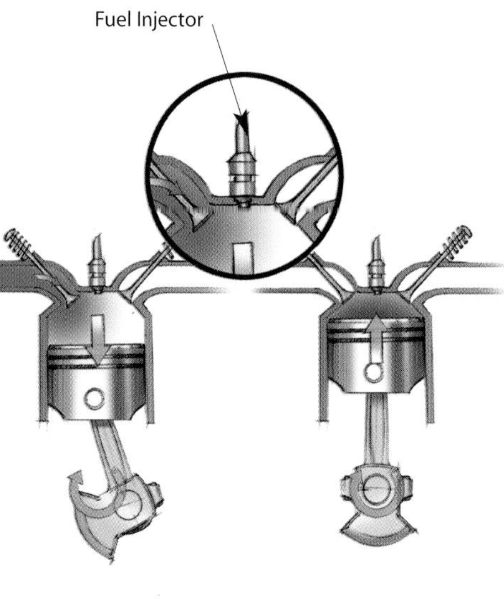

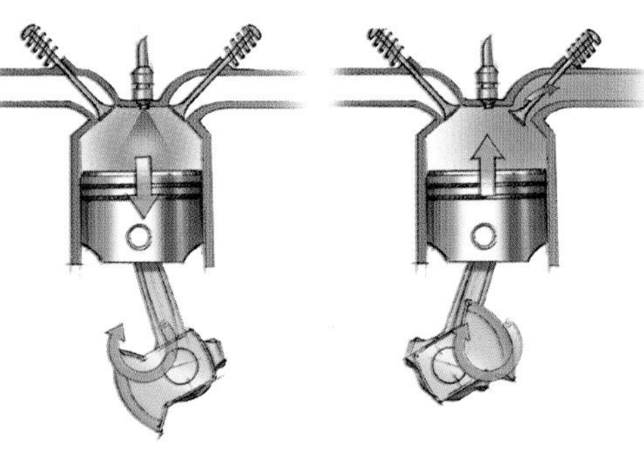

Intake   Compression   Combustion   Exhaust

How **ELECTRIC MOTORS** Work

Electric motors are everywhere! In your house, almost every mechanical movement that you see around you is caused by an AC (alternating current) or a DC (direct current) electric motor. By understanding how a motor works, you can learn a lot about magnets, electromagnets, and electricity in general.

An electric motor uses magnets to create motion. If you have ever played with magnets, you know that opposites attract and likes repel. So if you have two bar magnets with their ends marked north and south, then the north end of one magnet will attract the south end of the other. On the other hand, the north end of one magnet will repel the north end of the other, and similarly, south will repel south. Inside an electric motor, these attracting and repelling forces create rotational motion.

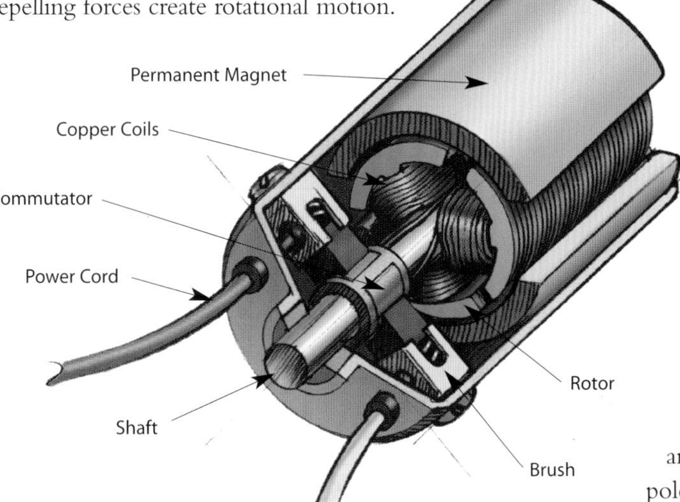

Electromagnets

An electromagnet is the heart of an electric motor. Imagine creating a simple electromagnet by wrapping 100 loops of wire around a nail and connecting it to a battery. The nail would become a magnet and have a north and south pole. Now say that you take your nail electromagnet, run an axle through the middle of it, and suspend it in the middle of a horseshoe magnet. The north end of the electromagnet would be repelled from the north end of the horseshoe magnet and attracted to the south end of the horseshoe magnet. The nail would move about half a turn and then stop.

The key to an electric motor is to go one step further so that, at the moment that this half-turn of motion completes, the field of the electromagnet flips.

You flip the magnetic field simply by changing the direction of the electrons flowing in the wire (you can do that by flipping the battery over). If the field of the electromagnet flipped at just the right moment at the end of each half-turn of motion, the electric motor would spin freely.

Parts of an Electric Motor

The armature takes the place of the nail in an electric motor. It is an electromagnet made of thin wire coiled around two or more poles of a metal core. The armature has an axle, and the commutator is attached to the axle.

The flipping of the electric field is accomplished by two motor parts: the commutator and the brushes. The commutator and brushes work together to let current flow to the electromagnet and also to flip the direction in which the electrons are flowing at just the right moment. The contacts of the commutator are attached to the axle of the electromagnet, so they spin with the magnet. The brushes are just two pieces of springy metal or carbon that make contact with the contacts of the commutator.

HSW Web Links

www.howstuffworks.com

How Power Door Locks Work
How Singing Fish Work
How Power Windows Work
How Wires, Fuses, and Connectors Work
How Electric Motors Work

Two-Pole vs. Three-Pole Motors

If you ever have the chance to take apart a small electric motor, you will find that it contains the same pieces described in this article: two permanent magnets, a commutator, two brushes, and an electromagnet made by winding wire around a piece of metal. Almost always, however, the motor will have three poles rather than two. In a two-pole motor, you can imagine the armature getting stuck or having trouble deciding which way to spin. That never happens in a three-pole motor.

How NUCLEAR POWER Works

Nuclear power plants provide about 17% of the world's electricity, and they also power nuclear submarines and aircraft carriers. A nuclear power plant is basically a controlled atom bomb, in which the power released by splitting atoms (fission) supplies immense heat to boil water that drives steam turbines and conventional electrical generators.

HSW Web Links

www.howstuffworks.com

How Atom Smashers Work
How Atoms Work
How Nuclear Power Works
How Nuclear Radiation Works
How Radon Works

Uranium is fairly common on earth. You find it as several isotopes (isotopes are atoms that have the same number of protons, but different numbers of neutrons):

- **Uranium-238 (U-238)**—This isotope is 99% uranium and has an extremely long half-life (4.5 billion years).
- **Uranium-235 (U-235)**—This isotope is 0.7% uranium.
- **Uranium-234 (U-234)**—This isotope is 0.3% uranium and is formed by the decay of U-238.

Some elements are naturally radioactive in all of their isotopes. Uranium is the best example and is the heaviest naturally occurring radioactive element, which means its atoms go through one of three common processes: alpha decay, beta decay, or spontaneous fission. For example, when an atom undergoes spontaneous fission, it spontaneously splits into two smaller atoms and gives off some radiation.

U-235 is one of the few materials that can undergo induced fission, so it is the isotope most commonly used in creating nuclear power. If a free neutron runs into a U-235 nucleus, the nucleus absorbs the neutron, becomes unstable, splits immediately, and produces two smaller atoms, two or three new free neutrons, and an immense amount of energy, in the form of gamma radiation and heat. This process occurs quickly (in about a picosecond, or one millionth of one millionth of a second). The free neutrons that are released cause another fission to occur if they run into other U-235 atoms.

For fission to be efficient, the uranium sample must be enriched so that it contains at least 2% to 3% U-235. Civilian nuclear power generation requires 3%. Weapons-grade uranium must be at least 90% U-235. In a nuclear bomb, you want the entire mass of uranium to undergo fission as quickly as possible, releasing all the energy in an instant and creating a gigantic explosion. In a nuclear power plant, you control the fission process to release that same energy over several months. However, exactly the same process is at work in the nuclear bomb and the nuclear power plant.

U-235 is not the only possible fuel for a power plant. Plutonium-239 can also undergo fission and can be created easily by bombarding U-238 with neutrons—something that happens all the time in a nuclear reactor. A breeder reactor is a reactor that runs on plutonium and also produces more plutonium as part of its operation.

Supercritical State

When a uranium atom splits, it throws off two or three neutrons. If, on average, exactly one of those neutrons splits another uranium atom, then the mass is critical and the mass will maintain the same temperature over time. If more than one of the neutrons splits another uranium atom, then over time the reaction rate grows and the uranium overheats and melts. This is a supercritical state. Plant operators insert neutronabsorbers to control the reaction rate and move it between subcritical, critical, and supercritical.

Inside a Nuclear Power Plant

To build a nuclear reactor, you need some mildly enriched uranium, usually formed into pellets with approximately the same diameter as a dime and a length of an inch or so. The pellets are arranged into long rods, which are collected together into bundles. The bundles are then typically submerged in water inside a pressure vessel. The water acts as a coolant. For the reactor to work, the bundle, submerged in water, must be slightly supercritical. If left to its own devices, the uranium in a supercritical bundle would eventually overheat and melt.

To prevent overheating, neutron-absorbing control rods are inserted into the bundle, by using a mechanism that can raise or lower the control rods. By absorbing neutrons, the control rods can cut down on the amount of induced fission that occurs in the bundle. The control rods allow operators to control the rate of the nuclear reaction. When an operator wants the uranium core to produce more heat, the rods are raised out of the

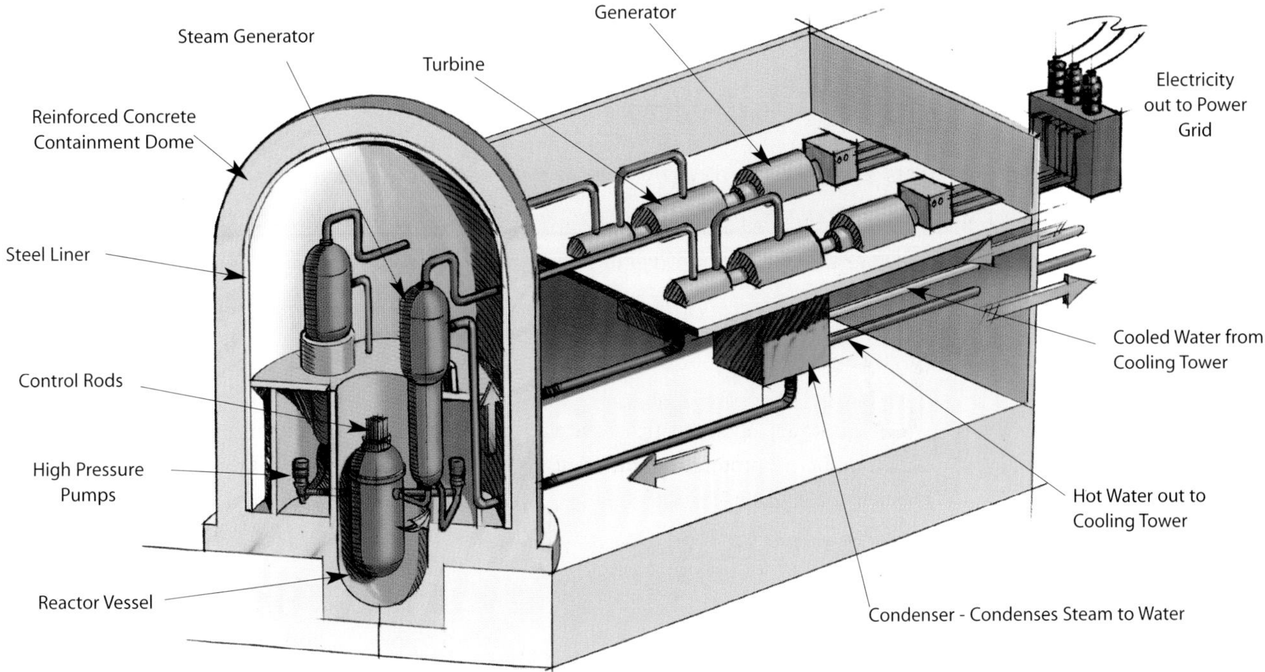

uranium bundle. To create less heat, the rods are lowered into the uranium bundle. The rods can also be lowered completely into the uranium bundle to shut down the reactor.

In most reactors, the uranium bundle heats the water into steam, which drives a steam turbine that spins a generator to produce power. In some reactors, the steam from the reactor goes through a secondary, intermediate heat exchanger to convert another loop of water to steam, which drives the turbine. The advantage to this design is that the radioactive water and steam never contact the turbine.

Once you get past the reactor itself, there is very little difference between a nuclear power plant and a coal-fired or an oil-fired power plant. All these power plants use heat to create steam that drives a turbine. The only difference between them is the source of heat.

The reactor's pressure vessel is typically housed inside a concrete liner that acts as a radiation shield. The liner is housed within a much larger steel containment vessel. This containment vessel encloses the reactor core as well as the cranes and other machinery that allow workers at the plant to refuel and maintain the reactor. It also prevents radioactive gases or fluids from leaking outside the plant. The containment vessel is protected by an outer concrete building, which is strong enough to survive catastrophes, such as 747s crashing right into the reactor building.

What Can Go Wrong

During normal operation, a well-constructed nuclear power plant is extremely clean compared to a coal-fired power plant. Nuclear power plants produce no air pollution or ash. Unfortunately, there are significant problems with them:

- Mining and purifying uranium has not, historically, been a very clean process.
- Improperly functioning nuclear power plants can leak radioactive substances. The Chernobyl disaster was caused by a poorly designed and improperly operated reactor that scattered a significant amount of radioactive dust into the atmosphere.
- Spent fuel from nuclear power plants is toxic for centuries and, as yet, there is no safe permanent way to store it.
- Transporting nuclear fuel to and from plants poses some risk, although to date the safety record in the United States has been good.

These problems, at least in the United States, have largely derailed the creation of new nuclear power plants. Society seems to have decided that the risks outweigh the rewards.

Energy from Fission

When a single atom splits, an incredible amount of energy is released. The resulting two atoms later release beta and gamma radiation of their own as well. Because the fission products and the neutrons weigh less than the original U-235 atom, the difference in weight is converted directly to energy at a rate governed by the equation $E = mc^2$. A pound of highly enriched uranium as used to power a nuclear submarine or nuclear aircraft carrier is equal to about a million gallons of gasoline. When you consider that a pound of uranium is smaller than a baseball and a million gallons of gasoline would fill a cube 50 feet per side (50 feet is as tall as a five-story building), you get an idea of the amount of energy available in just a little bit of U-235!

How SUBMARINES Work

You have probably seen submarines in movies and on TV; for example, the movies The Hunt for Red October, U-571, Crimson Tide, *and* Das Boot *all feature submarines. Military submarines can move around the oceans, virtually undetected, for months at a time, to conduct surveillance, disrupt shipping, and launch missiles. Research submarines can take scientists and explorers to incredible depths.*

HSW Web Links

www.howstuffworks.com

How Cruise Missiles Work
How Diesel Engines Work
How Electric Motors Work
How Nuclear Power Works
How Ultrasound Works

Submarines are very much like spaceships in that they are complete living environments, built so people can survive in a hostile environment. But when you simplify a submarine down to the basics, it is just a big, waterproof container that can float and sink on command, provide a livable environment for its crew for months at a time, move through the water to get from one place to another, and find its way around the deep ocean.

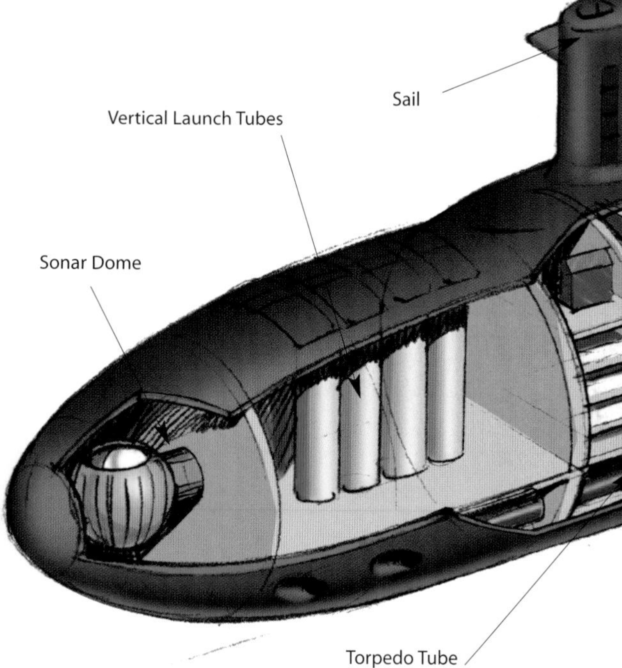

Diving and Surfacing

When you look at any boat floating on the water, you are seeing an example of buoyant force. As the boat sinks into the ocean, it displaces water. As soon as the amount of water displaced equals the weight of the entire boat, the boat stops sinking and a large part of the boat floats above the water. Water is heavy—it weighs about 62 pounds per cubic foot (1,000 kg/m^3). Water is much heavier than the combination of steel, people, and air that makes up a boat, so most of the boat is visible above the waterline.

A submarine floats for the same reason a boat does. But with a sub, you want to be able to float and sink on command. To make this possible, a submarine's designer pays attention to the sub's density. In other words, it is important for the weight of a submarine's steel, equipment, crew, and air to be close to the density of water. Then, to allow the sub to change density, the sub has ballast tanks—main ballast tanks in the hull and auxiliary, or trim, tanks on the interior.

The submarine's pilot can fill the ballast tanks with either water or air. When the tanks are full of air, the sub is light enough to float (positive buoyancy). When they are full of water, the sub is heavier than the water it displaces, and it sinks (negative buoyancy). If the amount of water and air in the tanks is perfectly balanced, a sub can hover in mid-ocean (neutral buoyancy). The sub has a supply of compressed air that it uses to blow the water out of the ballast tanks when it needs to. The submarine also has movable sets of short "wings" called hydroplanes on the bow (front) and/or stern (back) that help control the angle of the dive.

A pilot can steer a submarine in the water by using the tail rudder to turn to starboard (right) or port (left). A submarine may be equipped with a retractable motor and propeller that can swivel 360 degrees. This lets the sub act something like a helicopter in the water and move from side to side.

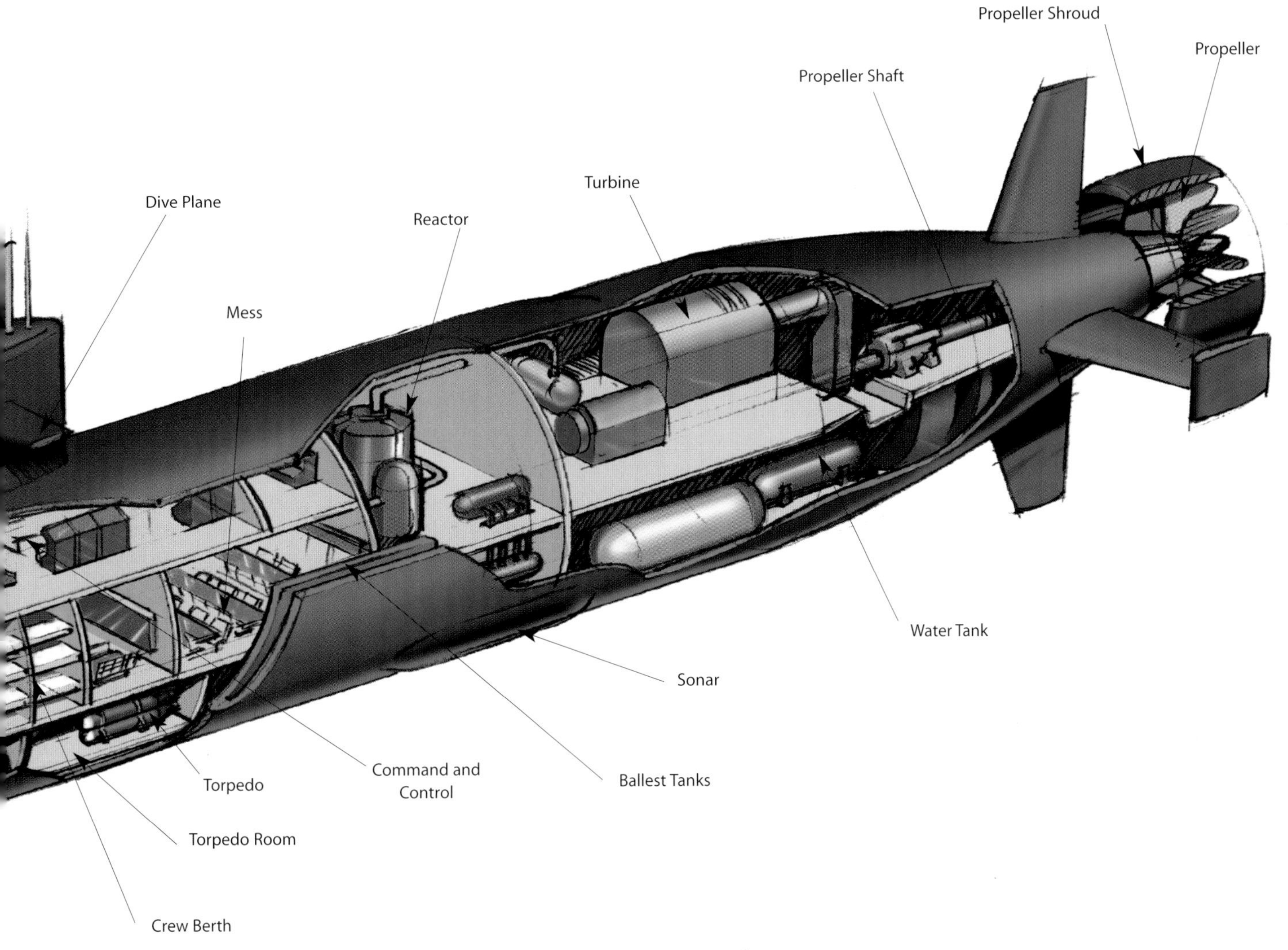

Life Support

The deep ocean is a hostile place. It is cold, the water pressure is very high, and there is no air for humans to breathe. This makes the deep ocean very much like outer space, and a sub has to have life-support systems just like a spaceship does. The life-support systems on a sub have to maintain the air quality, a fresh water supply, and temperature.

The air we breathe on earth is primarily made up of four gases:

- Nitrogen (78%)
- Oxygen (21%)
- Argon (0.94%)
- Carbon dioxide (0.04%)

When we breathe air, we consume oxygen and convert it to carbon dioxide. Exhaled air contains about 4.5% carbon dioxide. Our bodies do not do anything with the nitrogen or argon. Because a submarine is a sealed container that contains people and a limited supply of air, four things must happen to keep the submarine air breathable:

- Oxygen has to be replenished as it is consumed. If the percentage of oxygen in the air falls too low, a person suffocates.
- Carbon dioxide must be removed from the air. As the concentration of carbon dioxide rises, it becomes a toxin.
- The moisture that we exhale in our breath must be removed.
- Particulates (such as dust and smoke) must be removed.

A normal person at rest inhales and exhales about 2,906 gallons (11,000 l) of air every day and turns about 145 gallons (550 l) of oxygen into carbon dioxide every day.

How a Downed Submarine is Rescued

Rescue attempts from the surface must occur quickly, usually within 48 hours of the accident. Attempts typically involve trying to get some type of rescue vehicle down to remove the crew, or attaching some type of device to raise the submarine from the sea floor. Rescue vehicles include mini-submarines called Deep-Submergence Rescue Vehicles (DSRV) and diving bells. The DSRV can travel independently to the downed submarine, latch onto the submarine over a hatch (escape trunk), create an airtight seal so that the hatch can be opened, and load up to 24 crewmembers. A diving bell is typically lowered from a support ship down to the submarine, where a similar operation occurs. To raise the submarine, typically after the crew has been extracted, pontoons may be placed around the submarine and inflated to float it to the surface. Important factors in the success of a rescue operation include the depth of the downed submarine, the terrain of the sea floor, the currents in the vicinity of the downed submarine, the angle of the submarine, and the sea and weather conditions at the surface.

Oxygen on a sub comes either from pressurized tanks, an oxygen generator (which can form oxygen from electrolyzing water), or some sort of oxygen canister that releases oxygen through a very hot chemical reaction. On advanced subs, a computerized system senses the percentage of oxygen in the air and releases oxygen continuously as it is needed. On other subs, people release oxygen in batches when it is needed.

You can remove carbon dioxide from the air chemically by using soda lime (sodium hydroxide and calcium hydroxide) in a device called a scrubber. The carbon dioxide is trapped in the soda lime by a chemical reaction and removed from the air. Other similar reactions can accomplish the same goal.

The moisture in a submarine's air needs to be removed by using a dehumidifier or chemicals. If you don't remove the moisture, it condenses on the walls and equipment inside the ship. The smoke, dust, and other small particles are removed from a sub's air by passing the air through filters.

Most big submarines have a distillation plant that can take in seawater and produce fresh water. The distillation plant heats the seawater to water vapor, which removes the salts, and then cools the water vapor into a collecting tank of fresh water. The distillation plant on a large submarine can produce 10,000 to 40,000 gallons (37,850 to 151,399 l) of fresh water per day. The crew uses the water for drinking, cooking, and personal hygiene.

The temperature of the deep ocean surrounding the submarine is typically 39°F (4°C). The metal of the submarine conducts internal heat to the surrounding water, which makes a submarine like a giant refrigerator. So submarines must be heated to maintain a comfortable temperature for the crew. The power for the heaters comes from the nuclear reactor or diesel engine, or from batteries in an emergency.

Power Systems

The big problem for a submarine when it is under water is the lack of air. Any internal combustion engine breathes a huge amount of air and produces an equally huge amount of exhaust. This makes internal combustion engines useless when the sub is not floating on the surface.

A diesel submarine, therefore, is a very good example of a hybrid vehicle. Most diesel subs have two or more diesel engines. The diesel engines run generators that recharge a very large battery bank. The sub must surface (or cruise just below the surface, using a snorkel) to run the diesel engines. When the batteries are fully charged, the sub can head under water. The batteries power electric motors that drive the propellers. Battery operation is the only way a diesel sub can run while submerged. The limits of battery technology mean that the amount of time a diesel sub can stay under water is limited. (See "How Batteries Work," page 26.)

Because of the limitations of batteries, nuclear power in a submarine is a huge benefit. Nuclear generators need no oxygen, so a nuclear sub can stay under water for weeks at a time. (See "How Nuclear Power Works," page 30.) Also, because nuclear fuel lasts much longer than diesel fuel (years rather than days), a nuclear submarine does not have to come to the surface or to a port to refuel and can stay at sea for months at a time.

Nuclear subs (and aircraft carriers) are powered by nuclear reactors that are nearly identical to the reactors used in commercial power plants. The reactor produces heat to generate steam to drive a steam turbine. There are two big differences between commercial reactors and reactors in nuclear submarines:

- The reactor in a nuclear sub is smaller.
- The reactor in a nuclear sub uses highly enriched fuel to deliver a large amount of energy from a smaller reactor.

Nuclear submarines use nuclear reactors, steam turbines, and reduction gearing to drive the main propeller shaft, which provides forward and reverse thrust in the water (an electric motor drives the same shaft when docking or in an emergency). Submarines also need lots of electric power to operate the equipment on board, so steam turbines also drive electrical generators. Nuclear submarines also have batteries to supply electrical power in emergencies and when the reactor is offline.

Navigation

Small research subs and big military subs are completely different animals when it comes to navigation. In a research sub, the pilot has windows and lights and is normally moving around a very small area, such as an underwater shipwreck. The pilot navigates the sub visually. A big military sub, on the other hand, is cruising for long distances, deep in the open ocean. There is nothing to see, so there are no windows, and light does not penetrate very far into the ocean, so submarines must navigate through the water virtually blind.

When a military sub is on the surface, a global positioning system (GPS) receiver accurately determines latitude and longitude, but this system doesn't work when the submarine is under water. Under water, the submarine uses an inertial guidance system (IGS) that keeps track of the ship's motion from a fixed starting point, using gyroscopes and accelerometers. The IGSs are accurate for several days, but then small errors start to accumulate. A GPS, sonar, radio, radar, or satellites are used to realign the IGS every few days.

To locate a target, a submarine uses two types of sonar:

- **Active sonar**—Active sonar emits pulses of sound waves that travel through the water, reflect off the target, and return to the ship. By knowing the speed of sound in water and the time for the sound wave to travel to the target and back, the computers can quickly calculate distance between the submarine and the target. Whales, dolphins, and bats use the same technique—called echolocation—for locating prey. If the sub has a good underwater map, sonar systems can realign the inertial navigation systems by identifying known ocean floor features.
- **Passive sonar**—Passive sonar involves listening to sounds generated by the target. Passive sonar is used mostly because active sonar would give away the submarine's position to an enemy.

A big nuclear submarine can stay underwater for months at a time; it is a completely self-contained environment. Small research subs have taken people to the deepest corners of the ocean. It is very much like space travel, but a lot closer to home.

And Another Thing...

When a submarine goes down because of a collision or an on-board explosion, the crew will radio a distress call or launch a buoy that will transmit a distress call and the submarine's location. Depending upon the circumstances of the disaster, the nuclear reactors will shut down and the submarine may be on battery power alone. If this is the case, then the crew of the submarine has four primary dangers facing them:

- Flooding of the submarine must be contained and minimized.
- Oxygen use must be minimized so that the available oxygen supply can hold out long enough for possible rescue attempts.
- Carbon dioxide levels will rise and could produce dangerous, toxic effects.
- If the batteries run out, then the heating systems will fail and the temperature of the submarine will fall.

How **STEAM ENGINES** Work

Steam engines were the first type of engines to see widespread use. Thomas Newcomen first invented them in 1705, and James Watt (whom we remember each time we talk about 60-watt light bulbs and such) made big improvements to steam engines in 1769. Steam engines powered all early locomotives, steamboats, and factories, and therefore they acted as the foundation of the Industrial Revolution.

HSW Web Links

www.howstuffworks.com

How Car Engines Work
How Diesel Engines Work
How Diesel Two-Stroke Engines Work
How Gas Turbine Engines (and Jet Engines) Work
How Two-Stroke Engines Work

A steam engine is an external combustion engine. Fuel burns outside the engine to create high-pressure steam, and the pressure drives a piston. In a double-acting steam engine, steam can work alternately against both sides of the piston.

Exhaust Steam

The key to the steam engine is a sliding valve that controls the movement of steam in and out of the engine. The control rod for the valve is usually hooked into a linkage attached to the crosshead so that the motion of the crosshead slides the valve at the right times. On many steam engines, the exhaust steam simply vents into the air; in old locomotives, the exhaust steam ran through a smokestack to help draw more exhaust and therefore air through the coal. The way exhaust steam works explains why steam locomotives have to take on water at stations; the water is constantly being lost through the steam exhaust. It also explains where the "choo-choo" sound comes from. When the valve opens the cylinder to release the steam exhaust, the steam escapes under a great deal of pressure and makes a choo sound as it exits. When the train is first starting, the piston is moving very slowly, so the choo sounds are infrequent, and you hear "Choo . . . choo . . . choo." As speed builds the choo sounds blend together into a rhythmic puffing.

Steam in

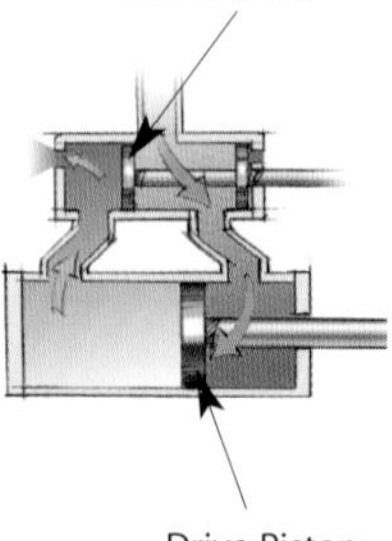

Boilers

The high-pressure steam in a steam engine comes from a boiler. The boiler's job is to apply heat to water to create steam. There are two types of boilers:

- **Fire tube boilers**—These boilers were common in the 1800s and are found in most remaining steam locomotives. A fire tube boiler consists of a tank of water perforated with pipes. The hot gases from a coal or wood fire run through the pipes to heat the water in the tank. In a fire tube boiler, the entire tank is under pressure, so if the tank bursts, it creates a major explosion.
- **Water tube boilers**—Water tube boilers are the most commonly used boilers today. In this type of boiler, water runs through a rack of tubes that are positioned in the hot gases from the fire.

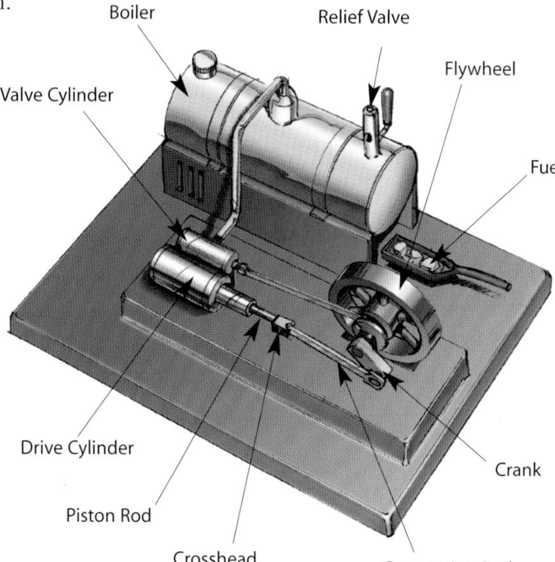

Steam engines are not very common today because they have been replaced by steam turbines. Most major power plants are essentially giant, stationary steam generators that extract power from heat with a steam turbine. The heat comes from coal, oil, natural gas, or a nuclear reactor.

How **TWO-STROKE ENGINES** Work

Two-stroke engines offer an incredible amount of power for their size. Because they are light, you find them in machines such as chain saws and weed trimmers.

Two-stroke engines have two important advantages over four-stroke engines: First, they do not have valves, which makes them simpler and lighter than four-stroke engines. Also, they fire once every revolution, whereas four-stroke engines fire once every other revolution; this difference gives two-stroke engines a significant power boost.

These two advantages make two-stroke engines lighter, simpler, and less expensive to manufacture than four-stroke engines. They also have the potential to pack about twice the power into the same space because there are twice as many power strokes per revolution than in a four-stroke engine. The combination gives two-stroke engines a great power-to-weight ratio.

The Two-Stroke Cycle

The two-stroke engine, as its name implies, has two strokes: one that handles intake and compression and another to handle combustion and exhaust. In a four-stroke engine there are four separate strokes: intake, compression, combustion, and exhaust.

The piston does three different things in a two-stroke engine:

- On one side of the piston is the combustion chamber. The piston is compressing the air/fuel mixture and capturing the energy released by ignition of the fuel.
- On the other side of the piston is the crankcase, where the piston is creating a vacuum to suck in air/fuel from the carburetor through the reed valve and then pressurizing the crankcase so that air/fuel is forced into the combustion chamber.
- Throughout the whole process, the sides of the piston are acting like the valves, covering and uncovering the intake and exhaust ports drilled into the side of the cylinder wall.

It's amazing to see the piston doing so many different things, and reusing the piston in three different ways like this is what makes two-stroke engines so simple and lightweight.

Disadvantages of Two-Stroke Engines

If two-stroke engines have such important advantages over four-stroke engines, why do all cars and trucks use four-stroke engines? There are four reasons:

- Two-stroke engines don't last nearly as long as four-stroke engines. The lack of a dedicated lubrication system means that two-stroke engine parts wear relatively quickly.
- Two-stroke oil is expensive and you need about 4 ounces (118 ml) of it for every 1 gallon (3.8 l) of gas. You would burn about a gallon of oil every 1,000 miles (1609 km) if you used a two-stroke engine in a car.
- Two stroke engines do not use fuel efficiently, so you would get poorer mileage.
- Two-stroke engines produce a lot of pollution—so much, in fact, that it is likely that you won't see them around too much longer. The pollution comes from two sources: the combustion of the oil and the little bit of raw fuel and oil that makes it out the exhaust port on every stroke.

These disadvantages mean that two-stroke engines are used only in applications where the motor is not used very often and the fantastic power-to-weight ratio of the two-stroke engine is important. To overcome the disadvantages of two-stroke engines, manufacturers have been working to miniaturize four-stroke engines.

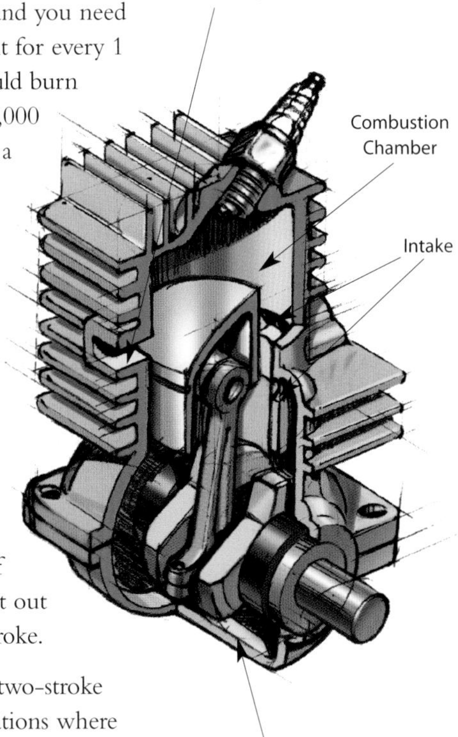

HSW Web Links

www.howstuffworks.com

How Car Engines Work
How Chain Saws Work
How Diesel Engines Work
How Diesel Two-Stroke Engines Work
How Steam Engines Work

How CHAIN SAWS Work

A chain saw is really just an engine with some cover plates bolted onto it, a handle to hold it, and a chain and bar to cut with. The engine is actually the main structural member that everything bolts onto. If you want to see a basic gasoline engine in its simplest application, then a chain saw is the best place to start!

HSW Web Links

www.howstuffworks.com

How Car Engines Work
How Electric Motors Work
How Fuel Injection Systems Work
How Gas Lanterns Work
How Oil Refining Works
How Smoke Detectors Work

The purpose of a chain saw is to spin the chain so you can cut wood. The chain looks a lot like the chain on a bicycle, but every second or third link has a saw tooth on it. By spinning the chain very fast, the chain becomes a powerful saw that is able to slice through logs very easily.

The rest of the saw is designed to make the chain go and to control the saw. The engine generates the power to spin the chain, and the handles and switches help the operator control the saw.

If you take the covers and handles off a chain saw, you get down to a small package about the size of a cantaloupe that is the heart of the machine. This package contains the engine itself, to which everything on the chain saw attaches, as well as a cooling fan and the following parts:

- **Centrifugal clutch**—The centrifugal clutch drives the chain.
- **Magneto**—The magneto provides the spark for the spark plug.
- **Carburetor**—The carburetor feeds air and fuel to the engine.

This whole package might weigh only 5 to 10 pounds (3 to 7 kg), but it is able to deliver around 3 horsepower, which is an amazing power-to-weight ratio. (See "How Two-Stroke Engines Work," page 37, for more information on why two-stroke engines have such a ratio.)

The Centrifugal Clutch

The centrifugal clutch is the link between the engine and the chain. The clutch disengages when the engine is idling so that the chain does not move. When the engine speeds up because the operator has pulled the throttle trigger, the clutch engages so that the chain can cut.

The clutch has three parts:

- **Outer drum**—The outer drum turns freely and includes a sprocket that engages the chain. When the drum turns, the chain turns.
- **Center shaft**—The center shaft is attached directly to the engine's crankshaft. If the engine is turning, so is the shaft.
- **Cylindrical clutch weights**—Two cylindrical clutch weights are attached to the center shaft, along with a spring that keeps them retracted against the shaft.

The center shaft and clutch weights spin as one unit. If they are spinning slowly, the weights are held against the shaft by the spring. If the engine spins fast enough, the centrifugal force on the weights overcomes the force being applied by the spring, and the weights sling outward. They come in contact with the inside of the drum, and the drum starts to spin. The drum, weights, and center shaft become a single spinning unit because of the friction between the weights and the drum. When the drum starts turning, so does the chain.

There are several advantages to a centrifugal clutch:

- It is automatic. In a car with a manual transmission, for example, you need a clutch pedal,but in a centrifugal clutch you don't.
- It slips automatically to avoid stalling the engine.
- When the engine is spinning fast enough, there is no slip in the clutch.

The Magneto

The magneto creates the electrical charge needed to fire the spark plug, and the spark

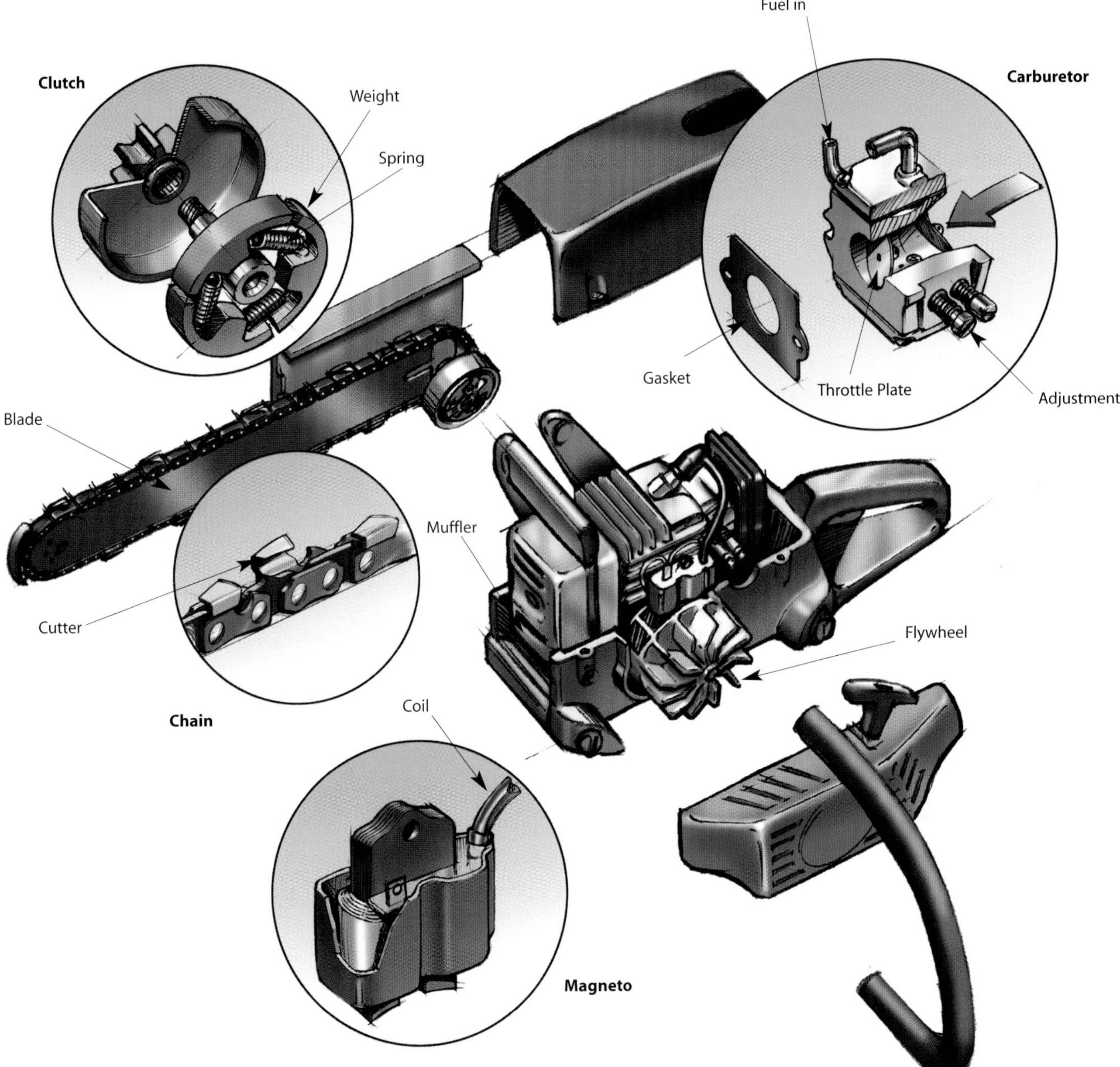

plug creates the spark inside the combustion chamber to ignite the gasoline. The job of the magneto is to create a surge of high voltage (between 10,000 and 20,000 volts) at just the right moment during each revolution of the crankshaft. This voltage arcs across the tip of the spark plug to ignite the gasoline.

The idea behind a magneto is simple. It is basically an electrical generator that has been tuned to create a periodic high-voltage pulse rather than continuous current. An electrical generator (or a magneto) is the opposite of an electromagnet. In an electromagnet there is a coil of wire around an iron bar (the armature). When you apply current to the electromagnet's coil (such as with a battery), the coil creates a magnetic field in the armature. In a generator, you reverse the process. You move a magnet past the armature to create electric current in the coil.

The magneto consists of five parts:

- **An armature**—The armature is normally shaped like a capital U. The two ends of the U point toward the magnets on the flywheel.
- **A primary coil**—The primary coil consists of about 200 turns of thick wire wrapped around one leg of the U.

And Another Thing...

If you've ever watched a high-end chain saw competition on TV, you've seen some amazing chain saws. The key components are the engine and the chain. Engines often come from snowmobiles and exceed 300 cc. With expansion chambers and other tweaks they generate massive amounts of horsepower!

Home and Professional Chainsaws

A typical chain saw manufacturer will offer models that range from small home saws to professional rigs. A typical home model will have a 14-inch or 16-inch bar, a 2.5-cubic inch (40 cc) engine producing about 2.5 horsepower, and will weigh about 10 pounds. The largest pro models can have up to a 36-inch bar, a 7.4-cubic inch (120 cc) engine generating almost 10 horsepower, and weigh in at 25 pounds or so.

One of the most amazing things about a chain saw is the power-to-weight ratio. A typical 3-horsepower lawn mower engine is twice as big and usually weighs more than twice as much. The difference comes from the fact that a chain saw uses a two-stroke engine while most lawn mowers use four stroke engines. A two-stroke engine fires twice as often and has no valves or camshaft to add weight.

- **A secondary coil**—The secondary coil consists of about 20,000 turns of very thin wire wrapped around the primary coil.
- **An electronic ignition**—The electronic ignition is a simple electronic control unit.
- **A magnets**—A pair of strong permanent magnets is embedded in the flywheel.

The Carburetor

The carburetor on a chain saw is extremely simple, as carburetors go, but it is certainly not uncomplicated. The job of the carburetor is to accurately meter extremely tiny quantities of fuel, and mix the fuel with the air entering the engine so that the engine runs properly. If there is not enough fuel mixed with the air, then the engine "runs lean," which can cause the engine not to run and can potentially damage the engine. If there is too much fuel mixed with the air, the engine "runs rich" and either does not run (it floods), runs with a lot of smoke, runs poorly (bogs down, stalls easily), or at least wastes fuel. The carburetor is in charge of getting the mixture just right.

A carburetor is essentially a tube. An adjustable plate across the tube, called the throttle plate, controls how much air can flow through the tube. At some point in the tube there is a narrowing, called the venturi, and in this narrowing a vacuum is created. In the venturi there is a hole, called a jet, that lets the vacuum draw in fuel.

At full throttle the throttle plate is parallel to the length of the tube, allowing maximum air to flow through the carburetor. The air flow creates a vacuum in the venturi, which draws in a metered amount of fuel through the jet. Most small carburetors have two adjusting screws. One of these screws controls how much fuel flows into the venturi at full throttle.

When the engine is idling, the throttle plate is nearly closed. There is not really enough air flowing through the venturi to create a vacuum. However, on the back side of the throttle plate there is a lot of vacuum action because the throttle plate is restricting the airflow. A tiny hole drilled into the side of the carburetor's tube, called the idle jet, makes it possible for fuel to be drawn into the tube by the vacuum. The other adjustment screw on the carburetor controls the amount of fuel that flows through the idle jet.

When the chain saw engine is cold and you try to start it with the pull cord, the engine is running at extremely low revolutions per minute, so it needs a very rich mixture to start. This is where the choke plate comes in. When activated, the choke plate completely covers the venturi. If the throttle is wide open and the venturi is covered, the engine's vacuum draws a lot of fuel through the main jet and the idle jet. Usually this very rich mixture allows the engine to fire once or twice or to run very slowly. If you then open the choke plate, the engine starts running normally.

Chain saws are great when you are interested in getting up close and personal with gasoline power. It is a miracle of modern manufacturing technology that you can buy a chain saw, containing several hundred parts, for about $100! And it will run great for years with little or no maintenance.

How GAS TURBINE ENGINES Work

When you go to an airport and see the commercial jets there, you can't help but notice the huge engines that power them. Most commercial jets are powered by turbofan engines, and turbofans are one example of a general class of engines called gas turbine engines. You may have never heard of gas turbine engines, but they are used in all kinds of unexpected places. For example, many of the helicopters you see, a lot of small power plants, and even M-1 tanks use gas turbines.

There are many different kinds of turbines, all of which capture the energy of a moving fluid:

- **Steam turbines**—Most power plants use coal, natural gas, oil, or a nuclear reactor to create steam. The steam runs through a huge and very carefully designed multistage turbine to spin an output shaft that drives the plant's generator.
- **Water turbines**—Hydroelectric dams use water turbines to generate power. The turbines used in a hydroelectric plant look completely different from steam turbines because water is so much denser (and slower moving) than steam, but both types of turbines use the same principle.
- **Wind turbines**—Wind turbines, also known as windmills, use the wind as their motive force. A wind turbine looks nothing like a steam turbine or a water turbine because wind is slow moving and very light, but again the principle is the same.
- **Gas turbines**—In a gas turbine, a pressurized gas spins the turbine. In all modern gas turbine engines, the engine produces its own pressurized gas, and it does this by burning something like propane, natural gas, kerosene, or jet fuel. The burning fuel and the heat create a high-speed rush of hot gas that spins the turbine.

Advantages and Disadvantages of Gas Turbines

An M-1 tank uses a 1,500 horsepower gas turbine engine instead of a diesel engine because it has two big advantages:

- Gas turbine engines have a great power-to-weight ratio compared to piston engines. That is, the amount of power you get out of the engine compared to the weight of the engine is very good.
- Gas turbine engines are smaller than their piston-based counterparts of the same power.

The main disadvantage of gas turbines is that, compared to reciprocating engines of the same size, they are expensive. Because they spin at such high speeds and because of the high operating temperatures, designing and manufacturing gas turbines is a tough problem from both the engineering and materials standpoints. Gas turbines also tend to use more fuel than piston-based engines when they are idling, and they prefer a constant load rather than a fluctuating load. That makes gas turbines great for things like transcontinental jet aircraft and power plants, but it explains why you don't have a gas turbine engine under the hood of your car.

Parts and Processes of Gas Turbine Engines

Gas turbine engines are, theoretically, extremely simple. They have three parts:

- **Compressor**—The compressor compresses the incoming air.
- **Combustion area**—The combustion area burns the fuel and produces high-pressure, high-velocity gas.
- **Turbine**—The turbine extracts the energy from the high-pressure, high-velocity gas flowing from the combustion chamber.

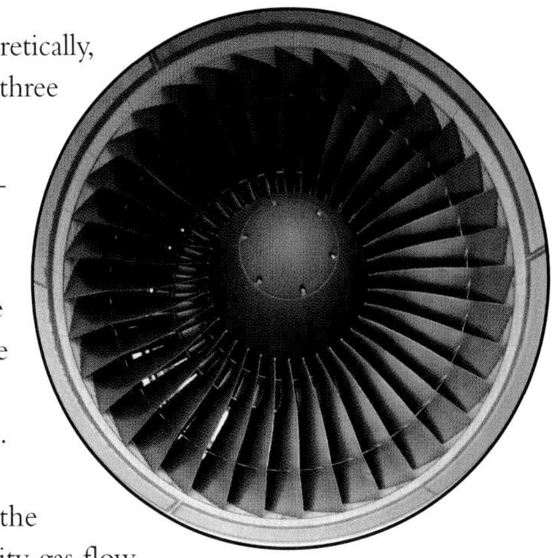

HSW Web Links

www.howstuffworks.com

How Air-Breathing Rockets Will Work
How Car Engines Work
How the Concorde Works
How Rocket Engines Work
How Space Shuttles Work

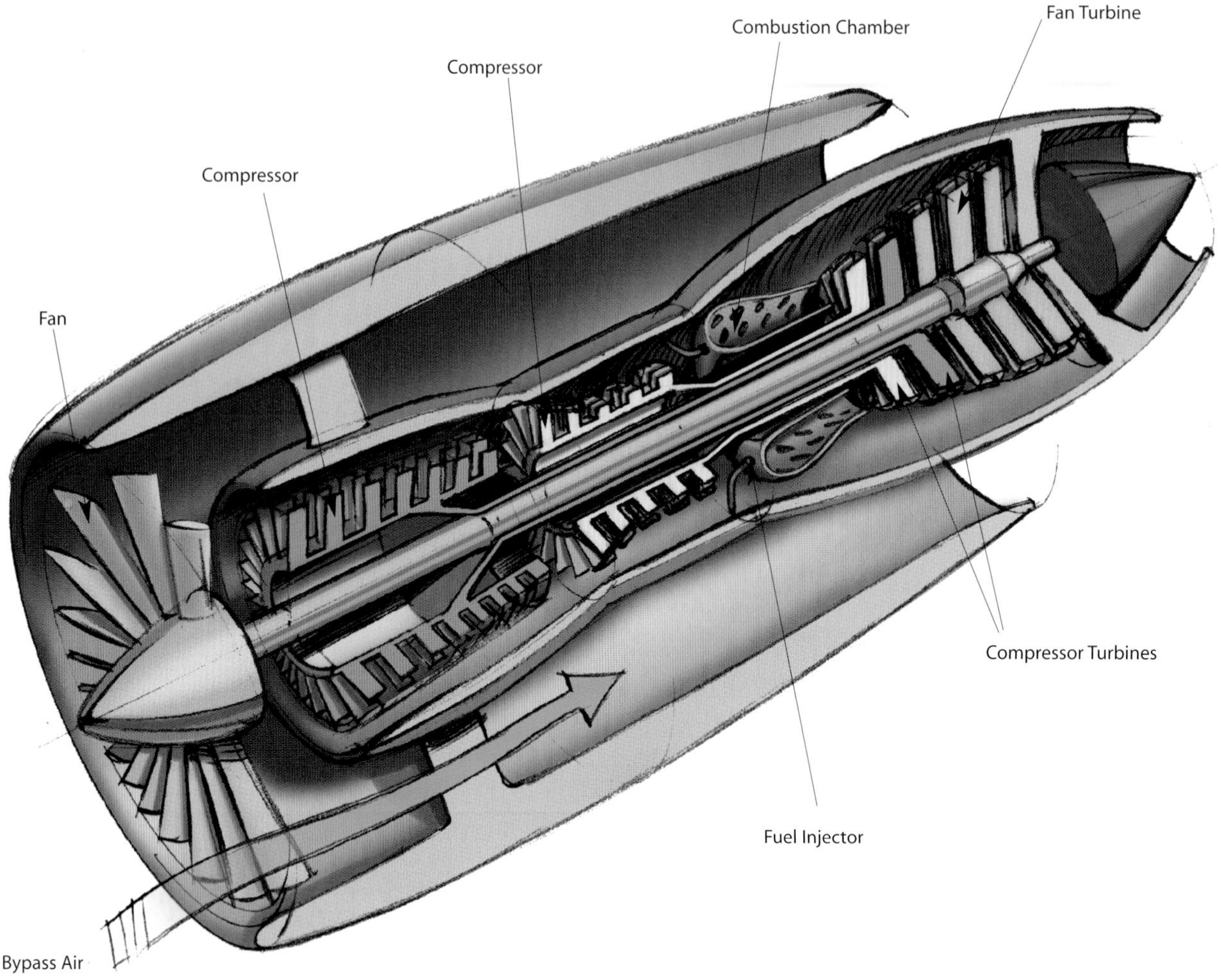

The compressor sucks air in. Generally, the compressor in a modern engine is basically a cone-shaped cylinder with small fan blades attached in rows. As air is forced through the compression stage, its pressure rises significantly. In some engines, the pressure of the air can rise by a factor of 30—for example, from 14.7 pounds per square inch (PSI) to 450 PSI.

This high-pressure air then enters the combustion area, where a ring of fuel injectors injects a steady stream of fuel. The fuel is generally kerosene, jet fuel, propane, or natural gas. If you think about how easy it is to blow out a candle, then you can see the design problem in the combustion area: High-pressure air is entering this area, moving at a very high velocity, and you need to keep a flame burning continuously in this environment. The device that solves this problem is called a flame holder, or sometimes a can, which is a hollow, perforated piece of heavy metal. Compressed air enters through perforations in the flame holder. Exhaust gases exit out the open end.

The gases next move through the turbines. The turbines are, essentially, sophisticated high-speed windmills. The hot gases pass through the turbines and spin them. In a jet airplane, the turbines feed their energy back up to drive the compressors and may also drive large fans or a propeller. In the gas turbine for the M-1 tank or a helicopter, a separate set of turbines drives an output shaft. This output shaft provides power the way the crankshaft on a reciprocating engine does. This final turbine stage and the output shaft are a completely standalone, freewheeling unit. They spin without any connection to the rest of the engine. And that is the amazing part about a gas turbine engine—there is enough energy in the hot gases blowing through the blades of that final output turbine to generate 1,500 horsepower and drive a 63-ton M-1 tank!

In the case of the turbine used in a tank or a power plant, there really is nothing to do with the exhaust gases but vent them through an exhaust pipe. Sometimes the exhaust runs through some sort of heat exchanger, either to extract the heat for some other purpose or to preheat air before it enters the combustion chamber.

Jet Engine Specifics

A jet engine for an airplane works slightly differently from other turbine engines. The whole point of the engine is to use exhaust to create thrust from the engine.

Large jetliners use turbofan engines, which are nothing more than gas turbines combined with large fans at the front of the engine. The core of a turbofan is a normal gas turbine engine. But the final turbine stage drives a shaft that makes its way back to the front of the engine to power the fan.

The purpose of the fan is to dramatically increase the amount of air moving through the engine and therefore increase the engine's thrust. When you look into the engine of a commercial jet at an airport, you see this fan at the front of the engine. It is huge (about 10 ft, or 3 m, in diameter on big jets), so it can move a lot of air. The air that the fan moves is called bypass air because it bypasses the turbine portion of the engine and moves straight through at a high speed to provide thrust.

A turboprop engine is similar to a turbofan engine, but instead of a fan there is a conventional propeller at the front of the engine. The output shaft connects to a gearbox to reduce the speed, and the output of the gearbox turns the propeller.

The goal of a turbofan engine is to produce thrust to drive the airplane forward. Thrust is generally measured in pounds of thrust in the United States (the metric system uses newtons [N], where 4.45 N is 1 lb of thrust). (See "How Rocket Engines Work," page 46, for a good discussion.)

Cabin Pressurization

One neat thing about having a gas turbine engine on an airplane is that it gives you a free air compressor to pressurize the cabin. The air pressure at an altitude of 30,00 feet is 4.3 PSI. If you've read the article "How Tire Pressure Gauges Work," page 72, then you know that the air pressure at sea level, 14.7 PSI, is what our bodies consider to be normal. According to Federal Aviation Regulations, passenger airplanes must be able to provide a cabin pressure of not more than that of air pressure at 8,000 feet, which means the cabin pressure has to be at least about 10.9 PSI. The air pressure in an unpressurized cabin at or above 30,000 feet is so low that the oxygen will diffuse out of a passenger's blood, causing a condition known as hypoxia. This lack of oxygen can cause a person to pass out, and even die. Passengers can actually begin to feel the effects of hypoxia at around 12,000 feet, so the FAA requires that passengers be provided with supplemental oxygen either through an oxygen mask or cabin pressurization.

Obviously, it doesn't make sense for airlines to require their passengers to wear oxygen masks for the duration of flights, so they use cabin pressurization. In order to properly pressurize the cabin, high-pressure air is used to "pump up" the cabin in much the same way a tire pump is used to pump up the air in a tire. The compression stage of gas turbine engines produces high-pressure air. Because the air has been compressed, it is very hot, so the air is sent through air-conditioning packs before it is pumped into the cabin. As new air is pumped into the cabin, air is released through an outflow valve so that the cabin pressure is maintained at a stable level.

The basic principles described here govern all gas turbine engines, and these engines have had a major impact on modern air travel, among other things.

Afterburners

The idea behind an afterburner is to inject fuel directly into the exhaust stream and burn it to create more thrust. This can increase the thrust of a jet engine by 50% or more.

The big advantage of an afterburner is that you can significantly increase the thrust of the engine without adding much weight or complexity to the engine. The disadvantage of an afterburner is that it uses a lot of fuel for the power it generates.

An afterburner is nothing but a set of fuel injectors in a tube at the back of the engine. A jet engine with an afterburner needs an adjustable nozzle so that it can work both with the afterburners on and off.

How FUEL CELLS Work

One of the biggest problems with a gasoline engine is that a lot of the energy in the gasoline is wasted as heat. Only about 20% of the power available in the gasoline actually makes it to a car's tires. Gasoline engines also create pollution (see "How Catalytic Converters Work," page 77). Fuel cell technology is extremely interesting because it can create power more efficiently and with less pollution than can gasoline engines.

HSW Web Links

www.howstuffworks.com

How Air-Breathing Rockets Will Work
How Asteroids Work
How Car Engines Work
How Gas Turbine Engines (and Jet Engines) Work
How Space Shuttles Work
How Spacesuits Work

A fuel cell converts hydrogen and oxygen into electricity and heat. It is sort of like a battery that can be recharged while you are drawing power from it. Instead of recharging using electricity, however, a fuel cell uses hydrogen and oxygen. Like a battery, a fuel cell produces a DC (direct current) voltage.

Types of Fuel Cells

There are several different types of fuel cells, each of which uses a different chemistry. Fuel cells are usually classified by the type of electrolyte they use. Some types of fuel cells show promise for use in power-generation plants. Others may be useful for small, portable applications or for powering cars. The proton exchange membrane fuel cell (PEMFC) is one of the most promising technologies for small fuel cells.

The PEMFC uses one of the simplest reactions of any fuel cell. The basic idea is this:

1) Hydrogen (H_2) molecules hit a catalyst on one side of the fuel cell. This strips off the electrons and each hydrogen molecule becomes a hydrogen ion that is missing electrons.
2) On the other side of the fuel cell, oxygen (O_2) molecules hit a catalyst and each splits into two oxygen atoms. The oxygen

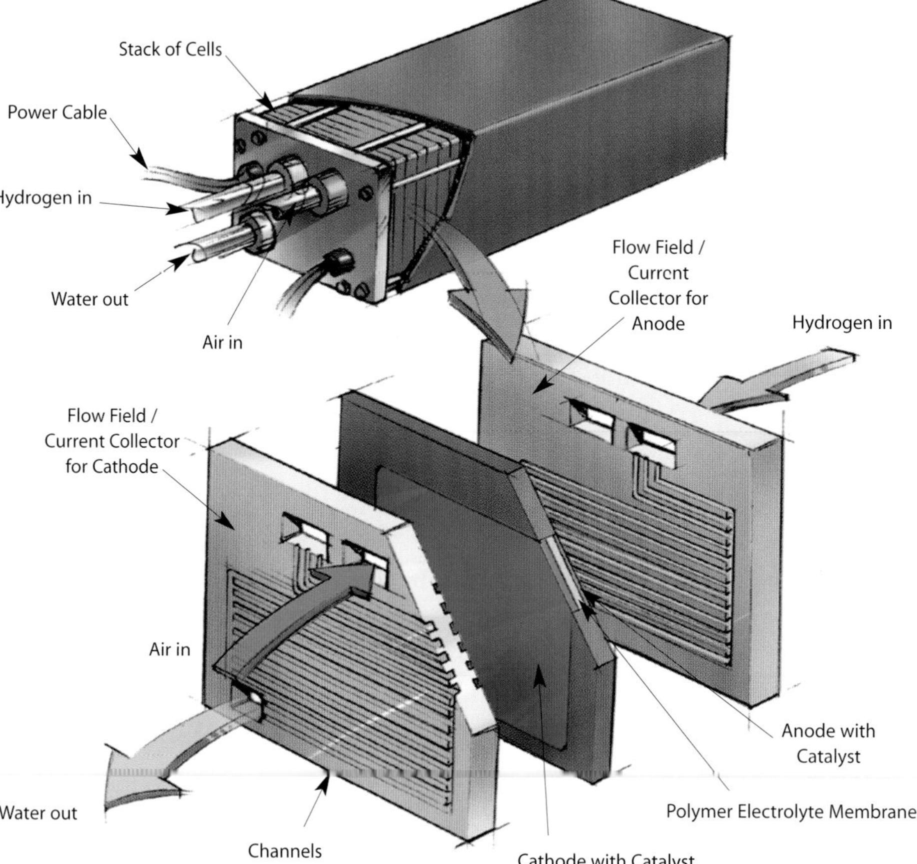

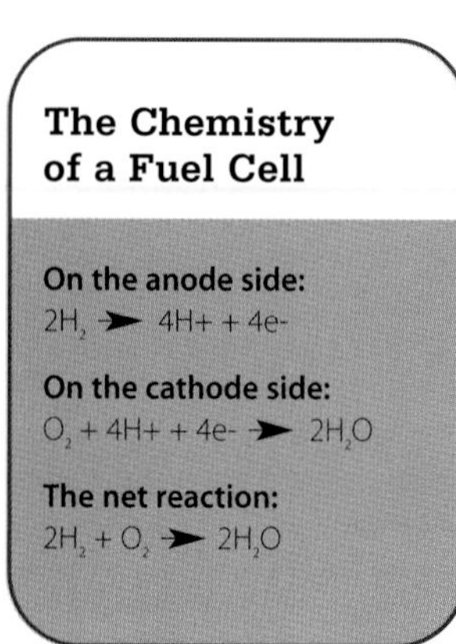

The Chemistry of a Fuel Cell

On the anode side:
$2H_2 \rightarrow 4H^+ + 4e^-$

On the cathode side:
$O_2 + 4H^+ + 4e^- \rightarrow 2H_2O$

The net reaction:
$2H_2 + O_2 \rightarrow 2H_2O$

atoms attract the hydrogen ions through the membrane and combine.

3) Meanwhile, the electrons freed from the hydrogen make their way through a circuit and arrive at the other side of the fuel cell. Hydrogen ions, oxygen atoms, and electrons combine to form water molecules.

This reaction in a single fuel cell produces only about 0.7 volts. To get this voltage up to a usable level, many fuel cells must be combined to form a fuel cell stack.

Parts of Fuel Cells

The anode, the negative post of the fuel cell, has several jobs. It conducts the electrons that are freed from the hydrogen molecules so that they can be used in an external circuit. It has channels etched into it that disperse the hydrogen gas equally over the surface of the catalyst.

The cathode is the positive post of the fuel cell. The cathode has channels etched into it that distribute the oxygen to the surface of the catalyst. It also conducts the electrons back from the external circuit to the catalyst, where they can recombine with the hydrogen ions and oxygen to form water.

The electrolyte is the proton-exchange membrane. This specially treated material, which looks something like ordinary kitchen plastic wrap, conducts only positively charged ions. The membrane blocks electrons.

The catalyst is a special material that facilitates the reaction of oxygen and hydrogen. It is usually made of platinum powder very thinly coated onto carbon paper or cloth. The catalyst is rough and porous so that the maximum surface area of platinum can be exposed to the hydrogen or oxygen.

Fuel cells use oxygen and hydrogen to produce electricity. The oxygen required for a fuel cell comes from the air. In fact, in the PEMFC, ordinary air is pumped into the cathode. Hydrogen, on the other hand, is difficult to store and distribute. A reformer addresses this problem by turning hydrocarbon fuel or alcohol fuel into hydrogen, which is then fed to the fuel cell.

The Efficiency of Fuel Cells

Fuel cells are more efficient than gasoline engines, but they are not perfect. Comparing a fuel cell–powered car to a gasoline engine–powered car and a battery-powered electric car helps us see the differences:

- **Fuel cell–powered electric car**— If the fuel cell is powered with pure hydrogen, it has the potential to be up to 80% efficient. It converts 80% of the energy content of the hydrogen into electrical energy. When you add a reformer to convert methanol to hydrogen, however, the overall efficiency drops to about 30% to 40%. You still need to convert the electrical energy into mechanical work by using a motor/inverter with about 80% efficiency. That gives an overall efficiency of about 24% to 32%.
- **Gasoline engine–powered car**— All the heat that comes out as exhaust or goes into the radiator is wasted energy. Only about 20% of the energy content of the gasoline is converted into mechanical work.
- **Battery-powered electric car**— An electric car has fairly high efficiency. The battery is about 90% efficient (most batteries generate some heat, but not much), and the electric motor/inverter is about 80% efficient. This gives an overall efficiency of about 72%. But this is not the whole story. The electricity used to power the car has to be generated somewhere. If it was generated at a power plant that used a combustion process (rather than nuclear, hydroelectric, solar, or wind power), then only about 40% of the fuel required by the power plant converts into electricity. The process of charging the car requires the conversion of AC power to DC power. This process has an efficiency of about 90%. That gives an overall efficiency of 26%.

Maybe you are surprised by how close these three technologies are. This example points out the importance of considering the whole system involved, not just the car itself.

Other Types of Fuel Cells

- **Alkaline fuel cell (AFC):** One of the oldest designs—in use since the 1960s— the AFC is very susceptible to contamination so it requires pure hydrogen and oxygen. It is also very expensive, so this type of fuel cell is unlikely to be commercialized.
- **Phosphoric acid fuel cell (PAFC):** The phosphoric acid fuel cell has potential for use in small stationary power generation systems. Because it operates at a higher temperature than PEM fuel cells, the PAFC has a longer warm-up time, making it unsuitable for use in cars.
- **Solid oxide fuel cell (SOFC):** These fuel cells are best suited for large-scale stationary power generators that could provide electricity for factories or towns. Because SOFCs operate at very high temperatures (around 1832°F, 1000°C), reliability is a problem, but it also has an advantage. The steam produced by the fuel cell can be channeled into turbines to generate more electricity, improving the overall efficiency of the system.
- **Molten carbonate fuel cell (MCFC):** Best suited for large stationary power generators, MCFCs operate at 1112°F (600°C), so they also generate steam that can be used to generate more power. They have a lower operating temperature than the SOFC, which means they don't need such exotic materials, making the design a little less expensive.

How **ROCKET ENGINES** Work

One of the most amazing endeavors humans have ever undertaken is the exploration of space, which can't happen without rocket engines. Rocket engines are on the one hand so simple that you can build and fly your own model rockets very inexpensively. On the other hand, rocket engines (and their fuel systems) are so complicated that only two countries have actually ever used them to put people in orbit. In order to get people and equipment into orbit, rocket engines turn massive amounts of energy into motion.

HSW Web Links

www.howstuffworks.com

How Air-Breathing Rockets Will Work
How Asteroids Work
How Gas Turbine Engines (and Jet Engines) Work
How Space Shuttles Work
How Spacesuits Work

Rocket engines are fundamentally different from electric motors and gasoline engines because they are reaction engines. The basic principle driving a rocket engine is the famous Newtonian principle that to every action, there is an equal and opposite reaction. A rocket engine throws mass out the back (the action) and gets pushed forward by the reaction.

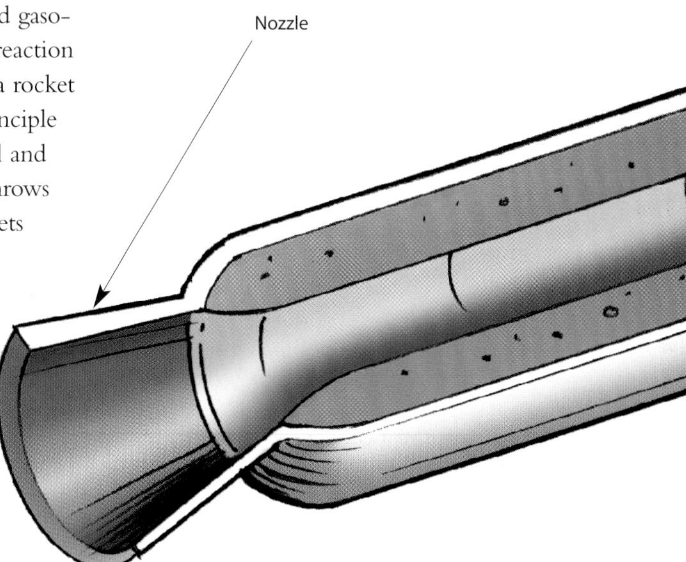

Action and Reaction

This concept of throwing mass backward and moving forward seems hard to grasp because rocket engines appear to be about flames, noise, and pressure, rather than about throwing things. So let's look at two examples:

- If you've ever fired a shotgun, especially a 12-gauge shotgun, then you know that it has a lot of "kick." That kick is a reaction. If you were wearing roller skates or standing on a skateboard when you shot such a gun, then the gun would be acting like a rocket engine, and you would react by rolling in the direction opposite the direction of the gunshot.
- When you blow up a balloon and let it go so that it flies all over the room, you also get an action/reaction response. In this case, the air molecules inside the balloon are being thrown. We tend to think of air as weightless, but air molecules do have mass. When you throw them out the nozzle of a balloon, the rest of the balloon reacts by flying in the opposite direction.

Let's say that you are in a spacesuit floating in space, and you have a baseball. If you throw the baseball, your body reacts by moving backward, in the opposite direction the ball is traveling. The mass *(m)* of the baseball and the amount of acceleration *(a)* that you apply to it control the speed at which your body moves. Mass multiplied by acceleration is force ($f = ma$).

The force that you apply to the baseball is equal to the reaction force that the baseball applies to you: ($m_{ball}a_{ball} = m_{body}a_{body}$). So let's say that the baseball weighs 1 pound (0.45 kg) and your body plus the spacesuit weighs 100 pounds (45 kg). You throw the baseball away with some acceleration. Your body reacts with an identical force ($m_{ball}a_{ball}$), but because you are 100 times more massive than the baseball, you move away at 1/100 the acceleration of the ball.

If you want to generate more thrust from the baseball, you have only two options:

- **Increase the mass**—Throw a heavier ball or a number of baseballs, one after another.

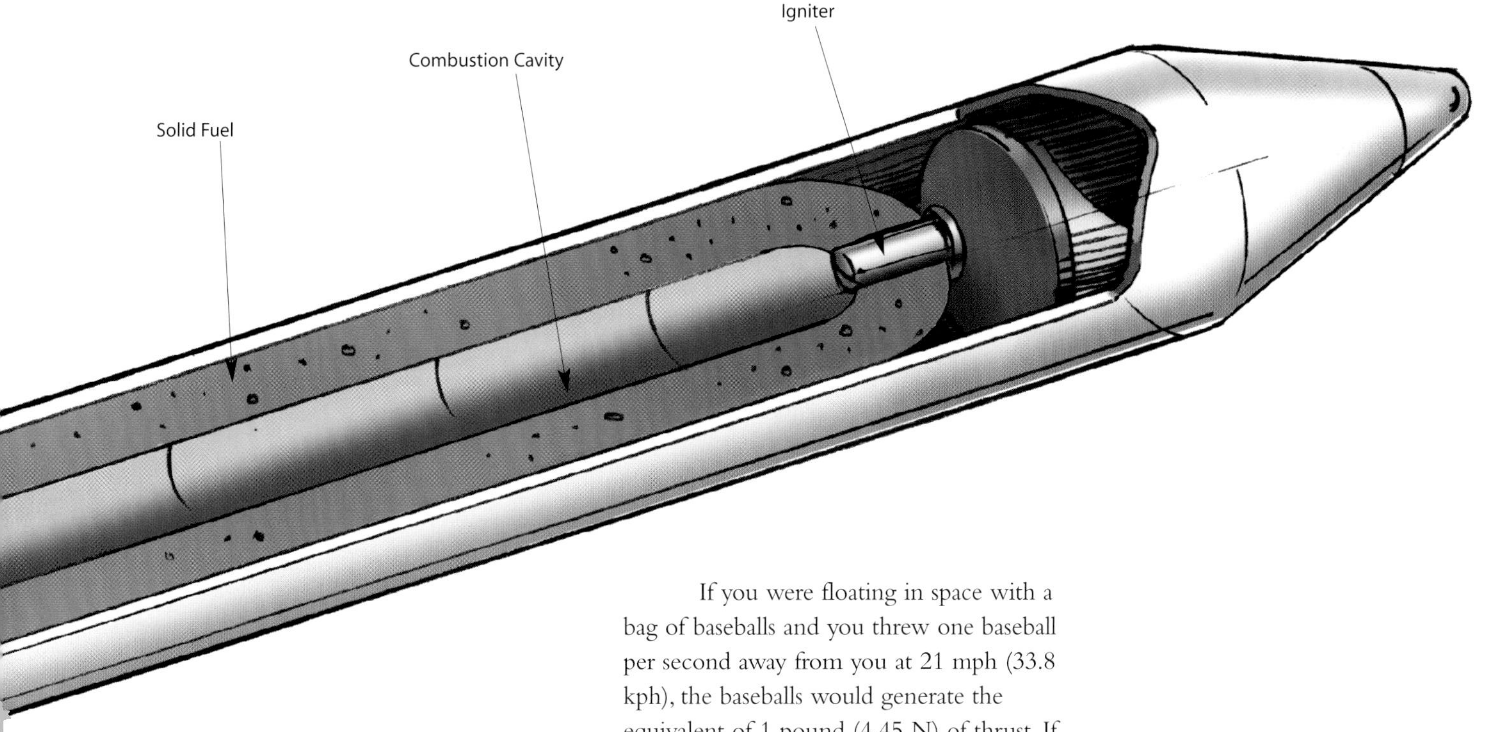

- **Increase the acceleration on the ball**—Throw the baseball faster.

A rocket engine generally throws mass in the form of high-pressure gas molecules (like many small baseballs, thrown very fast). The mass comes from the weight of the fuel that the rocket engine burns. The burning process accelerates the mass of fuel so that it comes out of the rocket nozzle at high speed. The fact that the fuel turns from a solid or liquid into a gas when it burns does not change its mass. If you burn a pound of rocket fuel, a pound of exhaust comes out the nozzle in the form of high temperature, high-velocity gas. The form changes, but the mass does not.

Thrust

The force of a rocket engine is called thrust. Thrust is measured in pounds in the English system and in newtons (N) in the metric system (4.45 N = 1 lb thrust). A pound of thrust is the force that it would take to keep a 1-pound object stationary against the force of gravity on earth. So on earth, the acceleration of gravity is 32 feet per second (21 mph per second), or 9.8 m per second.

If you were floating in space with a bag of baseballs and you threw one baseball per second away from you at 21 mph (33.8 kph), the baseballs would generate the equivalent of 1 pound (4.45 N) of thrust. If you were to throw the baseballs instead at 42 mph (67.6 kph), then you would generate 2 pounds (8.9 N) of thrust. If you were to throw them at 2,100 mph (3,379 kph)—perhaps by shooting them out of some sort of baseball gun—then you would generate 100 pounds (445 N) of thrust.

One of the funny problems rockets have is that what the engine wants to throw (the fuel) actually weighs something, and the rocket has to carry that weight around. So let's say that you want to generate 100 pounds (445 N) of thrust for an hour by throwing one baseball every second at a speed of 2,100 mph (3,379 kph). That means that you have to start with 3,600 1-pound baseballs (there are 3,600 seconds in an hour) or 3,600 pounds (16,020 N) of baseballs. Because you weigh only 100 pounds (445 N) in your spacesuit, you can see that the weight of your fuel (the baseballs) dwarfs the weight of the payload (you). In fact, the fuel weights 36 times more than the payload. That is why you have to have a huge rocket to get a person into space—you have to carry a lot of fuel.

You can see this weight equation very clearly on a space shuttle. At launch, there

> **Cool Facts**
>
> The Space Shuttle throws fuel backward at 6,000 mph (typical rocket exhaust velocities for chemical rockets = 5,000 - 10,000 mph). The SRBs burn for about 2 minutes and generate about 3.3 million lb of thrust each at launch (2.65 million lb average over the burn). The three main engines burn fuel from the external tank for about 8 minutes; thereby generating 375,000 lb of thrust each during the burn.

Did You Know?

- Solid rockets were invented hundreds of years ago in China.
- The line about "the rocket's red glare" in the National Anthem (written in the early 1800s) is talking about small military solid-fuel rockets used to deliver bombs or incendiary devices.
- In 1926, Robert Goddard tested the first liquid propellant rocket engine. His engine used gasoline and liquid oxygen. He also worked on and solved a number of fundamental problems in rocket engine design, including pumping mechanisms, cooling strategies and steering arrangements. These problems are what make liquid propellant rockets so complicated.

are three parts: the orbiter, the big external tank, and the two solid rocket boosters (SRBs). The entire vehicle totals 4.4 million pounds (9.7 million kg) at launch because the fuel weighs almost 20 times more than the shuttle. Therefore, it takes 4.4 million pounds to get 165,000 pounds (363,000 kg) in orbit!

Solid-fuel Rocket Engines

The idea behind a simple solid-fuel rocket is straightforward. You want to create something that burns very quickly but does not explode. Gunpowder, which is 75% nitrate, 15% carbon, and 10% sulfur, explodes. In a rocket engine, you don't want an explosion—you would like the power released evenly over a period of time. Therefore, you might change the mix to 72% nitrate, 24% carbon, and 4% sulfur, to create a simple rocket fuel. This sort of mix burns very rapidly, but it does not explode if loaded properly. In a small model rocket engine or in a tiny bottle rocket, the burn might last a second or less. In a space shuttle SRB containing more than a million pounds (500,000 kg) of fuel, the burn lasts about 2 minutes.

Solid-fuel rocket engines have two important advantages: simplicity and low cost. They also have two disadvantages: You cannot control the thrust and you cannot stop or restart the engine after you ignite it. The disadvantages mean that solid-fuel rockets are useful for short-lifetime tasks (such as deploying missiles) and for booster systems. When you need to be able to control the engine, you must use a liquid propellant system.

Liquid-propellant Rockets

The basic idea of a liquid-rocket engine is simple. A fuel and an oxidizer (for example, gasoline and liquid oxygen) are pumped into a combustion chamber, where they burn to create a high-pressure and high-velocity stream of hot gases. These gases flow through a nozzle, which accelerates them even more; typical exit velocity is 5,000 to 10,000 mph (8,000 to 16,000 kph).

The fuel and/or the oxidizer are often a cold liquefied gas, such as liquid hydrogen or liquid oxygen. One big problem in a liquid-propellant rocket engine is that the combustion chamber and nozzle need to be cooled; therefore, the cryogenic liquids are first circulated around the superheated parts to cool them. The pumps have to generate extremely high pressures to overcome the backpressure that the burning fuel creates in the combustion chamber. The main engines in a space shuttle use two pumping stages and burn fuel to drive the second-stage pumps. All this pumping and cooling makes a typical liquid-propellant engine extremely complex.

A variety of fuel combinations, such as the following, are used in liquid-propellant rocket engines:

- Liquid hydrogen and liquid oxygen are used in a space shuttle's main engines.
- Gasoline and liquid oxygen were used in early rockets.
- Kerosene and liquid oxygen were used in the Apollo Saturn V booster's first stage.
- Alcohol and liquid oxygen were used in German V2 rockets.
- Nitrogen tetroxide (NTO) and monomethyl-hydrazine (MMH) are used in a space shuttle's maneuvering engines.

Other Ways to Generate Thrust

There are many ways to generate thrust besides burning fuel. Any system that throws mass generates thrust. If you could figure out a way to accelerate baseballs to extremely high speeds, for example, you would have a viable rocket engine.

Many rocket engines are very small. For example, attitude thrusters on satellites don't need to produce much thrust. One common engine design used in satellites uses no "fuel" at all; pressurized nitrogen thrusters simply blow nitrogen gas from a tank through a nozzle.

Engine designers are trying to find ways to accelerate ions or atomic particles to extremely high speeds to create thrust more efficiently. For example, NASA's Deep Space-1 spacecraft was the first to use ion engines for propulsion.

How PC POWER SUPPLIES Work

If there is any one component that is absolutely vital to the operation of a computer, it is the power supply. Without it, a computer is just an inert box full of plastic and metal. The power supply converts the alternating current (AC) line from your home to the direct current (DC) needed by the personal computer (PC).

In a typical computer, the power supply is in a metal box found in a rear corner of the case. You can see the back of the power supply for many systems: It contains the power cord receptacle and the cooling fan.

Switcher Technology

Chips such as the microprocessor and random-access memory (RAM), as well as components such as hard disks and CD-ROM drives, depend on low-voltage DC power. A PC power supply has to convert the 120-volt AC power from a wall outlet to DC voltages. The typical voltages supplied are:

- 3.3 volts, 14 amps
- 5 volts, 30 amps
- 12 volts, 12 amps

The traditional way to make this conversion is with a transformer. The transformer accepts 120-volt AC and produces, for example, 12-volt AC. A rectifier converts the AC to DC, and a filtering capacitor smoothes out the DC voltage so that it runs at a very stable 12 volts.

The problem with transformers running at 60 Hertz (Hz), which is the frequency delivered by a wall outlet, is that they are heavy and bulky. So modern PC power supplies use switching technology to optimize things.

A switcher supply converts the 60-Hz AC current to a much higher frequency. This frequency conversion lets the power supply use a small, lightweight transformer. The high-frequency AC current provided by a switcher supply is also easier to filter than the original 60-Hz AC line voltage, which reduces the size of the filtering capacitors.

Turning on the Power Supply

If you have been around PCs for many years, you probably remember that the original PCs had large red toggle switches that had a good bit of heft to them. When you turned the PC on or off, you knew you were doing it. These switches actually controlled the flow of 120-volt power to the power supply.

Today you turn on the power with a little push button, and you turn off the machine through a menu option. The push button sends a 5-volt signal to the power supply to tell it when to turn on; the power supply also has a circuit that supplies 5 volts even when it is officially "off," so that the button will work. The operating system sends a signal to the power supply to tell it to turn off.

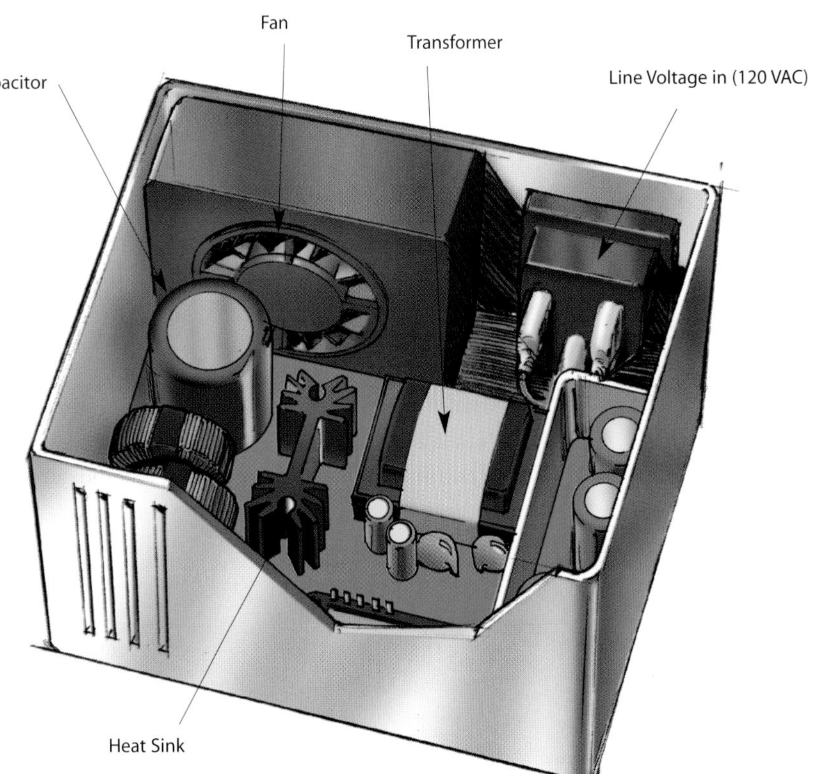

HSW Web Links

www.howstuffworks.com

How Computer Mice Work
How Digital Television Works
How Hard Disks Work
How Modems Work
How Surge Protectors Work

How the **POWER DISTRIBUTION GRID** Works

We take electrical power completely for granted. It's just there, constantly meeting your every need. Only during a power failure does anyone realize how important power is in daily life. Heating, air conditioning, cooking, refrigeration, light, sound, motion, computation, entertainment, and more depend on electricity!

HSW Web Links

www.howstuffworks.com

How Atoms Work
How Emergency Power Systems Work
How Solar Cells Work
How Solar Yard Lights Work

Power travels from the power plant to your house through an amazing system called the power distribution grid. The grid is quite public—if you live in a suburban or rural area, chances are it is right out in the open for all to see. It is so public, in fact, that you probably don't even notice it.

At the Power Plant

Electrical power starts at a power plant, which consists of a spinning electrical generator. The generator might be spun by a water wheel in a hydroelectric dam or by a large diesel engine or a gas turbine in a small power plant. Most often, a steam turbine spins the generator. The steam might be created by burning fossil fuel (coal, oil, natural gas) or by a nuclear reactor. All commercial electrical generators of any size generate three-phase AC (alternating current) power. To understand three-phase AC power, it is helpful to understand one-phase power first.

One-phase power is what you have in your house. A normal electrical outlet in the United States delivers one-phase 120-volt AC service. If you examine the power at a normal outlet in your house with an oscilloscope, you will find that it looks like a sine wave that oscillates between −170 volts and 170 volts at a rate of 60 cycles per second. The peaks are indeed at 170 volts; the average voltage is 120 volts. Oscillating power like this is called AC. The alternative to AC is DC (direct current). Batteries produce DC: A steady stream of electrons flow in one direction only, from the negative to the positive terminal of the battery.

AC has at least four advantages over DC in a power distribution grid:

- AC suffers from significantly less line loss than DC when transmitted over long distances because it is easy to step it up to extremely high voltages.
- Large electrical generators generate AC naturally, so conversion to DC would involve an extra step.
- Transformers must have AC to operate, and the power distribution grid depends on transformers.
- It is easy to convert AC to DC but fairly difficult and expensive to convert DC to AC.

For these reasons, the power plant produces AC. However, it produces three different phases of power simultaneously, and the three phases are offset 120 degrees from each other. Out of every power plant come four wires: one for each of the three phases plus a neutral, or ground, wire common to all three. There is nothing special or magical about three-phase power. It is simply three single phases synchronized and offsetby 120 degrees.

So why three phases? Why not one or two or four? One big

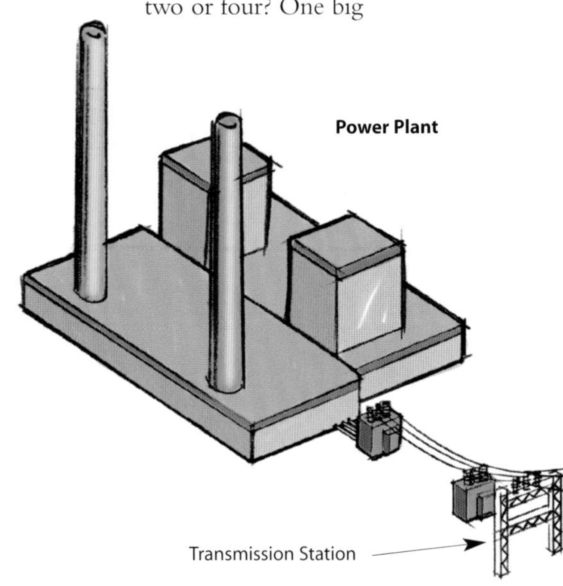

advantage that three-phase power has over one-phase or two-phase power is the fact that, at any given moment, one of the three phases is nearing a peak. In one-phase and two-phase power there are 120 moments per second when the sine wave(s) crosses zero volts. High-power three-phase motors (used in industrial applications) and things such as three-phase welding equipment have even power output. Four phases would not significantly improve the situation but would add a fourth wire, so three-phase is the natural settling point.

What about the fourth wire, the ground? The power grid essentially uses the earth as one of the wires in the power system. The earth is a pretty good conductor and it is huge, so it makes a good return path for electrons. (Car manufacturers do something similar; they use the metal body of the car as one of the wires in the car's electrical system and attach the negative pole of the battery to the car's body.) The "ground" in the power distribution grid is literally "the ground" all around you when you are walking outside. It is the dirt, rocks, groundwater, and so on of the earth.

The three-phase power leaves the generator and enters a transmission substation at the power plant where large transformers convert the generator's voltage (which is thousands of volts) up to extremely high voltages for long-distance transmission on the transmission grid. Typical voltages for long-distance transmission are in the range of 155,000 to 765,000 volts, to reduce line losses. A typical maximum transmission distance is about 300 miles (483 km).

You see huge steel towers with high-voltage transmission lines all over the place. All power towers like this always have three wires, one for each of the three phases. Many towers have extra wires running along their tops. These are ground wires.

Power Substations

For power to be useful in a home or business, it must be stepped down from the transmission lines to the distribution grid in a power substation. A power substation typically does two or three things:

- Transformers step transmission voltages (tens to hundreds of thousands of volts) down to distribution voltages (typically fewer than 20,000 volts).
- A bus splits the distribution power off in multiple directions.

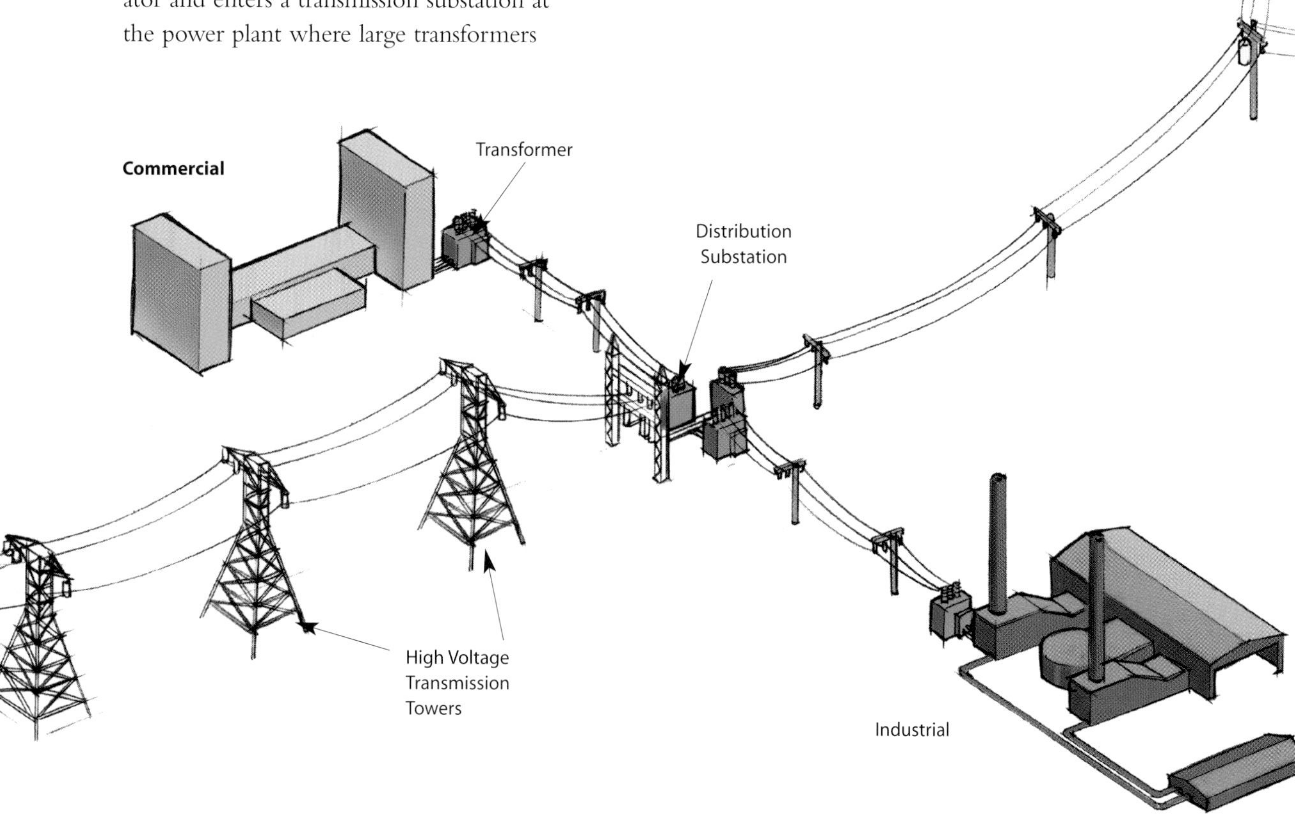

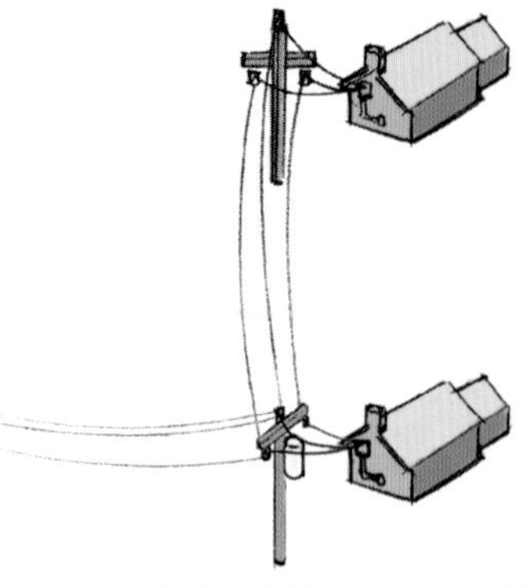

Did You Know?

Normal power in the United States and most of the Western hemisphere is 110 to 120 volts. Almost every other country uses 220 to 240 volts as a standard. The 110-volt systems have a 60-Hz cycle, and most of the 220-volt systems operate at 50 Hz. This difference in cycles per second can make certain items, such as electric clocks, run faster or slower. With a few exceptions, AC is the method used to deliver electricity. A few countries use DC, which can destroy any plugged-in equipment that wasn't made to operate in that system.

If you travel to another country and your equipment requires a specific voltage, you might need a converter or a transformer. A converter limits the amount of electrical output without reducing it, and a transformer reduces the voltage of the electricity going through it.

A converter uses an electronic switch to approximate 110-volt power by rapidly cutting on and off the current received from a 220-volt source. This is okay for some electrical items such as hair dryers, but it is not good for anything electronic (something with a computer chip). Also, a converter should not be used for anything that will be plugged in longer than a few minutes. You need to use a transformer with electronic items, and you need to use one when stepping up from 110-volt power to 220-volt power.

- When necessary, circuit breakers and switches can disconnect the substation from the transmission grid or distribution lines from the substation.

Three-phase power then leaves the substation and makes its way to individual neighborhoods.

At the House

A house needs only one of the three phases of electricity. In a neighborhood, therefore, one of two things happens. Either there is a single phase that is sent to a group of houses, or three phases run past all the houses, and each house taps into one of the phases.

At each house, a transformer drum is attached to the pole, or a transformer box is sitting in the yard. Here is some detail on what is going on at the transformer:

- The transformer's job is to reduce the thousands of volts running past the house down to the voltage that makes up normal household electrical service.
- There are two wires running out of the transformer and three wires running to the house. The two from the transformer are insulated, and each carries 120 volts, but they are 180 degrees out of phase, so the difference between them is 240 volts. The bare wire is the ground wire. This arrangement allows a homeowner to use both 120-and 240-volt appliances.
- The 240-volt electricity enters a house through a typical watt-hour meter so that the power company can charge you for putting up and maintaining all the wires.

Fuses and Circuit Breakers

Fuses and circuit breakers are safety devices. Their purpose is to prevent electrical fires when something goes wrong. The follow ing are examples of situations that fuses and circuit breakers prevent from becoming problems:

- A fan motor burns out a bearing, seizes, overheats, and melts, causing a direct connection between the power and the ground.
- A wire comes loose in a lamp and directly connects the power to the ground.
- A mouse chews through the insulation in a wire and directly connects the power to the ground.
- Someone accidentally sucks up a lamp wire with the vacuum cleaner, cutting it in the process and directly connecting the power to the ground.
- A person is hanging a picture in the living room and the nail used for the picture punctures a power line in the wall, directly connecting the power to the ground.

When a 120-volt power line from a device (such as a lamp) connects directly to the ground wire, it pumps as much electricity as possible through the connection. Either the device or the wire in the wall bursts into flames when that happens. The wire in the wall gets very hot, in the same way that the element in a toaster gets hot. A fuse is designed to overheat and burn out extremely rapidly in such a situation. In a fuse, a thin piece of foil or wire quickly vaporizes when an overload of current runs through it. This kills the power to the wire immediately, protecting it from overheating. A fuse must be replaced if it burns out. A circuit breaker does the same thing as a fuse, but it uses the heat from an overload to trip a switch; and a circuit breaker is therefore reusable—you simply reset it.

So the power starts at the power plant, it is transmitted and then distributed, it is transformed, it is fuse-protected, and it is finally wired into an outlet so you can use it. What an unbelievable path! It takes all that equipment, billions of dollars, and thousands of people to get power from the power plant to the light in your bedroom. The next time you drive down the road and look at the power lines, or the next time you flip on a light, you'll have a much better understanding of what is going on. The power distribution grid is truly an amazing system.

chapter three

ON THE ROAD

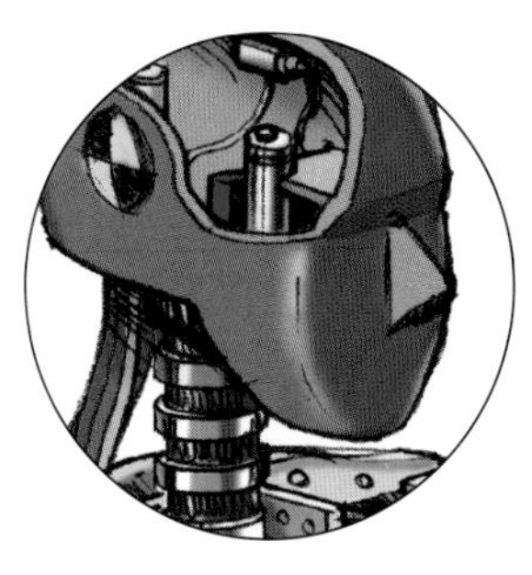

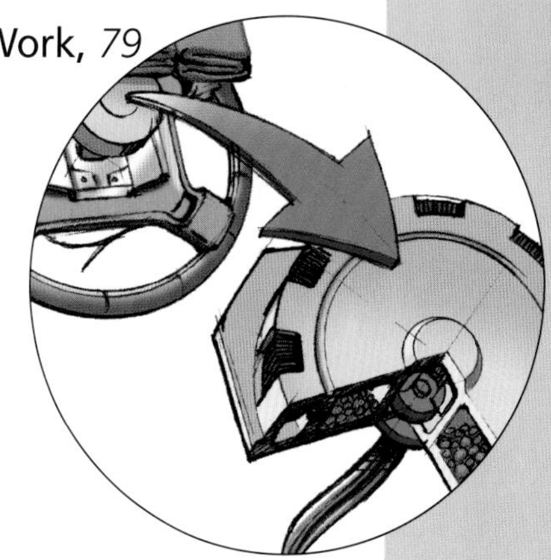

How **TURBOCHARGERS** Work

When people are talking about race cars, or high-performance sports cars, the topic of turbochargers inevitably comes up. Turbochargers, which are found on large diesel engines as well as in race cars, can significantly boost an engine's horsepower without significantly increasing its weight. This is a huge benefit when speed and power are the ultimate goals.

HSW Web Links

www.howstuffworks.com

- How Bearings Work
- How Car Engines Work
- How Caterpillar Backhoe Loaders Work
- How Champ Cars Work
- How Gas Turbine Engines Work
- How Mufflers Work

One of the best ways to get more power out of an engine is to increase the amount of air and fuel that it can burn. Generally, you do this by either adding cylinders or making the existing cylinders bigger. But sometimes these changes may not be feasible. A turbocharger can be a simpler, more compact way to add power.

Boosting Pressure

Turbochargers are a type of forced-induction system, meaning that they compress the air flowing into the engine. This air compression lets the engine squeeze more air into the cylinders; and because the air-to-fuel ratio is constant, more air means more fuel. With more air and fuel in each cylinder, you get more power from each combustion stroke. A turbocharged engine produces more overall power than would the same engine without a turbocharger; this significantly improves the engine's power-to-weight ratio because the turbocharger does not add very much weight.

The typical boost a turbocharger provides is 6 to 8 pounds per square inch (psi). Because normal atmospheric pressure is 14.7 psi at sea level, a turbocharger allows you to squeeze about 50% more air into the engine, so you would expect to get 50% more power. But turbocharging is not perfectly efficient, and you might in reality get a 30% to 40% improvement. A 200-horsepower engine becomes a 260- to 280-horsepower engine with the addition of a turbocharger.

A turbocharger is not perfectly efficient in part because of the fact that the power spinning the turbine is not free. Having a turbine in the exhaust flow increases the restriction in the exhaust, meaning that on the exhaust stroke, the engine has to push against a higher backpressure. This subtracts some power from the pistons that are pushing the exhaust out of the engine.

Step On the Gas!

When you first step on the gas pedal in any car, the throttle valve opens wide, allowing more air into the engine. The amount of fuel being injected into that air increases, thus producing more power. As power and engine speed increase, the volume and speed of the exhaust gases also increase and spin the turbine up to speed.

Parts of a Turbocharger

The turbocharger is bolted to the exhaust manifold of the engine. The exhaust from the cylinders spins the turbine. The turbine is connected by way of a shaft to the compressor, which is located between the air filter and the intake manifold.

The compressor pressurizes the air going into the cylinder. The exhaust from the cylinders passes through the turbine blades, causing the turbine to spin. The more exhaust that goes through the blades, the faster they spin. On the other end of the shaft that the turbine is attached to, the compressor is pumping air into the cylinders. The compressor is a type of centrifugal pump that draws air in at the center of its blades and flings it outward as it spins, forcing the air into the cylinders.

The turbine shaft has to be supported very carefully because it has to handle speeds of up to 150,000 RPM (that's about 30 times faster than most car engines can go). Most bearings would explode at speeds like this, so turbochargers usually use fluid bearings. This type of bearing supports the shaft on a thin layer of oil that is constantly being pumped around the shaft. This oil does two things: It cools the shaft and some of the other turbocharger parts, and it allows the shaft to spin without much friction.

The Intercooler

One especially important component that is used on many turbocharged engines, to improve their efficiency and performance, is the intercooler. When you compress air, it heats up, and when air heats up, it expands—so some of the pressure increase from a

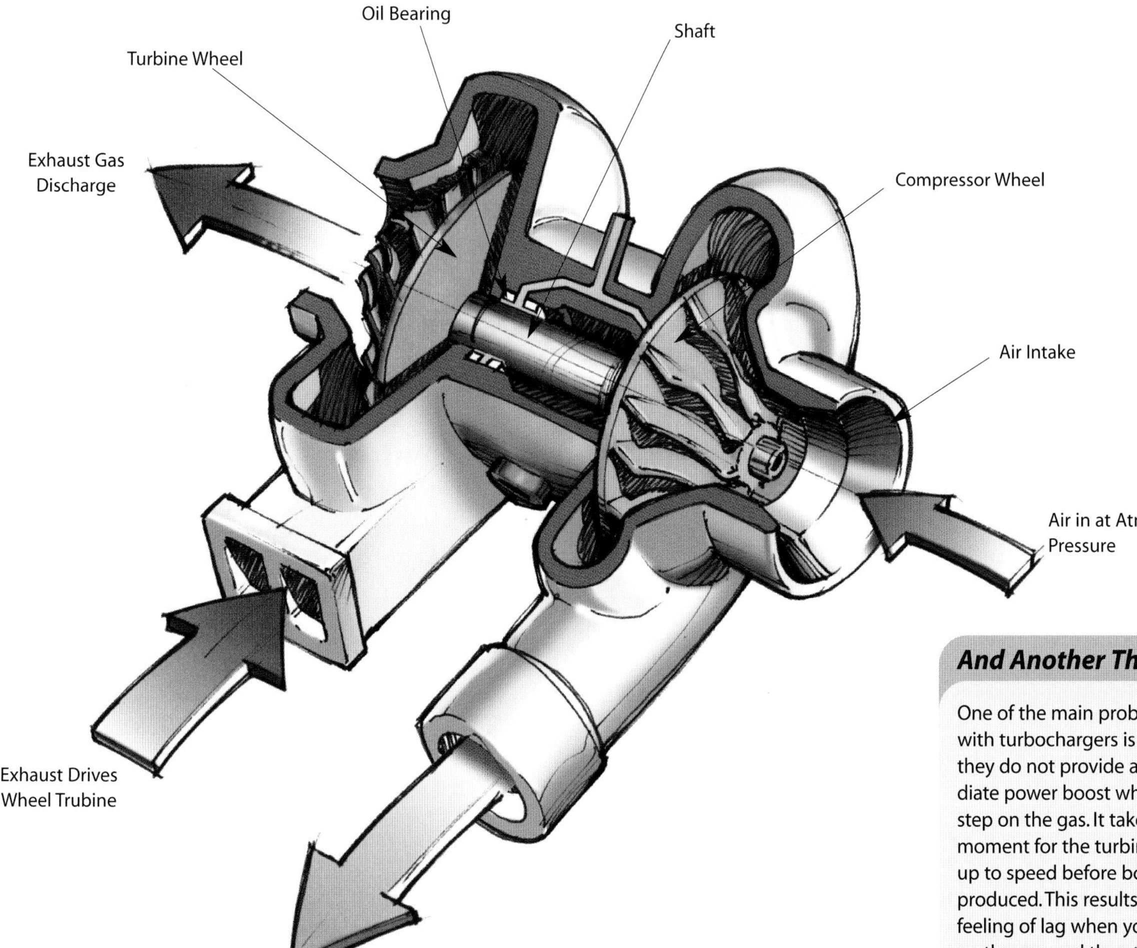

turbocharger is the result of heating the air before it flows into the engine.

To increase the power of the engine, you want to get more air molecules, not more air pressure, into the cylinder. This is where an intercooler comes in. It is a separate component that looks something like a radiator. When compressed air from the turbocharger passes through the sealed passageways inside the intercooler, colder air from outside is blown across fins by the engine-cooling fan.

Cool air is denser than warm air. By cooling the pressurized air coming out of the compressor before it goes into the engine, the intercooler produces denser air to be used in the cylinders. So if the turbocharger is operating at 7-psi boost, the intercooled system will put in 7 psi of cooler air, which contains more air molecules than warmer air. The more air, the more fuel and the more powerful the combustion.

Turbochargers use some very cool technology to make an engine more powerful, but as you can see, the concept is simple. By packing a little more air into the cylinders, turbochargers can make small, lightweight engines perform like much bigger engines. A turbocharger enables the small V-8 engines used in Champ Cars to produce around 900 horsepower — that's about triple the power of the big V-8 engines used on street cars, without the increased weight that would slow the car down. Not bad for adding some extra air.

And Another Thing...

One of the main problems with turbochargers is that they do not provide an immediate power boost when you step on the gas. It takes a moment for the turbine to get up to speed before boost is produced. This results in a feeling of lag when you step on the gas, and then the car lunges ahead when the turbo gets up to speed.

One way to improve turbo lag is to reduce the inertia of the rotating parts, mainly by reducing their weight. This allows the turbine and compressor to accelerate quickly and start providing boost earlier. One sure way to reduce the inertia of the turbine and compressor is to make it smaller. A small turbocharger provides boost more quickly at lower engine speeds, but it may not be able to provide much boost at higher engine speeds, when a really large volume of air is going into the engine.

Some car engines use two turbochargers of different sizes; this system is called sequential turbocharging. The smaller turbocharger spins up to speed very quickly, reducing lag, and the bigger one takes over at higher engine speeds to provide more boost.

How **BRAKES** Work

If a car has an engine, it needs to have brakes. The brake system on a car is simpler than the engine, but it is no less impressive. Brake systems also have some nice safety features because you can get into serious problems if they fail. An incredible amount of creativity has gone into the reliable brakes we have on cars today.

HSW Web Links

www.howstuffworks.com

How Anti-Lock Brakes Work
How Disc Brakes Work
How Drum Brakes Work
How Master Cylinders and Combination Valves Work
How Power Brakes Work

Any moving object has a certain amount of kinetic energy. The amount depends on the speed and mass of the object. In order to slow or stop an object, the kinetic energy has to be converted into another form of energy. Brakes use friction to convert kinetic energy into heat.

The brakes on a car work a lot like the brakes on a bicycle. When you squeeze the brake lever on a bike, a cable moves two pads that press against the rim of the wheel. Friction between the pads and the spinning wheel turns kinetic energy into heat and creates force that tends to slow the wheel down.

When you step on the brake pedal, it is a little more complicated than activating the brake lever on a bike, but approximately the same thing happens:

1) The brake pedal presses a piston in the master cylinder.
2) If the car has power brakes, the brake pedal also turns on the power assist to help you press the piston.
3) The master cylinder pressurizes the brake fluid.
4) The pressure in the brake fluid travels through the brake lines.
5) The pressure activates a slave cylinder at each wheel.
6) The slave cylinder moves the pads, and the pads press against the rotor or the drum.

To understand how the master and slave cylinders work, a good place to start is with hydraulics.

Hydraulics

The basic idea behind any hydraulic system is very simple: Force applied at one point is transmitted to another point by using a fluid that cannot be compressed. Hydraulic fluid is almost always an oil of some sort.

A hydraulic system makes force multiplication easy. Between the lever arm of the brake pedal and the multiplication in the hydraulic system, the pressure from your foot might be multiplied 40 times. If you put 10 pounds (44.5 N) of force on the pedal, 400 pounds (1,779 N) of force will be generated at the wheel to squeeze the brake pads.

The Master Cylinder

To increase safety, brake systems have two circuits, with two wheels on each circuit. If there is a leak in one circuit, only two of the wheels lose their brakes, so the car will still be able to stop when you press the brake pedal.

The master cylinder makes the two circuits possible. It has two pistons in the same cylinder and supplies pressure to both circuits at the same time, but it keeps the two circuits completely separate from each other.

When you press the brake pedal, the pedal in turn pushes on the primary piston. Pressure builds in the cylinder and lines as the brake pedal is pressed further. The pressure between the primary and secondary pistons forces the secondary piston to compress the fluid in its circuit. If the brakes are operating properly, the pressure is the same in both circuits.

If there is a leak in one of the circuits, one of two things happens. If the primary circuit loses pressure, then the primary piston runs into the secondary piston and pushes it normally. The same sort of thing happens if the secondary circuit leaks. If the secondary circuit leaks, the secondary piston runs into the end of the cylinder and the primary piston still works.

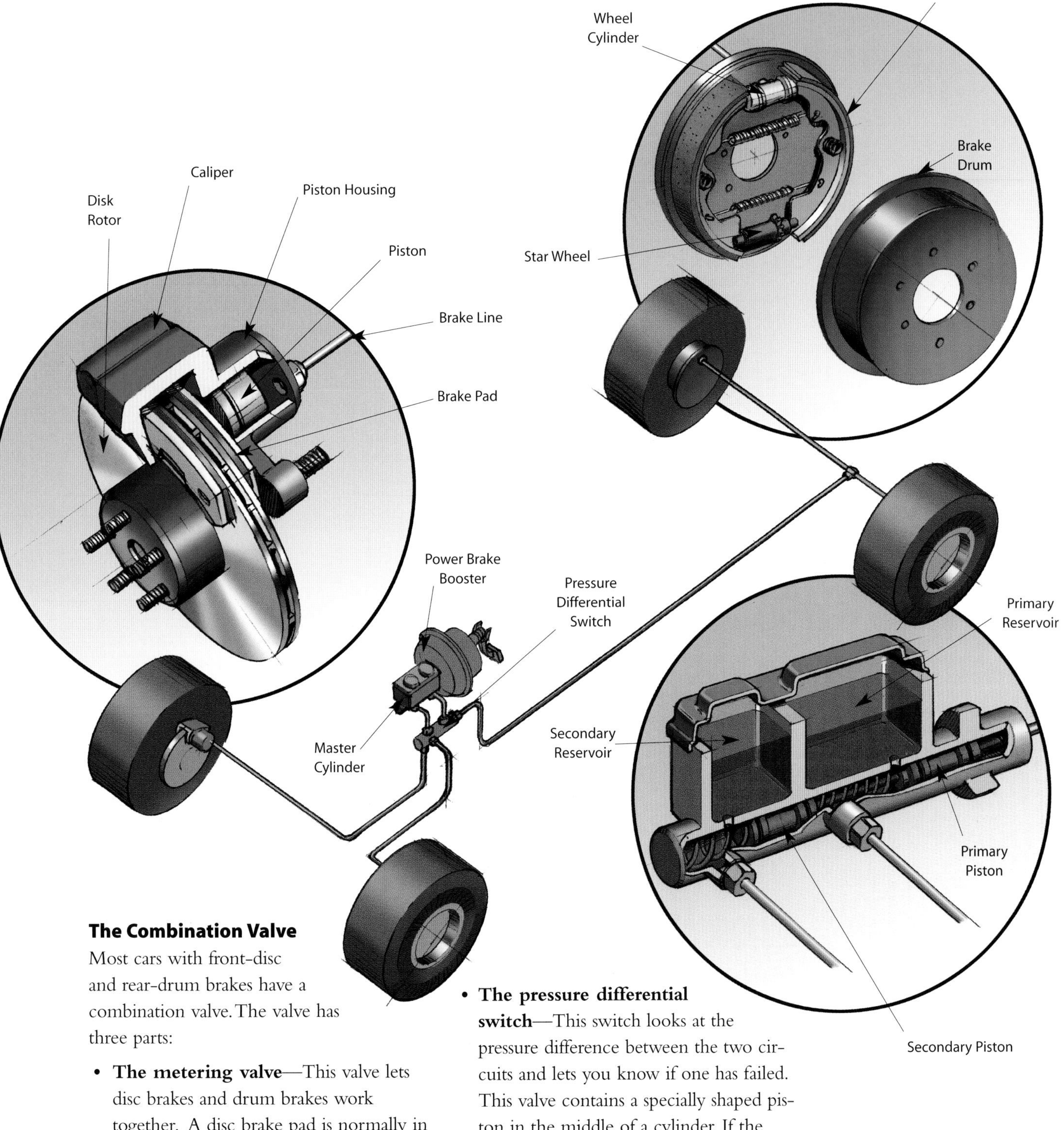

The Combination Valve

Most cars with front-disc and rear-drum brakes have a combination valve. The valve has three parts:

- **The metering valve**—This valve lets disc brakes and drum brakes work together. A disc brake pad is normally in contact with the disc, whereas the drum brake shoes are normally pulled away from the drum. The metering valve does not transfer any pressure to the disc brakes until a threshold pressure has been reached. The threshold pressure is low compared to the maximum pressure in the braking system, so the drum brakes just barely engage before the disc brakes kick in.
- **The pressure differential switch**—This switch looks at the pressure difference between the two circuits and lets you know if one has failed. This valve contains a specially shaped piston in the middle of a cylinder. If the pressure in both circuits is the same, the piston will stay centered in its cylinder. But if one side develops a leak, the pressure will drop in that circuit, forcing the piston off-center and closing a switch. This causes the instrument panel's brake light to turn on.
- **The proportioning valve**—This valve reduces the pressure to the rear brakes. Regardless of what type of brakes a car

has, the rear brakes require less force than the front brakes. The proportioning valve handles the adjustment.

Power Brakes

Most large cars have power brakes. Without them, you would have to apply a huge amount of force to the brake pedal to stop the car. A device called the brake booster also helps makes braking easy.

The brake pedal pushes a rod that passes through the booster into the master cylinder, actuating the master cylinder piston. The engine creates a partial vacuum inside the vacuum booster on both sides of the diaphragm. When you hit the brake pedal, the rod cracks open a valve, which allows air to enter the booster on one side of the diaphragm while sealing off the vacuum. This increases pressure on that side of the diaphragm and causes the diaphragm to help push the rod, which pushes the piston in the master cylinder.

As the brake pedal is released, the valve seals off the outside-air supply while reopening the vacuum valve. This restores vacuum to both sides of the diaphragm and allows everything to return to its original position.

Did You Know?

If you step on the brake pedal in a car after you turn off the engine, you might hear a hissing sound for a second. That is air getting sucked into the booster.

Disc Brakes

Disc brakes are a lot like the brakes on a bicycle. But instead of squeezing the wheel directly, a car's brake pads squeeze a rotor disc that is connected to the wheel. Friction between the pads and the disc slows the disc. Because a lot of heat is generated during a stop, disc brakes are vented; a set of vanes, between the two sides of the disc, pump air through the disc to provide cooling.

The single-piston floating-caliper disc brake is the most common type. This design is self-centering and self-adjusting. The caliper is able to slide from side to side, so it moves to the center each time the brakes are applied. Also, because there is no spring to pull the pads away from the disc, the pads always stay in light contact with the rotor. This is important because the pistons in the brakes are much larger in diameter than those in the master cylinder. If the brake pistons retracted into their cylinders, it might take several applications of the brake pedal to pump enough fluid into the brake cylinder to engage the brake pads.

Drum Brakes

Drum brakes work on the same principle as disc brakes: Shoes press against a spinning surface to slow the spinning. In this system, that surface is a drum. Like a disc brake, a drum brake has two brake shoes and a piston. But the drum brake also has an adjuster mechanism, an emergency brake mechanism, and lots of springs.

When you step on the brake pedal, the piston pushes the brake shoes outward, against the inside of the drum. Unlike disc brakes, the shoes in a drum brake cannot be left in contact with the drum. As soon as the shoes contact the drum, they wedge themselves more tightly against it. This increases the braking force, but it also means that there must be springs to retract the brake shoes from the drum.

As the pad wears down, extra space forms between the shoe and the drum. Each time the car stops while in reverse, the shoe is pulled tight against the drum by the wedging action. When the gap gets big enough, the adjusting lever rocks enough to advance the adjuster gear by one tooth. The adjuster has threads on it, like a bolt, so that it unscrews a little bit when it turns, lengthening to fill in the gap. When the brake shoes wear a little more, the adjuster can advance again. In this way, the shoes always stay close to the drum.

You probably don't give your brakes much thought—unless, of course, they malfunction. The next time you hit your brake pedal, you'll know exactly what happens to transfer the force from the sole of your shoe to the wheels, slowing you down or bringing you to a complete stop. In a car that does not have power-assisted brakes, your leg is supplying all the force needed to stop the car through an amazing system of force multiplication!

How ANTI-LOCK BRAKES Work

Stopping a car in a hurry on a slippery road can be a dangerous activity. An anti-lock braking system (ABS) can take a lot of the danger out of this situation. In fact, on slippery surfaces, even a professional driver with non-ABS brakes can't stop as quickly as an average driver with ABS brakes.

The theory behind anti-lock brakes is simple. A skidding wheel has much less traction than a nonskidding wheel. In addition, when the front wheels of your car lock up, you lose the ability to steer. ABS brakes improve traction by modulating the pressure in the brake lines. This benefits you in two ways: You stop faster, and you're able to steer while you stop.

Parts of an ABS

There are four main components of an ABS:

- **Speed sensors**—The ABS needs some way of knowing when a wheel is about to lock up. The speed sensors located at each wheel provide this information.
- **Valves**—A valve in the brake line of each brake is controlled by the ABS. The valve can release some of the pressure from the brake line.
- **A pump**—Because the valve is able to release pressure from the brakes, there has to be some way to put that pressure back when it's needed. When a valve reduces the pressure in a line, the pump is there to get the pressure back up.
- **A controller**—The ABS controller is one of the many computers in a car. It monitors the speed sensors and calculates the wheel accelerations. It uses this information to control the valves.

ABS Operation

In a simple ABS, the controller monitors the speed sensors at all times. It is looking for decelerations in the wheel that are out of the ordinary. Right before the wheel locks up, the wheel experiences a rapid deceleration. If left unchecked, the wheel would stop much more quickly than any car would. It might take a car 5 seconds to stop from 60 mph (96.6 kph) under ideal conditions, but a wheel that locks up could stop spinning in less than 1 second.

The ABS controller knows that such a rapid deceleration is impossible, so it reduces the pressure to that brake until it sees acceleration, and then it increases the pressure until it sees the deceleration again. When the ABS is in operation, you feel a pulsing in the brake pedal. This comes from the rapid opening and closing of the valves. Some systems can cycle up to 15 times per second.

Four-channel, four-sensor ABS is the best system. There is a speed sensor on each of the four wheels, and there is a separate valve for each wheel. With this setup, the controller monitors each wheel individually to make sure it is achieving maximum braking force.

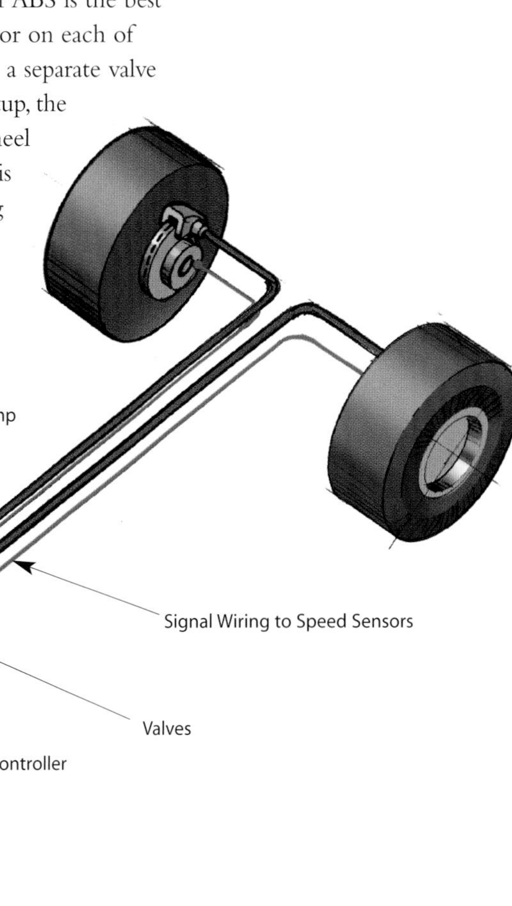

HSW Web Links

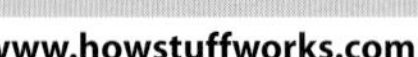

www.howstuffworks.com

How Brakes Work
How Disc Brakes Work
How Drum Brakes Work
How Master Cylinders and Combination Valves Work
How Power Brakes Work

How MANUAL TRANSMISSIONS Work

Cars need transmissions because of the physics of gasoline engines. Engines have narrow RPM ranges where they produce the most horsepower and torque. For example, an engine might produce its maximum horsepower at 5,500 RPM, but at 1,000 RPM the engine might make only a fraction of that power. By applying different gear ratios to the output of the engine, the transmission lets the engine stay in its powerband (the speed range where it makes usable power and torque), no matter what the vehicle's speed.

HSW Web Links

www.howstuffworks.com

How Automatic Transmissions Work
How Clutches Work
How Differentials Work
How Gear Ratios Work
How Gears Work
How Torque Converters Work

Inside the transmission are two shafts—the output shaft and the layshaft—each holding a set of gears of varying sizes. Each gear on one shaft mates with a gear on the other shaft, and each pair of gears forms a different gear ratio. The gears on the output shaft are not connected to the shaft; instead, they spin freely on bearings. On the layshaft, all the gears are permanently connected and the clutch drives this shaft. When the clutch is engaged, all the gears on the layshaft spin as one.

Gear Selection

The keyto the manual transmission is the mechanism that selects the gear ratio. This is where the shifter's "H" pattern on most manual transmissions comes from. When you

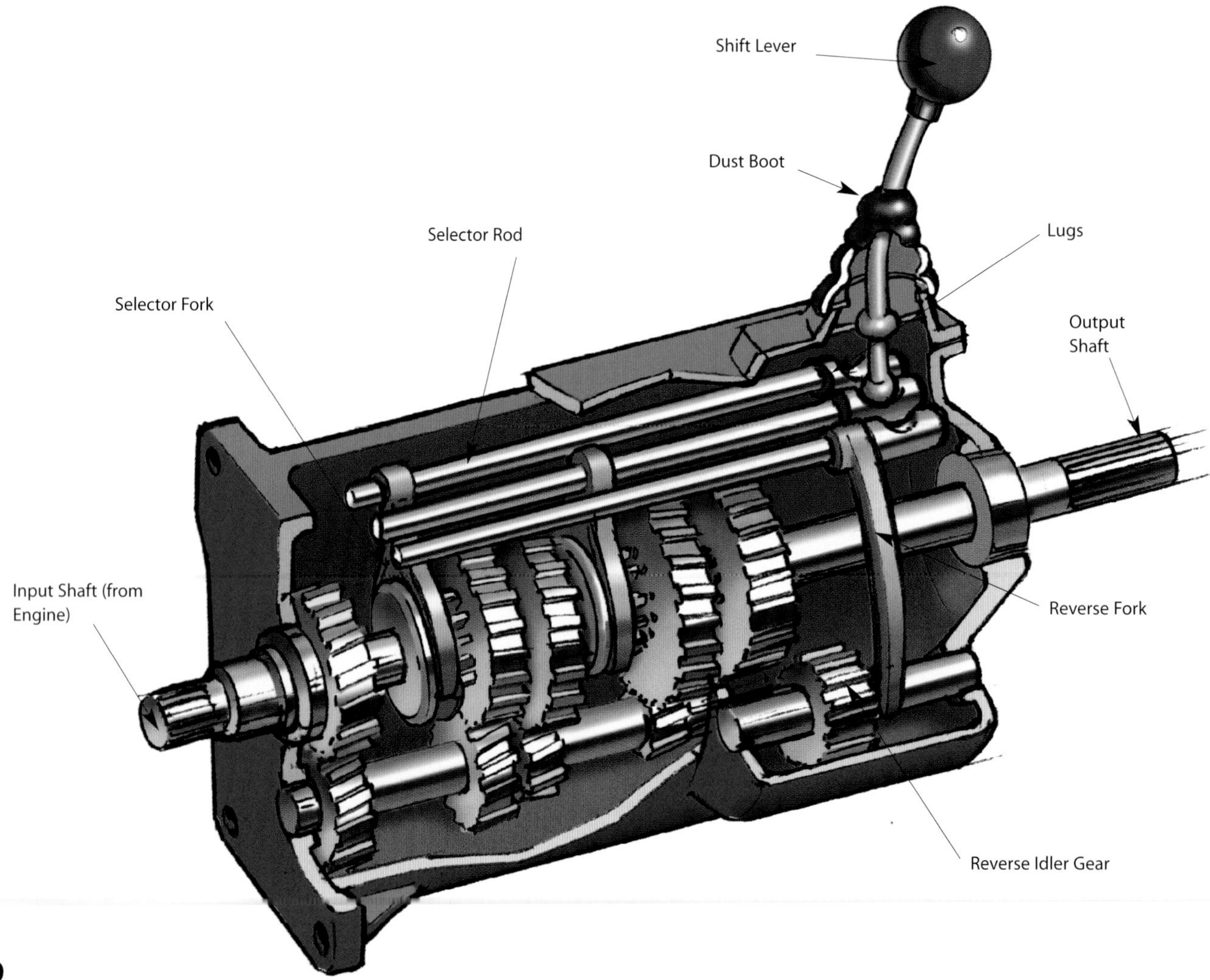

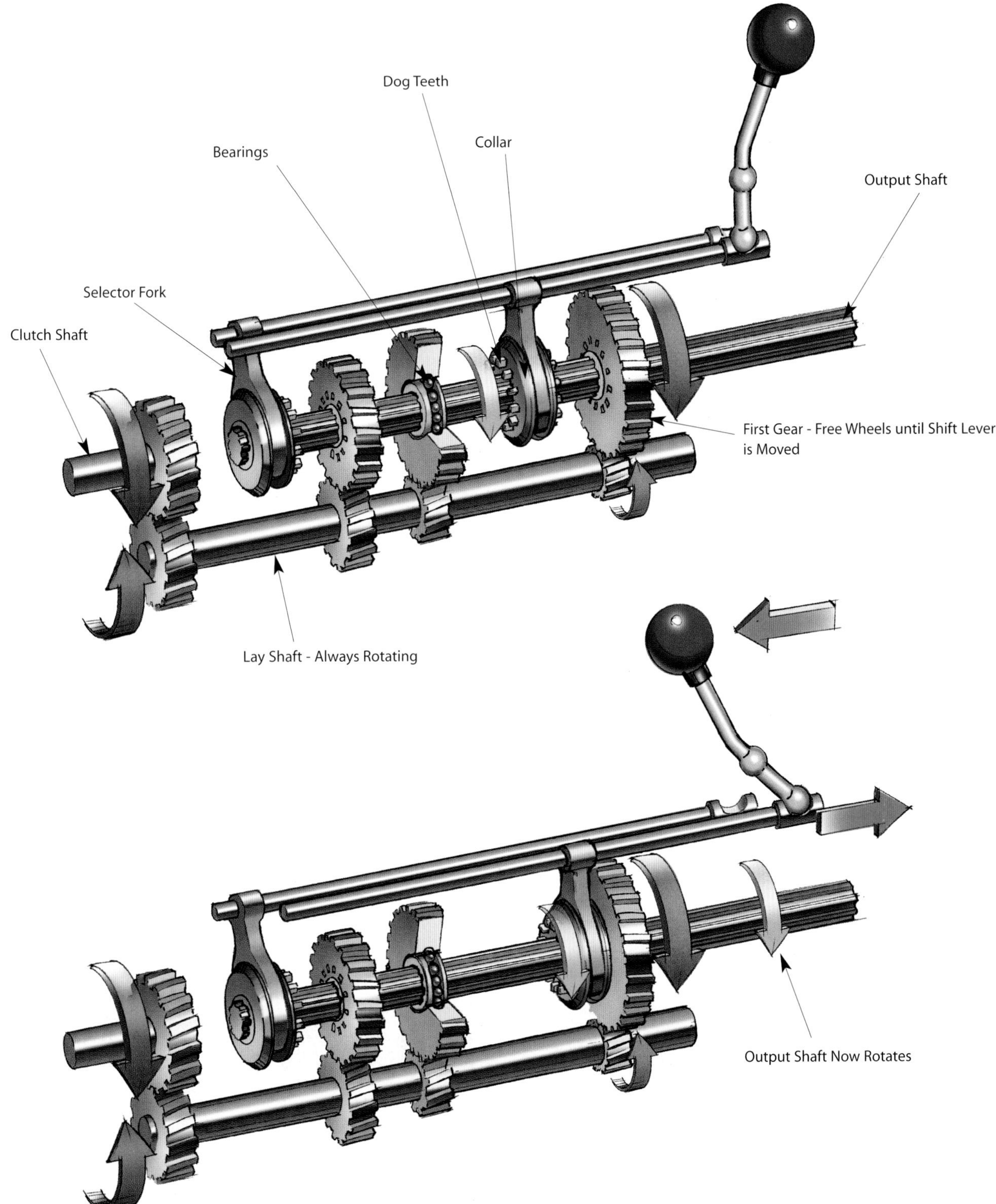

select a gear, the transmission engages one of the freewheeling gears to the output shaft, and this picks the gear ratio.

The output shaft is splined. Between each pair of gears on the output shaft is a collar that can slide along the splines. The collar has a set of teeth along the side that can mesh into the side of a freewheeling gear. In a five-speed transmission there are three of these collars. When you put the car into a gear, you are sliding one of these collars into one of the gears on the output shaft. As the collar slides into a gear, the teeth on the collar, called dog teeth, fit into holes on the sides of the gear, locking it to the output shaft. The gear that is locked to the output shaft determines the gear ratio of the transmission.

When the transmission is in neutral, none of the collars engage the gears, so all the gears freewheel on the output shaft at different rates. Because none of the gears is locked to the output shaft, no power from the engine is transmitted to the wheels.

The shift lever can control one of three rods, depending on its position. Each rod controls a fork that slides one of the collars back and forth. On most transmissions, reverse gear is handled by a small idler gear that fits between the gear on the layshaft and the gear on the output shaft, reversing the direction of rotation.

Advantages over Automatic Transmission

If you've ever looked at road tests in magazines, you may have noticed that cars with manual transmissions usually have faster acceleration times than the same cars with automatic transmissions. A manual transmission with five or six properly chosen gear ratios can outperform an automatic transmission by keeping the engine in its powerband better than an automatic transmission, which has fewer gears. A clutch is also more efficient than a torque converter at transmitting the engine's power to the wheels.

Synchronizers

Manual transmissions in modern passenger cars use synchronizers, or synchros, to make gear changing much smoother. A synchro's purpose is to allow the collar and the gear to make frictional contact before the dog teeth make contact so that the collar and the gear can synchronize their speeds before the teeth need to engage.

The cone on the gear fits into the cone-shaped area in the collar, and friction between the cone and the collar synchronize the collar and the gear. The outer portion of the collar then slides so that the dog teeth can engage the gear.

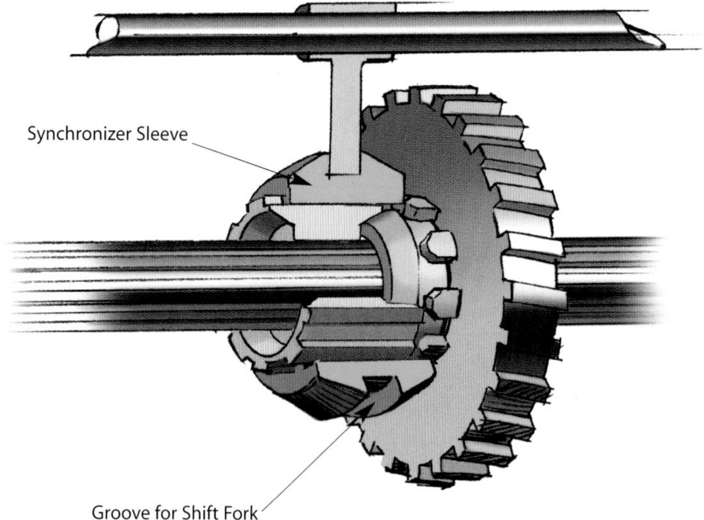

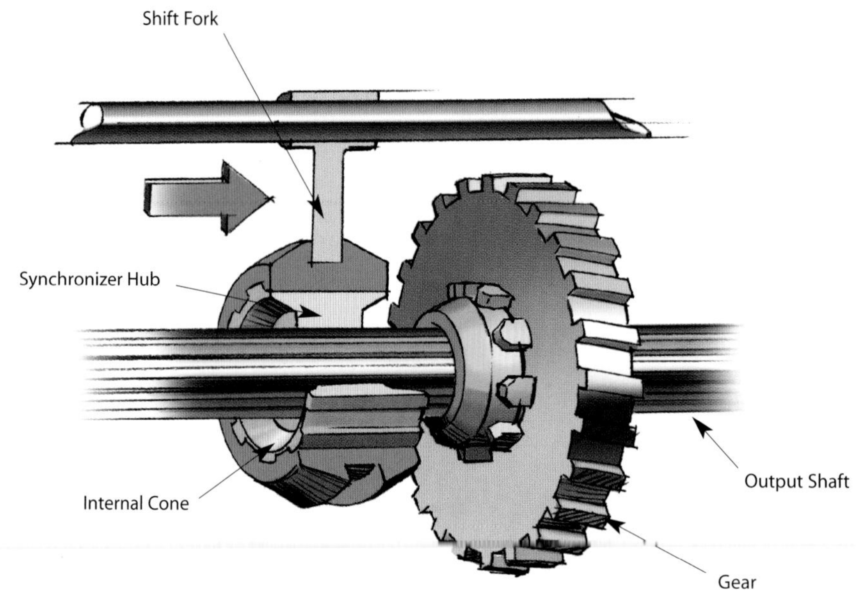

How CLUTCHES Work

When you first learn to drive a manual-transmission car, the clutch is the thing that drives you nuts. The skill you have to learn in order to get the car moving is called slipping: You half-engage the clutch so that the car can start without stalling the engine. If you have ever watched a new driver struggle with slipping, you know it is not an easy skill to master!

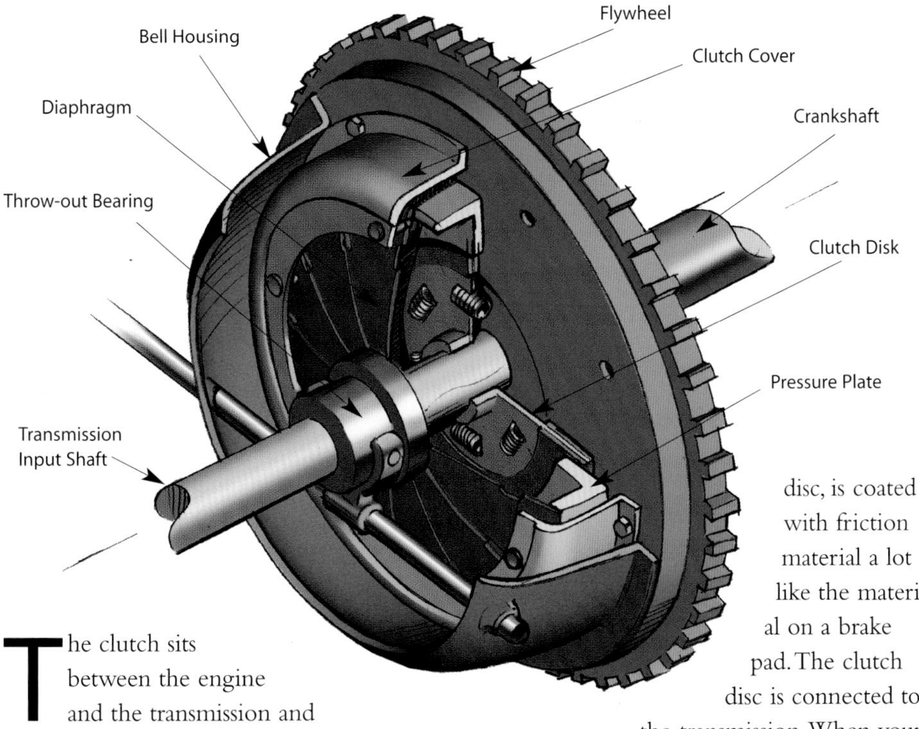

The clutch sits between the engine and the transmission and allows you to connect and disconnect them.

Starting,Stopping, and Switching Gears

The clutch in a manual transmission car helps you do three very important things:

- Come to a complete stop
- Start moving from a complete stop
- Switch gears while you're moving

When you start from a complete stop, you have to have a way to allow the engine to keep spinning when you connect it to a transmission that is not spinning. You also need the clutch in order to switch gears. When you change gears, the engine needs to quickly increase or decrease its speed to match up with the new gear ratio.

The Clutch and the Discs

The clutch works a lot like a disc brake. A spinning disc, called the flywheel, is attached to the engine. Another disc, called the clutch disc, is coated with friction material a lot like the material on a brake pad. The clutch disc is connected to the transmission. When your foot is off the clutch pedal, springs press the clutch disc against the flywheel and lock the engine to the transmission.

When you press the clutch pedal, a cable or hydraulic piston pushes on the release fork, which presses the throwout bearing against the middle of the diaphragm spring. As the middle of the diaphragm spring is pushed in, a series of pins near the outside of the spring cause the spring to pull the pressure plate away from the clutch disc. This releases the clutch from the spinning engine.

One important characteristic of the clutch is that when you engage it, you can allow the clutch disc to slip against the flywheel. This is important when you are starting out from a stop. It allows the engine to keep spinning while transmitting some torque to the transmission, which is not spinning yet. When the car gets moving, you can release the clutch pedal completely and lock the engine to the transmission.

HSW Web Links

www.howstuffworks.com

How Automatic Transmissions Work
How Car Engines Work
How Differentials Work
How Manual Transmissions Work
How Power Door Locks Work

Different Types of Clutches

Clutches are also used in many other places on a car. Automatic transmissions have several clutches inside them that allow them to switch gears. The air-conditioning compressor and cooling fan also have clutches so that they can go on and off while the engine keeps running. If your car has a limited-slip differential, then you might even find a clutch there, helping to improve traction.

How AUTOMATIC TRANSMISSIONS Work

If you have ever driven a car with an automatic transmission, then you know that there are two big differences between an automatic transmission and a manual transmission. In a car with an automatic transmission, there is no clutch pedal and there is no need to shift gears after you put the transmission into drive. An automatic transmission (with its torque converter) and a manual transmission (with its clutch) accomplish exactly the same thing, but they do it in totally different ways.

HSW Web Links

www.howstuffworks.com

How Clutches Work
How Gear Ratios Work
How Gears Work
How Manual Transmissions Work
How Torque Converters Work

Just like a manual transmission, an automatic transmission's primary job is to allow the engine to operate in its powerband while providing a wide range of output speeds for the wheels. Also like a manual transmission, an automatic accomplishes this by providing a number of different gear ratios. Unlike in a manual transmission, all the gear ratios in an automatic are produced by a set of planetary gears.

Planetary Gears

The planetary gearset is the device that makes the automatic transmission possible. The transmission creates the different gear ratios by changing the inputs and outputs of the planetary gearset. The gearset is not very large—it's about the size of a softball. Everything else in the transmission surrounds the planetary gearset and is involved in choosing the right gear.

A planetary gearset has three main components:

- The sun gear
- The planet gears and the planet gears' carrier
- The ring gear

Each of these three components can be the input or the output, or it can be held stationary. Choosing which piece plays which role determines the gear ratio for the gearset. For instance, a planetary gearset might have a ring gear with 72 teeth and a sun gear with 30 teeth. Let's take a look at a few of the gear ratios this gearset can produce:

- If the sun gear is driven by the torque converter, the ring gear is held stationary by a band, and the planet carrier is hooked up to the output, this planetary gearset will produce a 3.4:1 gear ratio (1 + [72/30]). This would make a good first gear. For every 3.4 turns of the crank shaft, the drive shaft turns once.
- If we let the torque converter drive the planet carrier, hold the sun gear stationary, and hook the output up to the ring gear, we get a 0.71:1 gear ratio ([1/1] + [(30/72]). This would make a good overdrive because the output spins faster than the input, which is what makes it an overdrive.
- Locking any two of the three components together locks up the whole device at a 1:1 gear reduction. This makes a good third gear. For every 0.71 turns of the crank shaft, the drive shaft turns once.

An automatic transmission normally has two planetary gearsets that are either separate, or folded into a single unit. Different combinations of the two gearsets can give the transmission three or four gears plus a reverse gear.

Parts of an Automatic Transmission

When you look inside an automatic transmission, you find an amazing assortment of parts arranged around the planetary gearset:

- **Clutches**—A set of three wet-plate clutches locks other parts of the gearset.
- **Bands**—A set of bands lock parts of the gearset.
- **Hydraulic control system**—An odd hydraulic control system controls the clutches and bands.
- **Pump**—A large pump moves transmission fluid around.
- **Governor**—A valve that tells the transmission how fast the car is going.

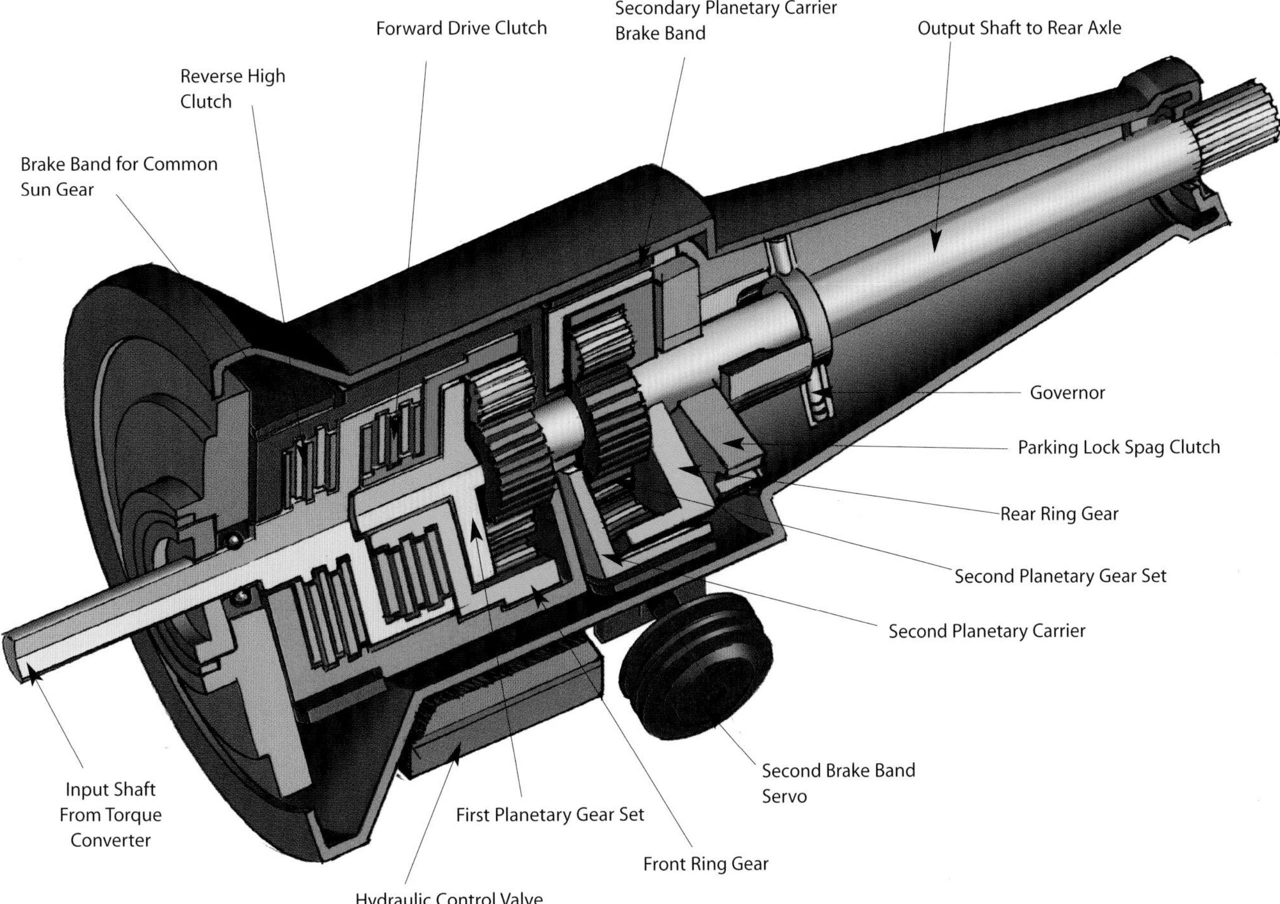

Clutches engage different shafts to control sun gears and planet gear carriers. Each clutch is actuated by hydraulic fluid that enters a piston inside the clutch. Springs make sure that the clutch releases when the hydraulic pressure is reduced. A wet clutch is made of alternating layers of clutch friction plates and steel plates. The friction plates are splined on the inside, where they lock to one of the gears or shafts. The steel plates are splined on the outside, where they lock to the clutch housing. When the clutch is pressurized, the clutch housing is locked to the gear or shaft; when the pressure is released, the clutch housing and gear or shaft are free to rotate separately.

The bands in a transmission are steel bands that wrap around sections of the gear train and connect to the housing. The bands wrap around the outside of the ring gear and can lock it. Hydraulic cylinders inside the case of the transmission actuate the bands.

The automatic transmission in a car has to do many different things, including the following:

- If the car is in overdrive (on a four-speed transmission), the transmission will automatically select the correct gear, based on vehicle speed and throttle pedal position.
- If you accelerate gently, shifts will occur at lower speeds than if you accelerate at full throttle.
- If you floor the gas pedal, the transmission will downshift to the next lower gear.
- If you move the shift selector to a lower gear, the transmission will downshift unless the car is going too fast for that gear. If the car is going too fast, the transmission will wait until the car slows down, and then it will downshift.
- If you put the transmission in second gear, it will never downshift or upshift out of second, even from a complete stop, unless you move the shift lever.

For many decades, an amazing hydraulic control system has provided all the intelligence needed to do these things. Computers are now taking over much of this work, but

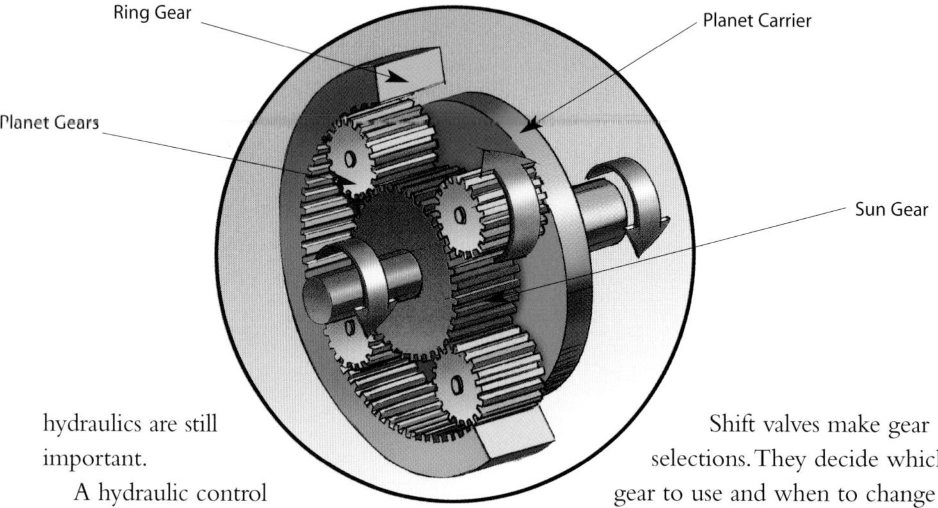

hydraulics are still important.

A hydraulic control system looks like some sort of organic creation. The passageways route fluid to all the different components in the transmission; without these passageways, dozens of hoses would be needed.

The hydraulic control system takes in information about how fast the car is going, how much load the engine is under, and what setting has been chosen for the transmission. From this it makes decisions about which gear to select when.

The governor is connected to the output shaft. The faster the car moves, the faster the governor spins. Inside the governor is a spring-loaded valve that opens in proportion to how fast the governor is spinning. The faster the governor spins, the more the valve opens. Fluid from the pump is fed to the governor through passageways in the output shaft. As the car speeds up, the fluid pressure from the governor increases.

An automatic transmission has to know how hard the engine is working. There are two different ways to get this information. Some cars have a simple cable linkage connected to a throttle valve in the transmission. The more the gas pedal is pressed, the more pressure is put on the throttle valve. Other cars use a vacuum modulator to apply pressure to the throttle valve. The modulator senses the manifold pressure, which drops when the engine is under a greater load.

The driver's shift lever hooks up to the manual valve. Depending on which gear is selected, the manual valve feeds hydraulic circuits that inhibit certain gears.

Shift valves make gear selections. They decide which gear to use and when to change gears. When a valve makes a decision, it controls the clutches and bands to select the proper gears. For instance, the one-to-two shift valve determines when to shift from first to second gear. This shift valve is pressurized with fluid from the governor on one side and the throttle valve on the other.

Imagine that the car's speed is increasing gently. The pressure from the governor builds slowly. This forces the shift valve over until the first-gear circuit is closed and the second-gear circuit opens. Because the car is accelerating at light throttle, the throttle valve does not apply much pressure against the shift valve.

When the car accelerates quickly, the throttle valve applies more pressure against the shift valve. This means that the pressure from the governor has to be higher (and therefore the vehicle speed has to be faster) before the shift valve moves over far enough to engage second gear.

Each shift valve responds to a particular pressure range. When the car is going faster, the two-to-three shift valve takes over because the pressure from the governor is high enough to trigger that valve.

Shifting gears doesn't seem like a complicated job because we do it every day without thinking about it. However, the automatic transmission and its control system is probably the most complicated component in a car. Hundreds of parts have to work together, using information from all over the car to decide when to shift and what to do during the shift.

Electronic Controls

Electronically controlled transmissions, which appear on some newer cars, still use hydraulics to actuate the clutches and bands, but each hydraulic circuit is controlled by an electric solenoid. This simplifies the plumbing on the transmission and allows for more advanced control schemes.

Electronically controlled transmissions can monitor parameters from all over the car. In addition to monitoring vehicle speed and throttle position, the transmission controller can monitor the engine speed, brake pedal position, and even parameters from the anti-lock braking system.

How TORQUE CONVERTERS Work

An engine is connected to a manual transmission by way of a clutch. Without this connection, a car would not be able to come to a complete stop without killing the engine. Cars with an automatic transmission use a device called a torque converter to do the same job as the clutch.

A torque converter is a type of fluid, rather than mechanical, coupling between the engine and transmission. This means that torque is transmitted from the engine to the transmission by a fluid. The engine drives a pump inside the torque converter, and the transmission is connected to a turbine that is powered by the fluid from the pump.

If the engine is turning slowly—for example, when the car is idling at a stoplight—the amount of torque passed through the torque converter is very small. A light pressure on the brake pedal is enough to keep the turbine from spinning and hold the car still.

Parts of the Torque Converter

There are four components inside the housing of the torque converter:

- Pump
- Turbine
- Stator
- Transmission fluid

The housing of the torque converter is bolted to the flywheel of the engine, so it turns at the same speed at which the engine is running. The fins that make up the pump of the torque converter are attached to the housing, so they also turn at the same speed as the engine.

The pump inside a torque converter is a type of centrifugal pump. As it spins, it flings fluid to the outside, much as the spin cycle of a washing machine flings water and clothes to the outside of the wash tub. As fluid is flung to the outside, a vacuum is created that draws more fluid in at the center.

The fluid leaves the pump and enters the blades of the turbine. The spinning turbine is connected to the transmission, and it causes the transmission and the wheels to spin.

The fluid exits the turbine at the center. It is moving opposite the direction that the pump and engine are turning. If the fluid were allowed to hit the pump going this direction, it would slow down the engine and waste power.

The stator is in the center of the torque converter. Its job is to redirect the fluid returning from the turbine before it hits the pump again. The stator has a very aggressive blade design and it lines up the fluid with the spin of the pump.

Torque Converters vs. Efficiency

The price of a torque converter is efficiency. Most of the time the engine spins faster than the transmission, so the torque converter wastes power. To counter this effect, some cars have a lockup clutch that locks the engine to the transmission at higher speeds, bypassing the torque converter completely.

By removing the clutch and clutch pedal, the driver does not have to worry about changing gears—the automatic transmission is automatic because of the torque converter.

HSW Web Links

www.howstuffworks.com

How Automatic Transmissions Work
How Car Engines Work
How Clutches Work
How Differentials Work
How Manual Transmissions Work

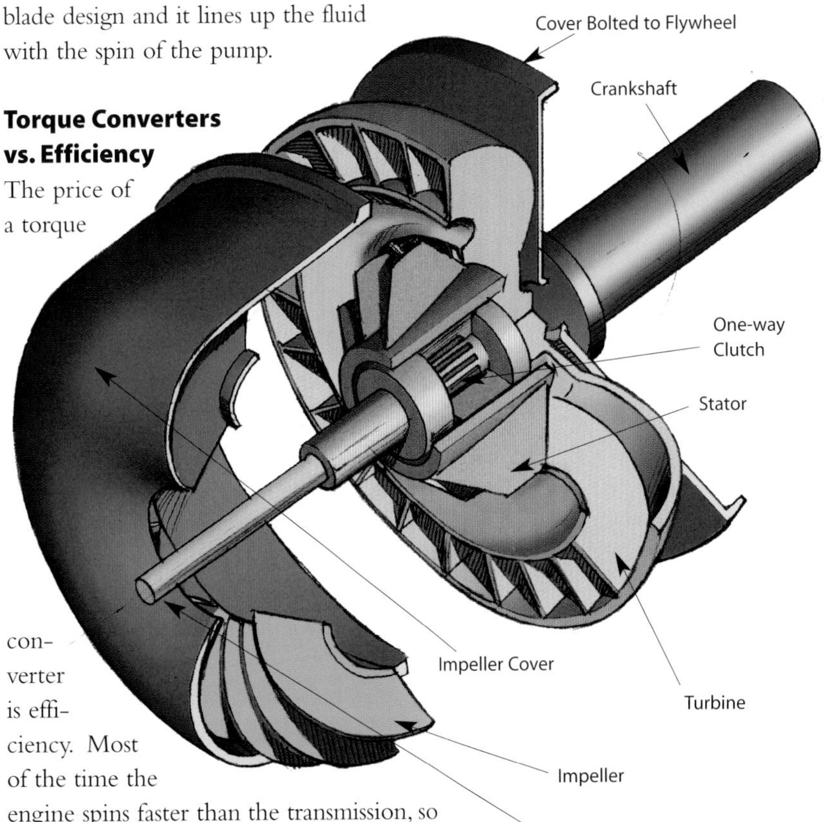

Cover Bolted to Flywheel
Crankshaft
One-way Clutch
Stator
Impeller Cover
Turbine
Impeller
Turbine Output Shaft

How **DIFFERENTIALS** Work

If a car is always traveling in a straight line, then it doesn't need a differential. During a turn is when the need pops up. When a car goes around a turn, all four wheels spin at different speeds. The simplest differential simply allows the speed differences to occur. The best differentials transmit torque evenly to the wheels despite this speed difference.

HSW Web Links

www.howstuffworks.com

How Automatic Transmissions Work
How Clutches Work
How Gears Work
How Manual Transmissions Work
How Torque Converters Work

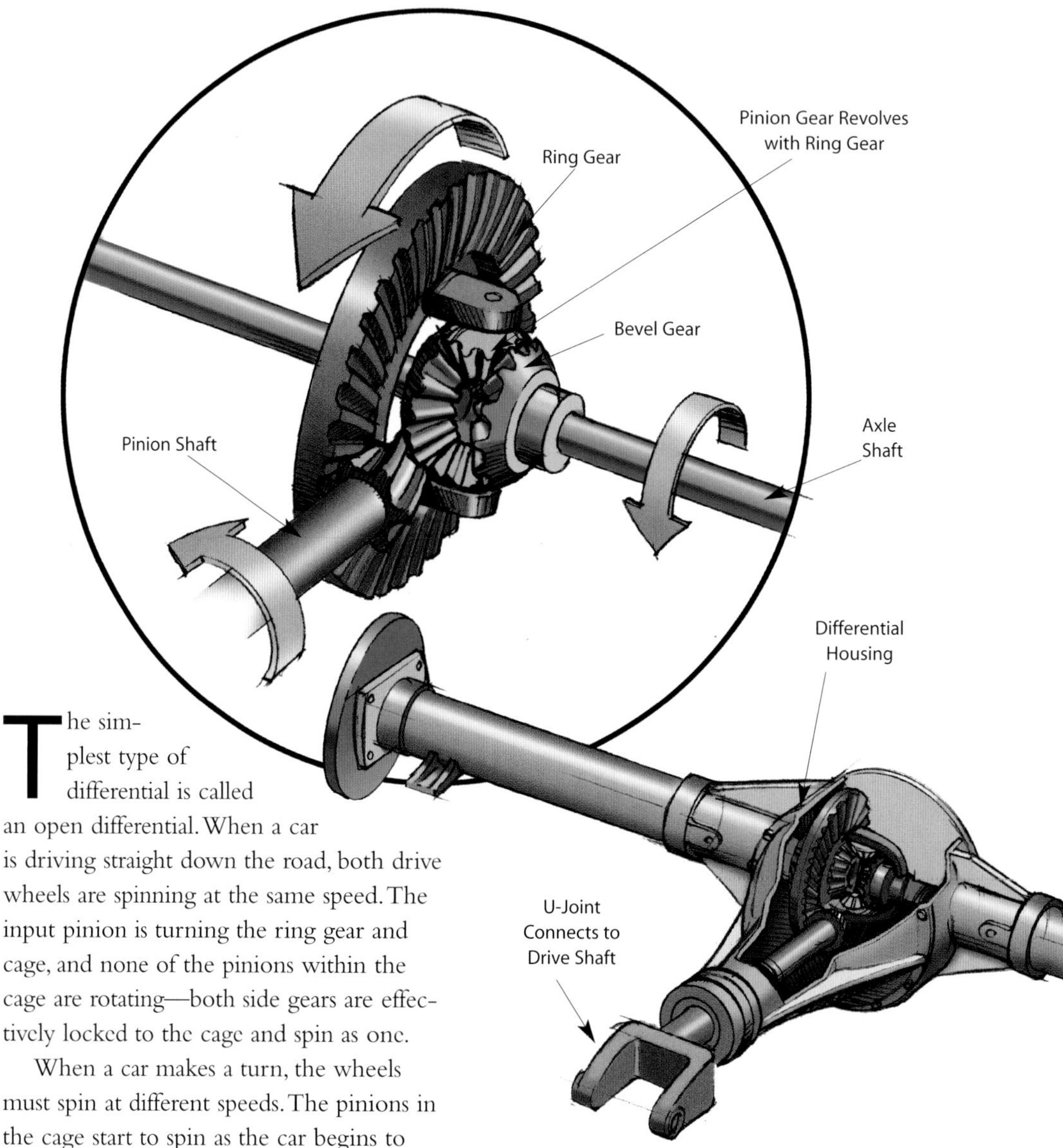

The simplest type of differential is called an open differential. When a car is driving straight down the road, both drive wheels are spinning at the same speed. The input pinion is turning the ring gear and cage, and none of the pinions within the cage are rotating—both side gears are effectively locked to the cage and spin as one.

When a car makes a turn, the wheels must spin at different speeds. The pinions in the cage start to spin as the car begins to turn. The inside wheel spins more slowly than the cage, and the outside wheel spins faster.

The problem with an open differential is that it always applies the same amount of torque to each wheel. If one wheel is on ice, it cannot absorb very much torque before it starts to slip. The differential cannot apply any more torque to the other wheel because both wheels get the same torque.

The solution to the problems that can occur with the open differential is the limited slip differential (LSD). LSDs can transfer more torque to the nonslipping wheel, while still allowing normal differential action when going around turns. Four different mechanisms are found in LSDs:

- Locking differential
- Clutch-type LSD

- Viscous coupling
- Torsen differential

Locking Differential

The locking differential is the brute-force way to solve differential problems and is useful in serious off-road vehicles. Many farm tractors also have this feature. This type of differential starts with an open differential, but it adds an electric, a pneumatic, or a hydraulic mechanism to lock the two output pinions together. This mechanism is usually activated manually by switch, and when it is activated, both wheels lock together and spin at the same speed. If one wheel ends up off the ground, the other wheel won't know or care. Both wheels will continue to spin at the same speed as if nothing had changed.

Clutch-type LSD

The clutch-type LSD is probably the most common version available. This type of LSD has all the components of an open differential, but it adds a spring pack and a set of clutches.

The spring pack pushes the side gears against the clutches, which are attached to the cage. When both drive wheels are moving at the same speed, both side gears spin with the cage, and the clutches aren't really needed. The only time the clutches step in is when something happens to make one wheel spin faster than the other, as in a turn. The clutches fight this behavior, wanting both wheels to turn at the same speed. If one wheel wants to spin faster than the other, it must first overpower the clutch. The stiffness of the springs, combined with the friction of the clutch, determine how much torque is required to overpower it. When the clutch is set properly, the differential still works in turns, and it provides some torque on slippery surfaces.

Viscous Coupling

The viscous coupling is often found in all-wheel-drive vehicles. It is commonly used to link the back wheels to the front wheels. When one set of wheels starts to slip, torque is transferred to the other set.

The viscous coupling has two sets of plates inside a sealed housing that is filled with a thick fluid. One set of plates is connected to each output shaft. Under normal conditions, both sets of plates and the viscous fluid spin at the same speed. When one set of wheels tries to spin faster than the other, perhaps because it is slipping, the set of plates corresponding to those wheels spins faster than the other set. The viscous fluid between the plates tries to catch up with the faster disks, dragging the slower disks along. This transfers more torque to the slower-moving wheels—the wheels that are not slipping.

Torsen Differential

The torsen (torque-sensing) differential is sometimes found on high-performance cars. It is a purely mechanical device—it has no electronics, clutches, or viscous fluids. The torsen differential works as an open differential when the amount of torque going to each wheel is equal. As soon as one wheel starts to lose traction, the difference in torque causes the gears to bind together. The design of the gears in the differential determines the torque bias ratio. For instance, if a particular torsen differential is designed with a 5:1 bias ratio, it is capable of applying up to five times more torque to the wheel with good traction than to the other wheels.

The differential is a very important component of a car. It has a great effect on the car's performance, especially in slippery conditions.

Did You Know?

NASCAR cars have permanently locked differentials. To compensate for the different wheel speeds when going around turns, a larger tire is mounted on the outside wheel so that the car can tolerate some wheel slippage on the straightaways.

The Input Pinion

The input pinion is smaller than the ring gear. This is the last gear reduction in the car. You may have heard terms like *rear axle ratio* and *final drive ratio*. These refer to the gear ratio in the differential. If the final drive ratio is 4.10:1, then the ring gear has 4.10 times as many teeth as the input pinion gear.

How TIRES Work

A car has gasoline, an engine, a clutch, a transmission, a differential, and tires. The tires are where the energy of the gasoline is converted into the motion of the car. The tires have a direct impact on acceleration, braking, cornering, comfort, and the car's performance in bad weather. Obviously, they are an important part of the system!

HSW Web Links

www.howstuffworks.com

- How Anti-Lock Brakes Work
- How Brakes Work
- How Champ Cars Work
- How Force, Power, Torque, and Energy Work
- How Tire Pressure Gauges Work

A tire is a rubber casing that holds air. The air helps tires handle uneven road surfaces so that the car's suspension does not have to handle every ripple and stone on the road. The rubber casing includes the tread and the sidewalls. The tread determines the amount of traction the tire have on wet and dry roads. The sidewalls handle the cornering forces. Designing a tire that does everything well, lasts a long time, and doesn't cost much is an interesting challenge.

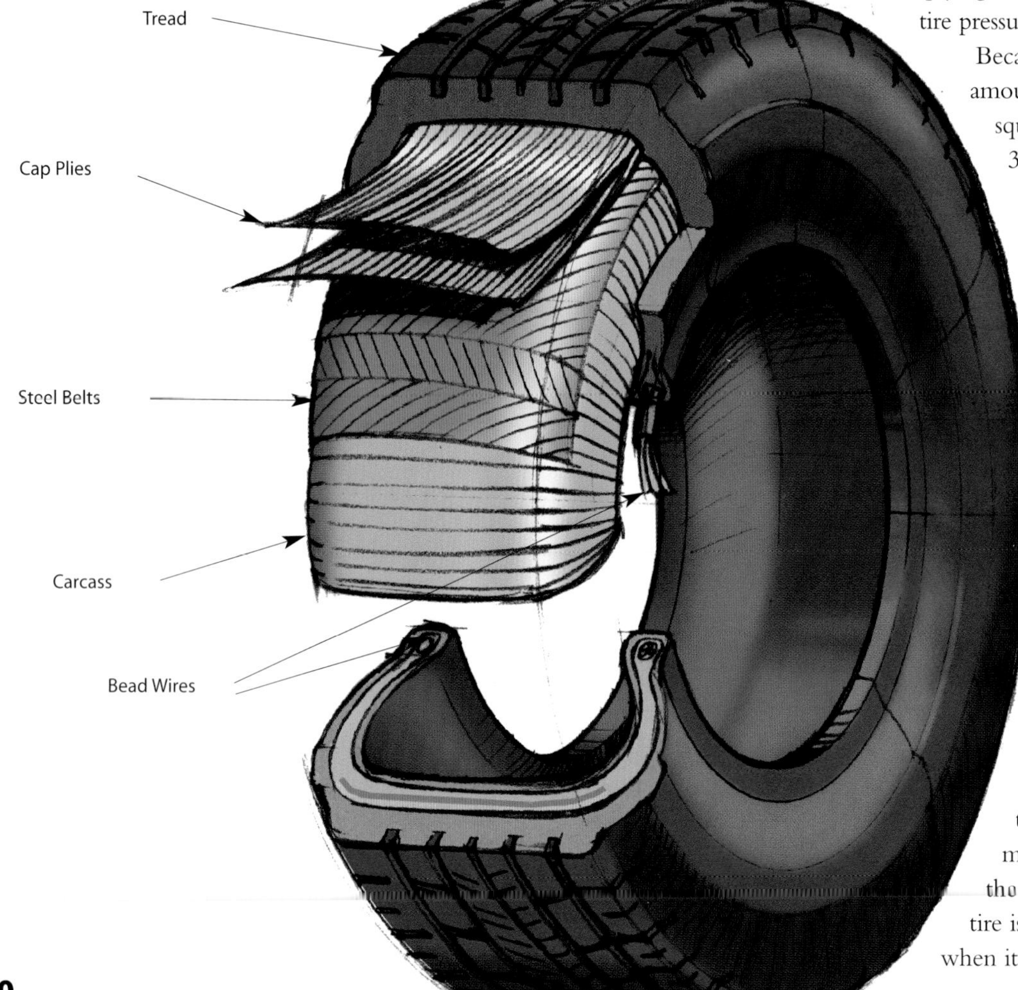

Support for the Car

The next time you are about to get into your car, take a close look at the tires, and you will notice that they are not really round. There is a flat spot on the bottom, where the tire meets the road. This flat spot is called the contact patch.

If you were looking up at a car through a glass road, you could measure the size of the contact patch. You could also make a pretty good estimate of the weight of your car by measuring the area of the contact patches of each tire, adding them together, and then multiplying the sum by the tire pressure.

Because there is a limited amount of pressure per square inch in the tire, say 30 pounds per square inch (psi), you need quite a few square inches of contact patch to carry the weight of the car. If you add more weight to the car or decrease the pressure in the tires, then you need even more contact area, so the flat spot gets bigger. When the tire is rolling, the contact patch must move around the tire to stay in contact with the road. It takes force to bend the tire, and the more the tire has to bend, the more force it takes. The tire is not perfectly elastic, so when it returns to its original shape, it does not return all the force that it took to bend it. Some of that force is converted to heat in the tire by the friction and work of bending all the rubber and steel in the tire. Because an underinflated or overloaded tire needs to bend more than another tire, it takes more force to push it down the road, so it generates more heat.

Tire Parts

A tire has a number of components, which are assembled in a tire-building machine:

- **Bead bundle**—The bead is a loop of high-strength steel cable coated with rubber. It gives the tire the strength it needs to stay seated on the wheel rim and to handle the forces applied by tire-mounting machines when the tires are installed on rims.
- **Body**—The body is made up of several layers of different fabrics, called plies. The most common ply fabric is polyester cord. The cords in a radial tire run perpendicular to the tread. The plies are coated with rubber to help them bond with the other components and to seal in the air. A tire's strength is related to the number of plies it has. Most car tires have 2 body plies. In comparison, large commercial jetliners often have tires with 30 or more plies.
- **Belts**—In steel-belted radial tires, belts made from steel are used to reinforce the area under the tread. These belts provide puncture resistance.
- **Cap plies**—Some tires have cap plies, an extra layer or two of polyester fabric to help hold everything in place. These cap plies are not found on all tires; they are mostly used on tires with higher speed ratings to help all the components stay in place at high speeds.
- **Sidewall**—The sidewall provides lateral stability for the tire, protects the body plies, and helps keep the air from escaping. It may contain additional layers to help increase the lateral stability during cornering.

Tread—The tread is made from a mixture of many different kinds of natural and synthetic rubbers. The tread and the sidewalls are extruded and cut to length, and then the tread patterns that give the tire traction are added.

Tire Assembly

The tire-building machine ensures that all the tire components are in the correct location and then forms the tire into a shape and size that are fairly close to the tire's finished dimensions.

At this point, the tire has all its pieces, but it's not held together very tightly and it doesn't have any markings or tread patterns. This is called a green tire. The next step is to run the tire into a curing machine. This machine functions something like a waffle iron, molding in all the markings and traction patterns. The heat also bonds all of the tire's components together in a process called vulcanizing. After a few finishing and inspection procedures, the tire is ready to roll.

Rolling Resistance

Tire manufacturers sometimes publish a coefficient of rolling friction (CRF) for their tires. You can use this number to calculate how much force it takes to push a tire down the road. The CRF has nothing to do with how much traction the tire has—it is used to calculate the amount of drag, or rolling resistance, caused by the tires. The CRF is just like any other coefficient of friction: The force required to overcome the friction is equal to the CRF multiplied by the weight on the tire.

Say your car weighs 4,000 pounds (1,814 kg) and the tires have a CRF of 0.015. The force is equal to 4,000 x 0.015, which equals 60 pounds (about 267 N). Now let's figure out how much power that is. Power is equal to force times speed. So the amount of power used by the tires depends on how fast the car is going. At 75 mph (120 kph), the tires use 12 horsepower, and at 55 mph (100 kph) they use 8.8 horsepower. All that power is turning into heat, most of which goes into the tires.

How TIRE PRESSURE GAUGES Work

You probably own an inexpensive, pen-sized tire pressure gauge. It's got a funny little spherical thing on one end and a sliding scale on the other end. Every time you use it a little voice in the back of your head probably asks, How the heck does this thing measure pressure? Why doesn't the little scale just blow out the end?"

HSW Web Links

www.howstuffworks.com

How Champ Cars Work
How Hot Air Balloons Work
How Hybrid Cars Work
How NASCAR Safety Works
How Tires Work

When you blow up a tire on a car or a bike, you use a pump to increase the number of atoms inside the tire. There is no magic here. The pump simply stuffs more air into a constant volume, so the pressure rises.

Air Pressure

Air creates pressure inside a container such as a tire or a balloon through atomic collisions—air atoms colliding with the inside of their container. The more gas atoms you put in the container, the more collisions you get and the greater the pressure they exert on the sides of the container. The other way to increase the pressure is to raise the temperature of the atoms inside the container. The hotter the atoms, the faster they move.

Parts of a Pressure Gauge

Inside the tube that makes up the body of the pressure gauge, there is a small, tight-sealing piston. The inside of the tube is polished smooth, the piston is made of soft rubber so it seals nicely against the tube, and the inside of the tube is lubricated with a light oil to improve the seal. A spring runs the length of the tube, between the piston and the stop.

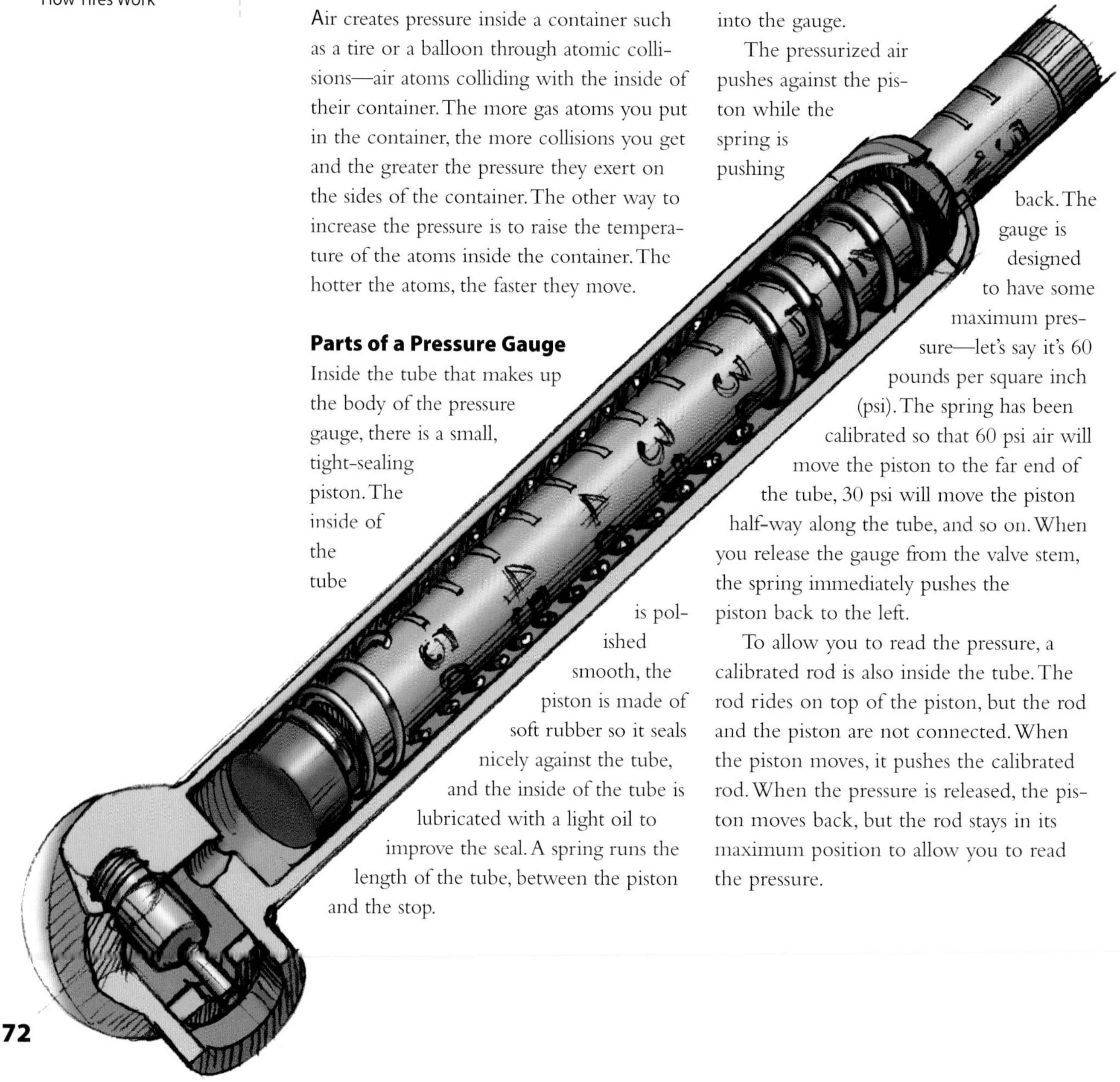

The funny spherical thing on the end of the gauge is hollow. If you look in the opening, you will be able to see a rubber seal and a small fixed pin. The rubber seal presses against the lip of the valve stem to prevent air from leaking during the measurement, and the pin depresses the valve pin in the valve stem to let air flow into the gauge.

The pressurized air pushes against the piston while the spring is pushing back. The gauge is designed to have some maximum pressure—let's say it's 60 pounds per square inch (psi). The spring has been calibrated so that 60 psi air will move the piston to the far end of the tube, 30 psi will move the piston half-way along the tube, and so on. When you release the gauge from the valve stem, the spring immediately pushes the piston back to the left.

To allow you to read the pressure, a calibrated rod is also inside the tube. The rod rides on top of the piston, but the rod and the piston are not connected. When the piston moves, it pushes the calibrated rod. When the pressure is released, the piston moves back, but the rod stays in its maximum position to allow you to read the pressure.

How CAR AIR BAGS Work

Air bags are one of those things that are completely hidden until the moment they are necessary. At the exact moment they're needed, they make their appearance in a matter of milliseconds, helping to cushion the impact of an accident and avoid injuries.

Since model year 1999, all new cars and light trucks in the United States have been required to have air bags on both driver and passenger sides. Statistics show that air bags reduce the risk of dying in a direct frontal crash by about 30%.

Stopping Momentum Safely

The goal of an airbag is to safely stop a moving object. Moving objects have momentum (the product of the mass and the velocity of an object), and the passengers in a car are objects with momentum of their own. Passengers continue moving at whatever speed the car is traveling, even if the car is stopped by a collision.

Stopping a passenger's momentum requires force acting over a period of time. The goal of any supplemental restraint system is to help stop the passenger with as little damage to the passenger as possible.

What an air bag is supposed to do is slow the passenger's speed to zero, with little or no damage. The air bag has only the space between the passenger and the steering wheel or dashboard and a fraction of a second to work with. Even that tiny amount of space and time is valuable, however, if the system can slow the passenger evenly rather than forcing an abrupt halt to the motion.

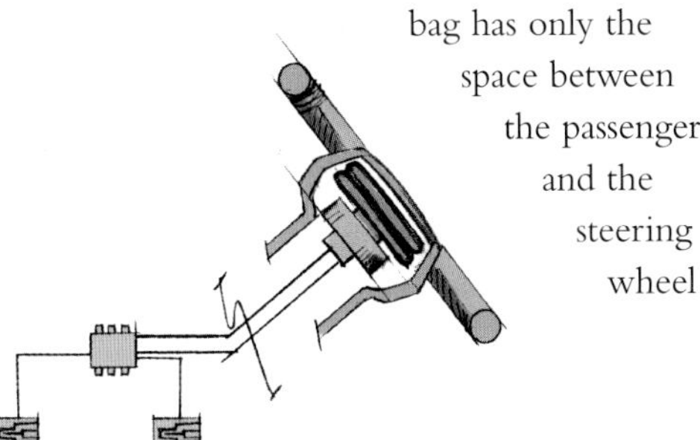

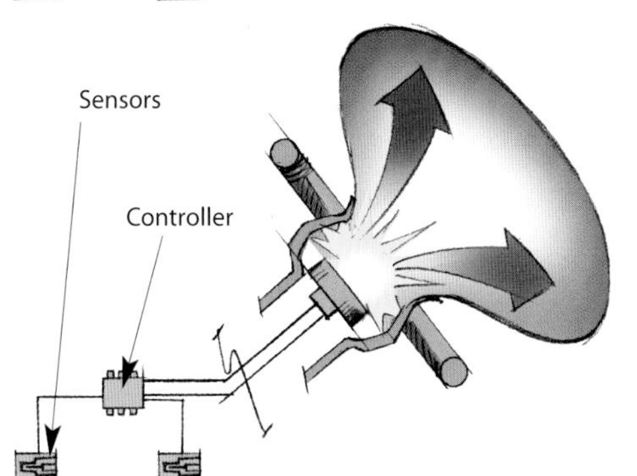

Parts of an Air Bag

To accomplish its magic, an air bag needs three parts:

- **Bag**—The bag itself is made of a thin nylon fabric that is folded into the steering wheel or dashboard.
- **Sensor**—The sensor tells the bag to inflate. Some sensors detect a crash by using a mechanical switch that closes when a mass shifts and an electrical contact is made. Electronic sensors use a tiny accelerometer that has been etched on a silicon chip.
- **Inflation system**—The air bag's inflation system uses the rapid pulse of hot nitrogen gas from the chemical reaction of sodium azide (NaN3) and potassium nitrate (KNO3) to inflate the bag.

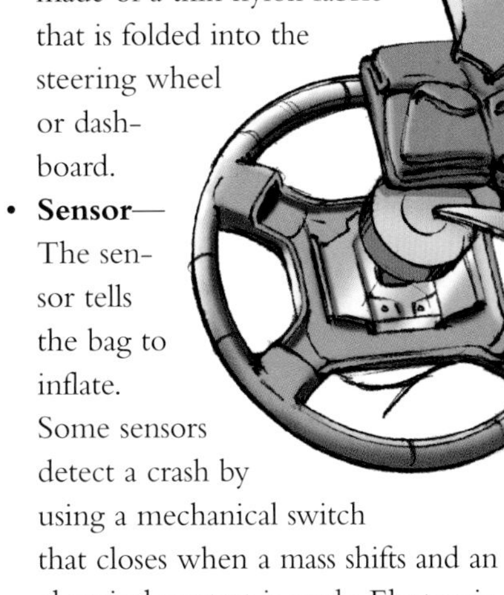

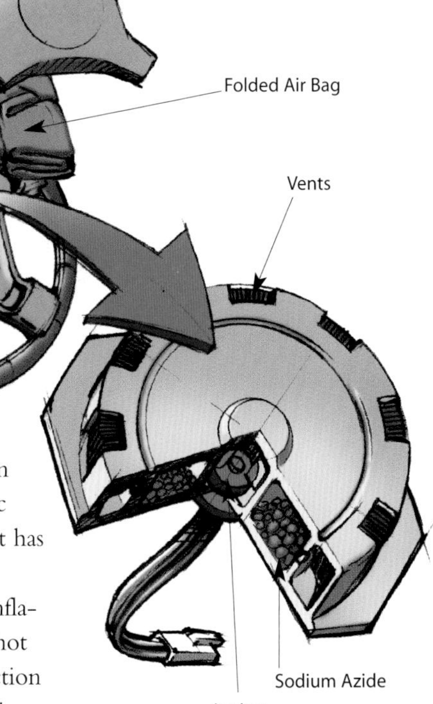

The inflation system could be compared to a solid rocket booster. The air bag system ignites a solid propellant, and it burns extremely rapidly to create a large volume of gas to inflate the bag. The bag then literally bursts from its storage compartment at up to 200 mph (about 322 kph). The gas then quickly dissipates through tiny holes in the bag to get out of the way. The whole process happens in only 40 milliseconds.

HSW Web Links

www.howstuffworks.com

- How Car Engines Work
- How Diesel Engines Work
- How Horsepower Works
- How Hybrid Cars Work
- How NASCAR Safety Works

Cool Fact

The powdery substance released from the air bag is regular cornstarch or talcum powder. Air bag manufacturers use it to keep the bags pliable and lubricated while they're in storage.

How **CRASH TESTING** Works

Crash-test dummies have been the subject of public service announcements, cartoons, and parodies—they're even the name of a band. Crash-test dummies are incredibly sophisticated devices that have saved thousands of lives by improving automotive safety.

HSW Web Links

www.howstuffworks.com

How Car Engines Work
How Diesel Engines Work
How Horsepower Works
How Hybrid Cars Work
How NASCAR Safety Works

The dummy's job is to simulate a human being during a crash. Dummies have instruments embedded throughout their bodies in order to collect data that would not be possible to collect any other way.

All crash tests in the United States use the Hybrid III dummy, which is built from materials that mimic the physiology of the human body. For example, the Hybrid III has a spine made from alternating layers of metal discs and rubber pads.

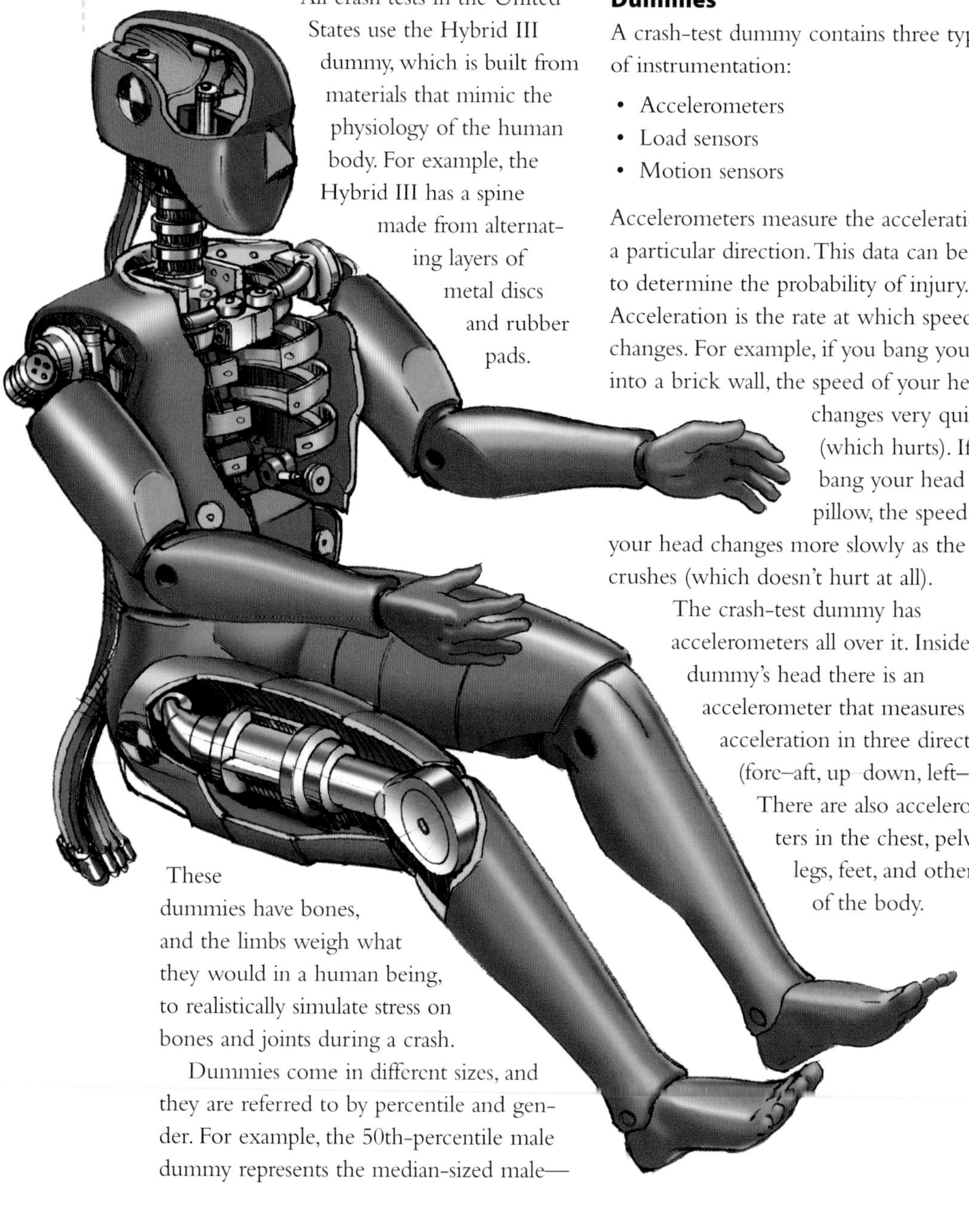

These dummies have bones, and the limbs weigh what they would in a human being, to realistically simulate stress on bones and joints during a crash.

Dummies come in different sizes, and they are referred to by percentile and gender. For example, the 50th-percentile male dummy represents the median-sized male—it is bigger than half the male population and smaller than the other half. It weighs 172 lbs (78 kg) and is 69 inches (5 ft 9 in, or 1.75 m) tall.

Instrumentation in Crash-test Dummies

A crash-test dummy contains three types of instrumentation:

- Accelerometers
- Load sensors
- Motion sensors

Accelerometers measure the acceleration in a particular direction. This data can be used to determine the probability of injury. Acceleration is the rate at which speed changes. For example, if you bang your head into a brick wall, the speed of your head changes very quickly (which hurts). If you bang your head into a pillow, the speed of your head changes more slowly as the pillow crushes (which doesn't hurt at all).

The crash-test dummy has accelerometers all over it. Inside the dummy's head there is an accelerometer that measures the acceleration in three directions (fore–aft, up–down, left–right). There are also accelerometers in the chest, pelvis, legs, feet, and other parts of the body.

During a crash test, all the accelerometers send data to recorders so that the results can be analyzed. Extremely high acceleration rates on different parts of the body indicate probable injuries. The highest values show up when something strikes a hard object.

Inside the dummy are load sensors that measure the amount of force on different body parts during a crash. The maximum load in the bone can be used to determine the probability of its breaking. Motion sensors are used in the dummy's chest. They measure how much the chest deflects during a crash. This data can indicate damage to the rib cage and internal organs.

Crash-test Paint

Before crash-test dummies are placed in a vehicle, researchers apply paint to them. Different colors of paint are applied to the parts of a dummie's body that are most likely to hit during a crash. The dummy's knees, face, and areas of the skull are each painted with a different color. The paint is designed to come off easily on the objects it touches.

After a crash, paint marks in the car indicate which part of the body hit which parts of the inside of the vehicle cabin. This information helps researchers develop improvements to prevent that type of injury in future crashes.

Running a Test

In the United States, the National Highway Traffic Safety Administration (NHTSA) conducts two types of crash tests as part of its New Car Assessment Program:

- **35 mph (56.3 kph) frontal impact**—The car runs straight into a solid concrete barrier. This test simulates a car moving at 70 mph hitting a stationary car of the same weight.
- **35 mph (56.3 kph) side impact**—A 3,015-pound (1,367.58-kg) sled with a deformable "bumper" runs into the side of the test vehicle. The test simulates a car running a red light and sideswiping a car that is crossing an intersection.

The test car is equipped with 15 high-speed cameras, including several under the car that are pointed upward. All the cameras shoot around 1,000 frames per second.

For the frontal-impact test, the test car is backed away from the barrier and prepared to crash. A pulley mounted in a track pulls the car down the runway. The car hits the barrier at 35 mph (56.3 kph). It takes only about 0.1 second from the time the car hits the barrier until it stops.

After a typical frontal-impact crash, the front of the car is completely crushed. This is good, as the car has to be crushed and collapse in order to absorb the kinetic energy and stop the car. A typical car might shrink by 2 feet (0.6 m) during impact. The engine is pushed under the car. The air bags deploy. The dummies and the cameras, during the instant of the crash, tell the crash story by producing reams of data.

Surviving a Crash

Crash testing has shown us that surviving a crash is all about kinetic energy. When your body is moving at 35 mph (56.3 kph), it has a certain amount of kinetic energy. After the crash, when you come to a complete stop, you have zero kinetic energy. To minimize risk of injury, you would like to remove the kinetic energy as slowly and evenly as possible. Some of the safety systems in your car help do this.

Many cars have seatbelt pretensioners and force limiters. They tighten the seatbelts very soon after your car hits the barrier, but before the air bag deploys. The seatbelt can absorb some of your energy as you move forward, toward the air bag. Milliseconds later, the force in the seatbelt holding you back would start to hurt you, so the force limiters kick in, making sure the force in the seatbelts doesn't become too great.

Next, the air bag deploys and absorbs more of your forward motion while protecting you from hitting anything hard. If you don't wear your seatbelt, then the first stage of your protection is lost, and it is going to hurt a lot more when you slam into the air bag.

How MUFFLERS Work

If you've ever heard a car engine running without a muffler, you know what a huge difference a muffler can make to the car's noise level. Without a muffler an engine can be painfully loud, but with a muffler the engine can be silent.

HSW Web Links

www.howstuffworks.com

How Automobile Ignition Systems Work
How Car Engines Work
How Catalytic Converters Work
How Speakers Work
How Turbochargers Work

A muffler dampens the sound of a car's operation by using several techniques together.

Pressure Pulse

When a balloon pops, pressure inside the balloon escapes in an instant and creates a loud noise. Every time an exhaust valve opens in an engine, it is like a very loud balloon popping—a burst of high-pressure gas suddenly enters the exhaust system—but the pressures inside a cylinder are much higher than those in a balloon.

If you let the air out of a balloon slowly, on the other hand, it does not create a noise. Similarly, a muffler reduces the spike of a pressure pulse by giving it a large area to expand into. Rather than coming out in pulses, the air comes out of the muffler in an even stream.

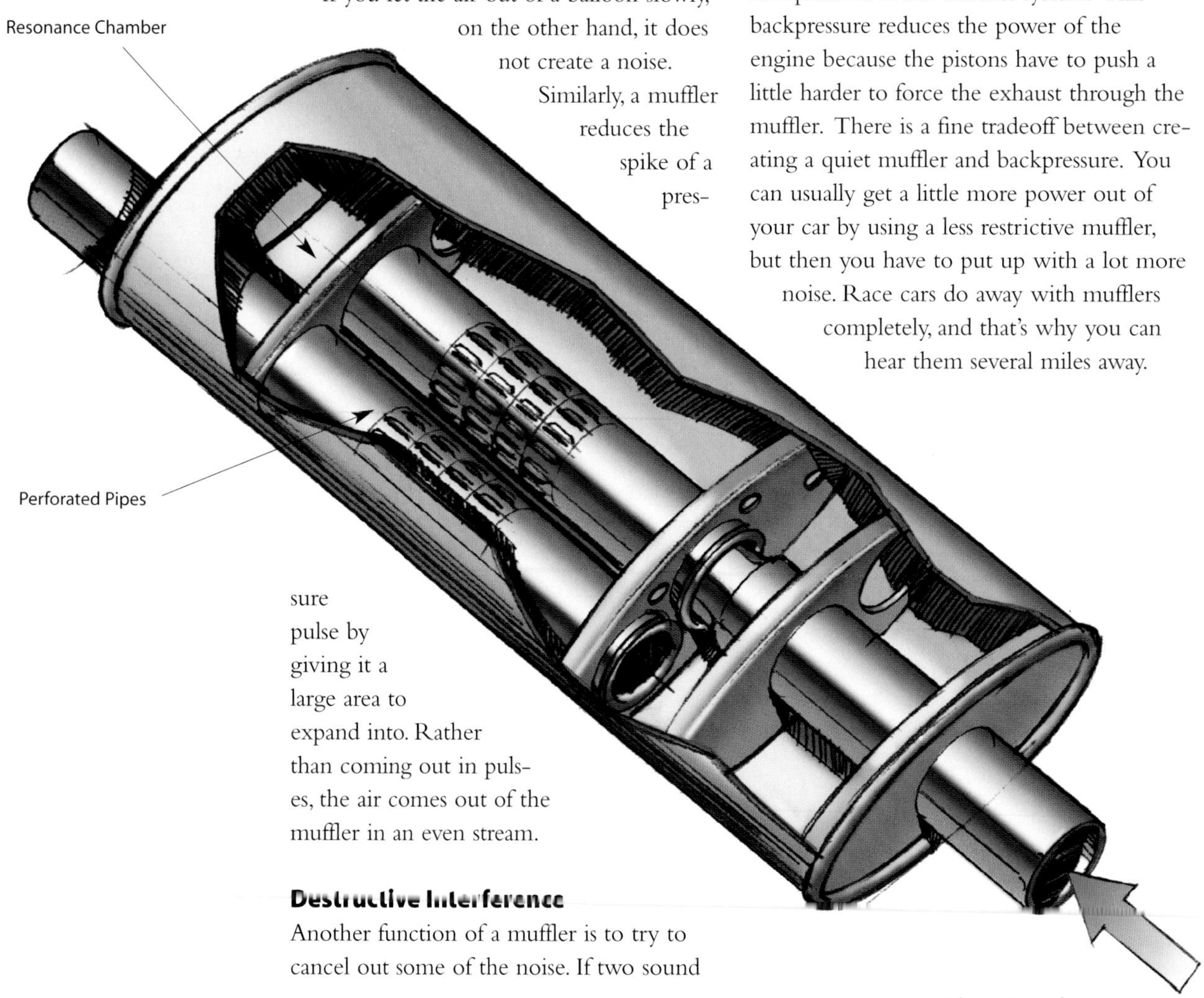

Destructive Interference

Another function of a muffler is to try to cancel out some of the noise. If two sound waves are perfectly out of phase (that is, the high-pressure part of one wave lines up perfectly with the low-pressure part of the other wave), then you will hear no sound at all. This is called destructive interference. The muffler is designed to create as much destructive interference as possible. Inside the muffler is a set of tubes that create reflected waves that interfere with each other.

Backpressure

All the tubes and turns in the muffler restrict the flow of exhaust gases and create some backpressure in the exhaust system. This backpressure reduces the power of the engine because the pistons have to push a little harder to force the exhaust through the muffler. There is a fine tradeoff between creating a quiet muffler and backpressure. You can usually get a little more power out of your car by using a less restrictive muffler, but then you have to put up with a lot more noise. Race cars do away with mufflers completely, and that's why you can hear them several miles away.

Resonance Chamber

Perforated Pipes

Exhaust Gases from Engine

How CATALYTIC CONVERTERS Work

There are millions of cars on the road in the United States, and each one is potentially a source of air pollution. Especially in large cities, the amount of pollution that all the cars produce together can create big problems. Catalytic converters eliminate harmful gases from car engine exhaust to reduce pollution as much as possible.

Modern car engines carefully control the amount of fuel they burn. They try to keep the air-to-fuel ratio very close to the stoichiometric point—the calculated ideal ratio of air to fuel. At this ratio, all the fuel will be burned using all the oxygen in the air. For gasoline, the stoichiometric ratio is about 14.7:1, which means that for each pound of gasoline, 14.7 pounds of air will be burned.

Emissions

A perfect car engine would produce nothing but nitrogen gas (N_2), carbon dioxide (CO_2), and water vapor (H_2O). The nitrogen is already in the air, and it would pass straight through the engine, and carbon and hydrogen in the fuel would bond with the oxygen in the air.

Because the combustion process is not perfect, car engines produce some undesirable gases as well as N_2, H_2O, and CO_2:

- **Carbon monoxide (CO)**—CO is a poisonous gas that is colorless and odorless.
- **Hydrocarbons or volatile organic compounds (VOCs)**—These are produced mostly from unburned fuel that evaporates. Sunlight breaks these down to form ground-level ozone (O_3), a major component of smog.
- **Oxides of nitrogen (NO and NO_2, together called NO_x)**—NO_x contributes to smog and acid rain, and it also cause irritation to human mucus membranes.

These gases are the three main regulated emissions, and they are also what the catalytic converter aims to reduce.

Types of Catalysts

A catalytic converter uses two different types of catalysts: a reduction catalyst and an oxidization catalyst. Both types appear as a ceramic honeycomb coated with a metal catalyst, usually platinum, rhodium, and/or palladium.

The reduction catalyst is the first stage of the catalytic converter. It uses platinum and rhodium to help reduce the NO_x emissions. When an NO or NO_2 molecule contacts the catalyst, the catalyst rips the nitrogen atom out of the molecule and holds on to it, freeing the oxygen to form O_2. The nitrogen atoms bond with other nitrogen atoms that are also stuck to the catalyst, forming N_2. Here's an example:

$$2NO \rightarrow N_2 + O_2$$

$$\text{or } 2NO_2 \rightarrow N_2 + 2O_2$$

The oxidation catalyst reduces the unburned hydrocarbons and CO by burning (oxidizing) them over a platinum and palladium catalyst. This catalyst assists the reaction of the CO and hydrocarbons with the remaining oxygen in the exhaust gas. Here's an example:

$$2CO + O_2 \rightarrow 2CO_2$$

The Control System

A control system monitors the exhaust stream and uses this information to control the fuel injection system. An oxygen sensor mounted between the engine and the catalytic converter tells the engine control unit (ECU) how much oxygen is in the exhaust. The ECU can increase or decrease the amount of oxygen in the exhaust by adjusting the air-to-fuel ratio.

For something so simple, the catalytic converter is an incredible advance in pollution control. Without it, it would be miserable to live in many of the larger cities of the world.

HSW Web Links

www.howstuffworks.com

How Automobile Ignition Systems Work
How Car Engines Work
How Fuel Injection Systems Work
How Hybrid Cars Work
How Mufflers Work

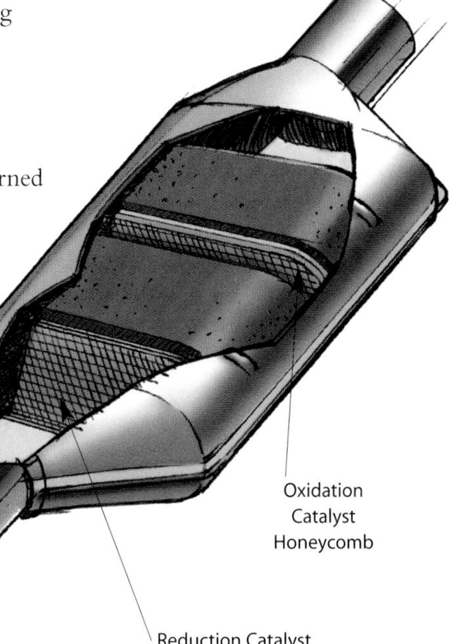

How **ODOMETERS** Work

On the dashboard of any car is an odometer. It is one of those things that you see every day of your life as a driver. If it is a mechanical odometer, it is a big event to watch it roll over 1,000 or 10,000 miles. The roll-over event is provided by one special feature—intermittent gear engagement—that gives the odometer the ability to fully display each of the digits.

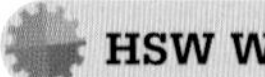
HSW Web Links

www.howstuffworks.com

How Cruise Control Systems Work
How Gear Ratios Work
How Gears Work
How Manual Transmissions Work
How Power Windows Work

An odometer, whether mechanical or computerized, measures the distance a car travels.

Mechanical Odometers

A mechanical odometer is simply a gear train with an incredible gear ratio. A typical odometer might have a 1,690:1 gear reduction. This means the input shaft of the odometer has to spin 1,690 times before the odometer will register 1 mile (1.6 km).

Mechanical odometers are turned by a flexible cable made from a tightly wound spring. The cable usually spins inside a protective rubber housing. The cable snakes its way from the transmission up to the instrument panel, where it is connected to the input shaft of the odometer.

A mechanical odometer uses a series of multiple worm gears to achieve its gear reduction. The input shaft drives the first worm, which drives a gear. Each full revolution of the worm turns the gear only one tooth. That gear turns another worm, which turns another gear, which turns the last worm and finally the last gear, which is hooked up to the tenths-of-a-mile indicator.

Each indicator has a row of pegs sticking out one side and a single set of two pegs on the other side. The two pegs provide intermittent engagement. When the set of two pegs come around, they mesh with a small helper gear. This helper gear also engages one of the pegs on the next higher indicator, turning it one-tenth of a revolution.

Computerized Odometers

If you make a trip to a bike shop, you won't find many mechanical odometers. Instead, you find bicycle computers. A magnet attaches to one of the wheels, and a pickup attaches to the bicycle's frame. Once per revolution of the wheel, the magnet passes by the pickup, generating a voltage in the pickup. The computer counts these voltage spikes, or pulses, and uses them to calculate the distance traveled.

If you have ever installed one of these bike computers, you know that you have to program them with the circumference of the wheel. The circumference is the distance traveled when the wheel makes one full revolution. Each time the computer senses a pulse, it adds another wheel circumference to the total distance and updates the digital display.

Many modern cars use a system very similar to this. Instead of a magnetic pickup on a wheel, though, they use a magnet mounted to the output of the transmission. Some cars use a slotted wheel and an optical pickup, like a computer mouse does.

The instrument panel contains a computer, and it updates the odometer with the new values. In cars with digital odometers, the dashboard simply displays the new value. Cars with analog odometers have a small stepper motor that turns the dials on the mechanical-looking odometer.

How POWER WINDOWS & POWER LOCKS Work

Power windows and power locks are two pure luxuries. Is it really that hard to roll down a window with a crank? Of course not. But when you make these things automatic and give the car's computer a way to control them, you can start doing some nice things. For example, the doors can automatically lock when the car starts rolling, and a little key-chain gizmo can unlock all the doors as you approach the car. It's a great example of how little touches affect our everyday lives.

Both power windows and power locks use small electric motors to drive the mechanisms. In both cases, the motors simply take over the actions that we used to do manually—the actual mechanism for raising the window or locking the door is exactly the same as it's always been.

Power Windows

The key to the lifting mechanism that raises and lowers the windows is a linkage that keeps the window level as it moves up and down. A long arm attaches to a bar that holds the bottom of the window. On the other end of the bar is a large plate that has gear teeth cut into it. The motor (or hand crank) turns a gear that engages these teeth. When the window reaches the top or bottom of its travel, a limit switch cuts the power to the motor.

For the sake of security, it should be impossible to force the glass down. A worm gear in the drive mechanism takes care of this. Most worm gears have a self-locking feature that means the worm can spin the gear, but the gear cannot spin the worm because friction between the teeth causes the gears to bind.

The motor can get its power directly from a switch, or the switch can send a signal to a computer, and the computer can operate the window. For example, in many cars you can turn off the engine and still operate the power windows as long as you don't open a door. The computer makes this sort of complex behavior possible.

Power Locks

The mechanism that unlocks a car's doors has to be very reliable because it is going to unlock the doors tens of thousands of times over the life of the car. It is also versatile because there are a number of ways you can unlock car doors:

- With a key
- By pressing an unlock button inside the car
- By using a combination lock on the outside of the door
- By pulling up the knob on the inside of the door
- With a keyless remote entry control
- By issuing a signal from a central control center

The actuator that locks and unlocks the door is quite simple. A small electric motor turns a series of spur gears that serve as a gear reduction. Then the last gear drives a rack-and-pinion gearset that is connected to the actuator rod. The rack converts the rotational motion of the motor into the linear motion needed to move the lock.

HSW Web Links 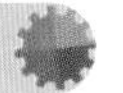

www.howstuffworks.com

How Clutches Work
How Electric Motors Work
How Gear Ratios Work
How Gears Work
How Radio Works

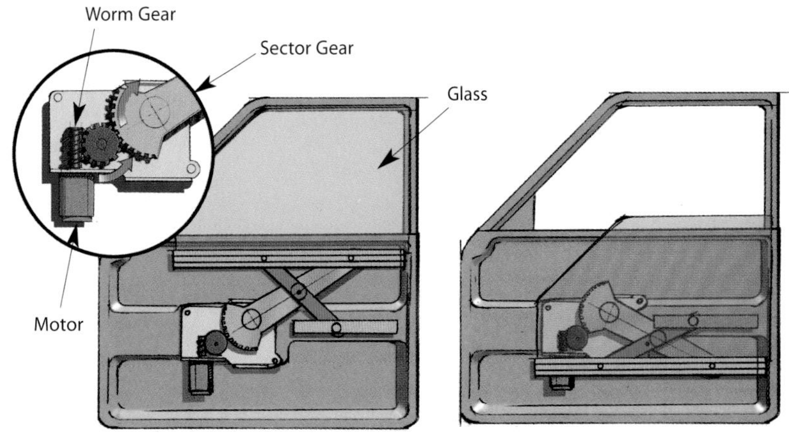

Did You Know?

On some cars the windows can be lowered by inserting the key in the driver's door, turning and holding it. This feature is controlled by the driver's door module, which monitors a switch in the door lock. If the key is held in the turned position for over a set amount of time, the driver's door module lowers the windows.

How CRUISE CONTROL SYSTEMS Work

Cruise control is one of those things that everyone could live without, but it's still nice to have it anyway, especially when it comes to avoiding speeding tickets. It turns out that these devices are a lot smarter than you might think. Try to maintain a constant speed next time you drive, and you'll realize that it actually takes a lot of concentration.

www.howstuffworks.com

How Automatic Transmissions Work
How Car Engines Work
How Fuel Injection Systems Work
How Manual Transmissions Work
How Odometers Work

A cruise control system compares the current speed of the car with the desired speed. If the current speed is less than desired, the system opens the throttle (which is what you do when you step on the gas pedal). If the current speed is more than desired, the system closes the throttle a little.

Controlling Cruise Control

Unlike a human driver, cruise control systems have no ability to anticipate. A cruise control system can only react to changes in speed; it cannot see hills and other changes in road condition that are coming up. In order to react quickly enough, cruise control systems use several control strategies.

The brain of a cruise control system is usually contained in the engine control unit. It connects to the throttle control as well as to several sensors and buttons. The most important input is the speed signal. The cruise control system does a lot with this signal.

Control Systems

There are two basic control systems: proportional control and proportional-integral-derivative (PID) control.

In a proportional control system, the cruise control adjusts the throttle proportionally to the error, where the error is the difference between the desired speed and the actual speed. So, if the cruise control is set at 60 mph (96.6 kph) and the car is going 50 mph (80.5 kph), the error is large and the throttle is opened quite a bit. When the car is going 55 mph (88.5 kph), the throttle position opening will be only half of what it was at 50 mph (80.5 kph). The closer the car gets to the desired speed, the slower it accelerates. On a steep enough hill, the car might not accelerate at all.

Most cruise control systems use PID control. The integral of speed is distance, and the derivative of speed is acceleration. A PID control system uses all three fators—proportional, integral, and derivative. The PID system calculates each individually and adds them to get the throttle position.

The integral factor is based on the difference between the distance your car actually travels and the distance it would have traveled if it were going the desired speed. Let's say your car starts to go up a hill and slows down. The proportional control increases the throttle a little, but you may still slow down. The integral control will start to increase the throttle, opening it more and more, because the longer the car maintains a speed that is slower than the desired speed, the larger the distance error will get.

The derivative factor helps the cruise control respond quickly to changes, such as hills. If the car starts to slow down, the cruise control can see this acceleration (slowing down and speeding up are both acceleration) before the speed can actually change much, and it responds by increasing the throttle position.

Advances in Cruise Control

Cruise control systems generally do a great job of maintaining a constant speed. But their inability to react to driving conditions tends to limit their use to wide-open roads. New systems, called adaptive cruise control, use forward-looking radar to sense the position of the car in front. They can be set to simply follow that car, which is exactly what most drivers do on congested roads.

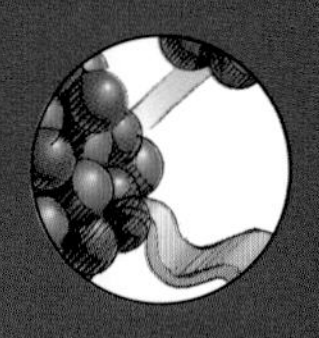

chapter four

SCIENCE 101

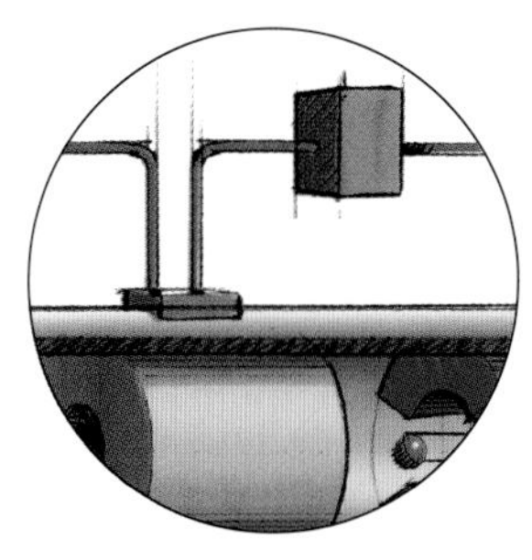

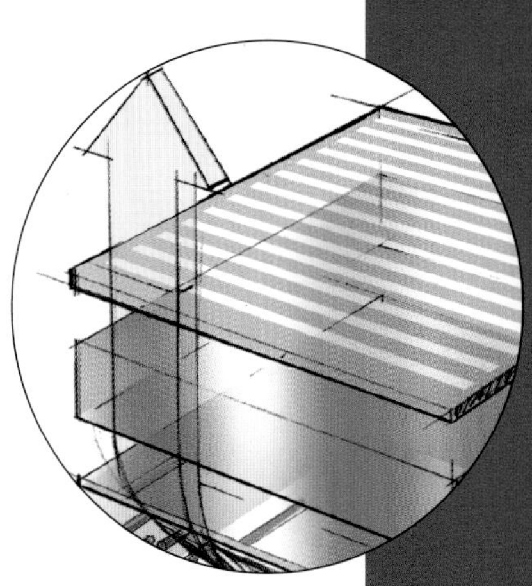

How **ATOMS** Work

We take atoms completely for granted, but there are atoms in our bodies, atoms in the furniture, atoms in the air... Everything is made of atoms. Everything we see and touch is either made of atoms or caused by atoms. Light, radioactivity, magnetism, and gravity all start with atoms.

HSW Web Links

www.howstuffworks.com

How Atom Smashers Work
How Lasers Work
How Nuclear Power Works
How Nuclear Radiation Works
How Photocopiers Work

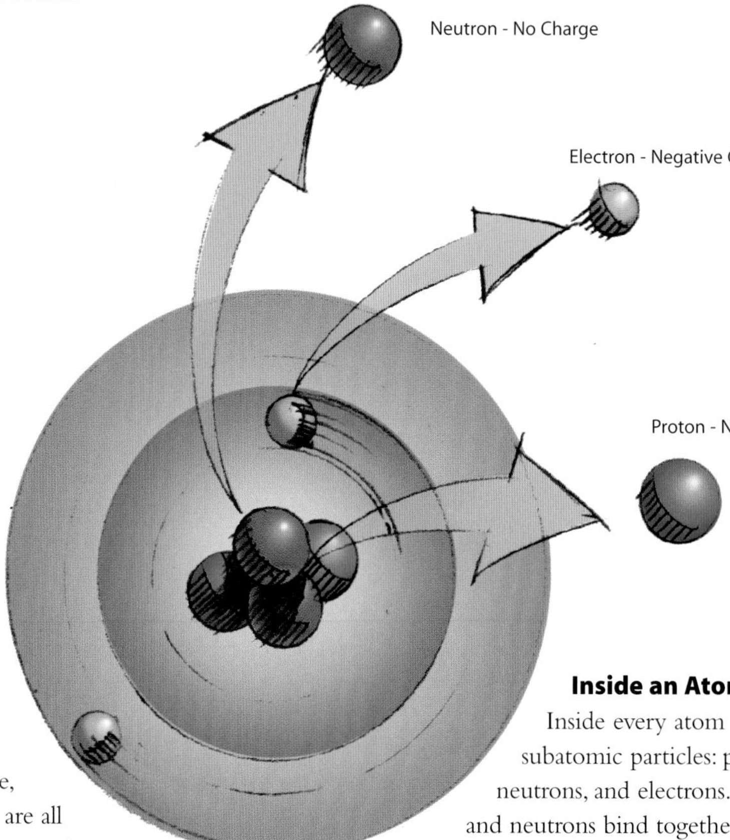

Neutron - No Charge

Electron - Negative Charge

Proton - No Charge

In nature, there are all sorts of atoms: iron atoms, oxygen atoms, carbon atoms, and so on. There are 92 different types of atoms found in nature, and they are known as elements.

The atom is the smallest particle of any element that still behaves like the element. So every substance on earth—metal, plastics, hair, clothing, leaves, glass, and everything else—is made up of combinations of the 92 atoms that are found in nature. The Periodic Table of Elements is a list of the elements found in nature plus a number of human-made elements.

Atoms bind together into molecules. A water molecule, for example, is composed of two hydrogen atoms and one oxygen atom. Elements and molecules have distinct behaviors, so Al (aluminum) is very different from H_2O (water), which is very different from FeO (iron oxide, or rust).

Inside an Atom

Inside every atom are three subatomic particles: protons, neutrons, and electrons. Protons and neutrons bind together to form the nucleus of the atom, and the electrons surround the nucleus. Protons and electrons have opposite charges: Electrons are negative and protons are positive. Because opposite charges attract, protons and electrons attract each other.

In most cases the number of electrons in an atom is the same as the number of protons in that atom, which makes the atom neutral in charge. If the atom has too many or too few electrons, the atom is called an ion and it has a positive or negative charge. (Ions are very useful in creating batteries, as described in "How Batteries Work," page 26.)

The neutrons in an atom are neutral. They bind protons together. Because the protons all have the same charge and would naturally repel one another, the neutrons act as "glue" to hold the protons tightly together in the nucleus.

The number of protons in the nucleus determines the behavior of an atom. For example, if you combine 13 protons with 14 neutrons to create a nucleus and then spin 13 electrons around that nucleus, you have an aluminum atom. If you group millions of aluminum atoms together, you get a substance that is aluminum, and you can form aluminum cans, aluminum foil, and cookware out of it. All aluminum that you find in nature is called aluminum-27. The 27 indicates the total number of neutrons and protons in the nucleus, which is called the atomic mass.

Atoms of the same element can have different forms. For example, copper has two stable forms: copper-63 (which accounts for about 70% of all natural copper) and copper-65 (which is the other approximately 30% of copper). The two forms are called isotopes. Atoms of both isotopes of copper have 29 protons, but a copper-63 atom has 34 neutrons and a copper-65 atom has 36 neutrons. The two isotopes act and look the same, and both are stable.

Light

Any light that you see is made up of a collection of one or more photons—packets of energy—propagating through space as electromagnetic waves. If you look around you right now, there is probably a light source in the room producing photons and objects in the room that reflect those photons. Your eyes absorb some of the photons flowing through the room, and that is how you see.

There are many different ways to produce photons, but all of them use the same mechanism inside an atom to do it. This mechanism involves the electrons orbiting each atom's nucleus. Electrons circle the nucleus in fixed orbits, sort of the same way satellites orbit the earth. There's a huge amount of theory about electron orbitals, but to understand light you need to understand just one key fact: An electron has a natural orbit that it occupies, but if you energize an atom, you can move electrons up to higher orbitals. A photon of light is produced whenever an electron in a higher-than-normal orbit falls back to its normal orbit. During the fall from high energy to normal energy, the electron emits a photon with very specific characteristics. The photon has a frequency, or color, that exactly matches the distance that the electron falls.

There are cases in which you can see this phenomenon quite clearly. For example, in lots of factories and parking lots you see sodium vapor lights, which give off very yellow light. A sodium vapor light energizes sodium atoms to generate photons. A sodium atom has 11 electrons, and because of the way that they are stacked in orbitals, 1 of those electrons is the most likely electron to accept and emit energy (this electron is called the 3s electron). The energy packets that this electron is most likely to emit have a wavelength of 590 nanometers, which corresponds to yellow light. If you run sodium light through a prism, you do not see a rainbow; instead, you see a pair of yellow lines.

Probably the most common way to energize atoms is with heat, and this is the basis of incandescence. If you heat up a horseshoe with a blowtorch, it will get red hot. When you apply enough heat to cause the light to appear white, you are energizing so many different electrons in so many different ways that all the colors of the spectrum are being generated, and they all mix together to look white.

Lasers

Laser light is very different from normal light, but it is produced by the same electron mechanism:

- The light released is monochromatic. It contains one specific wavelength of light (one specific color).
- The light released is coherent. The light is "organized"—all the photons have wave fronts that launch in unison.
- The light is very directional. A laser light has a very tight beam and is very strong and concentrated.

In a laser, the lasing medium is "pumped" to get the atoms into an excited state. Very intense flashes of light or electrical discharges pump the lasing medium and create a large collection of atoms with high-energy

Electrons

If an electron traveled as a wave, could you locate the precise position of the electron within the wave? A German physicist, Werner Heisenberg, answered no in what he called the uncertainty principle:

- To view an electron in its orbit, you must shine a wavelength of light on it that is smaller than the electron's wavelength.
- This small wavelength of light has a high energy.
- The electron will absorb that energy.
- The absorbed energy will change the electron's position.

We can never know both the momentum and position of an electron in an atom. Therefore, Heisenberg said that we shouldn't view electrons as moving in well-defined orbits about the nucleus!

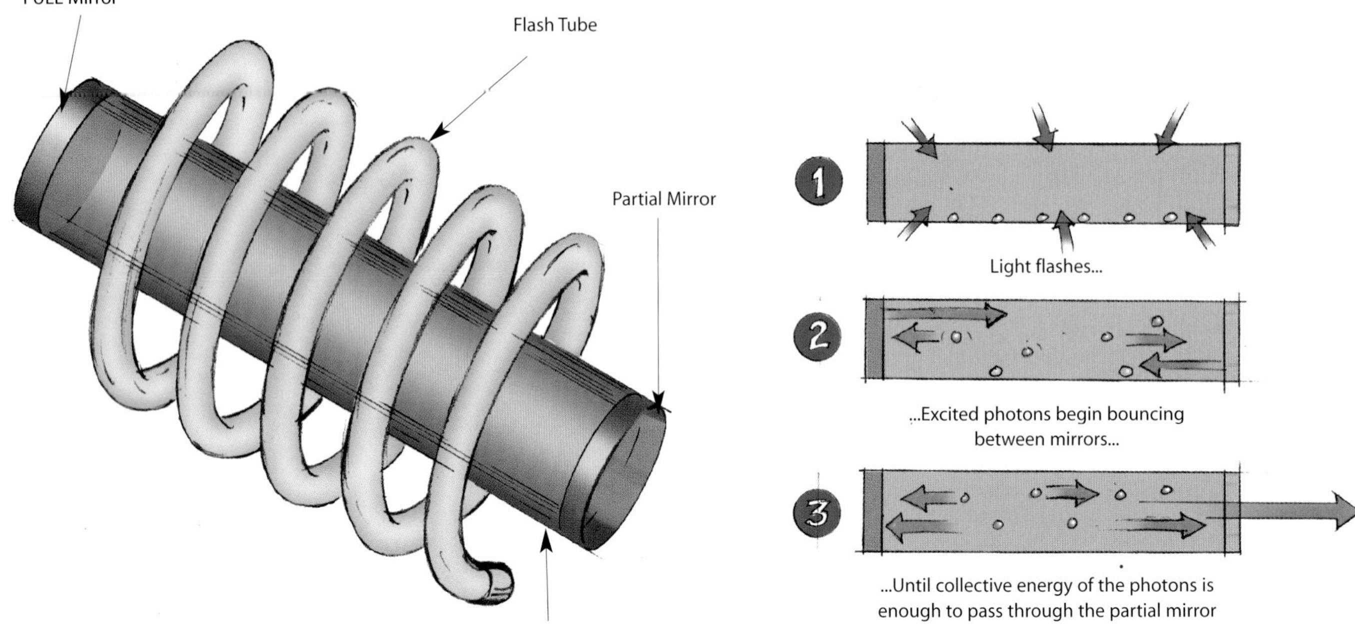

electrons. Two identical atoms with electrons in identical states will release photons with identical wavelengths.

The key to a laser is stimulated emission. The photon that any atom releases has a certain wavelength. If this photon has a certain energy and phase and encounters another atom that has an electron in the same excited state, stimulated emission can occur. The first photon can stimulate atomic emission so that the emitted photon from the second atom vibrates with the same frequency and direction as the incoming photon.

The other key to a laser is a pair of mirrors, one at each end of the lasing medium. Photons reflect off the mirrors to travel back and forth through the lasing medium. In the process, they stimulate more photons of the same wavelength and phase. A cascade effect occurs so that many photons of the same wavelength and phase exist together. The mirror at one end of the laser is half-silvered, meaning it reflects some light and lets some light through. The light that makes it through is the laser light.

And Another Thing...

Three types of emitted rays characterize radioactivity:

- Alpha particles are positively charged and massive. Ernest Rutherford showed that these particles were the nucleus of a helium atom.
- Beta particles are negatively charged and light (later shown to be electrons).
- Gamma rays are neutrally charged and have no mass (i.e., energy).

Radioactivity

Certain elements have isotopes that are radioactive. For example, hydrogen-3 (also known as tritium), has one proton and two neutrons. This isotope is unstable. If you have a container full of tritium and come back in a million years, you will find that it has all turned into helium-3 (two protons, one neutron), which is stable. The process by which it turns into helium is called radioactive decay.

Certain elements are naturally radioactive in all their isotopes. Uranium is a good example and is the heaviest naturally occurring radioactive element. There are eight other naturally radioactive elements: polonium, astatine, radon, francium, radium, actinium, thorium, and protactinium. All human-made elements that are heavier than uranium are radioactive as well.

An atom of a radioactive isotope will spontaneously decay into another element through one of three common processes: alpha decay, beta decay, or spontaneous fission. As this is happening, four different kinds of radioactive rays are produced: alpha rays, beta rays, gamma rays, and neutron rays.

Americium-241, a radioactive element best known for its use in smoke detectors, is a good example of an element that undergoes alpha decay. An americium-241 atom will spontaneously throw off an alpha particle. (An alpha particle is made up of two protons and two neutrons bound together,

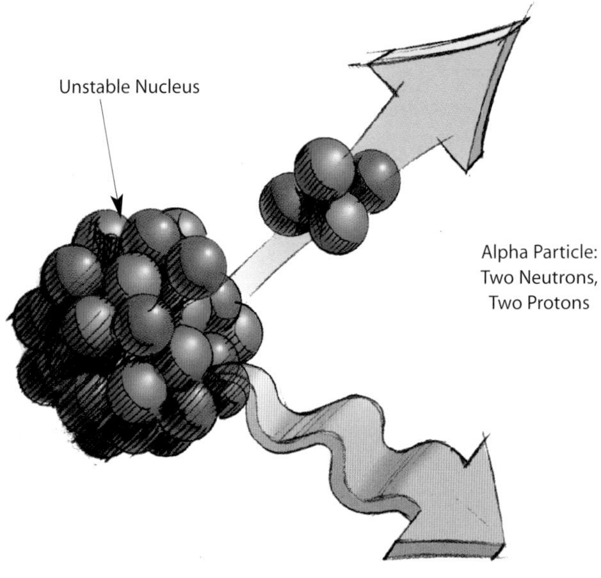

Alpha Decay

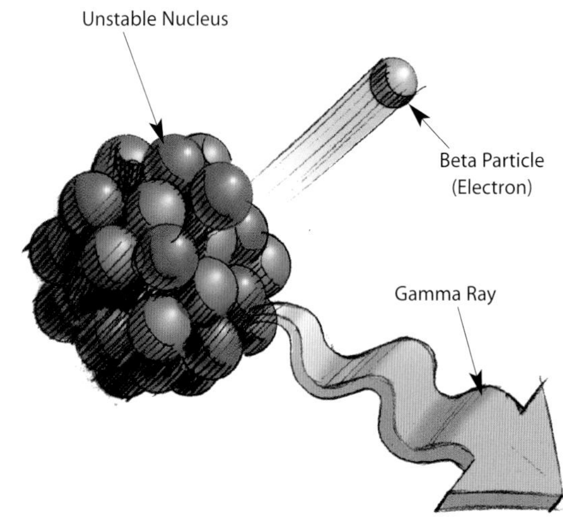

Beta Decay

which is the equivalent of a helium-4 nucleus.) In the process of emitting the alpha particle, the americium-241 atom becomes a neptunium-237 atom. The alpha particle leaves the scene at high velocity—perhaps 10,000 miles per second (16,000 kps).

If you are looking at an individual americium-241 atom, it would be impossible to predict when it would decay and throw off an alpha particle. However, if you have a large collection of americium atoms, then the rate of decay is quite predictable. For americium-241, it is known that half of the atoms decay in 458 years. Therefore, 458 years is the half-life of americium-241. Every radioactive element has a different half-life, ranging from fractions of a second to millions of years, depending on the specific isotope.

Tritium (hydrogen-3) is a good example of an element that undergoes beta decay. In beta decay, a neutron in the nucleus spontaneously turns into a proton, an electron, and a third particle, called an antineutrino. The nucleus ejects the electron (which is then referred to as a beta particle) and the antineutrino, and the proton remains in the nucleus. Therefore, the nucleus loses one neutron and gains one proton. A hydrogen-3 atom undergoing beta decay becomes a helium-3 atom.

In spontaneous fission, an atom actually splits instead of throwing off an alpha or a beta particle. The word *fission* means "splitting." A heavy atom such as fermium-256 undergoes spontaneous fission when it decays about 97% of the time. In the process, it becomes two atoms. For example, one fermium-256 atom may become a xenon-140 atom and a palladium-112 atom. In the process, it would eject four neutrons. These neutrons might be absorbed by other atoms, or they could collide with other atoms, like billiard balls, and cause gamma rays to be emitted.

In many cases, a nucleus that has undergone alpha decay, beta decay, or spontaneous fission is highly energetic and therefore unstable. This type of nucleus eliminates its extra energy as an electromagnetic pulse known as a gamma ray. Gamma rays are like X-rays in that they penetrate matter, but they are more energetic than X-rays. Gamma rays are made of energy, not moving particles, like alpha and beta particles.

Atoms have an amazing number of different properties, and we've only scratched the surface here. Everything we see in nature behaves the way it does because of these tiny particles.

How CELLS Work

At a microscopic level, you are composed of cells. Look at yourself in a mirror, and you'll see about 10 trillion cells of about 200 different types. Your muscles are made of muscle cells and your liver of liver cells; there are even very specialized types of cells that make the enamel for your teeth and the clear lenses in your eyes. Plants are also made of cells, and each bacterium is a creature that consists of one cell. If you want to understand how your body works, you need to understand cells. Everything from reproduction to infections to repairing a broken bone happens down at the cellular level.

HSW Web Links

www.howstuffworks.com

How DNA Evidence Works
How Mad Cow Disease Works
How Prenatal Testing Works
How Your Heart Works
How Your Immune System Works

The largest human cells are about the diameter of a human hair, but most human cells are smaller and might be one-tenth the diameter of a human hair. Run your fingers through your hair now and look at a strand of hair. It is not very thick—maybe 100 microns in diameter (a micron is one-millionth of a meter, so 100 microns is one-tenth of a millimeter). A typical human cell might be one-tenth the diameter of your hair (10 microns). Your little toe might be 2 or 3 billion cells, depending on how big it is. Imagine a whole house filled full of baby peas. If the house is your little toe, the peas are the cells. That's a lot of cells!

Bacteria

Bacteria are the simplest cells that exist. A bacterium is a single, self-contained, living cell. An Escherichia coli bacteria (or E. coli bacteria) is a typical bacterium—it is about one-hundredth the size of a human cell (maybe a micron long and one-tenth of a micron wide).

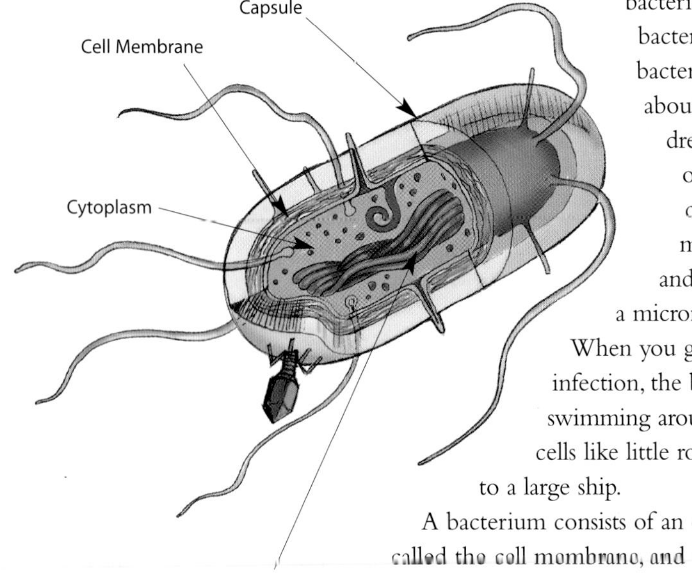

When you get a bacterial infection, the bacteria are swimming around your big cells like little rowboats next to a large ship.

A bacterium consists of an outer wrapper called the cell membrane, and watery fluid inside the cell membrane called the cytoplasm. Cytoplasm might be 70% water. The other 30% is proteins called enzymes that the cell has manufactured, along with smaller molecules such as amino acids, glucose molecules, and adenosine triphosphate (ATP). At the center of the cell is a ball of deoxyribonucleic acid (DNA) that is similar to a wadded up ball of string. If you were to stretch out this DNA into a single long strand, it would be incredibly long compared to the bacterium.

Human cells are much more complex than bacteria—they contain a special nuclear membrane to protect the DNA, other membranes and structures such as mitochondria and Golgi bodies, and a variety of other advanced features. However, the fundamental processes are the same in bacteria and human cells.

Enzymes

At any given moment, all the work being done inside any cell is being done by enzymes. A bacterium like E. coli has about 1,000 different types of enzymes floating around in the cytoplasm. Enzymes in a cell allow the cell to carry out chemical reactions very quickly. At the most basic level, a cell is really a little bag full of chemical reactions that are made possible by enzymes.

Enzymes are made from amino acids, which means that basically, they are proteins. Stringing between 100 to 1,000 amino acids together in a very specific and unique order makes an enzyme. The chain of amino acids then folds into a unique shape. That shape allows the enzyme to carry out specific chemical reactions. An enzyme acts as a very efficient catalyst for one specific chemical reaction—that is, the enzyme speeds up that reaction tremendously.

For example, the sugar maltose is made from two glucose molecules bonded together.

The enzyme maltase is shaped in such a way that it can break the bond and free the two glucose pieces. The only thing maltase can do is break maltose molecules, but it can do that very rapidly and efficiently. Other types of enzymes can put atoms and molecules together. Enzymes break molecules apart and put molecules together, and there is a specific enzyme for each chemical reaction needed to make a cell work properly.

There are hundreds or millions of copies of each different type of enzyme, and the number of copies depends on how important a reaction is to a cell and how often the reaction is needed.

Making DNA from Enzymes

As long as a cell's membrane is intact and it is making all the enzymes it needs to function properly, the cell is alive. The enzymes it needs to function properly allow the cell to create energy from glucose, construct the pieces that make up its cell wall, reproduce, and, of course, produce new enzymes. DNA is the key to creating enzymes. The whole purpose of DNA is to guide the cell in its production of new enzymes. The DNA in a cell is really just a pattern made up of four different parts, called nucleotides, or bases.

Imagine an alphabet that has only 4 different letters. DNA is a long sting of letters. In an E. coli cell the DNA pattern is about 4 million letters long. A human's DNA is about 3 billion letters long, or almost 1,000 times longer than an E. coli cell's DNA. Human DNA is so long that the wadded up approach does not work. Instead, human DNA is tightly wrapped into 23 structures called chromosomes to pack it more tightly and fit it inside a cell.

Amazingly, DNA is nothing more than a pattern that tells a cell how to make its enzymes. The 4 million bases in an E. coli cell's DNA tell the cell how to make the 1,000 or so enzymes that an E. coli cell needs to live its life. A gene is a section of DNA that acts as a template to form one enzyme.

You have probably heard of the human DNA molecule referred to as a double-helix. DNA is like two strings twisted together in a long spiral. DNA is found in all cells as base pairs made of four different nucleotides. Each base pair is formed from two complementary nucleotides bonded together. The 4 bases in DNA's alphabet are

Protein

A protein is a chain of amino acids. An amino acid is a small molecule that acts as the building block of any protein. If you ignore the fat, your body is about 20% protein and about 60% water. Most of the rest of your body is composed of minerals (for example, calcium in your bones).

Amino acids are called amino acids because they all contain an amino group (NH_2) and a carboxyl group (COOH), which is acidic. The human body is constructed of 20 different amino acids.

As far as your body is concerned, there are two different types of amino acids: essential and nonessential. Nonessential amino acids are amino acids that your body can create out of other chemicals found in your body. Essential amino acids cannot be created, and therefore the only way to get them is through food. Here are the different amino acids:

Nonessential amino acids:

- Alanine (synthesized from pyruvic acid)
- Arginine (synthesized from glutamic acid)
- Asparagine (synthesized from aspartic acid)
- Aspartic acid (synthesized from oxaloacetic acid)
- Cysteine (synthesized from homocysteine, which comes from methionine)
- Glutamic acid (synthesized from oxoglutaric acid)
- Glutamine (synthesized from glutamic acid)
- Glycine (synthesized from serine and threonine)
- Proline (synthesized from glutamic acid)
- Serine (synthesized from glucose)
- Tryosine (synthesized from phenylalanine)

Essential amino acids:

- Histidine
- Isoleucine
- Leucine
- Lysine
- Methionine
- Phenylalanine
- Threonine
- Tryptophan
- Valine

Protein in the diet comes from both animal and vegetable sources. Most animal sources (such as meat, milk, and eggs) provide "complete protein," meaning that they contain all the essential amino acids. The digestive system breaks all proteins down into their amino acids so that they can enter the bloodstream. Cells then use the amino acids as building blocks to build enzymes and structural proteins.

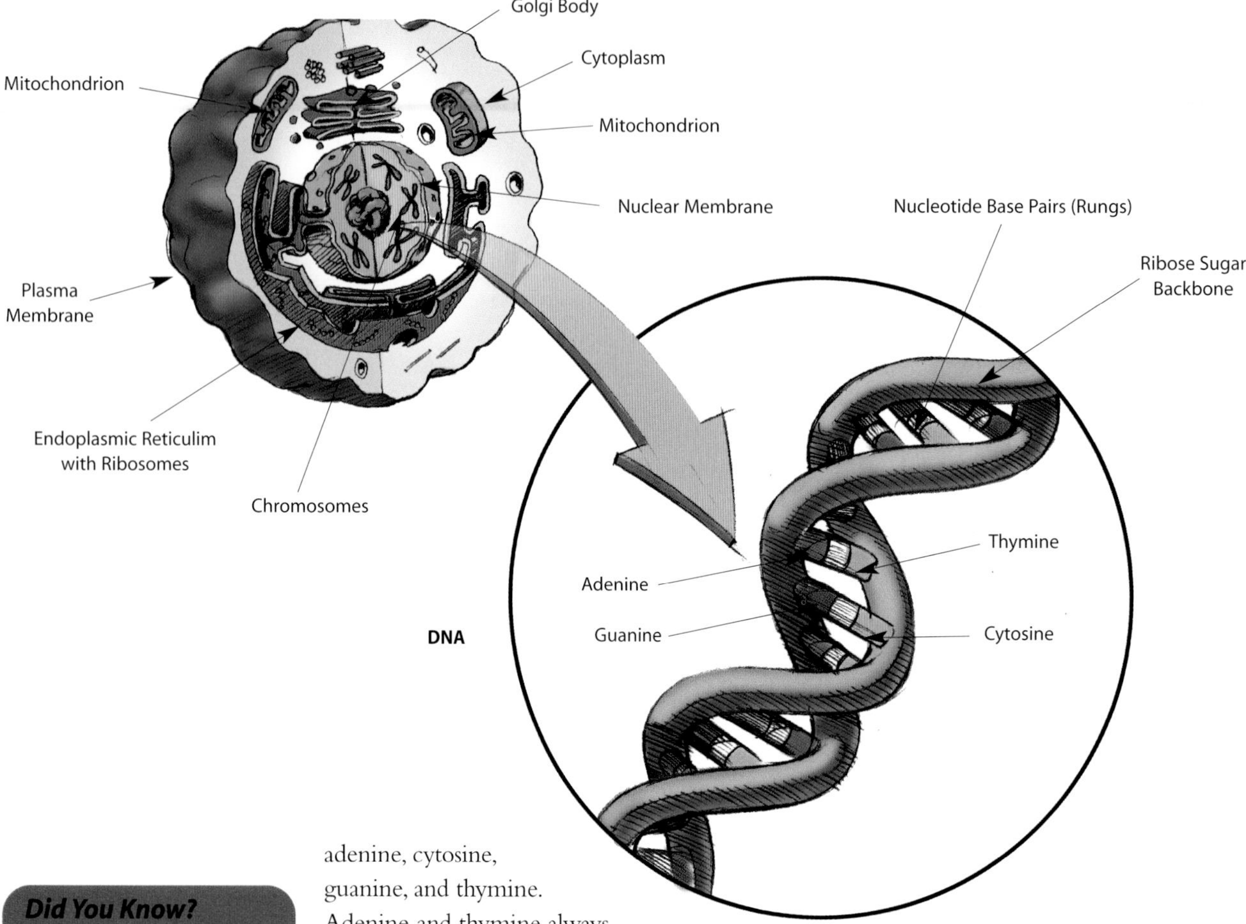

DNA

adenine, cytosine, guanine, and thymine. Adenine and thymine always bond together as a pair, and cytosine and guanine bond together as a pair. The pairs link together like rungs in a ladder.

In an E. coli bacterium, this ladder is about 4 million base pairs long. The two ends link together to form a ring, and then the ring gets wadded up to fit inside the cell. The entire ring is known as the genome, and scientists have completely decoded it. That is, scientists know all 4 million of the base pairs needed to form an E. coli bacterium's DNA exactly. The human genome project is the process of finding all 3 billion or so of the base pairs in a typical human's DNA and understanding what they all do.

Enzymes are formed from 20 different amino acids strung together in a specific order. How do you get from DNA, made up of only 4 nucleotides, to an enzyme containing 20 different amino acids? There are two answers to this question:

- An extremely complex and amazing enzyme called a ribosome reads messenger ribonucleic acid (RNA), produced from the DNA, and converts it into amino acid chains.
- To pick the right amino acids, a ribosome takes the nucleotides in sets of 3 to encode for the 20 amino acids.

This means that every 3 base pairs in the DNA chain encodes for 1 amino acid in an enzyme. The 3 nucleotides in a row on a DNA strand are therefore together referred to as a codon. Because DNA consists of 4 different bases, and because there are 3 bases in a codon, and because 4 X 4 X 4 = 64, there are 64 possible patterns for a codon. Because there are only 20 possible amino acids, there is some redundancy—several different codons can encode for the same amino acid. In addition, there is a stop codon that marks the end of a gene. So in a DNA strand there is a set of 100 to 1,000 codons (300 to 3,000 bases) that specify the amino acids to form a specific enzyme, and then there's a stop codon to mark the end of the chain. At the beginning of the chain is a

Did You Know?

As long as the enzymes in a cell are active and all the necessary enzymes are available, the cell is alive. If you take a bunch of yeast cells and put them in a blender to release the enzymes, the resulting soup will still do the sorts of things yeast cells do. For example, the soup will produce carbon dioxide and alcohol from sugar, like living yeast cells do, for some period of time. However, because the cells are no longer intact and therefore are not alive, no new enzymes are produced. Eventually, as the existing enzymes wear out, the soup stops reacting: The cells and the soup have "died."

section of bases that is called a promoter. A gene, therefore, consists of a promoter, a set of codons for the amino acids in a specific enzyme, and a stop codon.

To create an enzyme, the cell must first transcribe the gene in the DNA into messenger RNA (mRNA). The transcription is performed by an enzyme called RNA polymerase, which binds to the DNA strand at the promoter, unlinks the two strands of DNA, and then makes a complementary copy of one of the DNA strands into an RNA strand. RNA is very similar to DNA, except that it is happy to live in a single-stranded state (as opposed to DNA's desire to form complementary double-stranded helixes). So the job of RNA polymerase is to make a copy of the gene in DNA into a single strand of mRNA.

The strand of messenger RNA then floats over to a ribosome, which is possibly the most amazing enzyme in nature. A ribosome looks at the first codon in an mRNA strand, finds the right amino acid for that codon, holds the amino acid, and then looks at the next codon, finds its correct amino acid, stitches it to the first amino acid, and so on. In other words, the ribosome reads the codons, converts them to amino acids, and stitches the amino acids together to form a long chain. When the ribosome gets to the last codon—the stop codon—it releases the chain. The long chain of amino acids is an enzyme, and the enzyme folds into its characteristic shape and begins performing whatever reaction that enzyme performs.

As you can see, inside every cell, a variety of processes are keeping the cell alive.

Cell Reproduction

The hallmark of all living things is the ability to reproduce. In a bacterium, reproduction is one of many enzymatic behaviors. The enzyme DNA polymerase, along with several other enzymes that work alongside it, walks down the DNA strand and replicates the strand. DNA polymerase splits the double helix and creates a new double helix along each of the two strands. When it reaches the end of the DNA loop, there are two separate copies of the loop floating in the E. coli cell. The cell then pinches its cell wall in the middle, divides the two DNA loops between the two sides, and splits itself in half.

Under the proper conditions, an E. coli cell can split like this very rapidly. The enzymatic process of growing the cell, replicating the DNA loop, and splitting happens every 20 or 30 minutes.

Human beings have created some amazing machines, but it is very hard to compare them to cells in terms of design elegance and efficiency. Individual cells and multi-cellular structures are absolutely incredible in every way.

Poisons and Antibiotics

You can see that the life of a cell depends on a rich soup of enzymes that float in the cell's cytoplasm. Many different poisons work by disrupting the balance of the soup in one way or another.

For example, diphtheria toxin works by blocking the action of a cell's ribosomes, making it impossible for the ribosome to walk along the mRNA strand. The toxin in a death cap mushroom gums up the action of RNA polymerase and halts the transcription of DNA. In both cases, the production of new enzymes shuts down and cells affected by the toxin can no longer grow or reproduce.

An antibiotic is a poison that works to destroy bacterial cells, while leaving human cells unharmed. All antibiotics take advantage of the fact that there are many differences between the enzymes inside a human cell and the enzymes inside a bacterium. If a toxin is found, for example, that affects an E. coli ribosome but leaves human ribosomes unharmed, then it may be an effective antibiotic. Streptomycin is an example of an antibiotic that works in this way.

Penicillin was one of the first antibiotics. It foils a bacterium's ability to build cell walls. Because bacterial cell walls and human cell walls are very different, penicillin has a big effect on certain species of bacteria but no effect on human cells.

The unfortunate problem with any antibiotic is that it becomes ineffective over time. Bacteria reproduce so quickly that the probability for mutations is high. In your body there may be millions of bacteria that the antibiotic kills. But if just one of the bacteria has a mutation that makes it immune to the antibiotic, that one cell can reproduce quickly and then spread to other people. Most bacterial diseases have become immune to some or all of the antibiotics used against them through this process.

How ELECTROMAGNETS Work

The basic idea behind an electromagnet is extremely simple: By running electric current through a wire, you can create a magnetic field.

HSW Web Links

www.howstuffworks.com

How Batteries Work
How Electric Motors Work
How Magna Doodle Works
How Relays Work
How Speakers Work

Making an Electomagnet

To create your own electomagnet, you'll need:

- A normal flashlight battery
- A piece of wire that is about one or two feet long
- A compass

Put the compass on the table and, with the wire near the compass, connect the wire between the positive and negative ends of the battery for a few seconds. The compass needle swings because of the magnetic field created in the wire by the flow of electrons.

A circular magnetic field develops around the wire. The compass needle aligns itself with this field, perpendicular to the wire. If you flip the battery around and repeat the experiment, you will see that the compass needle points in the opposite direction.

Because the magnetic field around a wire is circular and perpendicular to the wire, an easy way to amplify the wire's magnetic field is to coil the wire. What you get is an electromagnet! If you wrap the wire around a piece of iron or steel (a nail would work nicely) and connect it to a battery, you can create a strong electromagnet.

Before talking about electromagnets, let's talk about normal "permanent" magnets, like the ones you have on your refrigerator.

Characteristics of Magnets

You know that all magnets have two ends, normally referred to as north and south, and they attract things made of steel or iron. You also know the fundamental law of all magnets: Opposites attract, and likes repel. So if you have two bar magnets with ends marked north and south, the north end of one magnet will attract the south end of the other. On the other hand, the north end of one magnet will repel the north end of the other, and the south end of one will repel the south end of the other. An electromagnet is the same way, except it is "temporary"—the magnetic field exists only when electric current is flowing in an electromagnet.

The Parts of an Electromagnet

An electromagnet starts with a battery (or some other source of power) and a wire. A battery produces electrons. If you look at a battery, say a normal D-cell battery from a flashlight, you can see that there are two ends, marked plus (+) and minus (-). Electrons collect at the negative end of the battery, and, if you let them, they will gladly flow to the positive end. The way you let them flow is with a wire. If you attach a wire directly between the positive and negative terminals of a D-cell battery, three things will happen:

- Electrons will flow from the negative side of the battery to the positive side as fast as they can.
- The battery will drain fairly quickly (in a matter of several minutes). For that reason it is generally not a good idea to connect the two terminals of a battery to one another directly.

- A small magnetic field is generated in the wire. This small magnetic field is the basis of an electromagnet.

You probably knew about the first two things on this list, but the third might be a surprise to you.

If you wrap a wire around a nail 10 times, connect the wire to a battery, and bring one end of the nail near a compass, you will find that it has a large effect on the compass. In fact, the nail behaves just like a bar magnet, and the magnet exists as long as the current is flowing from the battery.

You have created an electromagnet that is able to pick up small steel things like paper clips, staples, and thumbtacks. You can change the poles of the magnet by reversing the current—that is, by reversing the battery in the circuit.

By using the simple principles of electromagnetics, you can create all sorts of things, including motors, solenoids, speakers, and even read/write heads for disk and tape drives in cassette tape players, VCRs, and floppy and hard disks in a computer.

How GPS RECEIVERS Work

The global positioning system (GPS) is made up of 24 satellites orbiting the earth at an altitude of about 11,000 miles (17,703 km). These satellites send out radio signals that you can pick up with a GPS receiver. A GPS receiver uses the signals to determine its own location on earth, in latitude, longitude, and altitude, so when you use a GPS receiver, you always know where you are! For anyone who has ever been lost—while hiking in the woods, driving in an unfamiliar city, or flying a small plane at night—a GPS receiver is a miracle.

A GPS receiver uses measurements from satellite radio signals and basic geometry to calculate an exact location.

Trilateration

The geometric principle at work is called trilateration. The logic behind trilateration is easy to understand in two-dimensional space.

Let's say that you are somewhere in the United States and you are totally lost. You find a friendly looking person and ask, "Where am I?" The person says, "You are 625 miles (1,006 km) from Boise, Idaho." This piece of information isn't very useful by itself. You could be anywhere on a circle with Boise as the center and a radius of 625 miles (1,006 km).

So you ask another person the same question, and that person says, "You are 690 miles (1,110 km) away from Minneapolis, Minnesota." If you combine this information with the Boise information, you can draw two circles that intersect. You now know that you are at one of two points, but you don't know which one.

If a third person tells you that you are 615 miles (990 km) from Tucson, Arizona, you can figure out which of the two points is your location. So with three known distances, you can figure out that you are near Denver, Colorado!

This same concept works in three-dimensional space, but you have to use spheres instead of circles. By intersecting four spheres, you can find an exact location.

A GPS receiver uses the signals from a GPS satellite to figure out its exact distance from the satellite. The GPS receiver also knows exactly where the satellite is at any given moment because the satellites fly in extremely precise and well-known orbits. If the GPS receiver finds that it is *X* miles from one satellite with a known position, it knows it's somewhere on an imaginary sphere with the satellite as the center and the radius *X*.

If the GPS receiver can lock onto the signals of four satellites, it can find its point in space. Because the GPS receiver also knows about the sphere of the earth, the GPS receiver can tell you your longitude, latitude, and altitude. By storing up points over time, the GPS receiver can also calculate your current speed, average speed, distance traveled, and so on.

Location and Distance

For a GPS receiver to find your location, it needs to determine two things:

- The location of at least four GPS satellites above you
- The distance between you and each of those satellites

Finding the locations of the satellites is very simple. GPS satellites travel so high above the earth (about 11,000 miles, or 17,703 km)

HSW Web Links

www.howstuffworks.com

How Cell Phones Work
How Digital Television Works
How Radios Work
How Space Shuttles Work

Did You Know?

The Global Positioning System was developed by the U.S. Department of Defense, originally for military purposes. The first GPS satellite was launched in 1978, but the system didn't become fully operational until 1995. Initially, miscalculations were programmed into GPS transmissions to limit the accuracy of non-military GPS receivers. This operation was canceled in May 2000.

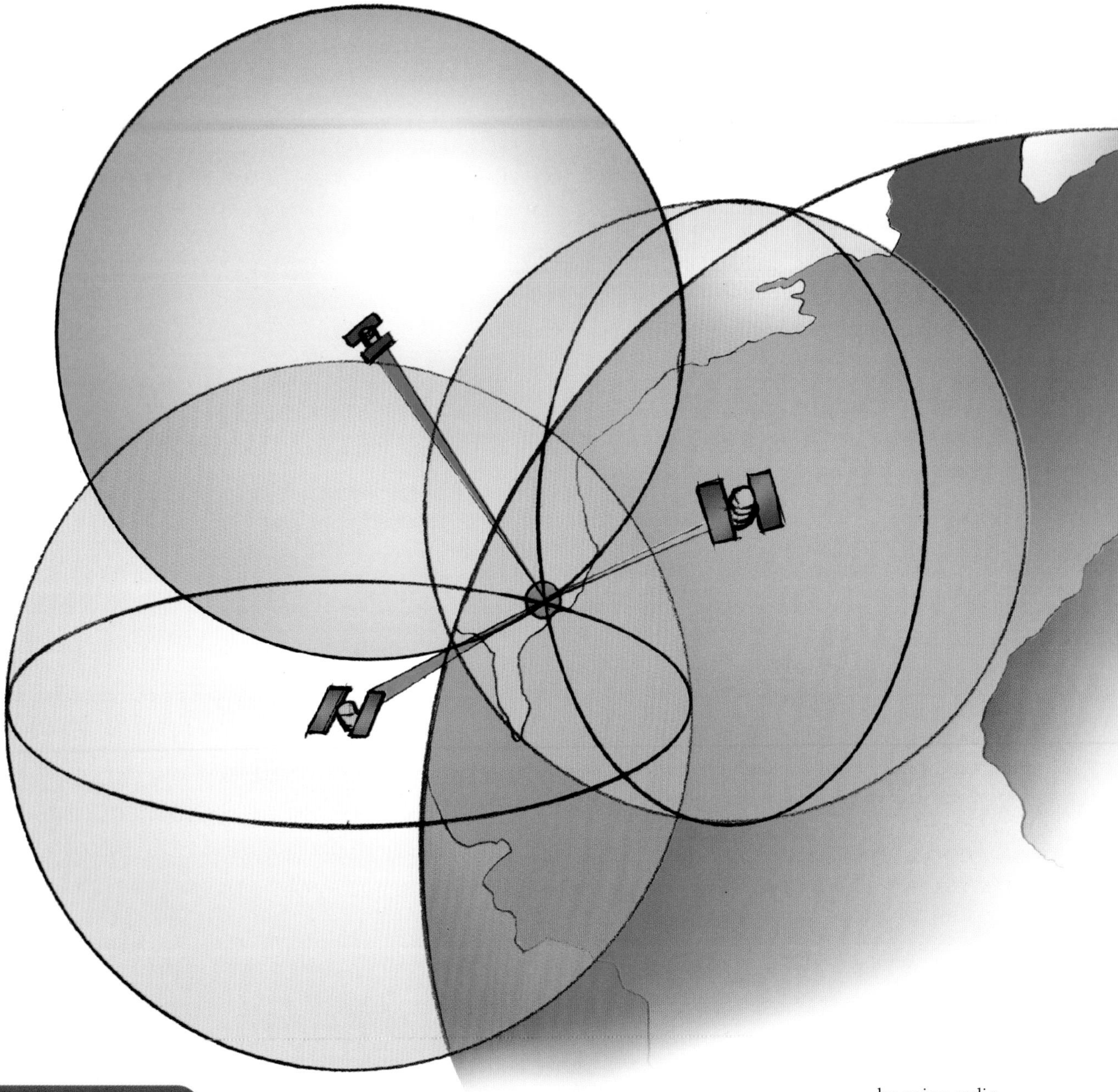

Cool Facts

- The 24 GPS satellites now in orbit cost an estimated $12 billion to build and launch.
- Each satellite weighs about 1,735 lbs (787 kg).
- A GPS satellite takes 12 hours to orbit the earth once.
- The Russians are working on a GPS called GLONASS that is not yet fully operational.

that the atmosphere does not affect them. This makes their orbits very predictable. The receiver has an almanac in its memory that lists where every satellite should be at any given time. Things like the pull of the moon and the sun do change the satellites' orbits very slightly, however. So the U.S. Department of Defense monitors exact satellite positions and transmits any adjustments to all GPS receivers, as part of the satellites' signals.

Measuring the distance to the satellites is a bit trickier than finding their locations. GPS satellites broadcast to all GPS receivers by using radio signals. To find the distance to the satellite, the receiver has to figure out how far a radio wave has traveled. We know how fast radio waves move; they are electromagnetic waves, so (in a vacuum) they travel at the speed of light—about 186,000 miles per second (299,338 kps), or about 1 foot (30 cm) per nanosecond. Because

Distance = Speed X Time

We can figure out how far away a satellite is by determining how long it took the satellite's signal to get to the receiver.

To figure out how long the signal takes to travel, GPS satellites and receivers use a fascinating scheme. Let's say that both the satellite and receiver had perfectly synchronized and very accurate clocks. At exactly midnight, they both started very quickly counting 1, 2, 3, 4, 5, and so on—say they are counting from 1 to 1 million in a second, so over the course of the day they count from 1 to 86 trillion. Let's also say that the satellite transmits its stream of numbers. Because the satellite and the receiver started counting at exactly the same time, the satellite's stream of numbers would lag slightly behind the receiver's internal count. The receiver would be able to calculate how much time it took for the signal to arrive by looking at the lag between the numbers.

This is exactly how the GPS system works, but the GPS system communicates by using a pseudorandom code rather than simple counting. In simple counting, the algorithm is "add 1 to form the next number." In a pseudorandom code, the algorithm is a bit more complicated, and it is designed so that the numbers coming out appear to be random. They are pseudorandom, however, because if you know the formula, you can know what each number will be. Each satellite transmits a different pseudorandom sequence, and this is how the receiver separates all the different satellite signals from one another.

Clock Synchronization

The clocks in the GPS receiver and the satellites have to be extremely accurate, so the time measurements are in nanoseconds. This means you need atomic clock–level accuracy on both the satellite and the receiver. The satellites do contain atomic clocks, but the receivers cannot. The problem is that atomic clocks typically cost between $50,000 and $100,000, making them prohibitively expensive as consumer items. The solution to this problem is fascinating.

A GPS receiver has to make do with an ordinary quartz clock. It gets atomic-level performance by checking the inaccuracy of its system and readjusting the calculations accordingly. The trick is to determine the distance of more than three satellites. The first three spheres (plus the earth) intersect at a specific point. If everything is measured accurately, the fourth sphere should also intersect with this point. But if the measurements are inaccurate because the clocks aren't synchronized correctly, the fourth sphere will not intersect. The distance of the fourth sphere from the point tells the receiver how far off its calculations were.

Because the receiver makes all its time measurements with the same clock, the distances are all proportionally incorrect. The receiver can calculate exactly what distance adjustment will cause the four spheres to intersect at one point. This provides an accurate location point and also tells the receiver how inaccurate its clock is. The receiver is constantly checking its measurements and therefore constantly readjusting its clock. For this reason, a GPS receiver keeps extremely accurate time, on the order of the actual atomic clocks in the satellites.

Another clock-related problem is determining the speed of the radio waves. As you saw earlier, electromagnetic signals travel through a vacuum at the speed of light. But of course the earth is not in a vacuum. Its atmosphere slows the transmission of the signal according to the particular conditions at that atmospheric location, the angle at which the signal enters it, and so on. A GPS receiver guesses the actual speed of the signal by using complex mathematical models of a wide range of atmospheric conditions. The satellites transmit relevant weather information as part of their radio signal.

You can see that a GPS receiver does a lot of calculating. It calculates exactly where each satellite is in its orbit, exactly how long the signals take to arrive at the receiver, exactly where the four (or more) spheres intersect, and exactly how inaccurate its internal clock is. Using this system of calculations, a receiver displays the latitude, longitude, and altitude of its current position on its LCD screen. Most receivers then combine this data with other information, such as maps, to make the receiver more user friendly.

Speed and Temperature Limits

Some GPS receivers have speed limits. GPS receiver manufacturers sometimes program speed limits into the devices, so that if the device is moving above a certain speed, it will not work properly. A receiver meant to be used in a car may not work on an airplane, which travels much faster than an automobile. This is more often the case in car, airplane, or boat-mounted receivers than in the hand-held models.

GPS receivers have temperature limits. Like most electronic devices, especially those with LCD screens, GPS receivers may not function properly above or below certain temperatures. If you plan to use your receiver in any extreme temperature situations, such as mountain climbing or hiking in the desert, you should check to make sure the receiver model can function in those conditions.

How COMPASSES Work

No matter where you stand on earth, you can hold a compass in your hand, and it will point toward the north pole. What an unbelievably neat and amazing thing! Imagine that you are in the middle of the ocean, and in every direction all you can see is water. It is overcast so you cannot see the sun. How in the world would you know which way to go? A compass completely solves this problem.

HSW Web Links

www.howstuffworks.com

How Electric Motors Work
How Electromagnets Work
How Gyroscopes Work
How Magna Doodle Works
How Magnetic Resonance Imaging (MRI) Works

Gyro Compass

Sometimes you have trouble using a magnetic compass on a moving platform, such as a ship or an airplane. A magnetic compass must be level, and it tends to correct itself rather slowly when the platform turns. Because of this tendency, most ships and airplanes use gyroscopic compasses instead.

If you support a spinning gyroscope in a gimbaled frame, it will maintain the direction it is pointing toward, even if the frame moves or rotates. In a gyrocompass this tendency is used to emulate a magnetic compass. At the start of the trip, the axis of the gyrocompass is pointed toward north by using a magnetic compass as a reference. A motor inside the gyrocompass keeps the gyroscope spinning, so the gyrocompass will continue pointing toward north and will do it swiftly and accurately, even if the boat is in rough seas or the plane hits turbulence. Periodically the gyrocompass is checked against the magnetic compass to correct any error it might pick up.

There are two main types of com-passes—magnetic and gyroscopic—and both are extremely simple devices.

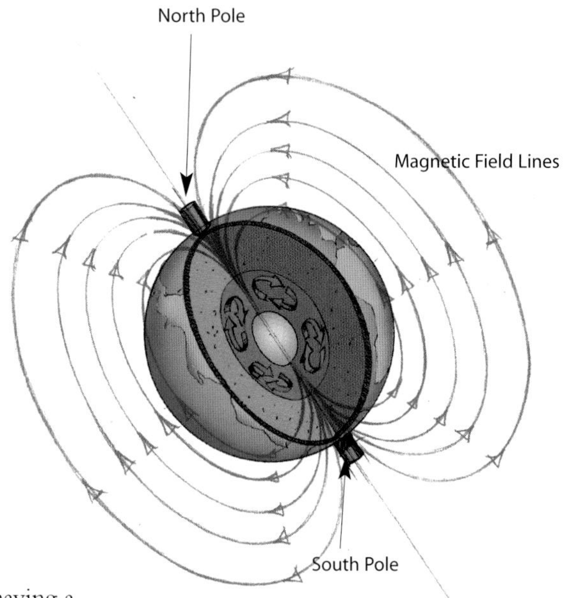

Magnetic Compasses

A magnetic compass consists of a small, lightweight magnet balanced on a nearly frictionless pivot point. The magnet is usually called a needle. One end of the needle is often marked N, for north, or colored in some way to indicate that it points northward. That's all there is to a compass.

You can think of the earth as having a gigantic bar magnet buried inside. In order for the north end of the compass to point toward the north pole, you have to assume that the buried bar magnet has its south end at the north pole. If you think of the world this way, then you can see that the normal opposites-attract rule for magnets would cause the north end of the compass needle to point toward the south end of the buried bar magnet—at the north pole.

The bar magnet does not run exactly along the earth's rotational axis. It is skewed slightly off center. For example, the rotational north pole and the magnetic north pole are more than 1,000 miles (1,609 km) apart. This skew is called the declination. Most good maps indicate what the declination is in different areas.

The earth's magnetic field is fairly weak on the surface of the planet. After all, the earth is almost 8,000 miles (12,875 km) in diameter. Therefore, a compass needs to have a lightweight magnet and a frictionless bearing. Otherwise, there just isn't enough strength in the earth's magnetic field to turn the needle.

The analogy of the big bar magnet buried in the core of the earth works to explain why the earth has a magnetic field, but obviously there is not really a bar magnet in the earth. One theory is that the earth's core consists largely of molten iron. Convection caused by heat radiating from the core, along with the rotation of the earth, causes the liquid iron to move in a rotational pattern. It is believed that these rotational forces in the liquid iron layer create weak magnetic forces around the earth's axis of spin. (For more information, see "How Electromagnets Work," on page 90.)

If the earth did not have a magnetic core, or if we had never figured out how to use it easily, it is safe to say that it would have slowed down scientific progress, map making, and early travel to some degree.

How **ATOMIC CLOCKS** Work

Atomic clocks are the most accurate time-keeping devices in the world. A normal quartz watch is remarkably good at keeping time and might be accurate to—that is, gain or lose—a second per day. The best atomic clocks are accurate to a second every several million years.

A clock's job is to keep track of the passage of time. All clocks do this by counting the ticks of a resonator. In a pendulum clock the resonator is a pendulum. In a quartz clock the resonator is a vibrating quartz crystal. An atomic clock uses the resonance frequencies of atoms as its resonator.

Atomic Resonance

The advantage of the atomic approach is that atoms resonate at extremely consistent frequencies. An atom of cesium will resonate at exactly the same frequency as any other atom of cesium. Cesium-133 oscillates at 9,192,631,770 cycles per second. This sort of accuracy is completely different from the accuracy of a quartz clock. In a quartz clock, the quartz crystal is manufactured so that its oscillating frequency is close to some standard frequency. Manufacturing tolerances cause every quartz crystal to be slightly different, however, and things like temperature can change the frequency. A cesium atom always resonates at the same known frequency, whatever the temperature or other atmospheric conditions—that is what makes atomic clocks so precise. The atomic clocks used in a GPS satellite are about the size of a loaf of bread and cost thousands of dollars each.

Detecting Excited Atoms

The basic idea in creating an atomic clock is to excite atoms at their resonance frequency and detect the atoms that actually do get excited. In the case of cesium, microwave radio waves are used to excite the atoms. A quartz oscillator that is close to the resonance frequency creates the micro-waves. The microwave frequency is adjusted up and down across a range that includes the resonance frequency of cesium. For example, imagine the microwave generator sweeping between 9,192,600,000 cycles per second and 9,192,700,000 cycles per second. At the point where the frequency is exactly 9,192,631,770 cycles per second, the maximum number of cesium atoms get excited. By detecting this maximum, the microwave generator can lock onto exactly 9,192,631,770 cycles per second, and then the frequency of the micro-wave generator can be used as a very precise resonator in a clock.

In a typical atomic clock, a stream of cesium atoms in a vacuum passes through the microwave field. A magnetic field separates the excited atoms from the unexcited ones and aims the excited atoms at a detector. When the detector detects the maximum number of atoms, it knows that the microwave generator is at the right frequency.

Atomic clocks and the incredibly accurate time they keep make possible things like GPSs satellite communications,and orbital calculations for space exploration. By being the basis of the world's time, they also keep planes on schedule and help you get up in the morning.

Clocks in the Home

You may have heard about "atomic clocks" for the home that automatically set themselves each day. These clocks and watches are not actually atomic clocks, but they truly do synchronize themselves with an atomic clock. This feature is made possible by a radio system set up and operated by the National Institute of Standards and Technology (NIST), located in Fort Collins, Colorado. NIST operates radio station WWVB, which is the station that transmits the time codes.

WWVB has high transmitter power (50,000 watts), a very efficient antenna, and an extremely low frequency: 60,000 Hz. In comparison, a typical AM radio station broadcasts at a frequency of 1,000,000 Hz. The combination of high power and low frequency gives the radio waves from WWVB a lot of bounce, and this single station can therefore cover all of the continental United States plus much of Canada and Central America as well. The "atomic clock" you might have in you home has a little radio receiver built in that receives these radio waves, decodes the time signal, and sets the clock.

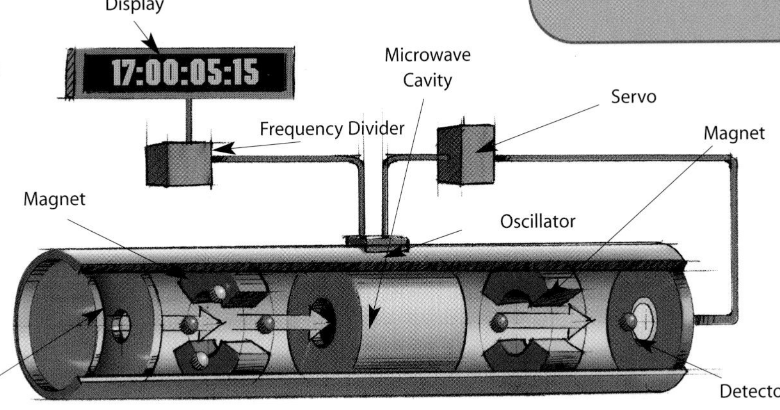

How SATELLITES Work

Not very long ago, satellites were exotic, top-secret devices. They were used mostly in military functions, for activities such as navigation and espionage. Now they are an essential part of our daily lives. We see and recognize their use in weather reports, global positioning system (GPS) signals, television transmissions by various satellite TV companies, and everyday telephone calls.

HSW Web Links

www.howstuffworks.com

How Cell Phones Work
How Digital Television Works
How GPS Receivers Work
How Radio Works
How Space Shuttles Work

Cost of Satellites

Satellites are expensive, costing from hundreds of millions of dollars into the billions. They are expensive because they are normally custom built, they often use exotic devices and technologies, they have to operate on extremely low power, and they have to be incredibly reliable. A satellite will stay live in orbit for a decade or more. Creating something that can operate reliably for that long, in the extreme environment of space, without repairs, is an expensive challenge.

Another important cost factor with satellites is the launch. A satellite launch can cost anywhere between $50 million and $500 million, depending on whether the satellite goes up on a disposable rocket or a space shuttle. You can see that building a satellite, getting the satellite into orbit, and then maintaining it from the ground control facility is a major financial endeavor!

A satellite is any object that revolves around a planet in a circular or an elliptical path. The moon is earth's original, natural satellite. Now there are many human-made, or artificial, satellites of the earth, many of them much closer to earth than the moon.

Orbital Velocity and Altitude

A rocket must accelerate to at least 25,039 mph (40,320 kph) to completely escape earth's gravity and fly into space. It takes much less escape velocity to put a satellite into orbit around the earth than to make a rocket escape earth's gravity. When you're sending a satellite into orbit, the goal is not for the satellite to escape earth's gravity, but for the satellite to balance it. If the satellite goes too slowly, gravity will pull it back to earth. So orbital velocity is the velocity needed to achieve balance between gravity's pull on the satellite and the inertia of the satellite's motion—the satellite's tendency to keep going—which is approximately 17,000 mph (27,359 kph) at an altitude of 150 miles (242 km).

The orbital velocity of a satellite depends on its altitude above earth. The nearer it is to earth, the faster the required orbital velocity. At an altitude of 150 miles (200 km), the required orbital velocity is just more than 17,000 mph (about 27,400 kph). To maintain an orbit that is 22,223 miles (35,786 km) above earth, the satellite must orbit at a speed of about 7,000 mph (11,300 kph). That orbital speed and distance permit the satellite to make one revolution in 24 hours. Because the earth also rotates once in 24 hours, a satellite at 22,223 miles (35,786 km) altitude stays in a fixed position relative to a point on the earth's surface. Because the satellite stays right over the same spot all the time, this kind of orbit is called a geostationary orbit. Geostationary orbits are ideal for weather satellites and communications satellites.

Types of Satellites

Satellites can have all shapes and sizes and play a variety of roles. For example, the following are common types of satellites:

- Weather satellites
- Communications satellites
- Broadcast satellites
- Scientific satellites
- Navigational satellites
- Rescue satellites
- Earth observation satellites
- Military satellites

The Hubble space telescope is actually a huge satellite that contains a massive telescope.

Parts of a Satellite

Despite the significant differences between all the different types of satellites, they all have some of the same basic parts:

- **A bus**—A satellite has a metal or composite frame and body, known as the bus. The bus holds everything together in space and provides enough strength for the satellite to survive the launch. The components of the satellite attach to the bus. Covers and hatches enclose the components and protect them from space debris and sunlight.
- **A power source**—The power source usually consists of solar cells as well as batteries for storage. Arrays of solar cells provide power to charge rechargeable batteries. Power on most satellites is precious and very limited. Nuclear power has been used on space probes to other

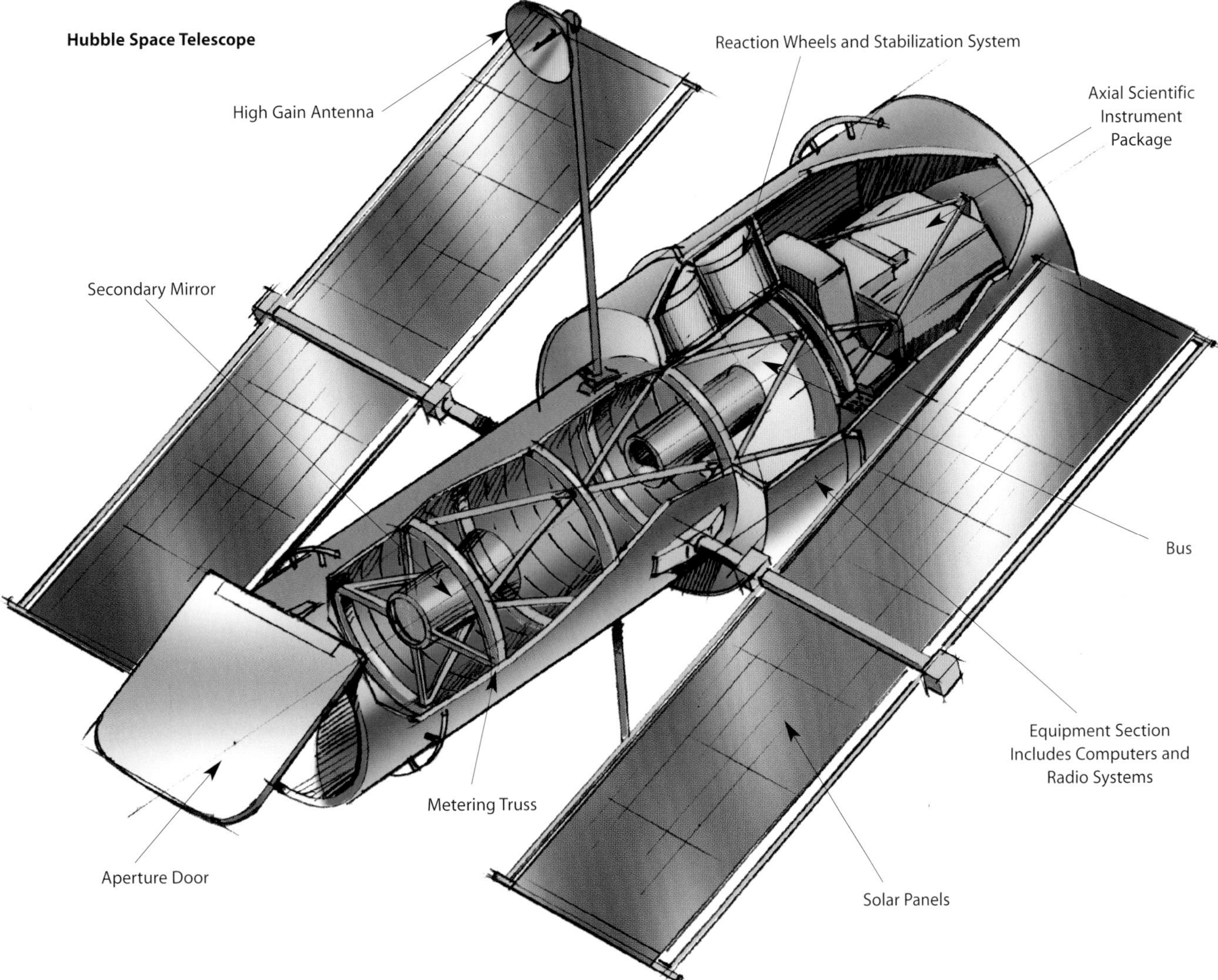

planets. Power systems are constantly monitored, and data on power and all other on-board systems is sent to earth stations in the form of telemetry signals.

- **An on-board computer**—An on-board computer controls and monitors the different systems.
- **A radio system and an antenna**—Most satellites have a radio transmitter/receiver so that the ground control crew can request status information from the satellite, monitor its health, and send adjustments. Many satellites can be controlled in various ways from the ground to do anything from change the orbit to reprogram the computer system.
- **An attitude control system (ACS)**—The ACS keeps the satellite pointed in the right direction. The Hubble space telescope has a very elaborate control system so that the telescope can point at the same position in space for hours or days at a time (even though the telescope travels at 17,000 mph, or 27,359 kph!). The system contains gyroscopes, accelerometers, a reaction wheel stabilization system, thrusters, and a set of sensors that watch guide stars to determine position.

Satellites can range in size from a device as big as a watermelon to something as big as a school bus.

If there were an easy way to see them, you could look up in the sky at any given moment and see hundreds of satellites. They do dozens of different things that make life better—everything from weather forecasting to cable TV depends on satellites!

How GYROSCOPES Work

Gyroscopes can be perplexing physical objects because they move in peculiar ways. They even seem to defy gravity. Their special abilities make gyroscopes extremely important in everything from bicycles to the advanced navigation systems on space shuttles. A typical airplane uses about a dozen gyroscopes in everything from its compass to its autopilot, and the Hubble space telescope uses a bunch of navigational gyroscopes as well. Gyroscopic effects are also central to items like yo-yos and Frisbees.

HSW Web Links

www.howstuffworks.com

How Boomerangs Work
How Compasses Work
How Force, Power, Torque, and Energy Work
How the Hubble Space Telescope Works
How Submarines Work

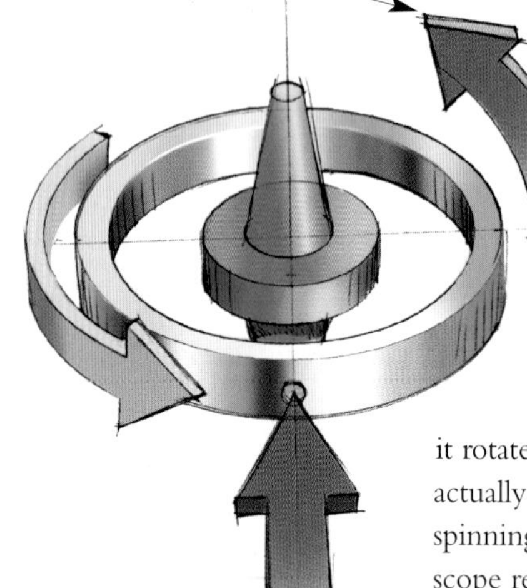

If you have ever played with toy gyroscopes, you know that they can perform all sorts of interesting tricks. They can balance on a string or a finger, and they resist motion around the spin axis in very odd ways. But the most interesting effect of a gyroscope is its gravity-defying part, known as precession.

Precession

Precession works like this: If you have a spinning gyroscope and you try to rotate its spin axis, the gyroscope will instead try to rotate around an axis at right angles to your force axis. This allows you to dangle a gyroscope horizontally on a string. Instead of falling like any normal object would, the gyroscope mysteriously hangs in the air and spins slowly.

If you think about what is actually happening to the different sections of the gyroscope as it rotates, you can see that its odd behavior actually makes sense! Because a gyroscope is spinning, the different sections of the gyroscope receive forces at one point, but then it rotates to a new position. Imagine a gyroscope hanging by a string:

1) The section that is at the top has a desire to move left.
2) When the section at the top of the gyro rotates 90 degrees to the side, it continues in its desire to move to the left.
3) Similarly, the section at the bottom wants to move right.
4) The bottom section rotates 90 degrees to the side and continues in its desire to move right.
5) As the top and bottom portions of the gyroscope rotate another 90 degrees, the left and right forces cancel each other out.
6) With the forces that normally make the gyroscope fall canceling each other out, the gyrocospe cannot fall. Instead, it turns, or precesses. So, the gyroscope's axle hangs in the air. You can see that precession isn't mysterious at all—it is the way things have to be!

Uses of Gyroscopes

After you spin a gyroscope, its axle wants to keep pointing in the same direction. If you mount a gyroscope in a set of gimbals so that it can continue pointing in the same direction, it will. This is the basis of the gyrocompass. If you mount two gyroscopes with their axles at right angles to one another on a platform and place the platform inside a set of gimbals, the platform will remain completely rigid as the gimbals rotate anywhere they please. This is this basis of an inertial navigation system (INS). In an INS, sensors on the gimbals' axles detect when the platform rotates. The INS uses those signals to understand the vehicle's rotations relative to the platform. If you add to the platform a set of three sensitive accelerometers, you can tell exactly where the vehicle is heading and how its motion is changing in all three directions. With this information, an airplane's autopilot can keep the plane on course, and a rocket's guidance system can put the rocket into a desired orbit.

Bikes, Frisbees, and yo-yos all depend on gyroscopes, and so do space shuttles and airplanes. Now, when you ride a bike or watch the space shuttle launch, you'll know how and why gyroscopes affect their inner workings!

How **SPACESUITS** Work

Imagine that you have to go to work outside, but outside there is no air to breathe—in fact it is a complete vacuum —and it is 248°F (120°C) in sunlight and –148°F (–100°C) in the shade. These are the conditions in outer space. Surviving in these conditions requires some extremely specialized and expensive clothing. A spacesuit is basically a miniature spacecraft that has a price tag of about $12 million.

In order for a spacesuit to protect you in outer space, it has to provide you with oxygen to breathe, and it has to eliminate the carbon dioxide and moisture you exhale. It also has to provide a pressurized environment for your body because the fluids in the human body boil quickly in a vacuum. It must insulate you from the heat and cold of the sun and shade, and it has to protect you from the sun's rays, which are totally unfiltered in outer space. In addition, a spacesuit needs to provide communications systems, maneuvering systems, computer systems, and displays to check all the components. It also needs some comforts of home, such as water and restroom facilities. All these requirements make modern spacesuits extremely complex machines.

Today's Spacesuit

Whereas early spacesuits were made entirely of soft fabrics, the spacesuit in use today, the extravehicular mobility unit (EMU), has a combination of soft and hard components plus a backpack to provide support, mobility, and comfort. The soft part of the suit has 13 layers of material, including an inner cooling garment (2 layers), a pressure garment (2 layers), a thermal micrometeroid garment (8 layers), and an outer cover (1 layer). All the layers are sewn and cemented together to form the suit. In contrast to early spacesuits, which were individually tailored for each astronaut, the EMU has component pieces of varying sizes that can be put together to fit any given astronaut.

The EMU has the following parts:

- **Maximum absorption garment**—A diaper that collects the astronaut's urine.
- **Liquid cooling and ventilation garment**—Water-cooled, tricot/spandex long underwear that removes excess body heat. Water circulates from this garment through the backpack to remove heat.
- **EMU electrical harness (EEH)**—A harness that provides connections for communications and bioinstruments in the suit's backpack.
- **Communications carrier assembly (CCA)**—A fabric cap that contains microphones and earphones for communications.
- **Lower torso assembly (LTA)**—The lower half of the EMU, including pants, knee and ankle joints, boots, and the lower waist.
- **Hard upper torso (HUT)**—The hard fiberglass vest shell that supports several structures, including the arms, torso, helmet, life-support backpack, and control module.
- **Arms**—Sleeves that contain shoulder, upper arm, and elbow joints.
- **Gloves**—Outer and inner gloves that have wrist bearings and rubberized fingertips to help move and grip things.
- **Helmet**—A clear, impact-resistant polycarbonate plastic helmet that is padded for comfort and has a purge valve to remove carbon dioxide if the backup oxygen supply is used.

HSW Web Links

www.howstuffworks.com

How Robonauts Will Work
How Space Planes Will Work
How Space Shuttles Work
How Space Stations Work
How Space Tourism Will Work

Did You Know?

If you were to step outside a spacecraft or into a world with little or no atmosphere (such as the moon or Mars) without a spacesuit, the following would happen:

- You would become unconscious within 15 seconds because there is no oxygen.
- Your blood and body fluids would "boil" and then freeze because there is little or no air pressure.
- Your tissues, including skin, heart, and other internal organs, would expand because of the boiling fluids.
- You would face extreme changes in temperature, and you would be exposed to various types of radiation, such as cosmic rays and solar wind.
- You could be hit by small particles of dust or rock that move at high speeds (micrometeoroids) or orbiting debris from satellites or spacecraft.

- **Extravehicular visor assembly (EVA)**—The helmet's outer covering, with a metallic visor and an impact-resistant thermal cover that protects the astronaut from bright sunlight. It also contains a TV camera and four headlamps.
- **In-suit drink bag (IDB)**—A 32-ounce (0.9-l) water bag with a straw that provides water for the astronaut. A fruit bar is also located inside the helmet.
- **Primary life support subsystem (PLSS)**—A backpack attached to the HUT that provides oxygen, battery power, carbon dioxide removal, cooling water, ventilation, radio equipment, and a warning system. It provides up to 7 hours of oxygen supply and carbon dioxide removal.
- **Secondary oxygen pack (SOP)**—A small oxygen tanks that fits below the PLSS and provides enough emergency oxygen for about 30 minutes.
- **Display and control module (DCM)**—A chest-mounted unit that contains displays and controls to run the PLSS.

To prepare for a spacewalk, a crewmember must do the following:

1) Reduce the pressure in the shuttle to 0.7 atm and increase the oxygen.
2) Pre-breathe 100% oxygen for 30 minutes to remove nitrogen from the blood and tissues. This helps prevent the bends.
3) Put on the MAG.
4) Enter the airlock.
5) Put on the LCVG.
6) Attach the EEH to the HUT.
7) Attach the DCM to the HUT (the PLSS is preattached to the HUT).
8) Attach the arms to the HUT.
9) Rub the helmet with anti-fog compound.
10) Place a wrist mirror and checklist on the sleeves.
11) Insert a food bar and water-filled IDB inside the HUT.
12) Check the lights and cameras on the EVA.
13) Place the EVA over the helmet.
14) Connect the CCA to the EEH.
15) Step into the LTA and pull it above the waist.
16) Plug the SCU into the DCM and into the shuttle.
17) Squirm into the upper torso portion of the suit.
18) Attach the cooling tubes of the LVCG to the PLSS.
19) Attach the EEH electrical connections to the PLSS.
20) Lock the LTA to the HUT.
21) Put on the CCA and eyeglasses (if they are to be worn).
22) Put on comfort gloves.
23) Lock on the helmet and EVA.
24) Lock on the outer gloves.
25) Check the EMU for leaks by increasing the pressure inside the suit to 0.20 atm above the airlock pressure.

If there are no leaks, then the airlock is depressurized to the vacuum of space. The EMU automatically depressurizes to its operating pressure, and the outer airlock door is opened.

The astronaut steps out of the airlock, into the shuttle's cargo bay, and the spacewalk begins. At this point, the EMU is a spacecraft that's independent of the shuttle or the space station. After the spacewalk, these steps are reversed to get out of the suit and back into the spacecraft.

Future Spacesuits

When working on the moon, Apollo astronauts had difficulties moving around in their spacesuits. The Apollo suits were not nearly as flexible as the EMU used today; however, the EMU weighs almost twice as much as the Apollo suit—but that's not really a problem because the EMU was designed for work in microgravity, not on a planet's surface. For future space missions to Mars, NASA is developing "hard suits" that are more flexible, more durable, lighter-weight, and easier to don than current spacesuits.

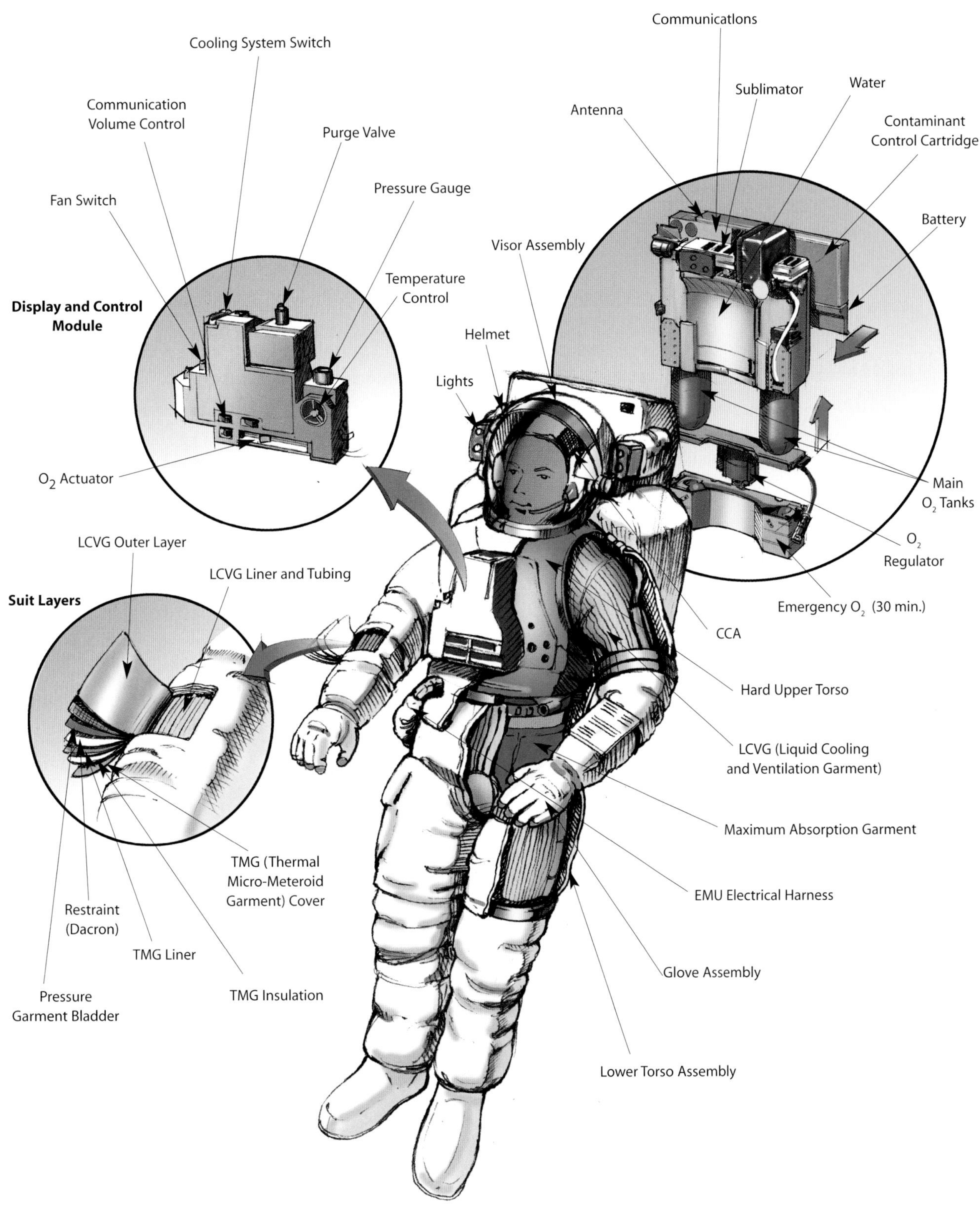

Communications
Cooling System Switch
Sublimator
Water
Antenna
Communication
Volume Control
Contaminant
Control Cartridge
Purge Valve
Pressure Gauge
Fan Switch
Battery
Visor Assembly
Temperature
Control
Display and Control
Module
Helmet
Lights
O_2 Actuator
Main
O_2 Tanks
O_2
Regulator
LCVG Outer Layer
LCVG Liner and Tubing
Suit Layers
Emergency O_2 (30 min.)
CCA
Hard Upper Torso
LCVG (Liquid Cooling
and Ventilation Garment)
Maximum Absorption Garment
TMG (Thermal
Micro-Meteroid
Garment) Cover
EMU Electrical Harness
Restraint
(Dacron)
TMG Liner
Glove Assembly
Pressure
Garment Bladder
TMG Insulation
Lower Torso Assembly

How LIQUID CRYSTAL DISPLAYS (LCDs) Work

You probably use items containing liquid crystal displays (LCDs) every day. An LCD is like a window with tiny shutters to let in light in a precise pattern. LCDs are common because they offer some big advantages over other display technologies. They are thinner and lighter and draw much less power than any other type of display.

HSW Web Links

www.howstuffworks.com

How Computer Monitors Work
How Laptop Computers Work
How PDAs Work
How Projection Television Works

Matter can have three common states: solid, liquid, and gaseous. Solids act the way they do because their molecules always point the same way and stay in the same position with respect to one another. The molecules in liquids are just the opposite: They can point in any direction and move anywhere in the liquid. But some substances can exist in an odd state that is something like a liquid and something like a solid. When substances are in this state, their molecules tend to point the same way, like the molecules in a solid, but they can also move around to different positions, like the molecules in a liquid.

Liquid Crystals

Liquid crystals are closer to a liquid state than they are to a solid state. It takes a fair amount of heat to change most substances from a solid into a liquid. It takes only a little heat to turn a liquid crystal into a real liquid. This explains why liquid crystals are very sensitive to temperature and why they are used to make thermometers and mood rings. It also explains why a laptop computer's display may act funny in cold weather or on a hot day at the beach.

One feature of liquid crystals is that electric current affects them. Applying a voltage to liquid crystals twists them to

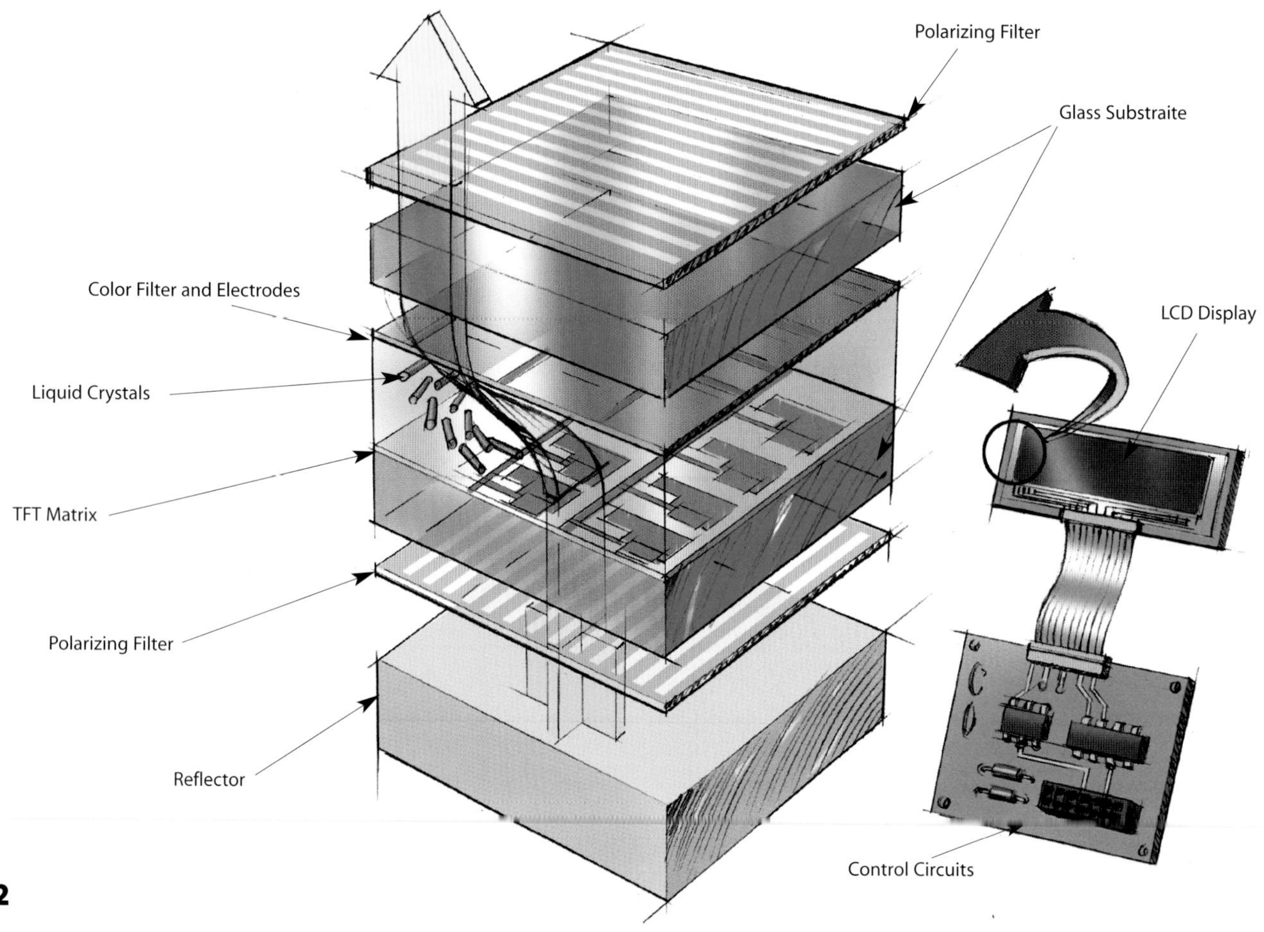

varying degrees. Depending on the voltage applied to a liquid crystal in an LCD, the liquid crystal will either transmit light or block light.

Liquid crystal materials emit no light of their own: They simply block or transmit light. Small and inexpensive LCDs are usually reflective, and to display anything, they must reflect light from external light sources. Most computer displays are lit with built-in fluorescent tubes above, beside, and sometimes behind the LCD. A white diffusion panel behind the LCD redirects and scatters the light evenly to ensure a uniform display.

Making an LCD

Creating an LCD begins with two polarized pieces of glass. A special polymer that creates microscopic grooves in the surface is rubbed on the side of the glass that does not have the polarizing film on it. The grooves must be in the same direction as the polarizing film. Then a coating of nematic liquid crystals is added to one of the filters. The grooves cause the first layer of molecules to align with the filter's orientation. The second piece of glass is added, with the polarizing film at a right angle to the first piece. Each successive layer of molecules gradually twists, until the uppermost layer is at a 90-degree angle to the bottom, matching the polarized glass filters.

As light strikes the first filter, it is polarized. The molecules in each layer then guide the light they receive to the next layer. As the light passes through the liquid crystal layers, the molecules also change the light's plane of vibration to match their own angle. When the light reaches the far side of the liquid crystal substance, it vibrates at the same angle as the final layer of molecules. If the final layer is matched up with the second polarized glass filter, then the light passes through.

If you apply an electric charge to liquid crystal molecules, they untwist. When they straighten out, they change the angle of the light passing through them so that it no longer matches the angle of the top polarizing filter. No light can pass through that area of the LCD, which makes that area darker than the surrounding areas.

Passive- and Active-matrix LCDs

Two main types of LCDs are used in computers: passive matrix and active matrix. Passive-matrix LCDs use a simple grid to supply the charge to a particular pixel on the display. The grid starts with two glass layers called substrates. One substrate is given columns and the other is given rows made from a transparent conductive material, usually indium tin oxide. The rows or columns connect to integrated circuits (ICs) that control the voltage sent down a particular column or row. The layer of liquid crystals is sandwiched between the two glass substrates, and a polarizing film is added to the outer side of each substrate. To turn on a pixel, the IC sends a charge down the correct column of one substrate, and a ground is activated on the correct row of the other. The row and column intersect at the designated pixel, and that delivers the voltage to untwist the liquid crystals at that pixel.

Active-matrix LCDs depend on thin film transistors (TFTs). TFTs are tiny switching transistors and capacitors that are arranged in a matrix on a glass substrate. To address a particular pixel, the proper row is switched on, and then a charge is sent down the correct column. Because all the other rows that the column intersects are turned off, only the capacitor at the designated pixel receives a charge. The capacitor is able to hold the charge until the next refresh cycle. By carefully controlling the amount of voltage supplied to a crystal, you can make it untwist only enough to allow some light through, and doing this in very exact increments creates a grayscale.

Color LCDs

An LCD that can show colors must have three subpixels with red, green, and blue color filters to create each color pixel. Through the careful control and variation of the voltage applied, the intensity of each subpixel can range more than 256 shades. Combining the subpixels produces a possible palette of 16.8 million colors (256 shades of red x 256 shades of green x 256 shades of blue). Color displays take an enormous number of transistors. For example, a typical laptop computer supports resolutions up to 1,024 x 768. If you multiply 1,024 columns x 768 rows x 3 subpixels, you get 2,359,296 transistors etched onto the glass!

How GEARS Work

Gears are used in just about every mechanical device. They do several important jobs, but most often they provide gear reduction in motorized equipment. In many situations a small motor spinning very fast can provide enough power for a device, but it cannot provide enough torque. Gears turn the motor's speed into torque. Gears also adjust the direction of rotation. For instance, in the differential between the rear wheels of a car, the power is transmitted by a shaft that runs down the center of the car, and the differential has to turn that power 90 degrees to apply it to the wheels.

HSW Web Links

www.howstuffworks.com

How Differentials Work
How Gear Ratios Work
How Manual Transmissions Work
How Odometers Work
How Power Windows Work

Each device has its own unique requirements, which can determine the type of gear used in that device. For instance, a device whose main goal is quiet operation will use a different gear than a device that requires a large gear reduction and transmits lots of power.

Types of Gears

Let's take a look at the different types of gears and see where they are used:

- **Spur gears**—Spur gears are the most common type of gears. They have straight teeth and are mounted on parallel shafts. Sometimes, many spur gears are used at once to create very large gear reductions. Spur gears are used in many devices, such as electric screwdrivers, drills, toys, oscillating sprinklers, windup alarm clocks, washing machines, and power company meters. But you won't find many in your car. This is because the spur gear can be really loud. Each time a gear tooth engages a tooth on the other gear, the teeth collide, and this impact makes a noise. It also increases the stress on the gear teeth. To reduce the noise and stress in the gears, most of the gears in a car are helical.
- **Helical gears**—The teeth on helical gears are cut at an angle to the face of the gear. When two teeth on a helical gear system engage, the contact starts at one end of the tooth and gradually spreads as the gears rotate, until the two teeth are in full engagement. This gradual engagement makes helical gears operate much more smoothly and quietly than spur gears. For this reason, helical gears are used in almost all car transmissions. Because of the angle of the teeth on helical gears, they create a thrust load on the gear when they mesh. Devices that use helical gears have bearings that can support this thrust load. One interesting thing about helical gears is that if the angles of the gear teeth are correct, helical gears can be mounted on perpendicular shafts, adjusting the rotation angle by 90 degrees.
- **Bevel gears**—Bevel gears are useful when the direction of a shaft's rotation needs to be changed. They are usually mounted on shafts that are 90 degrees apart, but they can be designed to work at other angles as well. The teeth on bevel gears can be straight, spiral, or hypoid. Straight bevel gear teeth have the same problem as straight spur gear teeth: as each tooth engages, it affects the corresponding tooth all at once. Just like with spur gears, the solution to this problem is to curve the gear teeth. These spiral teeth engage the same way as helical teeth: The contact starts at one end of the gear and progressively spreads across the whole tooth. On straight and spiral bevel gears, the shafts must be perpendicular to each other, but they must also be in the same plane. If you were to extend the two shafts past the gears, they would intersect. The hypoid gear, on the other hand, can engage with the axes in different planes.
- **Worm gears**—Worm gears are used when large gear reductions are needed. It is common for worm gears to have reductions of 20:1, and even up to 300:1 or greater. Many worm gears have an

interesting property that no other gear set has: The worm can easily turn the gear, but the gear cannot turn the worm. The angle on the worm is so shallow that when the gear tries to spin it, the friction between the gear and the worm holds the worm in place. This feature is useful for machines such as conveyor systems, in which the locking feature can act as a brake for the conveyer when the motor is not turning.

- **Rack-and-pinion gears**—Rack-and-pinion gears are used to convert rotation into linear motion. A perfect example of this is the steering system on many cars. The steering wheel rotates a gear, which engages the rack. As the gear turns, it slides the rack either to the right or left, depending on which way you turn the wheel. Rack-and-pinion gears are also used in some scales to turn the dial that displays your weight.

Gear Ratio

Understanding the concept of gear ratio is easy if you understand the concept of the circumference of a circle. Keep in mind that the circumference of a circle is equal to the diameter of the circle multiplied by pi (pi is equal to approximately 3.14159). Therefore, if you have a circle or a gear with a diameter of 1 inch (2.54 cm), the circumference of that circle will be 3.14159 inches.

Most gears have teeth, and the teeth have three main purposes:

- They prevent slippage between the gears. Therefore, axles connected by gears are always synchronized exactly with one another.
- They make it possible to determine exact gear ratios. You just count the number of teeth in the two gears and divide the larger number by the smaller. So if one gear has 60 teeth and another has 20, the gear ratio when these two gears are connected together is 3:1.
- They make it so slight imperfections in the actual diameter and circumference of two gears don't matter. The gear ratio is controlled by the number of teeth, even if the diameters are a bit off. See "How Multiplying Force Works," page 109, for more information.

To create large gear ratios, gears are often connected together in gear trains. A small gear and a larger gear are connected together, one on top of the other. Gear trains often consist of multiple gears in the train.

If you look inside a clock, a car, or a VCR, you can immediately see how important gears are in today's society—they are everywhere.

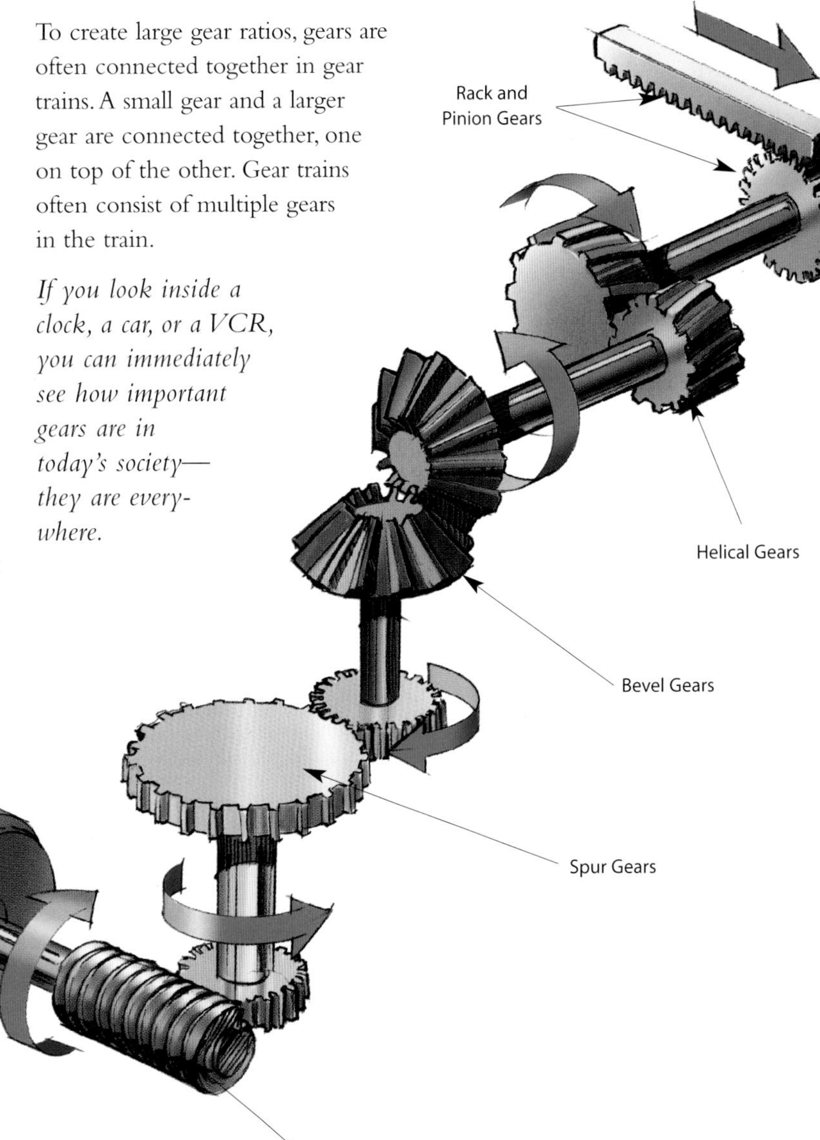

How BEARINGS Work

Bearings are amazingly simple and work behind the scenes, but they make many things about the modern world possible. Without bearings we would be constantly replacing parts worn from friction. The concept behind a bearing is very simple: Things roll better than they slide.

HSW Web Links

www.howstuffworks.com

How Ballpoint Pens Work
How Bicycles Work
How Brakes Work
How Earthquakes Work
How Smart Structures Will Work

Sliding is bad because when things slide, the friction between them causes them to slow down. If the two surfaces can roll over each other instead of sliding, then the friction is greatly reduced. Bearings let surfaces roll by providing smooth metal balls or rollers and a smooth inner and outer metal surface for the balls to roll against. These balls or rollers "bear" the load, allowing the device to slide smoothly.

Types of Loads

Bearings typically have to deal with two kinds of loading: radial and thrust. Depending on where the bearing is being used, it may see all radial loading, all thrust loading, or a combination of both:

- A radial load is transmitted through the bearing to the shaft in a direction that is perpendicular to the shaft. For instance, the bearings in the alternator in a car experience mostly radial loads that come from the tension in the belt that drives the pulley.
- A thrust load, on the other hand, acts in a direction that is parallel to the shaft. For instance, the bearing in a barstool is loaded mainly in thrust. The entire load comes from the weight of the person sitting on the stool.
- Some bearings are loaded both radially and in thrust. The hub of a car wheel is a good example: The radial load comes from the weight of the car, and the thrust load comes from the cornering forces when you go around a turn.

Types of Bearings

There are many types of bearings, each used for different purposes. Here are some of the most common types of bearings:

- **Ball bearings**—Ball bearings are probably the most common type of bearing. They are found in everything from inline skates to hard drives. These bearings can handle both radial and thrust loads, and they are usually used in applications where the load is relatively small. In a ball bearing, the load is transmitted from the outer race to the ball and from the ball to the inner race. Because the ball is a sphere, it contacts the inner and outer

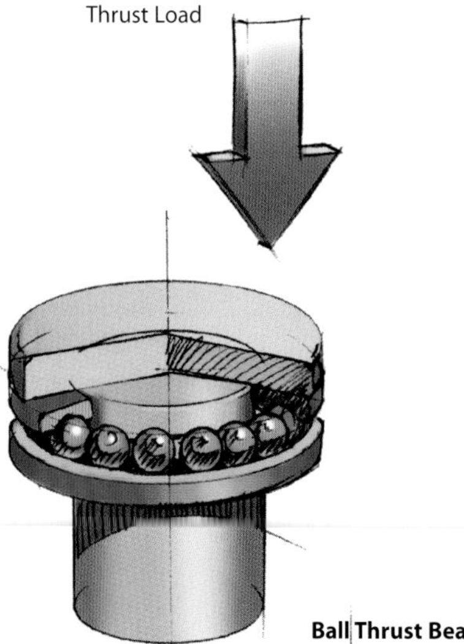

Ball Thrust Bearings

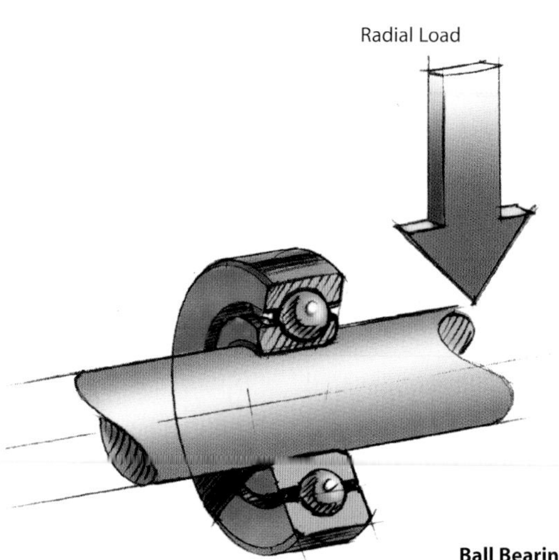

Ball Bearings

races at only a very small point, which helps it spin very smoothly. But this also means that there is not very much contact area holding that load, so if the bearing is overloaded, the balls can become deformed, cracked, or otherwise ruined.

- **Roller bearings**—Roller bearings are used in applications where they must hold heavy radial loads, such as in conveyer belts. In these bearings, the roller is a cylinder, so the contact between the inner and outer race is not a point but a line. This spreads the load out over a larger area, allowing the bearing to handle much greater loads than a ball bearing. However, this type of bearing is not designed to handle much thrust loading.

  A variation of the roller bearing, called a needle bearing, uses cylinders with a very small diameter. This allows the bearing to fit into tight places.
- **Ball thrust bearings**—Ball thrust bearings are mostly used for low-speed applications and cannot handle much radial load. Barstools and Lazy Susan turntables use this type of bearing.
- **Roller thrust bearings**—Roller thrust bearings can support large thrust loads. They are often found in gearsets, such as in car transmissions between gears and between the housing and the rotating shafts. The helical gears used in most transmissions have angled teeth, which cause a thrust load that must be supported by a bearing.
- **Tapered roller bearings**—Tapered roller bearings can support large radial and large thrust loads. Tapered roller bearings are used in car hubs, where they are usually mounted in pairs facing opposite directions so that they can handle thrust in both directions.

Interesting Uses of Bearings

Small ball bearings are hidden all around us in our cars and various appliances. You also find more exotic uses of bearings:

- **Magnetic bearings**—Some very high-speed devices, such as advanced flywheel energy storage systems, use magnetic bearings. These bearings allow the flywheel to float on a magnetic field created by the bearing. Some flywheels run at speeds upward of 100,000 revolutions per minute (rpm), and normal bearings with rollers or balls would melt or explode at these speeds. The magnetic bearing has no moving parts, so it can handle these incredible speeds.
- **Giant roller bearings**—Probably the first use of bearings occurred when the Egyptians were building the pyramids. They put round logs under the heavy stones so that they could roll them to the building site. This method is still used today when large, very heavy objects—such as the Cape Hatteras lighthouse in North Carolina—need to be moved.
- **Bearings to earthquake-proof buildings**—The San Francisco International Airport uses many advanced building technologies to help it withstand earthquakes. One of these technologies involves giant ball bearings. Each of the 267 columns that support the weight of the airport rides on a ball bearing that is 5 feet (1.5 m) in diameter. The ball rests in a concave base that is connected to the ground. In the event of an earthquake, the ground can move 20 inches in any direction. The columns that rest on the balls move somewhat less than this as they roll around in their bases, and this helps isolate the building from the motion of the ground. When the earthquake is over, gravity pulls each column back to the center of its base.

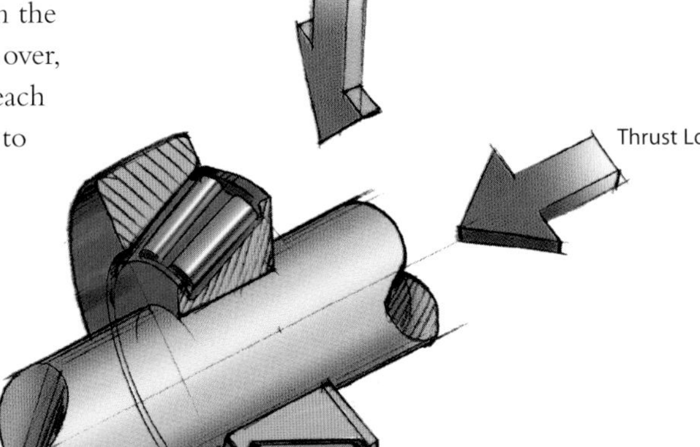

Tapered Roller Bearing

How MULTIPLYING FORCE Works

If you have ever looked at the end of a crane, used an engine hoist or a come-along, or seen the rigging on a sailboat, then you have witnessed a force multiplication at work. A block and tackle is an arrangement of rope and pulleys that allows you to trade force for distance. Levers, gears, and hydraulics are also used to multipy force.

HSW Web Links

www.howstuffworks.com

How Electromagnets Work
How Gears Work
How Gyroscopes Work
How Horsepower Works
How Relays Work

Let's look at how a block and tackle works, and also examine several other force multiplying devices.

Block-and-Tackle Arrangements

Imagine that you have a 100-pound (45.4-kg) weight suspended from a rope. If you are going to suspend the weight in the air, then you have to apply an upward force of 100 pounds (445 N) to the rope. If the rope is 100 feet (30.5 m) long and you want to lift the weight 100 feet (30.5 m), you have to pull in 100 feet (30.5 m) of rope to do it. This is simple and obvious. Now imagine that you add a pulley to the mix.

Does the pulley change anything? Not really. The only thing that changes is the direction of the force you have to apply to lift the weight. You still have to apply 100 pounds (445 N) of force to keep the weight suspended, and you still have to reel in 100 feet (30.5 m) of rope in order to lift the weight 100 feet (30.5 m).

So now let's add a second pulley. This two-pulley arrangement changes things in an important way. Now the weight is suspended by two ropes rather than one, so the weight is split equally between the two ropes. Each rope holds only half the weight, or 50 pounds (22.7 kg). That means that if you want to hold the weight suspended in the air, you only have to apply 50 pounds (224.5 N) of force (the ceiling exerts the other 50 pounds, or 224.5 N, of force on the other end of the rope). If you want to lift the weight 100 feet (30.5 m) higher, then you have to reel in twice as much rope—200 feet (61 m). This demonstrates a force–distance tradeoff. The force has been cut in half, but the distance the rope must be pulled has doubled.

A block and tackle can contain as many pulleys as you like, although at some point the amount of friction in the pulley shafts begins to become a significant source of resistance. You could, in theory, wrap the rope through pulleys 100 times. You would only have to apply 1 pound (4.5 N) of force, but you would have to reel in 10,000 feet (609.6 m) of rope, which would be almost 2 miles (3.2 km) of rope!

Levers

Imagine that you have a lever whose left end is twice as long as its right end. If you apply a force (F) to the left end of the lever, then the right end of the lever has an available force of 2F. Changing the relative lengths of the left and right ends of the lever will change the multipliers. For example, if the left end were three times as long as the right, then the force of the right end would be 3F.

Gears

Say that you have two gears—a left and a right gear—and that the left-hand gear has twice the diameter of the right-hand gear. For every turn of the left-hand gear, the right-hand gear turns twice. If you apply a certain amount of torque to the left-hand gear through one rotation, the right-hand gear will exert half as much torque, but it will turn two revolutions.

Hydraulics

Assume that you have two cylinders full of water, a pipe connecting the two cylinders, and two pistons pushing down on the water in each cylinder. The left cylinder is 1 inch (2.54 cm) in diameter, and the right cylinder

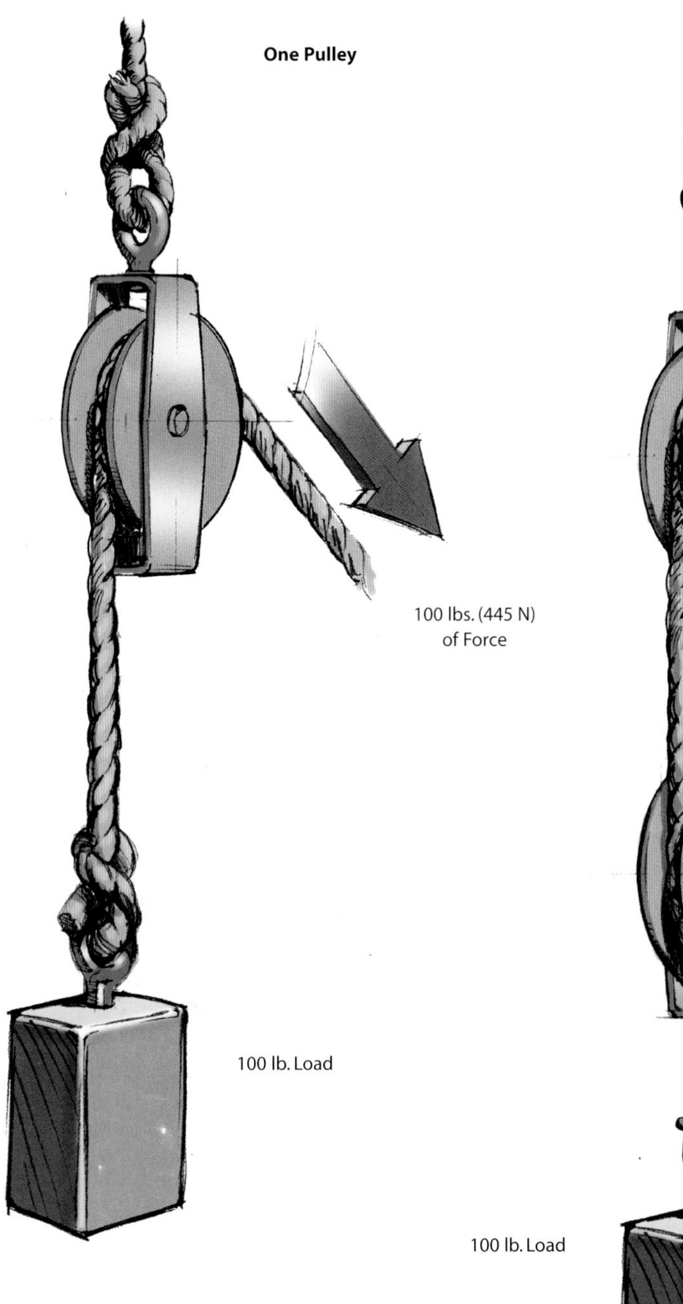

is 4 inches (10.16 cm) in diameter. If you apply a force to the left-hand piston, it creates a pressure in the left-hand cylinder. Let's say you push the left-hand piston down through 16 inches (40.64 cm) with a force of 10 pounds (44.5 N). The right-hand piston will rise 1 inch (2.54 cm) with a force of 160 pounds (712 N). Hydraulic cylinders of all sorts take advantage of this simple force-multiplying effect every day.

You can see that a block and tackle, a lever, a gear train, and a hydraulic system all do the same thing: They let you magnify a force by proportionally diminishing the distance through which the magnified force can act. This sort of force multiplication is an extremely useful capability!

Simple Examples

Here are some devices that use the principles of multiplying force:

- **Car jack** (lever or threaded gear)
- **Fingernail clippers** (lever) **Automobile transmission** (gears)
- **Come-along** (block and tackle and gear)
- **Can opener** (gear and lever)
- **Crowbar** (lever)
- **Hammer claw** (lever)
- **Bottle opener** (lever)
- **Car brakes** (hydraulics) **Hydraulic shop lift** (hydraulics)
- **Elevator** (block and tackle or hydraulics)

How RELAYS Work

Relays may be invisible, but they are everywhere: There are relays in your car, your thermostat, your refrigerator, and even your clock radio. They are in so many places because relays act as an important interface between digital control systems and high-power devices.

HSW Web Links

www.howstuffworks.com

How Boolean Gates Work
How Electromagnets Work
How Electronic Gates Work

A relay is basically like a regular wall switch that turns on a light when you flip it with your finger. But rather than being activated by your finger, a relay is activated by an electromagnet that is supplied with a small amount of electricity.

Relays have been around for more than a century. They are so versatile that several early computers were made with relays instead of vacuum tubes. Even so, relays are incredibly simple: A relay is a simple electro-mechanical switch made up of an electromagnet and a set of contacts.

Parts of a Relay

Relays are simple devices. There are four parts in every relay:

- An electromagnet
- A metal armature that can be attracted by the electromagnet
- A spring
- A set of electrical contacts

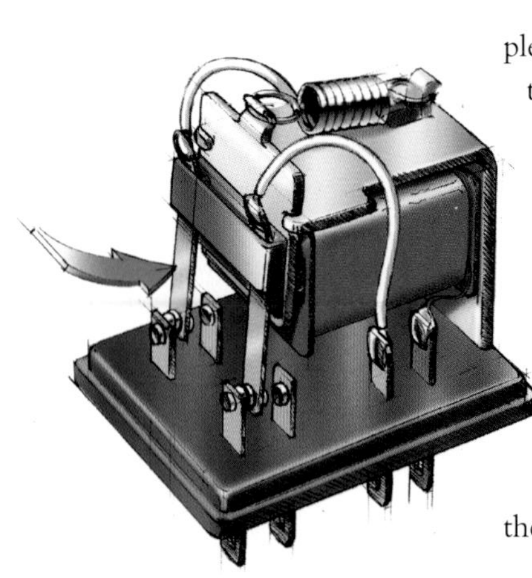

A relay consists of two separate and completely independent circuits. The first drives the electromagnet. When power in this circuit is on, the electromagnet is on and it attracts the armature. The armature acts as a switch in the second circuit. When the electromagnet is energized, the armature closes the second circuit. When the electromagnet is not energized, the spring pulls the armature away, and the circuit is open. Depending on the relay's design, it is also possible to reverse the opening and closing.

When you purchase relays, you generally have control over several different variables:

- The voltage and current that are needed to activate the electromagnet
- The maximum voltage and current that can run through the armature and the armature contacts
- The number of armatures (generally one or two)
- The number of contacts for the armature (generally one or two)
- Whether the contact in the second circuit is normally open (NO) or normally closed (NC)

Designers choose the best relay for a given application based on these parameters.

Relay Applications

In general, the point of a relay is to use a small amount of power in the electromagnet to move an armature that is able to switch a much larger amount of power. The electromagnet's power might come from a small dashboard switch or a low-power electronic circuit. For example, you might want the electromagnet to energize using 5 volts and 50 milliamps (250 milliwatts), and the armature might be able to support 120 volts of alternating current at 2 amps (240 watts).

Relays are quite common in home appliances, where there is an electronic control turning on something like a motor or a light. They are also common in cars, where the 12-volt supply voltage means that just about everything needs a large amount of current. Car manufacturers have started combining relay panels into the fuse box to make maintenance easier.

In places where a large amount of power needs to be switched, relays are often cascaded. So a small relay switches the power needed to drive a much larger relay, and that second relay switches the power to drive the load.

Every day relays touch your life in many different ways. Cars, computers, thermostats, and many other devices all use either solid-state or mechanical relays. Relays have been around for over a century, but as things get more electronic, relays become more important.

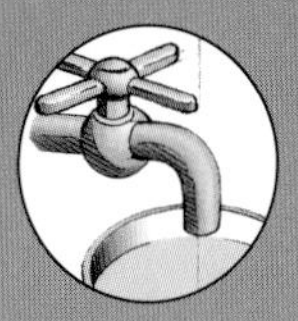

chapter five

Of MICROPROCESSORS, MICE, and MODEMS

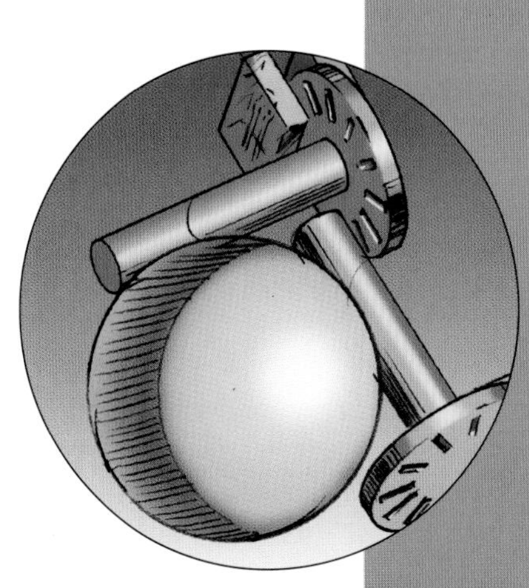

How BYTES & BITS Work

Random-access memory (RAM), hard disk space, and file sizes are all measured in bytes. You hear about bytes all the time. You also hear about bits—for example, when you read about a 32-bit processor or a 64-bit data bus. When you understand bytes and bits, you understand quite a lot about how computers work.

HSW Web Links

www.howstuffworks.com

How Boolean Logic Works
How Electronic Gates Work
How File Compression Works
How Microcontrollers Work
How Microprocessors Work

The easiest way to understand bits is to compare them to something you know: digits. In the standard decimal number system, a digit is a single place that can hold a numeric value between 0 and 9.

Digits are normally combined together in groups to create larger numbers. For example, the number 6,357 has four digits. It is understood that in the number 6,357, the 7 is filling the 1s place, the 5 is filling the 10s place, the 3 is filling the 100s place, and the 6 is filling the 1,000s place. So you could express the number this way if you wanted to be explicit:

(6 x 1000) + (3 x 100) + (5 x 10) + (7 x 1)
= 6,000 + 300 + 50 + 7 = 6,357

Another way to express it would be to use powers of 10. If you use the symbol ^ to represent the concept of raising to the power of (for example, 10 squared is written as 10^2), you could say this:

(6 x 10^3) + (3 x 10^2) + (5 x 10^1) +
(7 x 10^0)
= 6,000 + 300 + 50 + 7
= 6,357

You can see that each digit is a placeholder for the next higher power of 10.

Bits

Whereas humans think in terms of the base-10, or decimal, number system, computers operate using the base-2 number system, also known as the binary number system. Computers use binary numbers, and therefore use binary digits, in place of decimal digits. The word *bit* is a shortening of the words *binary digit*. Whereas decimal digits have 10 possible values ranging from 0 to 9, bits have only two possible values: 0 and 1.

A binary number is composed of only 0s and 1s, like this: 1011. How do you figure out the value of the binary number 1011? You do it in the same way you did it for 6,357, but you use a base of 2 instead of a base of 10. Here's what it looks like:

(1 x 2^3) + (0 x 2^2) + (1 x 2^1)
+ (1 x 2^0)
= 8 + 0 + 2 + 1
= 11

You can see that in binary numbers, each bit holds the value of increasing powers of 2. That makes counting in binary pretty easy. Starting at 0 and going though 10, counting in decimal and binary looks like this:

Decimal	Binary
0	0
1	1
2	10
3	11
4	100
5	101
6	110
7	111
8	1000
9	1001
10	1010

Bytes

Bits are rarely seen alone in computers. They are almost always bundled together into 8-bit collections called bytes. Why are there 8 bits in a byte? A similar question is, "Why are there 12 eggs in a dozen?" The 8-bit byte is something people have settled on through trial and error over the past 50 years.

With 8 bits in a byte, you can represent 256 values ranging from 0 to 255, as shown here:

Decimal	Binary
0	0
0	00000000
1	00000001
2	00000010
...	...
254	11111110
255	11111111

ASCII Codes

One common use of computers is to hold and edit text documents. Bytes are frequently used to hold individual characters in a text document. In the ASCII character set, each binary value between 0 and 127 is given a specific character.

Computers store text documents, both on disk and in memory, using these codes. For example, if you create a text file that contains the words, "Four score and seven years ago," the text editor would use 1 byte of memory per character (including 1 byte for each space character between the words). When stored in a file on disk, the file will also contain 1 byte per character and space. You will find that the file has a size of 30 bytes on disk—1 byte for each character. If you add another word to the end of the sentence and resave the file, the file size will jump to the appropriate number of bytes—each additional character will consume another 1 byte.

Lots of Bytes

When you start talking about lots of bytes, you get into prefixes like kilo, mega, and giga, as in kilobyte, megabyte, and gigabyte (also shortened to K, M, and G, as in Kbytes, Mbytes, and Gbytes or KB, MB, and GB). The table to the right shows the multipliers. You can see that kilo is about a thousand, mega is about a million, giga is about a billion, and so on. So when someone says, "This computer has a 100-gig hard drive," what the person means is "100 gigabytes," which means approximately 100 billion bytes, or exactly 107,374,182,400 bytes.

Name	Abbreviation	Size
Kilo	K	2^10 = 1,024
Mega	M	2^20 = 1,048,576
Giga	G	2^30 = 1,073,741,824
Tera	T	2^40 = 1,099,511,627,776
Peta	P	2^50 = 1,125,899,906,842,624
Exa	E	2^60 = 1,152,921,504,606,846,976
Zetta	Z	2^70 = 1,180,591,620,717,411,303,424
Yotta	Y	2^80 = 1,208,925,819,614,629,174,706,176

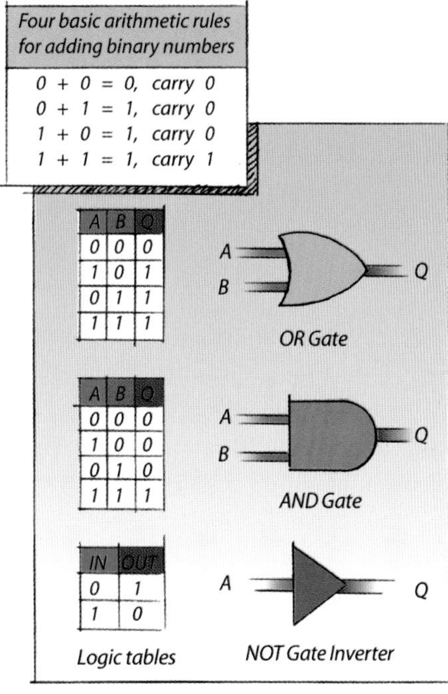

There are three basic logic gates. Each one is made of transistors organized so as to obey basic rules of binary logic, as shown in the corresponding logic tables.

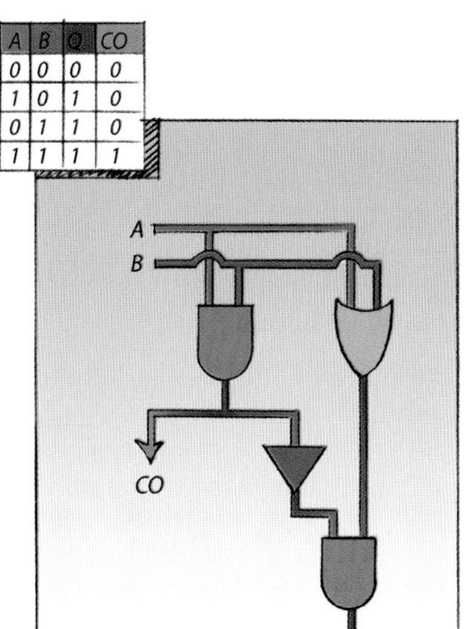

This is an example of a HALF ADDER. This configuration of gates can be combined with other half adders to create full adders.

How OPERATING SYSTEMS Work

The operating system defines the computing experience. It's the first software you see when you turn on a computer, and it's the last software you see when you turn off the computer. It's the software that enables all the programs you use. The operating system organizes and controls all the hardware on your desk and in your hands.

HSW Web Links

www.howstuffworks.com

How Computer Mice Work
How Hard Disks Work
How Microprocessors Work
How Modems Work
How USB Ports Work

All desktop computers have operating systems. The most common are the Windows family of operating systems (Windows 3.1, 95, 98, 2000, NT, CE, and so on), the UNIX family of operating systems (which includes Linux, BSD UNIX, and many other derivatives), and the Macintosh operating systems. There are hundreds of other operating systems available for special-purpose applications, including mainframes, robotics, manufacturing, and real-time control systems.

At the simplest level, an operating system does two things:

- It manages the hardware and software resources of the computer system. These resources include things such as the processor, memory, and disk space.
- It provides a stable, consistent way for applications to deal with the hardware without having to know all the details of the hardware.

Providing a consistent application interface is especially important if more than one of a particular type of computer will be using the operating system, or if the hardware that makes up the computer is ever open to change. A consistent application programming interface (API) allows a software developer to write an application on one computer and have a high level of confidence that it will run on another computer of the same type. The Windows operating systems are a great example of the flexibility an operating system provides. Windows runs on hardware from thousands of vendors. It can accommodate thousands of different printers, disk drives, and special peripherals in many combinations.

Types of Operating Systems

There are four broad types of operating systems, categorized based on the types of computers they control and the sort of applications they support:

- **Real-time operating system**—Real-time operating systems are used to control machinery, scientific instruments, and industrial systems. A real-time operating system typically has very little user interface capability and no end-user utilities because the system is a "sealed box" when delivered for use. A very important part of a real-time operating system is managing the resources of the computer so that a particular operation executes in precisely the same amount of time every time it occurs. In a complex machine, having a part move more quickly just because system resources are available may be just as catastrophic as having it not move at all because the system is busy.
- **Single-user, single-task operating system**—As the name implies, a single-user, single-task operating system is designed to manage the computer so that one user can effectively do one thing at a time. The Palm operating system for Palm computers is a good example of a modern single-user, single-task operating system.
- **Single-user, multitasking operating system**—This is the type of operating system most people use on their desktop and laptop operating systems today. Windows and Macintosh are examples of operating systems that let a single user have several programs in operation at the same time. For example, it's possible for a Windows user to be writing a note in a word processing program while downloading a file from the Internet while printing the text of an email message.
- **Multiuser operating system**—A multiuser operating system allows many different users to take advantage

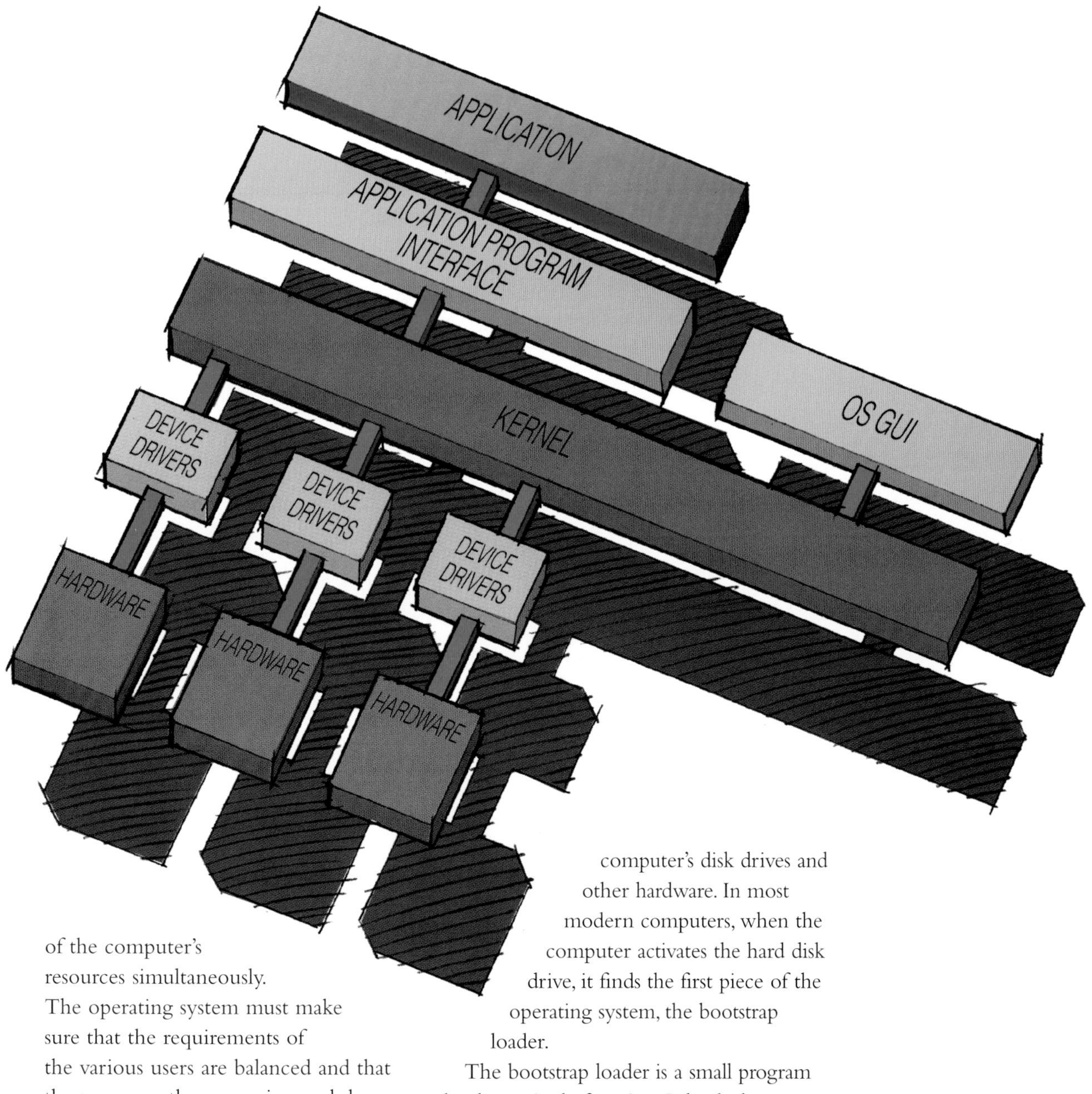

of the computer's resources simultaneously. The operating system must make sure that the requirements of the various users are balanced and that the programs they are using each have sufficient and separate resources so that a problem with one user doesn't affect the entire community of users. UNIX, VMS, and mainframe operating systems, such as MVS, are examples of multiuser operating systems.

Tasks of an Operating System

When you turn on a computer, the first program that runs is usually a set of instructions kept in the computer's read-only memory (ROM). This program examines the system hardware to make sure everything is functioning properly. The software loaded in ROM then begins to activate the computer's disk drives and other hardware. In most modern computers, when the computer activates the hard disk drive, it finds the first piece of the operating system, the bootstrap loader.

The bootstrap loader is a small program that has a single function: It loads the operating system into memory and allows it to begin operation. The bootstrap loader sets up the small driver programs that interface with and control the various hardware subsystems of the computer and loads the entire operating system into RAM. Finally, it turns the control of the computer over to the operating system.

The operating system's tasks generally fall into five categories:

- Processor management
- Memory and storage management
- Device management
- API
- User interface

Multiple Processors

In many high-end machines, there can be two, four, or more CPUs on the same motherboard. In most multi-CPU systems, each CPU has its own cache but they all access the same RAM space. The operating system is in charge of taking maximum advantage of all the CPUs.

In asymmetric operating systems, the operating system consumes one of the CPUs, and then it uses all of the other CPUs to run applications. In symmetric operating systems, the operating system itself runs as a group of independent tasks (usually called threads) that spread out across all of the CPUs. Each application is also a set of threads, and they spread out as well. The operating system is in charge of allocating hundreds of threads to all of the processors available.

Because of contention in RAM and other overhead, you do not get a perfect doubling of performance if you double the number of CPUs, but multiple-CPU machines with a good operating system can work surprisingly well up to about 16 CPUs.

Managing the processor involves two related issues:

- Ensuring that each process and application receives enough of the processor's time to function properly.
- Using as many processor cycles for real work as possible.

The basic unit of software that the operating system deals with in scheduling the work done by the processor is either a process or a thread, depending on the operating system.

The operating system controls and schedules processes for execution by the central processing unit (CPU). In a single-tasking system, the schedule is straightforward. The operating system allows the application to begin running, suspending the execution only long enough to deal with interrupts (special signals sent by hardware or software to the CPU) and user input.

The job of the operating system is much more complicated in a multitasking system, where the operating system must arrange the execution of applications so that you believe that there are several things happening at once. This is complicated because the CPU can do only one thing at a time. In order to give the appearance of lots of things happening at the same time, the operating system has to switch between different processes hundreds or thousands of times per second.

A process occupies a certain amount of RAM. In addition, the process makes use of registers, stacks, and queues within the CPU and operating system memory space. When two processes are multitasking, the operating system gives a certain number of CPU execution cycles to one process. After that number of cycles, the operating system makes copies of all the registers, stacks, and queues used by the process and notes the point at which the process paused in its execution. It then loads all the registers, stacks, and queues used by the second process and allows it a certain number of CPU cycles. All the information needed to keep track of a process when switching is kept in a data package called a process control block.

When an operating system manages the computer's memory, there are two tasks to be accomplished. First, each process must have enough memory in which to execute, and it can neither run into the memory space of another process nor be run into by another process. Next, the different types of memory in the system must be used properly, so that each process can run most effectively. The first task requires the operating system to set up memory boundaries for types of software and for individual applications.

The operating system must balance the needs of the various processes with the availability of the different types of memory, moving data in blocks (called pages) between available memory as the schedule of processes dictates. (See "How the Memory Hierarchy Works," page 120, for details.)

The path between the operating system and all hardware connected to the computer goes through a program called a driver. A driver is a translator between the electrical signals of the hardware subsystems and the high-level programming languages of the operating system and application programs.

One reason that drivers are separate from the operating system is so that new functions can be added to a driver without requiring the operating system itself to be modified, recompiled, and redistributed. Through the development of new hardware device drivers, the manufacturer of the subsystem, rather than the publisher of the operating system, pays for development.

Operating systems also manage the API and the user interface. The user interface is what you see on the screen; it lets you control the computer by using menus, icons, and other user-friendly features rather than by entering commands from a command prompt. Each operating system's user interface is unique and offers its own set of features.

The API is a set of thousands of function calls that developers use when they want to access all the capabilities of the operating system.

The advanced operating systems in use today have made hardware and software easier to use and much less expensive. They also define much of the user experience you have when working with your computer.

How COMPUTER MEMORY Works

When you buy a computer, an MP3 player, or a digital camera, you have to think about and decide on the amount of "memory" you will get. The memory stores data in a digital format so that it is quickly and easily accessible. There are many different types of memory, each one suited for a particular application.

A microprocessor cannot operate without memory. Memory stores the instructions and data that the microprocessor must have in order to operate properly. Computer memory can be split into two main categories: volatile and nonvolatile. Volatile memory loses any data as soon as it loses power; it requires constant power to remember anything. Random-access memory (RAM)—including both dynamic RAM (DRAM) and static RAM (SRAM)—fall into this category. Nonvolatile memory can remember things for years. Read-only memory (ROM), erasable programmable ROM (EPROM), and flash memory all fall into this category.

DRAM

Similar to a microprocessor, a memory chip is an integrated circuit (IC) made of millions of transistors and capacitors. The most common form of computer memory is DRAM. *Dynamic* means that it must be dynamically refreshed periodically in order to maintain its memory. *Random access* means that the microprocessor can access any location instantly. When you buy a computer with "128 megabytes of RAM," it is DRAM.

DRAM is popular because it is the densest, least expensive form of RAM. A transistor and a capacitor are paired to create a memory cell, which represents a single bit of data. The capacitor holds the bit of information—a 0 or a 1.

A capacitor is like a small bucket that is able to store electrons. To store a 1 in the memory cell, the bucket is filled with electrons. To store a 0, the bucket is emptied. The problem with the capacitor's bucket is that it has a leak. In a matter of a few milliseconds, a full bucket becomes empty. Therefore, for dynamic memory to work, either the central processing unit (CPU) or the memory controller has to come along and recharge all the capacitors holding a 1 before they discharge. To do this, the memory controller reads the memory and then writes it back. This refresh operation happens automatically hundreds of times per second.

Memory cells are etched onto a silicon wafer in an array of columns (bitlines) and rows (wordlines). The intersection of a bitline and wordline constitutes the address of the memory cell. DRAM works by sending a charge through the appropriate column to activate the transistor at each bit in the column. When writing, the row lines contain the state the capacitor

HSW Web Links

www.howstuffworks.com

How Flash Memory Works
How Magnetic RAM Will Work
How RAM Works
How ROM Works
How Virtual Memory Works

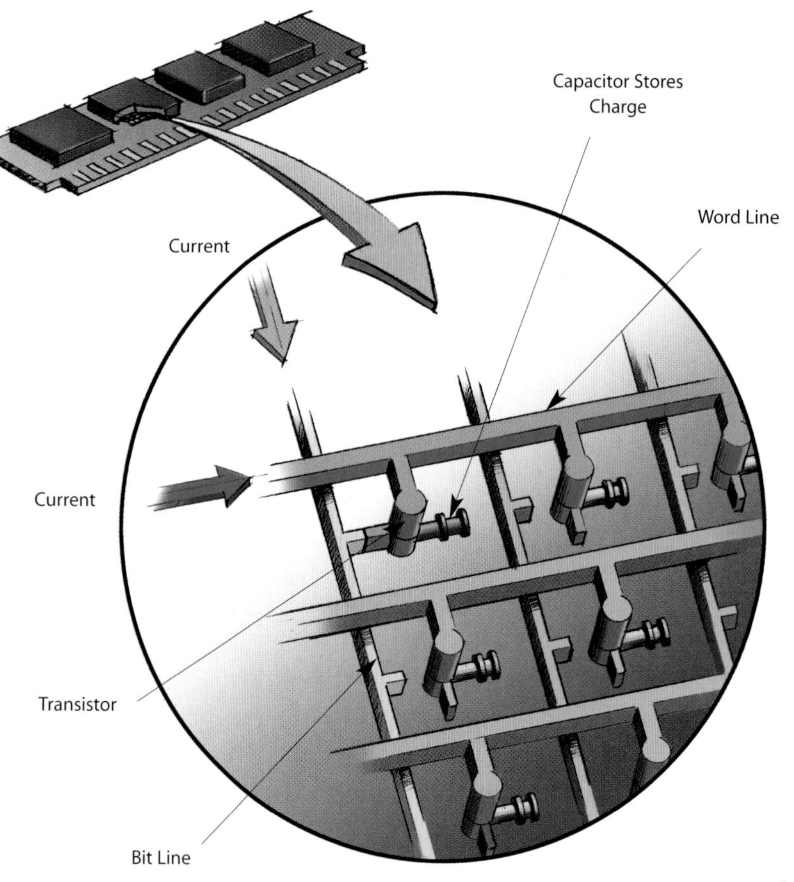

should take on. When reading, the sense-amplifier determines the level of charge in the capacitor. If the charge is more than 50%, it reads as a 1; otherwise, it reads as a 0. The length of time necessary to do all this is so short that it is expressed in nanoseconds (that is, billionths of a second). A memory chip rating of 70 ns means that it takes 70 nanoseconds to completely read or charge each cell.

DRAM Prices

One of the most interesting things about dynamic RAM is the way prices have fallen over the years.

For example, the Apple Macintosh was first released in 1984, and the IBM PC had been out about two years at that point. You purchased dynamic RAM as individual chips and plugged them into the motherboard in groups of eight or nine. Chips held 64K bits and cost about $6 each, so you could buy 64K bytes of RAM for about $50.

By 1991, RAM came packaged in SIMMs. One megabyte of RAM cost about $50, so RAM prices fell by a factor of 16 in 7 years.

By 2001, you could buy 256 megabyte SIMMs for about $100. Compared to 1984, RAM prices fell by a factor of 2,000 times!

SRAM

SRAM uses a completely different technology from DRAM. In SRAM, a form of flip-flop holds each bit of memory. A flip-flop is a memory cell that takes four or six transistors along with some wiring to remember a bit, but it never has to be refreshed. This makes SRAM significantly faster than DRAM. However, because it has more parts, a static memory cell takes a lot more space on a chip than a dynamic memory cell. Therefore, you get less memory per chip, and that makes SRAM more expensive than DRAM.

SRAM is fast and expensive, and DRAM is less expensive but slower. Therefore, SRAM is used to create the CPU's speed-sensitive cache and DRAM forms the larger system RAM space.

Types of ROM

Nonvolatile memory does not lose its data when the system or device is turned off. ROM, also known as firmware, is an IC that is programmed with specific data when it is manufactured. Other forms of ROM can be programmed in the field.

There are five basic ROM types:

- ROM
- Programmable ROM (PROM)
- EPROM
- Electrically erasable programmable ROM (EEPROM)
- Flash memory

Each type of ROM has unique characteristics. Data stored in ROM chips is either unchangeable or requires a special operation to change—unlike RAM, which can be changed as easily as it is read. This means that removing the power source from a ROM chip will not cause it to lose any data.

Similarly to RAM, ROM chips contain a grid of columns and rows. But where the columns and rows intersect, ROM chips are fundamentally different from RAM chips. Whereas RAM uses transistors to turn on or off access to a capacitor at each intersection, ROM uses a diode. Where there is a diode, the memory state is 1, and where there is no diode, it is 0. The diodes are manufactured as part of the chip, and they can never change.

A PROM chip has a grid of columns and rows, just as any other ROM chip does. But every intersection of a column and row in a PROM chip has a fuse connecting the column and row. A charge sent through a column will pass through the fuse in a cell to a grounded row indicating a value of 1. Because each cell has a fuse, the initial, or blank, state of a PROM chip is all 1s. To change the value of a cell to 0, you use a programmer to break the connection between the column and row by burning out the fuse. This process is known as burning the PROM and is a one-time operation.

EPROM addresses the one-time writing issue; EPROM chips can be rewritten many times. Erasing an EPROM chip requires ultraviolet (UV) light. EPROM chips are configured using an EPROM programmer.

In an EPROM chip, the cell at each intersection has two transistors, and a thin oxide layer separates the two transistors from each other. One of the transistors is known as a floating gate and the other one is the control gate. The floating gate's only link to the row, or wordline, is through the control gate. As long as this link is in place, the cell has a value of 1.

Changing the cell's value to a 0 requires a process called Fowler-Nordheim tunneling. Tunneling alters the placement of electrons in the floating gate and allows an electrical charge to be applied to the floating gate. This charge causes the floating-gate transistor to act like an electron gun. The excited electrons are pushed through and trapped on the other side of the thin oxide layer, giving it a negative charge. These negatively charged electrons act as a barrier between

Analogy of a Capacitor Leak

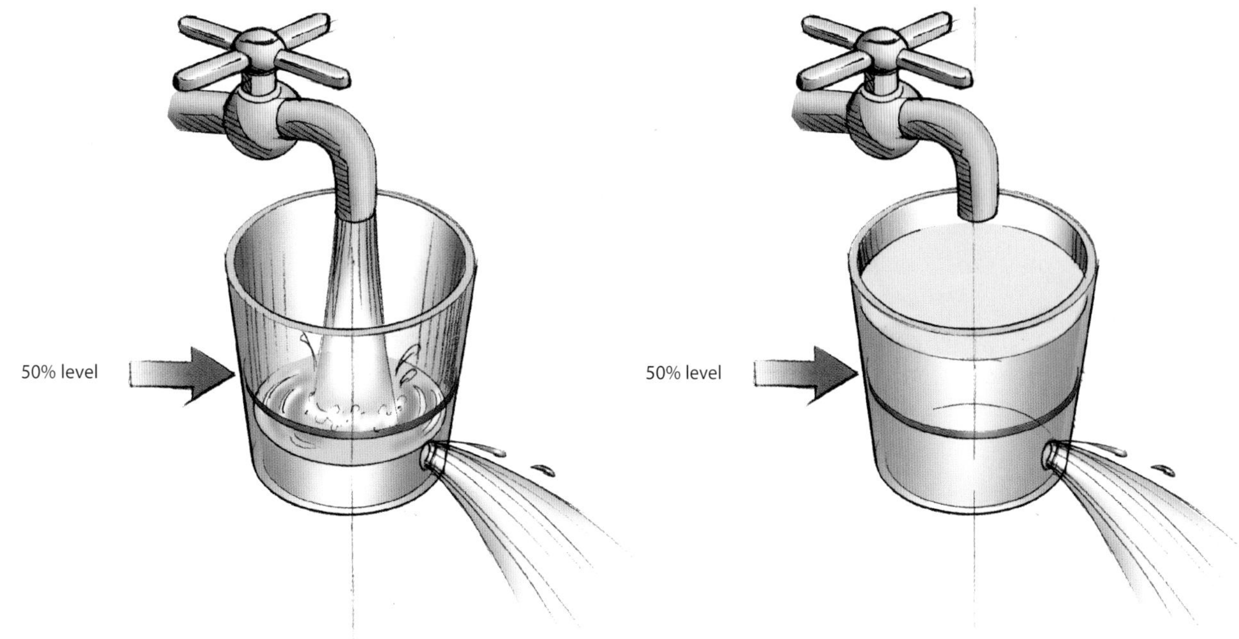

the control gate and the floating gate. A device called a cell sensor monitors the level of the charge passing through the floating gate. If the flow through the gate is greater than 50% of the charge, it has a value of 1. When the charge passing through drops below the 50% threshold, the value changes to 0. A blank EPROM chip has all the gates fully open, giving each cell a value of 1.

Although EPROM chips are a big step up from PROM chips in terms of reusability, they still require dedicated equipment and a labor-intensive process to remove and reinstall them each time a change is necessary. Also, changes cannot be made incrementally to an EPROM chip; the whole chip must be erased. EEPROM chips remove the biggest drawbacks of EPROM chips. An EEPROM chip does not have to be removed from the computer to be rewritten, an entire chip does not have to be completely erased to change a specific portion of it, and changing the contents of a chip does not require additional dedicated equipment.

Instead of using UV light, you can return the electrons in the cells of an EEPROM chip to normal with the localized application of an electric field to each cell. This erases the targeted cells of the EEPROM chip, which can then be rewritten. EEPROM chips are changed a byte at a time, which makes them versatile but slow. In fact, EEPROM chips are too slow to use in many products that make quick changes to the data stored on the chip.

Manufacturers responded to this EEPROM limitation with flash memory, which is a type of EEPROM. Flash memory works much faster than traditional EEPROM because it writes data in chunks, usually 512 bytes in size, instead of 1 byte at a time. Flash memory storage devices such as CompactFlash and SmartMedia cards are today's most common form of electronic nonvolatile memory. The electrons in the cells of a flash memory chip can be returned to normal by the application of an electric field generated on the chip. Flash memory can reset the entire chip or predetermined sections of a chip, known as blocks.

Because memory is so important to microprocessors, new types of memory and improvements to existing types are being developed all the time. That's why, when the original IBM PC came out in the early 1980s, it shipped with just 16 kilobytes of RAM. Today you can buy hundreds of megabytes of much faster RAM for the same price that those early 16 kilobytes cost!

How the MEMORY HIERARCHY Works

You know that a desktop computer or a laptop computer has memory. You hear about it when you buy the computer, and then every so often your computer complains because it runs out of memory. Your computer contains a complete memory hierarchy that is fascinating.

HSW Web Links

www.howstuffworks.com

How Flash Memory Works
How Magnetic RAM Will Work
How RAM Works
How ROM Works
How Virtual Memory Works

A modern microprocessor that acts as the central processing unit (CPU) of a desktop machine is an amazing device. The fastest processors can perform billions of operations every second. Of course, in order to process data that quickly, the processor has to access the instructions and data at the same pace, so a modern microprocessor needs billions of bytes streaming in and out of it at all times. All those bytes are stored in some form of memory. From the moment you turn on your computer until the time you shut it down, your CPU is constantly using memory.

The problem is that memory that can match the native speed of a microprocessor is extremely expensive. If everything that a computer had to remember were stored in high-performance memory like this, no one could afford a computer. To solve this problem, computer designers have developed the idea of a memory hierarchy, where layers of memory interact with each other to keep the price of computers low.

The CPU accesses memory according to a distinct hierarchy. Whether it comes from permanent storage (such as the hard drive) or input (such as the keyboard), most data goes in random-access memory (RAM) first. The CPU then stores pieces of data it will need to access often in a cache and maintains certain special instructions in the register.

Levels of the Hierarchy

Computers need many memory systems. A typical computer has these forms of memory:

- Level 1 and Level 2 memory caches
- Normal system RAM
- Virtual memory
- A hard disk

Each of these different forms of memory has a specific price per byte and a specific performance level, and each one fits into a different level of the memory hierarchy.

Fast, powerful CPUs need quick and easy access to large amounts of data in order to maximize their performance. If the CPU cannot get to the data it needs, it literally stops and waits for it.

The cheapest form of read/write memory in wide use today is the hard disk. Hard disks provide large quantities of inexpensive, permanent storage. You can buy hard disk space for pennies per megabyte, but it can take a good bit of time (approaching a second) to read a megabyte from a hard disk. Because storage space on a hard disk is so cheap and plentiful, it forms the final stage of a CPU's memory hierarchy: It holds files that the computer can access when it needs them. It is also the backbone of a system called virtual memory.

The data rate that a hard disk supports—on the order of a megabyte per second—is way too slow for a CPU. The CPU needs data about a thousand times faster than that. That's where RAM comes in.

Locality of Reference

The reason caches and virtual memory work so well and make the memory hierarchy possible is because of a principle called locality of reference. At any given nanosecond, a microprocessor can only access one address in RAM. It turns out that, in general, the next area of RAM to be accessed will be near the last area, or it will have been recently accessed. On average, during any given second, out of all the megabytes of RAM that the computer has loaded, because of locality of reference, very little of it is actually used.

If this were not the case, caching and virtual memory would not work very well. It is possible to write programs, for example, which access large amounts of data randomly. In that case the cache is no good and the hard disk runs constantly -- it can bring a computer to its knees.

The next level of the hierarchy is RAM, which usually comes in the form of dynamic RAM (DRAM). This is a rough approximation, but you can think of DRAM as being about 100 times more expensive, and 100 times faster, than hard disk space. It also is volatile, meaning that it loses its memory when the power goes out. Every time you turn on a computer, the operating system loads itself into RAM from files on the hard disk.

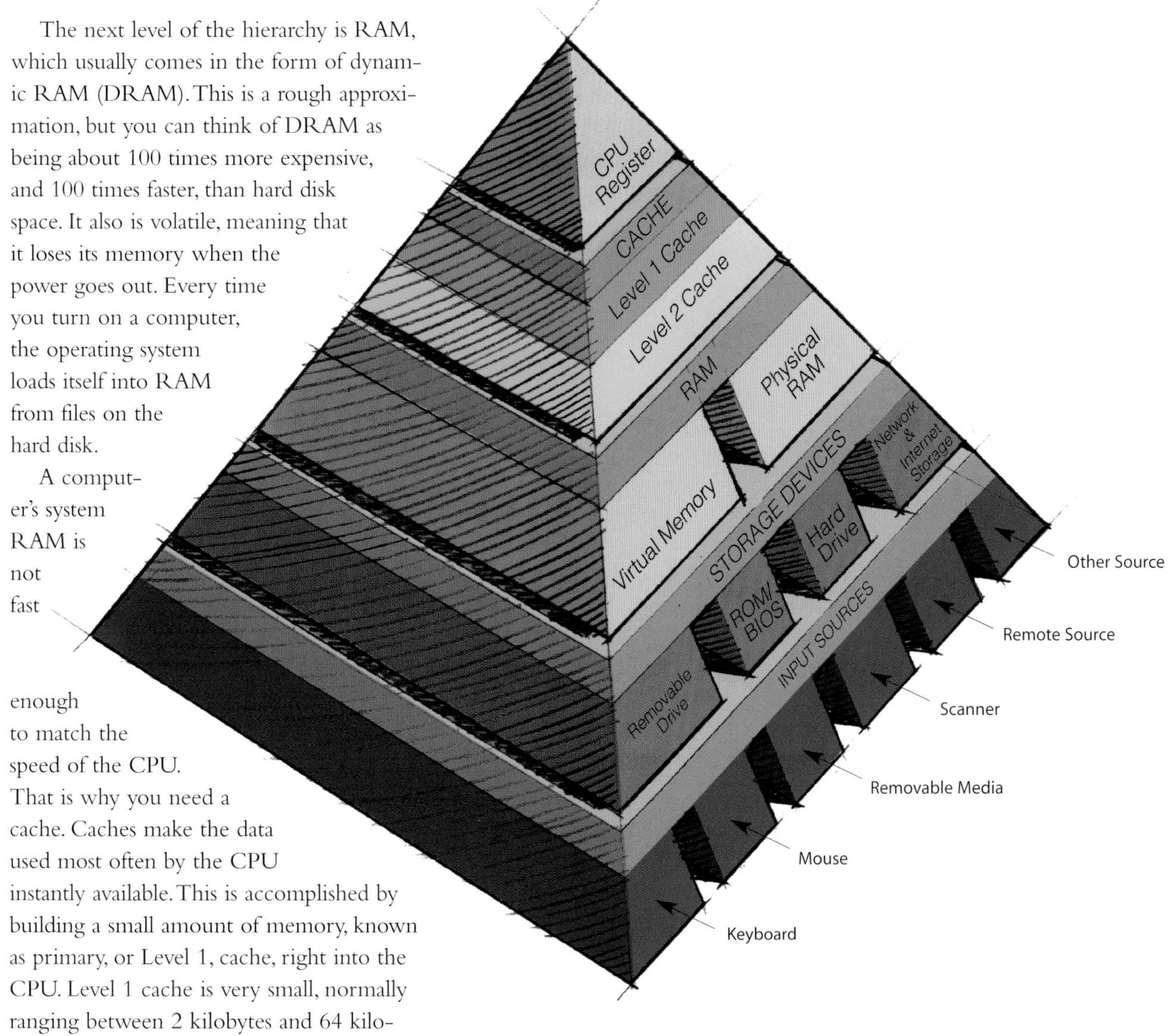

A computer's system RAM is not fast enough to match the speed of the CPU. That is why you need a cache. Caches make the data used most often by the CPU instantly available. This is accomplished by building a small amount of memory, known as primary, or Level 1, cache, right into the CPU. Level 1 cache is very small, normally ranging between 2 kilobytes and 64 kilobytes.

The secondary, or Level 2, cache typically resides on a memory card located near the CPU. The Level 2 cache has a direct connection to the CPU. A dedicated integrated circuit on the motherboard, the L2 controller, regulates the CPU's use of the Level 2 cache. Depending on the CPU, the size of the Level 2 cache ranges from 256 kilobytes to 2 megabytes. In most systems, data needed by the CPU is accessed from the cache approximately 95% of the time, greatly reducing the overhead needed when the CPU has to wait for data from the main memory.

Some inexpensive systems dispense with the Level 2 cache altogether. Many high-performance CPUs now have the Level 2 cache actually built into the CPU chip itself. Therefore, the size of the Level 2 cache and whether it is on-board (that is, on the CPU) are major determining factors in the performance of a CPU.

Caches use static RAM (SRAM). SRAM uses multiple transistors, typically four to six of them, for each memory cell. This means that it does not have to be continually refreshed, as DRAM does. Each cell maintains its data as long as it has power. Without the need for constant refreshing, SRAM can operate extremely quickly and keep pace with the CPU. But the complexity of each cell makes SRAM extremely expensive. It is roughly 10 times faster, and 10 times more expensive, than DRAM.

The SRAM in the cache can be asynchronous or synchronous. Synchronous SRAM is designed to exactly match the

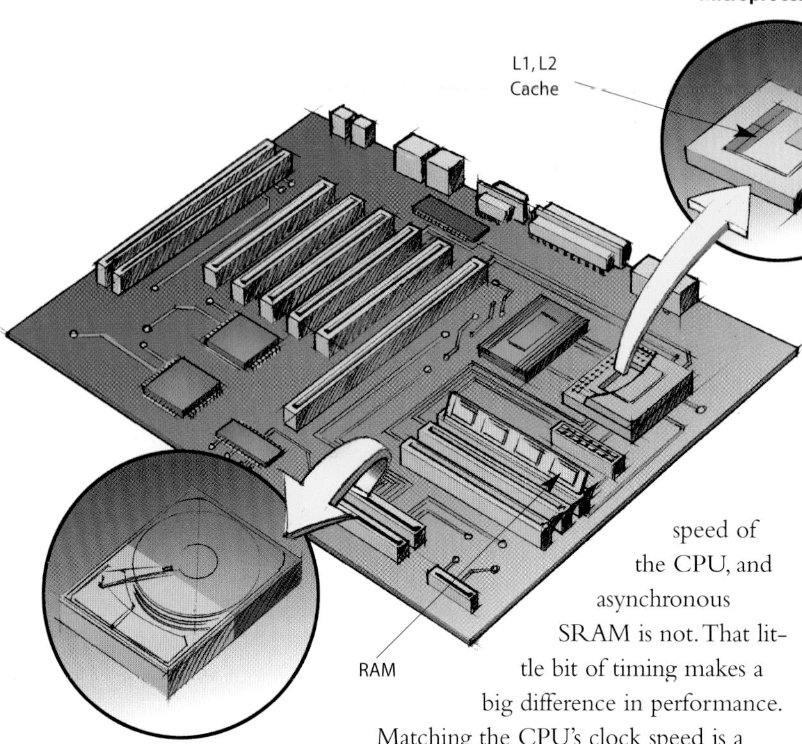

speed of the CPU, and asynchronous SRAM is not. That little bit of timing makes a big difference in performance.

Matching the CPU's clock speed is a good thing, so you generally want synchronized SRAM in the cache.

The final level of memory is the registers. The registers are memory cells built right into the CPU that contain specific data needed by the CPU, particularly the arithmetic/logic unit. An integral part of the CPU itself, the registers are controlled directly by the compiler that sends information for the CPU to process.

Virtual Memory

When you first turn on the computer, it goes to the hard disk and loads all the files for the operating system into RAM. When you double-click on a word processing document, the computer goes to the hard disk and loads all the files for the word processing application into RAM, and then the application loads the document file into RAM as well. As you open other documents and applications, they are all loaded into RAM.

Unfortunately, the amount of RAM in most computers is not sufficient to run all the programs that most users expect to run at once. For example, if you load the operating system, an email program, a Web browser, and a word processor into RAM simultaneously, all this will probably not fit into the RAM. If there were no such thing as virtual memory, your computer would have to say, "Sorry, you cannot load any more applications. Please close another application to load a new one." When RAM is full, there is no room for anything else. With virtual memory, the computer can look at RAM for areas that have not been used recently and copy them onto the hard disk. This frees up space in RAM to load the new application. Because it happens automatically, you don't even know it is happening, and it makes your computer feel like is has unlimited RAM space, even though there is only some small amount of RAM installed. Because hard disk space is so much cheaper than RAM chips, it also has a nice economic benefit.

Of course, the read/write speed of a hard drive is much slower than RAM, and the technology of a hard drive is not geared toward accessing small pieces of data at a time. If your system has to rely too heavily on virtual memory, you will notice a significant performance drop. The hard disk will be churning all the time as data swaps between RAM and the hard disk.

The key is to have enough RAM to handle everything you tend to work on simultaneously. Then the only time you feel the slowness of virtual memory is a slight pause when you change tasks. When that's the case, virtual memory is perfect. When it is not the case, virtual memory makes your computer feel very sluggish.

The area of the hard disk that stores the RAM image is called a page file. It holds pages of RAM on the hard disk, and the operating system moves data back and forth between the page file and RAM. On a Windows machine, page files have an .swp extension.

Using all these different levels, your computer can have a huge amount of memory that runs very quickly but costs very little. The memory hierarchy has played a big part in keeping computer prices low.

How **HARD DISKS** Work

Nearly every desktop computer and server in use today contains one or more hard disk drives. Every mainframe and supercomputer is normally connected to hundreds of them. Some VCR-type devices and camcorders even use hard disks instead of tape. These billions of hard disks do one thing well: They store changing digital information in a relatively permanent form. Thus, they give computers the ability to remember things when the power goes out.

Hard disks were invented in the 1950s. They started as large disks up to 20 inches in diameter that held just a few megabytes of data each. They were originally called fixed disks or Winchesters (a code name used for a popular IBM product). They later became known as hard disks to distinguish them from floppy disks. A hard disk has a hard platter made of aluminum or glass that holds the magnetic medium, as opposed to the flexible plastic film found in tapes and floppies.

Advantages of Hard Disks

Hard disks are important to computers because they have three big advantages over other storage devices that currently exist:

- Hard disks store data magnetically rather than electronically, so they retain memory even if the power is out. A hard disk provides permanent storage. Random-access memory (RAM) must have power at all times to retain data.
- Many memory technologies offer permanent storage—for example, magnetic tape, floppy disks, and rewritable CDs. Compared to all these technologies, hard disks are much faster.
- On a cost-per-byte basis, hard disks are incredibly inexpensive. Hard disks have been getting denser and cheaper at an amazing rate for decades. In the early 1980s, a 10-megabyte hard disk cost more than $1,000. Today you can buy gigabytes of storage space in a smaller, faster package for a few hundred dollars.

Hard disks are currently the fastest permanent storage medium available, and they have a reasonable cost per byte.

Hard Disks and Cassette Tapes

At the simplest level, a hard disk is not that different from a cassette tape. Hard disks and cassette tapes use the same magnetic recording techniques. Hard disks and cassette tapes also share the major benefits of magnetic storage: The magnetic medium can be easily erased and rewritten, and it can remember the magnetic flux patterns stored onto the medium for many years.

There are a few big differences between cassette tapes and hard disks:

- The magnetic recording material on a cassette tape is coated onto a thin plastic strip. In a hard disk, the magnetic recording material is layered onto a high-precision aluminum or glass disk. The hard disk platter is then polished to mirror smoothness.
- With a tape, you have to fast-forward or reverse through the tape to get to any particular point on the tape. This can take several minutes with a long tape. On a hard disk

HSW Web Links

www.howstuffworks.com

How Bits and Bytes Work
How CDs Work
How Floppy Disk Drives Work
How IDE Controllers Work
How Tape Recorders Work

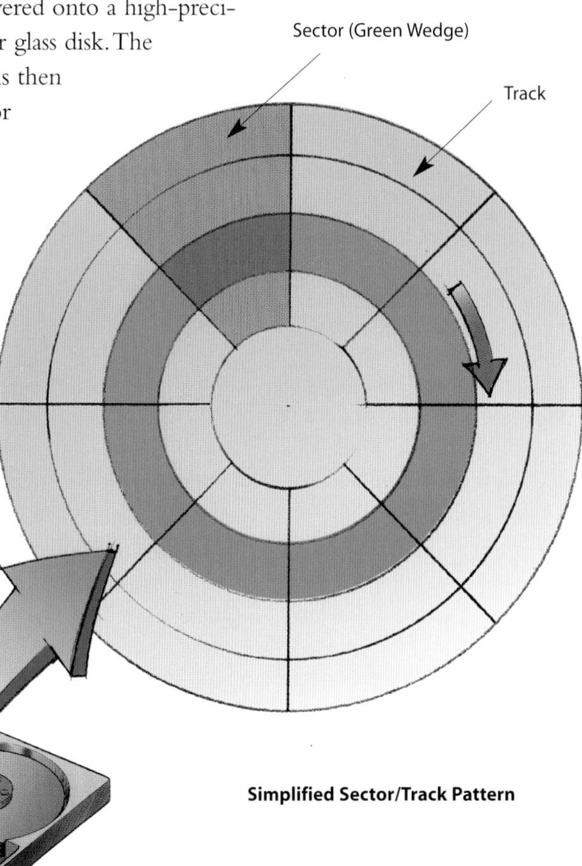

Simplified Sector/Track Pattern

RAID

On high-end systems, hard disks are grouped together under a system called RAID (Redundant Arrays of Inexpensive Disks). "Mirroring" is one example of RAID. The same data is written onto two disks, so if one fails the other can take over. In another configuration, perhaps four disks are combined with a parity disk; if any of the five fails, the other four have everything they need to continue.

you can move the read/write head to any point on the surface of the disk in a matter of milliseconds.

- In a cassette tape deck, the read/write head touches the tape directly, which causes wear. In a hard disk the read/write head flies over the disk, never actually touching it.
- The tape in a cassette tape deck moves over the head at about 2 inches (5.08 cm) per second. A hard disk platter can spin underneath its head at speeds up to 3,000 inches per second (170 mph, or 272 kph). This means that data can move on and off the disk at high speed.
- Compared to in a cassette tape, the information on a hard disk is stored in extremely small magnetic domains. The precision of the platter and the speed of the medium make the size of these domains possible. This means that you can store data very densely on a hard disk.

Most hard disks have two or more platters and a read/write head on each side of each platter, which improves the access time even more.

A modern hard disk is able to store an incredible amount of information in a very small space, and it can access any of its information in a fraction of a second.

Parts of a Hard Disk

When you look inside a hard disk drive, the first thing you notice is the disks themselves, which are also called platters. There are normally two or more platters on the same spindle. Platters typically spin at 3,600 or 7,200 RPM when the drive is operating. They are manufactured to have high tolerances and are mirror smooth.

The second thing you notice is the read/write heads. These are tiny rectangles mounted on lightweight arms. When the platters spin up, a very thin cushion of air forms between the read/write head and the platter, so the head does not actually touch the magnetic surface. However, the head is only microns away from the surface. Even the tiniest particle can cause the head to crash into the platter and ruin it, so the inside of a hard disk drive is a very clean place.

The arms that hold the read/write heads attach to a voice coil that moves the arms to different positions on the disk very precisely and at a very high speed. The arm that holds the read/write heads can move from the hub to edge of the drive and back up to 50 times per second—and it is an amazing thing to watch!

The surface of the platter is divided into tracks and sectors to store the data. Tracks are concentric rings, like the rings of a tree. A typical hard disk has thousands of tracks. Each track contains a group of sectors, and each sector holds a fixed number of bytes. For example, one sector on one track might hold 512 bytes.

Low-level formatting, which is done at the factory, establishes the tracks and sectors on the platter. The starting and ending points of each sector are written onto the platter. This process prepares the drive to hold blocks of bytes.

A computer operating system uses the tracks and sectors to hold files, which are named collections of bytes, and the bytes in turn are stored in one or more sectors on the disk.

High-level formatting creates the file-storage structures such as the file allocation table on the disk. The file allocation table contains the name of each file and remembers all the different tracks and sectors that hold the file. When the computer wants to access a file, it takes the following steps:

1) The computer looks up the file by name in the file allocation table.
2) The computer retrieves the track and sector locations of all the different pieces of the file.
3) The computer moves the read/write head to the right track and then waits for the right sector to spin into position to read each sector of the file.

Ideally, all the sectors of the file are stored sequentially in contiguous tracks, and the read/write head has to move very little to read the file. If many files have been deleted on the disk, however, files are sometimes

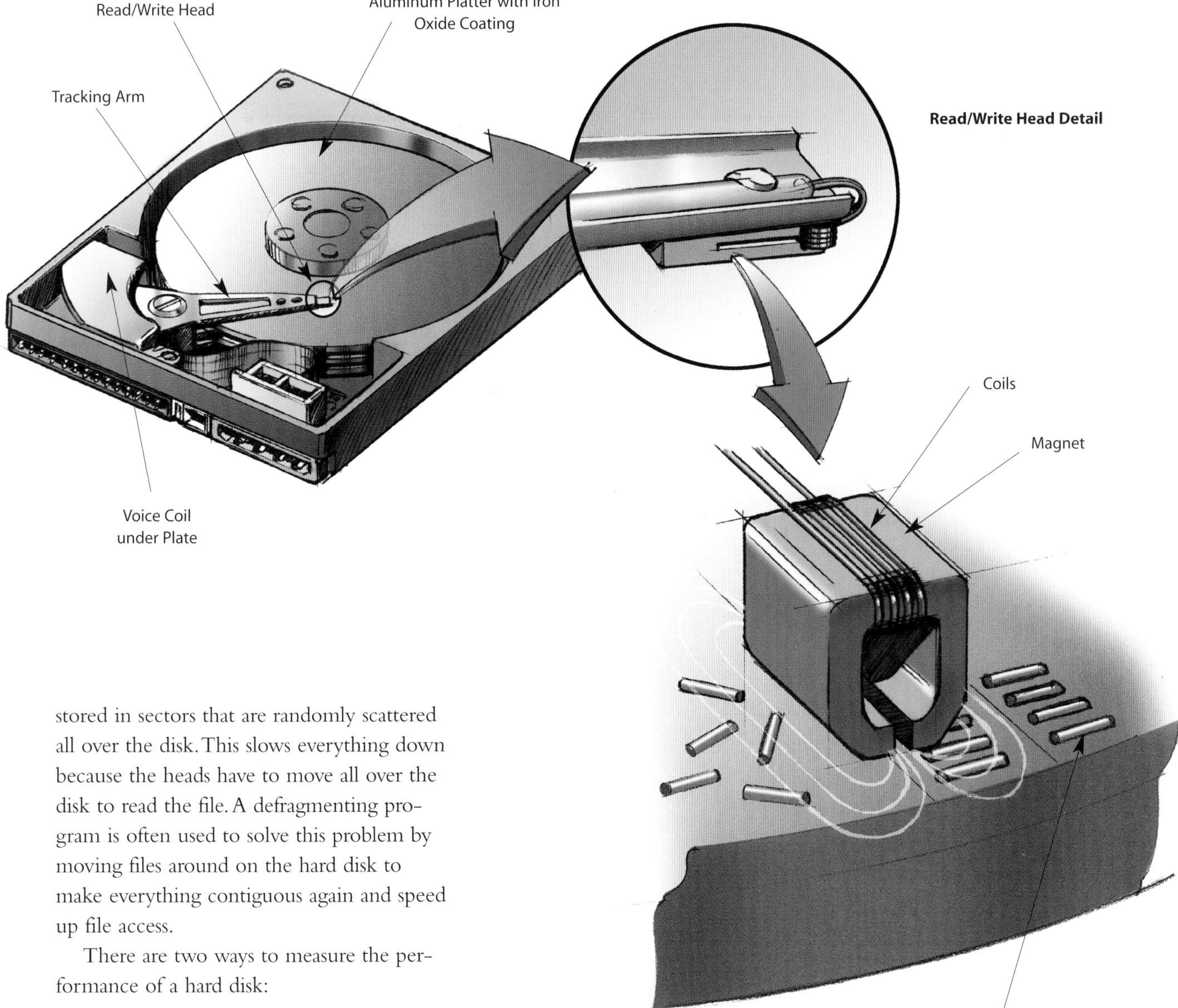

stored in sectors that are randomly scattered all over the disk. This slows everything down because the heads have to move all over the disk to read the file. A defragmenting program is often used to solve this problem by moving files around on the hard disk to make everything contiguous again and speed up file access.

There are two ways to measure the performance of a hard disk:

- **The data rate**—The number of bytes per second that the drive can deliver to the central processing unit
- **The seek time**—The amount of time it takes for the heads to move to different tracks and for a sector to spin into position

Another important parameter is the capacity of the drive—the number of bytes it can hold. This number has been increasing steadily for decades.

Engineers have been able to do remarkable work with the technology of memory storage. Not long ago, what you can store on today's average laptop took up several large rooms filled with disks!

How **FILE COMPRESSION** Works

If you download many programs and files off the Internet, you've probably encountered compressed files with extensions such as .zip or .gz. Compression systems let you shrink the number of bytes in a file so that the file takes less space on a disk or less transmission time on a modem. When you open the file, the computer uses a software application to expand it back to its original size.

HSW Web Links

www.howstuffworks.com

How Bits and Bytes Work
How Digital Cameras Work
How MP3 Files Work
How MP3 Players Work
How Email Works

File compression seems very mysterious. How can the expanded file be identical to the original file before it was compressed? Where do all the extra bytes go? The basic concept is actually very simple: The compression program simply eliminates redundancy in the file in such a way that the expansion program can put it back in. As an example, let's look at a type of information we're all familiar with: words.

Take John F. Kennedy's famous line: "Ask not what your country can do for you. Ask what you can do for your country." The quote has 61 letters, 16 spaces, and 2 periods. If each letter, space, and punctuation mark takes up one memory unit, you get a file size of 79 units. To compress the file, you need to look for redundancies.

Right away you notice that most of the words appear twice, so roughly half the phrase is redundant. Nine words—*ask, not, what, your, country, can, do, for, you*—give us almost everything we need for the entire quote. To construct the second half of the phrase, we simply point to the words in the first half and fill in the spaces and punctuation.

The Dictionary System

Most compression systems use the LZ adaptive dictionary-based algorithm scheme. *LZ* refers to Lempel and Ziv, the algorithm's creators, and *dictionary* refers to the method of cataloging pieces of data.

The system for arranging dictionaries varies, but it could be as simple as a numbered list. When you go through the Kennedy quote, you pick out the words that are repeated and put them into the numbered index. Then, you just write the number instead of writing out the whole word.

So if this is the dictionary:

1=ask
2=what
3=your
4=country
5=can
6=do
7=for
8=you

The sentence now reads:

1 not 2 3 4 5 6 7 8. 1 2 8 5 6 7 3 4.

If you knew the system, you could easily reconstruct the original phrase, using only this dictionary and number pattern. This is what the expansion program does when it expands a downloaded file.

But how much space have you actually saved? *1 not 2 3 4 5 6 7 8. 1 2 8 5 6 7 3 4.* is certainly shorter than *Ask not what your country can do for you. Ask what you can do for your country*. But keep in mind that we need to save the dictionary itself along with the file.

You already saw that the full phrase takes up 79 units. The compressed sentence (including spaces) takes up 35 units, and the dictionary (words and numbers) also takes up 37 units. This gives us a file size of 74, so we haven't reduced the file size by very much.

But if the compression program worked through the rest of Kennedy's speech, it would find these words and others repeated many more times. And it would also keep rewriting the dictionary to get the most efficient organization possible.

New Patterns

To humans, picking out repeated words is the most obvious way to write a dictionary.

But a compression program doesn't have any concept of separate words; it looks for patterns instead. To reduce the file size as much as possible, it carefully selects which patterns to include in the dictionary. If you were to approach the phrase from this perspective, you would end up with a completely different dictionary.

One pattern the program might notice is the letters *ou*, which appear in both *your* and *country*. If this were a longer document, writing this pattern to the dictionary could save a lot of space because it is a fairly common combination in the English language. But as the compression program worked through this sentence, it would quickly discover a better choice for a dictionary entry: Not only is *ou* repeated, but the entire words *your* and *country* are both repeated, and they are actually repeated together, in the phrase *your country*. In this case, the program would overwrite the dictionary entry for *ou* with an entry for *your country*.

The phrase can do for is also repeated, one time followed by *your* and one time followed by *you*, giving us a repeated pattern of *can do for you*. This lets us write 15 characters (including spaces) with one dictionary entry, whereas *your country* lets us write only 13 characters (including spaces), so the program would overwrite the *your country* entry as just *r country* and then write a separate entry for *can do for you*. The program proceeds in this way, picking up all repeated bits of information and then calculating which patterns it should write to the dictionary. This ability to rewrite the dictionary is the "adaptive" part of the LZ adaptive dictionary-based algorithm.

So, using the new patterns and adding __ for spaces, we come up with this larger dictionary:

1=ask__
2=what__
3=you
4=r__country
5=__can__do__for__you

And we have this smaller sentence:

1 not__2345.12354.

The sentence now takes up 16 units of memory, and the dictionary takes up 41 units. So we've compressed the total file size from 79 units to 57 units! And this is just one way of compressing the phrase—not necessarily the most efficient one. A computer is the perfect machine for finding and optimizing patterns like this.

In a long text document, there is an immense amount of duplication. Therefore, most text documents are reduced by half when you run them through a compression process. Certain text documents—especially long lists of similar items—compress even more, depending on how many repeated patterns the compression system finds when it looks at the file.

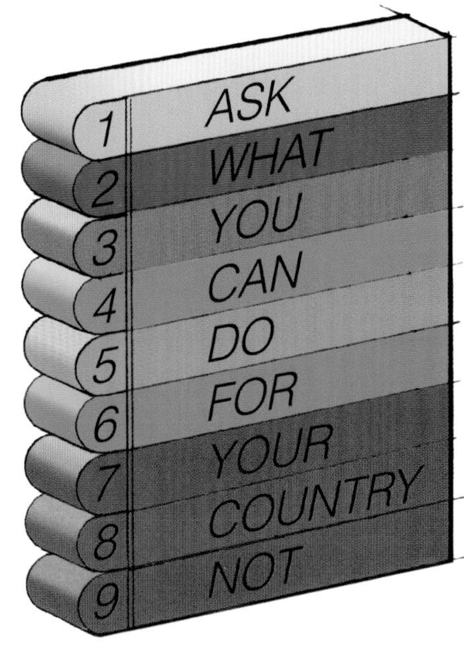

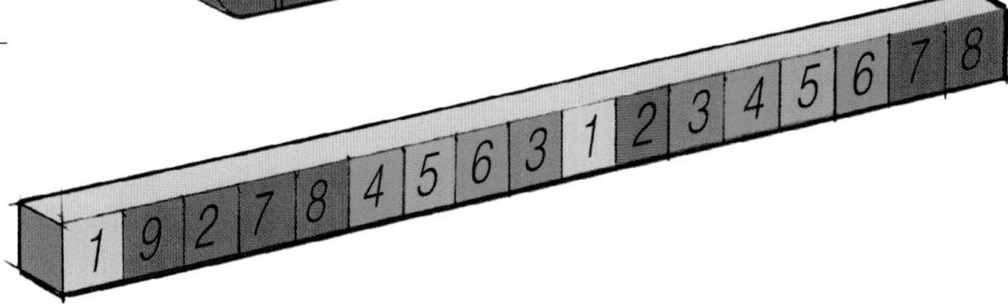

How **MOTHERBOARDS** Work

The motherboard is the heart of a computer. It holds the computer's central processing unit (CPU), random-access memory (RAM), and basic input/output system (BIOS). Everything from the video card to the sound card plugs into sockets on the motherboard. The motherboard provides all the connections that move the data between the different chips and cards. The motherboard also takes the power from the power supply and distributes it to everything that needs it.

HSW Web Links

www.howstuffworks.com

How BIOS Works
How Computer Memory Works
How IDE Controllers Work
How Microprocessors Work
How PC Power Supplies Work

Did You Know?

The consumer can upgrade functions that are integrated on the motherboard (such as audio and video controllers), as long as the motherboard manufacturer provides a way to disable the integrated function.

A motherboard may still have voltages present on it, even if the computer is switched off, due to recent advances in power management and power controls. Always make sure that the power cord is unplugged before you touch a motherboard!

The original IBM PC contained the original PC motherboard. This design premiered in 1982. The motherboard itself was a large printed circuit card that contained the 8088 microprocessor, the BIOS, sockets for the CPU's RAM, and a collection of slots that auxiliary cards could plug into. If you wanted to add a floppy disk drive, a parallel port, or a joystick, you bought a separate card and plugged it into one of the slots. This approach was originally pioneered in the mass market by the Apple II machine. By making it easy to add cards, Apple and IBM accomplished two things:

- They made it easy to add new features to the machine over time.
- They opened the computer to creative opportunities by third-party vendors.

Several years after the PC's debut, companies started creating compatible motherboards that were faster and cheaper and had many more capabilities. There is now an entire motherboard marketplace where hundreds of companies compete with different products. Motherboards have gotten progressively more complicated. Processors and chips are hundreds of times faster than those in the original PC, and most of the features that were on auxiliary cards now live right on the motherboard. This integration has made computers faster and less expensive.

Motherboard Sizes and Shapes

Motherboards of different vintages typically have different form factors. Form factor essentially means the size and shape of the actual motherboard. There are more than a half-dozen standard form factors for motherboards. The ATX form factor has been

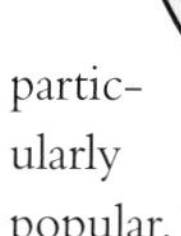

particularly popular. By buying a computer with a true ATX motherboard, you know that you will have the ability to upgrade by being able to reuse the personal computer (PC) case with a more recent replacement ATX board design. An ATX board is about 12 by 9.6 inches (30.4 by 24.3 cm).

Common Motherboard Parts

A motherboard is a multilayered printed circuit board. Copper circuit paths called traces carry signals and voltages across the motherboard. A motherboard has many layers; some layers of a motherboard can carry data for the input/output (I/O), processor, and memory buses, and other layers can carry voltage and ground. The insulated layers are manufactured into one complete, complex sandwich.

These are some of the parts that are commonly on a motherboard:

- Chips and sockets
- One or more microprocessors
- A BIOS chip
- Memory slots
- A chipset that adds features such as I/O ports and controllers
- Peripheral component interconnect (PCI) adapter card slots
- Industry standard architecture (ISA) adapter slots
- Accelerated graphics port (AGP) video card slots
- Universal serial bus (USB) ports

A typical motherboard has sockets for one or two microprocessor chips, sockets for the cache RAM for the microprocessors, sockets for RAM and the BIOS chip, and a number of chips that make up the chipset. The motherboard also has sockets for three to five PCI slots and an AGP socket for the video card.

The design of the motherboard controls the maximum data rate for data moving around on the board. For example, a motherboard might support 133-MHz bus speeds.

Motherboards today often contain all the circuitry for the sound card, the video card, the hard disk controller card and all the I/O functions. In many computers, all that's needed is a power supply, a hard disk drive, and a CD-ROM drive to build a working machine. The motherboard contains everything you need!

The transformation that has occurred since the first PC motherboard came out in the early 1980s has been amazing. Motherboards are now hundreds of times faster and have dozens of capabilities integrated within them. With hundreds of manufacturers competing, the price of computing power has fallen steadily for many years.

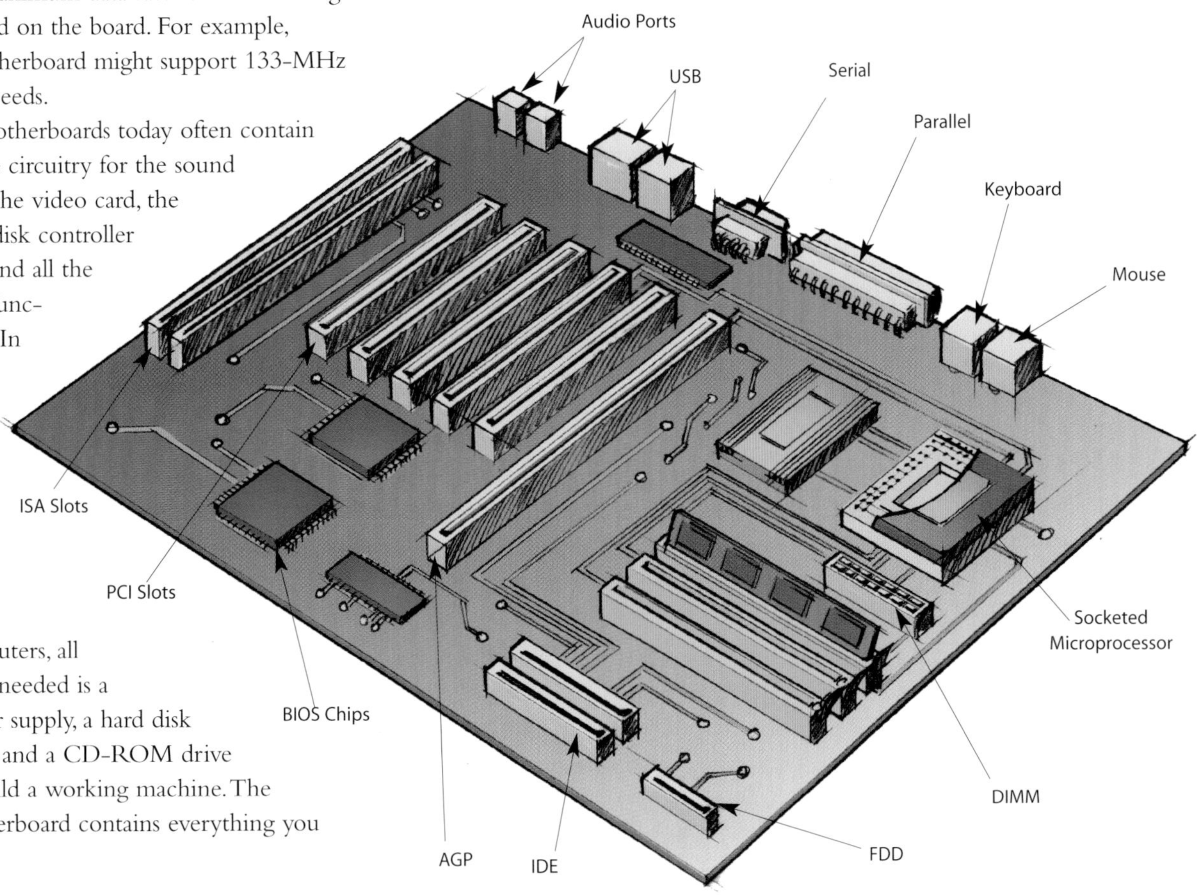

Data Bus Width

Modern Pentium-class motherboards have a data bus that is 64 bits wide. On the original PC motherboard the data bus was 8 bits wide. Moving data on a wider bus is one of the things that makes today's motherboards so fast. Typical bus names and widths (in bits) are:

- **ISA**—8 or 16 bits
- **Extended ISA (EISA)**—8 or 16 bits
- **Microchannel architecture (MCA)**—16 or 32 bits
- **VESA local bus (VLB)**—32 bits
- **PCI**—32 or 64 bits
- **AGP**—32 bits

Chipsets

Chipsets provide the support for the processor chip on the motherboard. The Intel 440BX is the dominant chipset in the non-Apple PC. The chipset is the heart of the computer because it controls and determines how fast and which type of processor, memory, and slots are used.

Another chip on the motherboard is called the super I/O controller. Its main function is to control the floppy disk drive, keyboard, mouse, and serial and printer ports.

Recent motherboard designs include additional chips to support USB, sound cards, video adapters, computer hosts, and network adapters. These chips save the cost of an adapter slot.

How **MICROPROCESSORS** Work

Have you ever wondered what the microprocessor in your computer is doing? The microprocessor is the brain of any normal computer, whether it is a desktop machine, a server, or a laptop. It uses fairly simple digital logic techniques to do its job, whether it's playing a game, calculating a spreadsheet, or spell checking a document.

HSW Web Links

www.howstuffworks.com

How Computerized Clothing Will Work
How Disposable Cell Phones Will Work
How Microcontrollers Work
How Motherboards Work
How Printable Computers Will Work

A microprocessor, also known as a central processing unit (CPU), is a complete computation engine that is fabricated on a single chip. The first microprocessor was the Intel 4004, introduced in 1971. The 4004 was not very powerful: All it could do was add and subtract, and it could only do that 4 bits at a time. But it was astounding that everything was on one chip. The 4004 powered one of the first portable electronic calculators.

A chip, also called an integrated circuit, is a small, thin piece of silicon onto which the transistors making up the microprocessor have been etched. A chip for a complex modern processor might be as large as an inch on a side and can contain many millions of transistors. Simple processors like the 4004 consist of just a few thousand transistors.

Modern processors can do much more than the original 4004. A 4004 can only add integers, whereas a modern ALU can perform hundreds of different floating-point operations, the CPU recognizes thousands of different instructions, and the CPU can handle things like virtual memory.

Inside a Microprocessor

To understand how a microprocessor works, it is helpful to look inside and learn about the logic used to create one.

A microprocessor executes a collection of machine instructions that tell the processor what to do. Based on the instructions, a microprocessor does three basic things:

- Using its ALU, a microprocessor can perform mathematical operations such as addition, subtraction, multiplication, and division. Modern microprocessors contain complete floating-point processors that can perform extremely sophisticated operations on large floating-point numbers.
- A microprocessor can move data from one memory location to another.
- A microprocessor can make decisions and jump to a new set of instructions based on those decisions.

A microprocessor may do other very sophisticated things, but these three are its basic activities. To fulfill these duties, the microprocessor uses random-access memory (RAM) and read-only memory (ROM). ROM holds a set of instructions, and it often has a way (for example, a hard disk) to load data and more instructions into RAM from files.

A microprocessor executes instructions one at a time. It fetches one instruction from memory (ROM or RAM), decodes it, and executes it. The decoding process tells the different parts of the microprocessor (for example, the ALU and the registers) what to do. Then the microprocessor increments its instruction counter and fetches the next instruction from memory. Some instructions force the microprocessor to change the instruction counter rather than incrementit. This is how a microprocessor can make decisions.

Parts of a Microprocessor

A simple microprocessor has the following components:

- An address bus that sends an address to memory.
- A data bus that can retrieve instructions from memory or move data into and out of memory.
- An RD (read) and a WR (write) line to tell the memory whether the processor wants to set or get the addressed location.

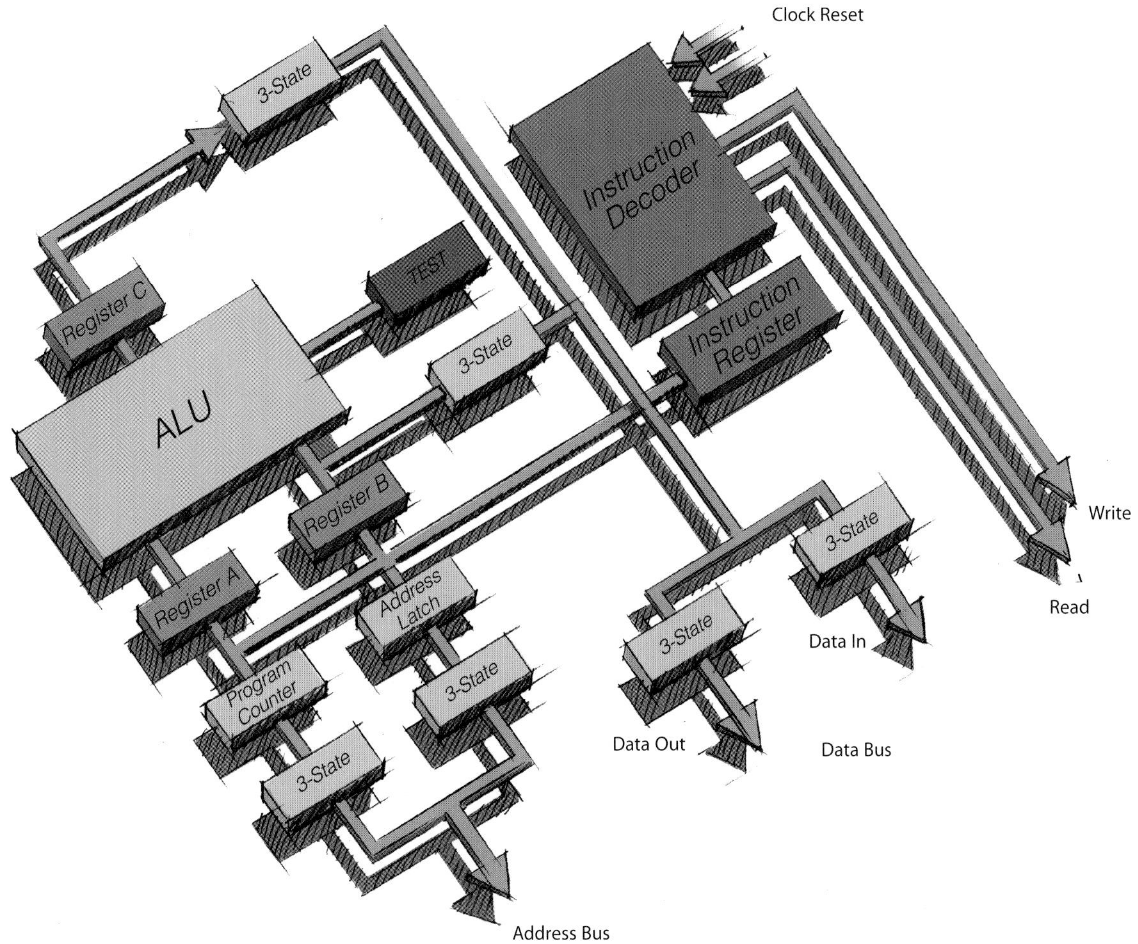

- A clock line that lets a clock pulse sequence the processor.
- A reset line that resets the program counter to zero (or whatever) and restarts execution.

Let's assume that both the address and data buses are 8 bits wide in this example.

Here are the components inside this simple microprocessor:

- Registers A, B, and C are simply latches made out of flip-flops. Each can remember a single byte.
- The address latch is just like registers A, B, and C.
- The program counter is a latch with the extra ability to increment by one when told to do so and also to reset to zero when told to do so.
- The ALU could be as simple as an 8-bit adder, or it might be able to add, subtract, multiply, and divide 8-bit values.
- The test register is a special latch that can hold values from comparisons performed in the ALU. An ALU can normally compare two numbers and determine whether they are equal, if one is greater than the other, and so on. The test register can also normally hold a carry bit from the last stage of the adder. It stores these values in flip-flops, and then the instruction decoder can use the values to make decisions.
- The six boxes marked "3-State" are tristate buffers. A tristate buffer can pass a 1 or a 0, or it can disconnect its output. A tristate buffer allows multiple outputs to connect to one wire, but it allows only one of the outputs to actually drive a 1 or a 0 onto the line.
- The instruction register and instruction decoder are responsible for controlling all the other components.

Address and data buses, as well as the RD and WR lines, connect either to RAM or ROM, and generally to both. Nearly all computers contain some amount of ROM. It is possible to create a simple computer that contains no RAM (many microcontrollers do this by placing a handful of RAM bytes on the processor chip itself), but it is generally impossible to create one that contains no ROM.

Microprocessor Instructions

Even an incredibly simple microprocessor can perform a fairly large set of instructions. The collection of instructions is implemented as bit patterns, each one of which has a different meaning when loaded into the instruction register. Humans are not particularly good at remembering bit patterns, so a set of short words are defined to represent the different bit patterns. This collection of words is the assembly language of the processor. An assembler can translate the words into their bit patterns very easily, and then the output of the assembler is placed in memory for the microprocessor to execute.

Here's a sample set of assembly language instructions (or opcodes) that a designer might create for the simple microprocessor described earlier:

- **LOADA mem**—Load register A from memory address.
- **LOADB mem**—Load register B from memory address.
- **SAVEB mem**—Save register B to memory address.
- **SAVEC mem**—Save register C to memory address.
- **ADD**—Add A and B and store the result in C.
- **COM**—Compare A and B and store result in test.
- **STOP**—Stop execution.

The instruction decoder needs to turn each of the opcodes into a set of signals that drive the different components inside the microprocessor. For example, this is what the ADD instruction needs to do:

- During the first clock cycle the microprocessor needs to actually load the instruction. Therefore, the instruction decoder needs to:
  1) Activate the tristate buffer for the program counter.
  2) Activate the RD line.
  3) Activate the data-in tristate buffer.
  4) Latch the instruction into the instruction register.
- During the second clock cycle the ADD instruction is decoded. It needs to do very little:
  1) Set the operation of the ALU to addition.
  2) Latch the output of the ALU into the C register.
- During the third clock cycle, the program counter is incremented (in theory, this could be overlapped into the second clock cycle).

Every instruction can be broken down as a set of sequenced operations, like these, that manipulate the components of the microprocessor in the proper order.

And Another Thing...

On a PC, the ROM is called the BIOS (basic input/output system). When the microprocessor starts, it begins executing instructions it finds in the BIOS. The BIOS instructions do things such as test the hardware in the machine and go to the hard disk to fetch the boot sector. This boot sector is another small program, and the BIOS stores it in RAM after reading it off the disk. The microprocessor then begins executing the boot sector's instructions from RAM. The boot sector program tells the microprocessor to fetch something else from the hard disk into RAM, which the microprocessor then executes, and so on. This is how the microprocessor loads and executes the entire operating system.

Microprocessor Performance

The number of transistors available has a huge effect on the performance of a processor. A typical instruction in a processor such

Clock Speed

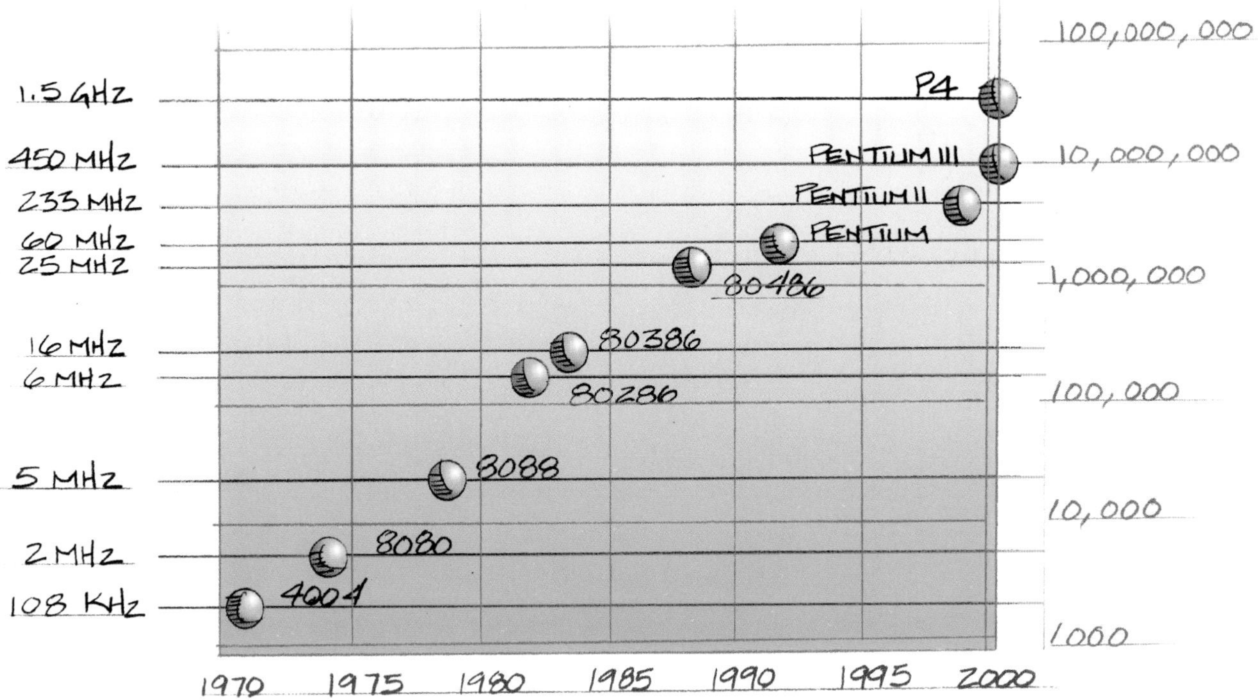

as an 8088 (the first processor used in an IBM PC) took 15 clock cycles to execute. Because of the design of the multiplier, it took approximately 80 cycles just to do one 16-bit multiplication on the 8088. With more transistors, much more powerful multipliers that are capable of single-cycle speeds become possible.

More transistors also allow a technology called pipelining. In a pipelined architecture, instruction execution overlaps. So even though it might take five clock cycles to execute each instruction, there can be five instructions in various stages of execution simultaneously, and it looks like one instruction completes every clock cycle.

Many modern processors have multiple instruction decoders, each with its own pipeline. This allows multiple instruction streams, which means more than one instruction can complete during each clock cycle. This technique can be quite complex to implement, so it takes lots of transistors.

Hardware virtual memory support and Level 1 caching have also been added to processor chips. These types of trends push up the transistor count, leading to the multi-million-transistor powerhouses that are available today and can execute billions of instructions per second.

You can create a very simple microprocessor with perhaps 1,000 transistors. It will be extremely slow, but you could program it to do anything a modern microprocessor can do. Today's microprocessors use millions of transistors to increase the speed dramatically, so that they can perform the most complex operations in a few nanoseconds. In the process, these high-speed machines are in the process of completely transforming our world.

How MODEMS Work

The modem appeared out of the ether to solve a problem. Imagine people in the 1960s first starting to think about connecting computers to terminals in remote offices. The obvious question is the wiring: How do you connect a terminal to a computer that's 100 miles away? Because the phone lines already connected everything together, they were an obvious way to go. So then another question came up: How do you move the computer's digital data across an analog phone line? That question inspired the birth of the modem.

HSW Web Links

www.howstuffworks.com

How Cable Modems Work
How Cell Phones Work
How Serial Ports Work
How Telephones Work
How Virtual Private Networks Work

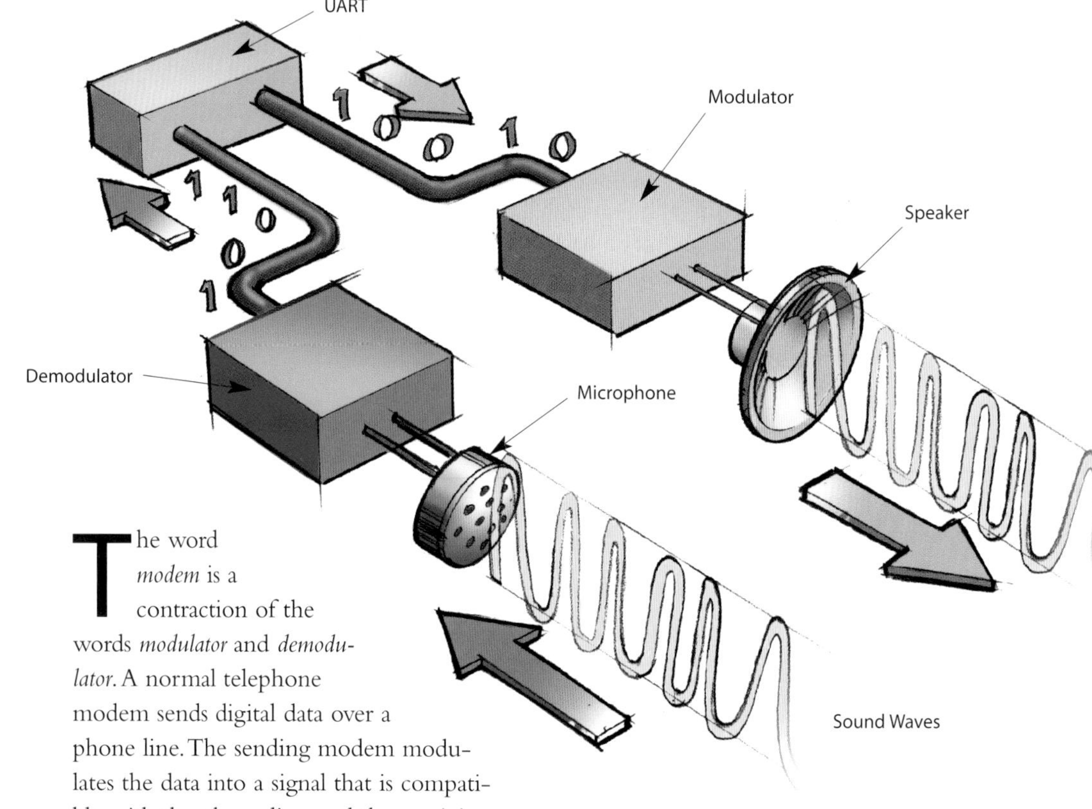

The word *modem* is a contraction of the words *modulator* and *demodulator*. A normal telephone modem sends digital data over a phone line. The sending modem modulates the data into a signal that is compatible with the phone line, and the receiving modem demodulates the signal back into digital data. Wireless modems are also frequently used to convert data into radio signals and back.

Modems came into existence in the 1960s as a way to allow terminals to connect to computers over phone lines. A dumb terminal at an off-site office or store could dial in to a large, central computer. The 1960s were the age of time-shared computers, so a business would often buy computer time from a time-share facility and connect to it via a 300 bits-per-second (bps) modem.

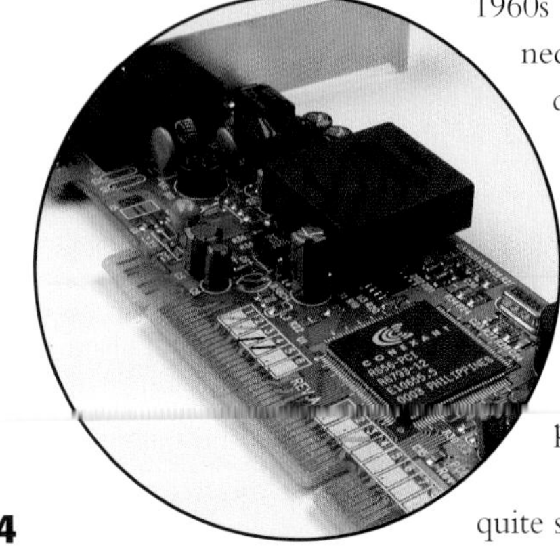

People got along at 300 bps for quite some time. This speed was tolerable because 300 bps represents about 30 characters per second, and that is much faster than a person can type characters or read.

300-bps Modems

A 300-bps modem is a device that uses frequency shift keying (FSK) to transmit digital information over a telephone line. In FSK, a different tone (frequency) is used for each different bit.

When a terminal's modem dials a computer's modem, the terminal's modem is called the originate modem. It transmits a 1,070-Hertz (Hz) tone for a 0 and a 1,270-Hz tone for a 1. The computer's modem is called the answer modem, and it transmits a 2,025-Hz tone for a 0 and a

2,225-Hz tone for a 1. Because the originate and answer modems transmit different tones, they can both use the line simultaneously. This is known as full-duplex operation. Modems that can transmit in only one direction at a time are known as half-duplex modems, and they are rare.

Let's say that two 300-bps modems are connected, and the user at the terminal types the letter *a*. The ASCII code for this letter is 97 in decimal, or 01100001 in binary. A device inside the terminal, called a universal asynchronous receiver/transmitter (UART), converts the byte into its bits and sends them out one at a time through the terminal's RS-232 port (also know as a serial port). The terminal's modem is connected to the RS-232 port, so it receives the bits one at a time, and its job is to send them over the phone line by converting the bits into tones. The modem at the other end recognizes the tones, converts them into bits, and sends them back.

Faster Modems

FSK is simple to implement and easy to understand, but its speed is limited. In order to create faster modems, modem designers had to use techniques more sophisticated than FSK. First they moved to phase shift keying (PSK) and then quadrature amplitude modulation (QAM). These techniques allow an incredible amount of information to be crammed into the 3,000 Hz of bandwidth available on a normal voice-grade phone line.

56,000-bps modems are about the limit of these techniques over normal phone lines, but many phone lines are so old that they cannot support anything near 56,000 bps. These high-speed modems incorporate a concept of gradual degradation, meaning that they can test the phone line and fall back to slower speeds if the line cannot handle the modem's fastest speed.

DSL Modems

The newest digital subscriber line (DSL) modems let you connect a home computer to the Internet at amazing speeds—up to 20 times faster than a standard modem. This is surprising because a DSL modem uses the same wiring a normal phone line uses. To make these high speeds possible over a normal phone line, the phone company tests your phone line to make sure it can handle the speed and puts the receiving device—the device that is talking to your DSL modem—within a mile or two of your home. Therefore, the DSL modem has a clear line and a known distance that it has to transmit.

To get the higher speeds, a DSL modem takes an interesting approach. It takes advantage of the fact that a dedicated copper wire can carry far more data than the 3,000-Hz signal needed for the phone's voice channel. The section of copper wire between your house and the phone company can act as a purely digital high-speed transmission channel. The capacity is something like 10 million bps. The same line can transmit both a phone conversation and the digital data, and the phone conversation is insignificant compared to the digital data.

The approach a DSL modem takes is very simple in principle. Imagine dividing the phone line's bandwidth that is between 24,000 Hz and 1,100,000 Hz into 4,000-Hz bands. This gives you the equivalent of hundreds of analog phone lines. Then imagine attaching a virtual 56,000-bps modem to each band. Each of these virtual modems would test its band and do the best it could with its allocated slice of bandwidth. The aggregate of the hundreds of virtual modems is the total speed of the pipe.

It's interesting that the technology of one 56,000-bps modem can be recycled in this way to create a system with incredible speed and capacity.

The modem gives computers the same capability that the telephone gives human beings—the ability to work together even when separated by a distance. In addition, modems that connect computers to the Internet have given people new ways to communicate in groups using email, newsgroups, and chat rooms.

Point-to-Point Protocol

Today no one uses dumb terminals or terminal emulators to connect to an individual computer. Instead you use a modem to connect to an Internet service provider (ISP), and the ISP connects you to the Internet. The Internet lets you connect to any machine in the world. Because of the relationship between your computer, the ISP, and the Internet, it is no longer appropriate to send individual characters. Instead, your modem routes Transmission Control Protocol/Internet Protocol (TCP/IP) packets between you and your ISP.

The standard technique for routing these packets through a modem is called Point-to-Point Protocol (PPP). The basic idea is simple: Your computer's TCP/IP stack forms its TCP/IP datagrams normally, but then it hands the datagrams to the modem for transmission. The ISP receives each datagram and routes it appropriately to the Internet. The same thing happens to get data from the ISP to your computer. The modem acts like a very slow network card; as far as your computer is concerned, it is on the Internet and is a full participant in all the Internet's features.

How GRAPHICS CARDS Work

The graphics card plays an important role in any personal computer (PC). It turns the digital information that the computer produces into something a human being can see and use. It then sends signals to the monitor so that you can actually view the information. Without a graphics card, the monitor would have nothing to display.

HSW Web Links

www.howstuffworks.com

How 3-D Graphics Work
How Computer Monitors Work
How Liquid Crystal Displays Work
How Random Access Memory Works
How Television Works

When you look at the screen of a typical PC, you see lots of different things. Windows that the operating system uses to display applications can contain text, drawings, and 3-D scenes. Around the windows there can be icons and menus.

If you look at the screen very closely, you can see that all the different things on the screen are made up of individual dots. These dots are called pixels, and each pixel has a color. On some screens (for example, on the original Macintosh), the pixels could have just two colors: black or white. On some screens today, a pixel can have one of 256 colors. On many screens, the pixels are full-color (also known as true color) and can have 16.7 million possible colors. Because the human eye can only discern about 10 million different colors, 16.7 million is more than enough.

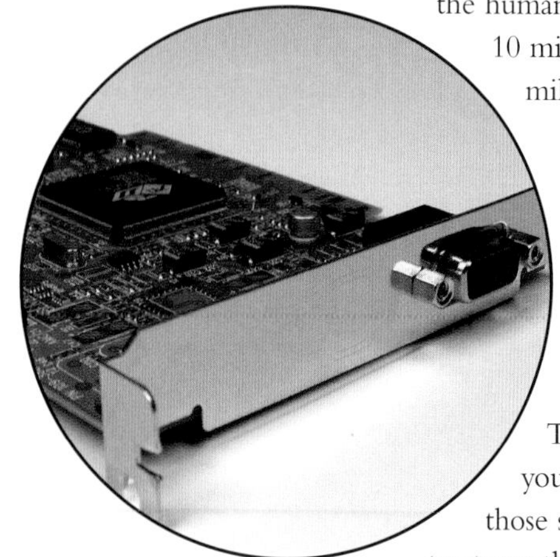

The goal of a graphics card is to create a set of signals that display the dots on the computer screen. (If you have read "How Computer Monitors Work," page 146, and "How Television Works," page 190, you have a good sense of what those signals are and how a monitor turns them into light.)

Simple Graphics Cards

You can understand the essence of a graphics card by looking at the simplest possible implementation. This card would be able to display only black or white pixels, and it would do that on a 640 X 480 pixel screen.

The first thing that a graphics card needs is memory. The memory holds the color of each pixel. In the simplest case, because each pixel is only black or white, you need just 1 bit to store each pixel's color. (See "How Bytes & Bits Work," page 112, for details.) Because 1 byte holds 8 bits, you need 640 ÷ 8 = 80 bytes to store the pixel colors for one line of pixels on the display. You need 480 X 8 = 3,480 bytes of memory to hold all the pixels visible on the display.

The second thing a graphics card needs is a way for the computer to change the graphics card's memory. This is normally done by connecting the graphics card to the card bus on the motherboard. The computer can send signals through the bus and change the memory.

The next thing that the graphics card needs is a way to generate the signals for the monitor. The card must generate color signals that drive the cathode ray tube electron beam as well as synchronization signals for horizontal and vertical sync. (See "How Television Works," page 190, for details.) Let's say that the screen is refreshing at 60 frames per second. This means that the graphics card scans the entire memory array 1 bit at a time, 60 times per second. It sends signals to the monitor for each pixel on each line, and then it sends a horizontal sync pulse. It does this repeatedly for all 480 lines, and then it sends a vertical sync pulse.

Color Graphics Cards

When a graphics card handles color, it does it in one of two ways. A true color card devotes 3 or 4 bytes per pixel (4 bytes allows an extra byte for an alpha channel). On a 1,600 X 1,200 pixel display, this adds up to about 8 million bytes of video memory. The other alternative is to use 1 byte per pixel and then use these bytes to index into a color look-up table (CLUT). The CLUT contains 256 entries, with 3 or 4 bytes per entry. The CLUT is loaded with the 256 true colors that the screen will display.

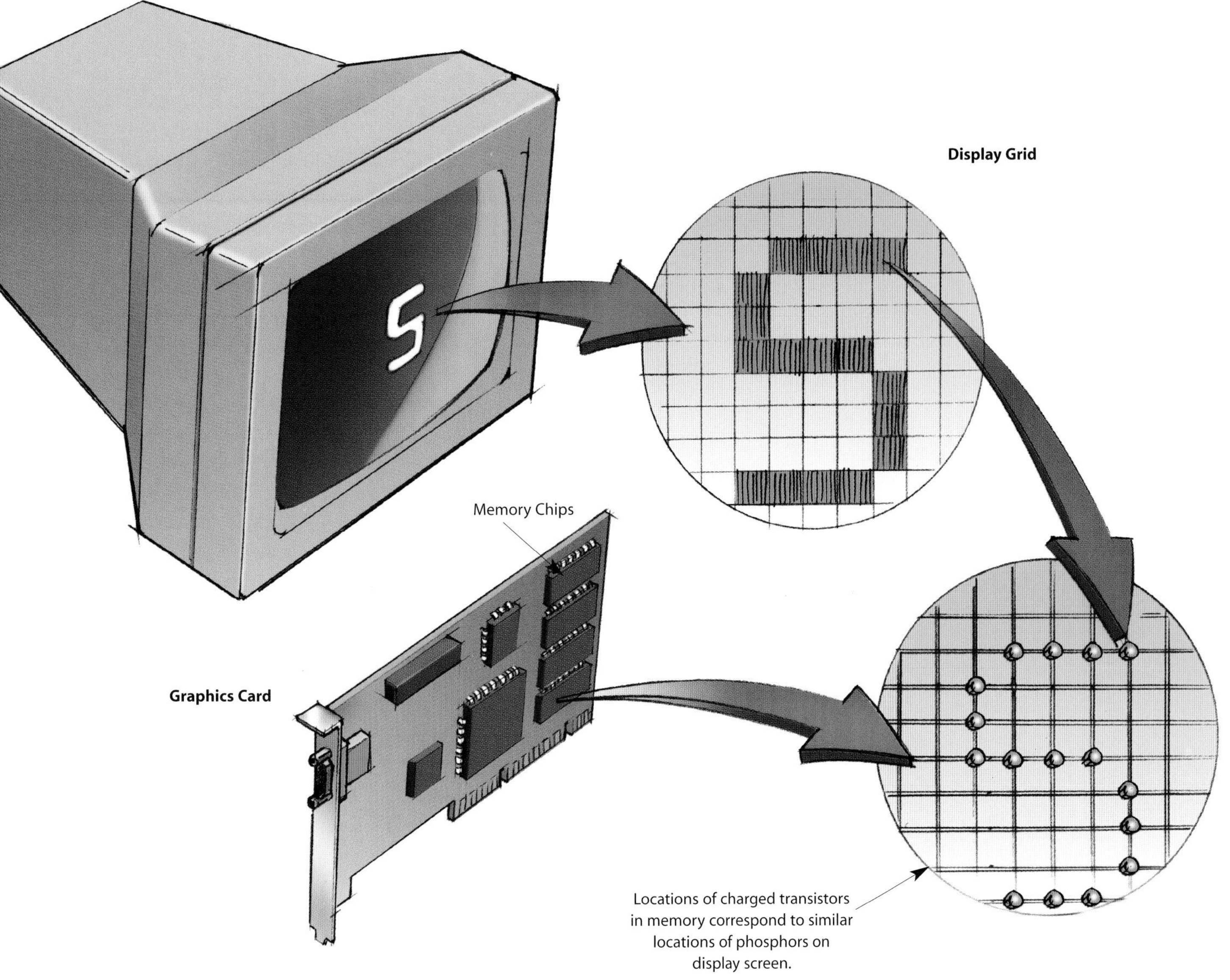

Graphics Coprocessors

A simple graphics card is called a frame buffer. The card holds a frame of information that is sent to the screen. The computer's microprocessor does all the work of changing every byte of video memory. The problem with frame buffers is that, on complex graphics operations, the microprocessor ends up spending all its time updating video memory, and it can't get any real work done. For example, if a 3-D image contains 10,000 polygons, the microprocessor has to draw and fill each polygon in the video memory, one pixel at a time.

Modern graphics cards have evolved to take this load off the microprocessor. A modern card contains its own high-power central processing unit (CPU) that is optimized for graphics operations. The computer's CPU tells the graphics coprocessor to draw a polygon with three specific vertices, and the graphics coprocessor does all the work of painting the pixels of the polygon into video memory. More and more complex graphics operations have moved to the graphics coprocessor, including shading, texturing, and antialiasing.

As graphics cards and coprocessors continue to evolve, the capabilities increase dramatically. Modern cards can draw millions of polygons per second which enables designers to create extremely realistic games and simulations.

How PARALLEL PORTS Work

If you have a printer connected to your computer, there is a very good chance that it uses a parallel port. Although universal serial bus (USB) ports are becoming increasingly popular, parallel ports are still the most used interface for printers and some other peripherals.

HSW Web Links

www.howstuffworks.com

How FireWire Works
How Microprocessors Work
How Modems Work
How USB Ports Work

Parallel ports are used to connect a host of popular computer peripherals, such as printers, scanners, CD writers, external hard drives, removable drives, network adapters, and tape backup drives. These devices use parallel ports because this is a relatively fast way to move data into and out of a computer without having to open the computer's case and install a new card in an input/output (I/O) slot. On newer computers and peripherals, this sort of connection is now handled by USB ports. (For more information on USB ports, see "How USB Ports Work," page 140.)

Parallel Port Connections

Parallel ports were originally developed by IBM as a way to connect a printer to a PC. When IBM was in the process of designing the PC, the company wanted the computer to work with printers offered by Centronics, a top printer manufacturer at the time. At that time Centronics used the same large and rather unwieldy connector on both ends of its printer cables. IBM decided not to use that connector on the computer side. Instead, IBM engineers coupled a 25-pin connector, a DB-25, with a 36-pin Centronics connector to create a special cable to connect the printer to the computer. Other printer manufacturers ended up adopting the Centronics interface, making this strange hybrid cable an unlikely de facto standard.

When a PC sends data to a printer or another device by using a parallel port, it sends 8 bits of data (1 byte) at a time. These 8 bits are transmitted parallel to each other, as opposed to the 1-bit-at-a-time approach used by serial ports. A standard parallel port is capable of sending 50 to 100 kilobytes (KB) of data per second.

Here's what each pin on a printer's parallel connector does:

- Pin 1 carries the strobe signal. It maintains a level of 5 volts but drops to 0 volts whenever the computer sends a byte of data. This drop in voltage tells the printer that data is being sent.
- Pins 2 through 9 carry the 8 data bits: 5 volts indicates a value of 1, and 0 volts indicates a value of 0. This is a simple but highly effective way to transmit digital information over an analog cable in real time.
- Pin 10 sends the acknowledge signal from the printer to the computer. Like Pin 1, it maintains 5 volts and drops to 0 volts to let the computer know that the data was received.
- If the printer is busy, it sends 5 volts on Pin 11, and then it drops the voltage to 0 volts to let the computer know it is ready to receive more data.
- The printer lets the computer know it is out of paper by sending 5 volts on Pin 12.
- As long as the computer is receiving 5 volts on Pin 13, it knows that the device is online.
- The computer sends an auto feed signal to the printer through Pin 14 by using a 5-volt signal.
- If the printer has any problems, it drops the voltage to 0 volts on Pin 15 to let the computer know that there is an error.
- Whenever a new print job is ready, the computer sends 0 volts on Pin 16 to initialize the printer.
- The computer uses Pin 17 to remotely take the printer offline. This is accomplished by sending 5 volts to the printer and maintaining it as long as the printer is supposed to be offline.
- Pins 18 to 25 are grounds.

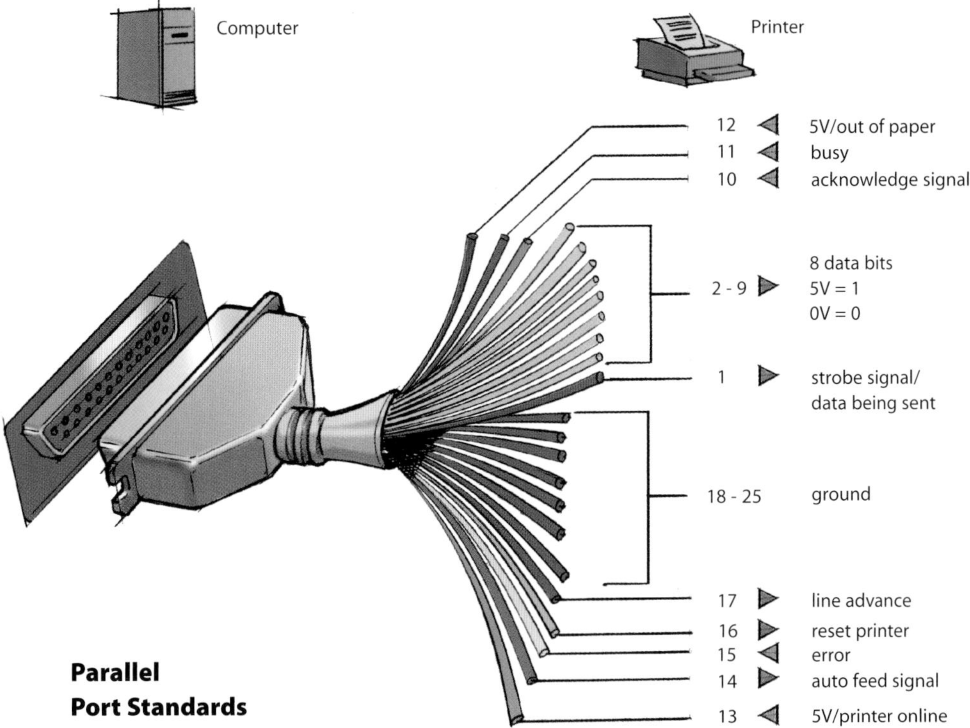

Parallel Port Standards

The original specification for parallel ports was unidirectional, meaning that data traveled in only one direction for each pin. With the introduction of the PS/2 in 1987, IBM offered a new bidirectional parallel port design, commonly known as Standard Parallel Port (SPP), which has completely replaced the original design. Bidirectional communication allows each device to receive data as well as transmit it. Many devices use the eight pins (Pins 2 through 9) that were originally designated for data. Using the same eight pins limits communication to half-duplex, meaning that information can travel in only one direction at a time. But pins 18 through 25, originally used as grounds, can also be used as data pins. This allows for full-duplex (both directions at the same time) communication.

Enhanced Parallel Port (EPP) was created by Intel, Xircom, and Zenith in 1991. EPP allows for much more data, 500 KB to 2 MB, to be transferred each second. It was targeted specifically toward non-printer devices that would attach to the parallel port—particularly storage devices that need the highest possible transfer rate.

Close on the heels of the introduction of EPP, Microsoft and Hewlett Packard jointly announced a specification called Extended Capabilities Port (ECP) in 1992. Whereas EPP was geared toward other devices, ECP was designed to provide improved speed and functionality for printers.

In 1994 the IEEE 1284 standard was released. It included the two specifications for parallel port devices, EPP and ECP. In order for them to work, both the operating system and the device must support the required specification. This is seldom a problem today because most computers sold support SPP, ECP, and EPP and are able to detect which mode needs to be used, depending on the attached device. If you need to manually select a mode, you can do so through the basic input/output system on most computers.

Parallel ports are an extremely good example of adaptation and evolution in the computer marketplace. A standard used by one manufacturer morphed into an industry wide standard used by all manufacturers. Then the standard morphed again to handle all sorts of non-printing activities. Finally, the standard was overshadowed by a new standard, USB ports, which are faster and cheaper. That kind of thing happens all the time in the fast-moving computer industry.

Serial Ports

Serial ports provide a standard connector and protocol to let you attach devices such as modems to a computer. When a PC sends data to a modem or other device by using a serial port, it sends 8 bits of data (1 byte) at a time. The name *serial* refers to the fact that a serial port serializes data. That is, it takes a byte of data and transmits the 8 bits in the byte one at a time. The advantage is that a serial port needs only one wire to transmit the 8 bits (whereas a parallel port needs eight wires). The disadvantage is that it takes eight times longer to transmit the data with one wire than it would if there were eight wires. Serial ports lower cable costs and make cables smaller.

Serial ports are bidirectional. They rely on a controller chip, the universal asynchronous receiver/transmitter (UART), to function properly. The UART chip transforms the parallel output of the computer's system bus into serial form for transmission through the serial port. In order to function faster, most UART chips have a built-in buffer that allows the chip to cache data coming in from the system bus while it is processing data going out to the serial port.

Serial ports, like parallel ports, are rapidly being replaced by USB ports. USB ports are, essentially, extremely high-speed serial ports that operate at about 12 million bits per second.

How USB PORTS Work

Just about any computer that you could buy for your home or office today comes with one or more universal serial bus (USB) connectors on the back. Essentially, USB uses a standard connector and protocol to let you attach everything from mice to printers to your computer quickly and easily.

HSW Web Links

www.howstuffworks.com

How Computer Mice Work
How FireWire Works
How Laptop Computers Work
How Microprocessors Work
How Modems Work
How Parallel Ports Work

The operating system supports USB so the installation of the device drivers is quick and easy. Compared to other ways of connecting devices to a computer (including using parallel ports, serial ports, and special cards that you install inside the computer's case), USB devices are incredibly simple. USB gives you a single, standardized, easy-to-use way to connect up to 127 devices to a computer. Each device can use up to a maximum of 6 megabits per second (Mbps) of bandwidth, which is fast enough for the vast majority of peripheral devices that most people want to connect to their machines. Most printers, scanners, digital cameras, joy sticks—almost any peripheral that you connect to a computer—now use USB ports to move data around.

Connecting a USB device to a computer is simple. You find the USB connector on the back of the machine and plug the USB connector into it. If it is a new device, the operating system autodetects it and asks for the driver disk. If the device has already been installed, the computer activates the device and starts talking to it. USB devices can be connected and disconnected at any time, whether or not the computer is running.

Behind the Scenes

You can think of USB as a small network that devices connect into to talk to the computer. The computer acts as the host for the network and controls all the devices. Up to 127 devices can connect to the host, either directly or by using USB hubs.

USB cables act as the wiring for the USB network. A USB cable has two wires for power (+5 volts and ground) and a twisted pair of wires to carry the data. On the power wires, the computer can supply up to 500 milliamps of power at 5 volts. Hubs can supply additional power. Small, low-power devices such as mice get their power from the USB cable to keep the device simple. High-power devices such as printers have their own power supplies and draw minimal power from the bus.

Individual USB cables can run as long as 16.4 feet (5 m), and with hub devices can stretch as far as 98.4 feet (30 m—using six cables) from the host.

The USB has a maximum data rate of 12 Mbps. That is, all the devices connected to one computer can transmit and receive at most 12 Mbps of data. Any individual device can request as much as 6 Mbps. Obviously, you can't really have more than one device requesting 6 Mbps, or you would exceed the 12-Mbps maximum for the bus.

USB devices are hot-swappable, meaning you can plug them into the bus and unplug them any time. The host computer can put many USB devices to sleep when it enters a power-saving mode. When the host computer powers up, it queries all the devices connected to the bus and assigns each one an address. This process is called enumeration. Devices are also enumerated when they connect to the bus. At this point the host also finds out from each device what type of data transfers it wants to perform:

- **Interrupt**—A device such as a mouse or a keyboard, which will be sending very little data, would choose the interrupt mode.
- **Bulk**—A device such as a printer, which receives data in one big packet (for example, per page), uses the bulk transfer mode. A block of data is sent to the printer and verified to make sure it is correct.
- **Isochronous**—A streaming device such as speakers uses isochronous mode. Data moves between the device and the host

in real time, and there is no error correction. The key is to keep the data moving at a steady rate so that the device never runs out of data.

The host can also send commands or query parameters to devices by using control packets. As devices are enumerated, the host is keeping track of the total bandwidth that all the isochronous and interrupt devices are requesting. They can consume up to 90% of the 12 Mbps of bandwidth that is available. After 90% is used up, the host denies access to any other isochronous or interrupt devices. Control packets and packets for bulk transfers use any bandwidth that is left over (at least 10%).

The USB divides the available bandwidth into frames, and the host controls the frames. A frame contains 1,500 bytes, and a new frame starts every millisecond. During a frame, isochronous and interrupt devices get a slot so that they are guaranteed the bandwidth they need. Bulk and control transfers use whatever space is left.

Running Out of Ports

Most computers that you buy today come with one or two USB sockets. With so many USB devices on the market today, you can run out of sockets very quickly. For example, a computer might have a USB printer, a USB scanner, a USB webcam, and a USB network connection. If the computer has only one USB connector on it, how do you hook up all the devices?

The easy solution to the problem is to use a USB hub. The USB standard supports up to 127 devices, and USB hubs are a part of the standard. A hub normally supports four or eight new USB ports, but some have many more. You plug the hub into your computer, and then you plug your devices (or other hubs) into the hub. By chaining hubs together, you can build up dozens of available USB ports on a single computer.

Hubs can be powered or unpowered. Obviously, a high-power device such as a printer or scanner will have its own power supply, but low-power devices such as mice and digital cameras get their power from the USB cable so that they can be as simple as possible. A powered hub makes sure there is enough power available for lots of USB devices.

USB ports have made it easy to connect almost any device to a computer. It's amazing to go behind the scenes and find out that all the devices are communicating with the host computer on a small private network!

Comparing Speeds

A USB port can transfer up to 12 megabits per second (Mbps), and any individual device can use a maximum of 6 megabits per second. Here are some other common port speeds for comparison:

- **Serial ports**—Up to 115 Kbps
- **USB 1.0**—Up to 6 Mbps per device
- **Parallel ports**—8 Mbps
- **Ethernet card**—10 Mbps
- **USB 2.0**—Proposed over 100 Mbps
- **Fast Ethernet**—100 Mbps
- **FireWire**—Up to 400 Mbps
- **IDE connection**—500 Mbps
- **AGP graphics bus**—Up to 8 Gbps

With a typical USB device like a mouse or a digital camera, USB has plenty of speed. A fast hard disk can stream something like 80 Mbps, so it needs something like an IDE connection to handle its data.

How **COMPUTER KEYBOARDS** Work

The part of the computer that we come into most contact with is probably the piece that we think about the least. But the keyboard is an essential piece of technology. For instance, the keyboard on a typical computer system is actually its own little standalone computer.

HSW Web Links

www.howstuffworks.com

How Computer Mice Work
How Laptop Computers Work
How Microprocessors Work
How Operating Systems Work
How USB (Universal Serial Bus) Ports Work

At its simplest, a keyboard is a series of switches connected to a microprocessor that monitors the state of each switch and initiates a specific response to a change in that state. The keyboard communicates with the host computer every time a key is pressed.

Keyboards have changed very little in layout since their introduction on manual typewriters many years ago. In fact, the most common change has simply been the natural evolution of adding more keys that provide additional functionality. A typical keyboard has four basic types of keys:

- Character keys
- Numeric keypad keys
- Function keys
- Control keys

The character keys are the section of the keyboard that contains the letters, generally laid out in the same style that was common for typewriters. This layout, known as QWERTY (for the first six letters in the layout), was originally designed to slow down fast typists by making the arrangement of the keys somewhat awkward.

The numeric keypad is a part of the natural evolution mentioned previously. As the use of computers in business environments increased, so did the need for data entry. Because a large part of the data was numbers, a set of 17 keys was added to the keyboard. These keys are laid out in the same configuration used by most adding machines and calculators.

The function keys were added because of their use on mainframe keyboards. Most software today does not really use the function keys that much.

Inside the Keyboard

Different keyboards use different designs to detect when someone presses a key. The simplest technique is ridiculously simple. Imagine two wires with bare ends. Imagine that the two bare ends are close to each other but do not touch. Now imagine that on the bottom of the key is a piece of metal that connects the two bare ends. That is approximately how the simplest keyboards work.

The key matrix is the grid of circuits underneath the keys. Pressing a key bridges the gap in the circuit, allowing current to flow through. The keyboard's processor monitors the key matrix for signs of current at any point on the grid. When it finds a circuit that is closed, it compares the location of that circuit on the key matrix to the character map in its read-only memory (ROM). The character map is basically a comparison chart for the processor that tells it what the key at each *x,y* coordinate in the key matrix represents.

One of the big reasons for the separate microprocessor in the keyboard is the problem of bounce. When you connect two bare wires together, dozens of connections and disconnections actually occur in a fraction of a second as the two wires press firmly together. The microprocessor has to look at all the interference and interpret all that bouncing as a single keystroke.

If more than one key is pressed at the same time (for example, Alt and Tab together), the keyboard processor has to understand that as well. It checks to see if that combination of keys has a designation in the character map.

A different character map provided by the computer can override the character map in the keyboard. This allows you, for

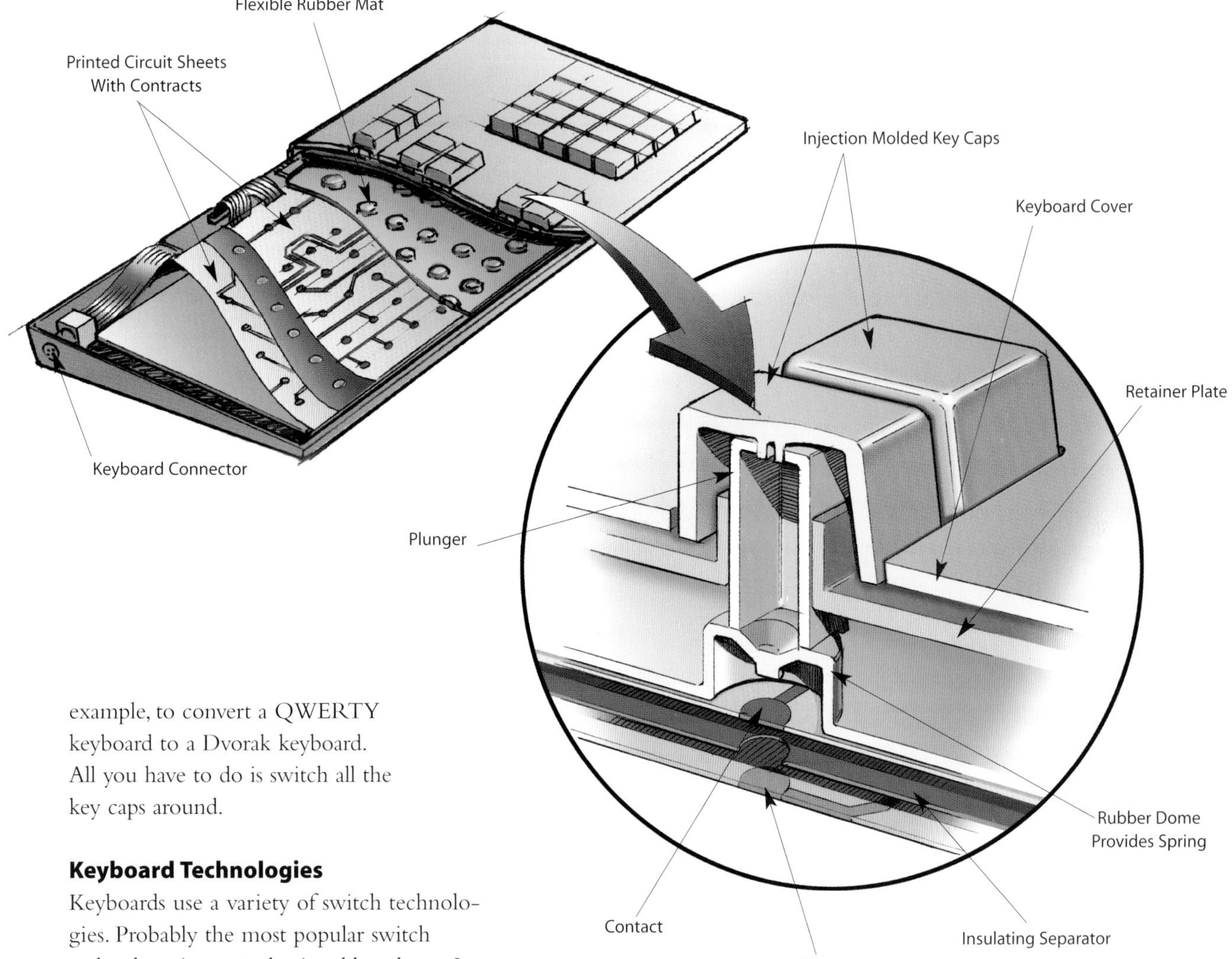

example, to convert a QWERTY keyboard to a Dvorak keyboard. All you have to do is switch all the key caps around.

Keyboard Technologies

Keyboards use a variety of switch technologies. Probably the most popular switch technology in use today is rubber dome. In rubber dome keyboards, each key sits over a small, flexible rubber dome with a hard carbon center. When a key is pressed, a plunger on the bottom of the key pushes down against the dome. This causes the carbon center to push down also, until it presses against a hard, flat surface beneath the key matrix. As long as the key is held, the carbon center completes the circuit for that portion of the matrix. When the key is released, the rubber dome springs back to its original shape, forcing the key back up to its at-rest position.

Basically every idea, literary passage, email message, and document that any human being has thought and recorded in computerized form has come into the digital realm through a keyboard. The keyboard is a simple yet powerful device.

Why QWERTY?

The reason typewriter manufacturers began using the QWERTY layout was because of design problems in early typewriters. The mechanical arms that imprinted each character on the paper could jam together if the typist pressed the keys too rapidly. Because the QWERTY configuration has been long established as a standard and people have become accustomed to it, manufacturers have developed keyboards for computers using the same layout, even though jamming is no longer an issue. Critics of the QWERTY layout have adopted another layout, Dvorak, that places the most commonly used letters in the most convenient arrangement.

How **COMPUTER MICE** Work

Every day of your computing life, you reach for your mouse whenever you want to move the cursor or click on something. The mouse senses your motion and your clicks, and it sends the motions to the computer so that it can respond appropriately.

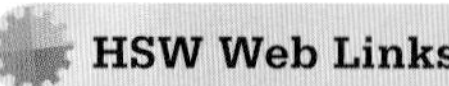

HSW Web Links

www.howstuffworks.com

How Ethernet Works
How Hard Disks Work
How Microprocessors Work
How Modems Work
How USB Ports Work

A mouse converts the motion of your hand into a form that your computer can use to control the cursor's position. In most mice, a small rubber ball on the bottom captures the motion. Many new mice use optical laser technology, and all sorts of other input devices accomplish the same goal, but the ball-based optomechanical mouse is still the most popular pointing device.

The Evolution of the Mouse

The mouse is simple and effective, but took a while to become a part of everyday life. In the beginning there was no need to point because computers—for example, Teletype machines and punch cards for data entry—used crude interfaces. Early text terminals did nothing more than emulate a Teletype machine (using the screen to replace paper), so it was many years (well into the 1960s and early 1970s) before arrow keys were found on most terminals. Full screen editors first took advantage of the cursor keys, and they offered people the first crude way to point.

Three technologies were competing with mice in the 1970s and early 1980s: the light pen, the touch screen, and the graphics tablet. Light pens work well, but the user has to hold it up to the screen to use it, and this gets uncomfortable after an hour or two. Touch screens don't have enough precision because our fingers are too big. And graphics tablets never really caught on.

When Apple introduced the Macintosh with an attached mouse, it was an immediate success. Something about a mouse device is completely natural—it is precise, inexpensive, and comfortable, allowing your hand and arm to rest on the desk all day.

Optomechanical Translation

The main goal of a mouse is to translate the motion of your hand into signals that the computer can use. Almost all mice today do the translation as follows:

1) A ball inside the mouse touches the desktop and rolls when the mouse moves.
2) Two rollers inside the mouse touch the ball. One of the rollers is oriented so that it detects motion in the *x* direction, and the other is oriented 90 degrees to the first roller so that it detects motion in the *y* direction. When the ball rolls, one or both of these rollers roll as well.
3) Each roller connects to a shaft, and the shaft spins a disk with holes in it. When a roller rolls, its shaft and disk spin.
4) On either side of the disk is an infrared light-emitting diode (LED) and an infrared sensor. The holes in the disk break the beam of light coming from the LED so that the infrared sensor sees pulses of light. The rate of the pulsing is directly related to the speed of the mouse.
5) An on-board processor chip inside the mouse reads the pulses from the infrared sensors and turns them into binary data that the computer can understand. The chip sends the binary data to the computer through the mouse's cord.

Most of the mice on personal computers use this optomechanical arrangement—the ball and disks move mechanically, and an optical system counts pulses of light.

Each encoder disk has two infrared LEDs and two infrared sensors, one on each side of the disk. This arrangement allows the processor to detect the disk's direction of rotation. There is a piece of plastic with a small, precisely located hole that sits between the encoder disk and each infrared sensor. This piece of plastic provides a window through which the infrared sensor can "see." The window on one side of the disk is located slightly higher than it is on the other; it is one-half the height of one of the holes in the encoder disk, to be exact. That difference causes the two infrared sensors to see pulses of light at slightly different times. At times, one of the sensors sees a pulse of light and the other does not. The mouse's processor chip decodes these differences to figure out the direction of rotation.

Mouse Data Interfaces

Mice typically have either PS/2- or USB-type connectors. (For more information on USB, see "How USB Ports Work," page 140.) Whenever the mouse moves or the user clicks a button, the mouse sends 3 bytes of data to the computer. The first byte's 8 bits contain the following:

Byte	Contents
1	Left button state (0 = off, 1 = on)
2	Right button state (0 = off, 1 = on)
3	0
4	1
5	x direction (positive or negative)
6	y direction
7	x overflow, if the mouse moved more than 255 pulses in the x direction in 1/40 second
8	y overflow

The next 2 bytes contain the *x* and *y* movement values, respectively. These two bytes contain the number of pulses detected in the *x* and *y* direction since the last packet was sent.

With a PS/2 mouse, there are four wires in the cable for +5 volts, ground, data, and a clock signal. The data is sent from the mouse to the computer serially on the data line, with the clock line pulsing to tell the computer where each bit starts and stops. Eleven bits are sent for each byte (1 start bit, 8 data bits, 1 parity bit, and 1 stop bit). The PS/2 mouse sends on the order of 1,200 bits per second, which allows it to report mouse position to the computer at a maximum rate of about 40 reports per second.

The mouse has completely changed the face of computing, enabling computers to go from text-based to completely graphical computing. Operating systems have gone from command-based to menu-driven point-and-click systems. Everything about the computer has gotten friendlier in the process. Not bad for a little peripheral that costs about $10!

How COMPUTER MONITORS Work

Computer monitors are so important that, without a monitor, a typical desktop computer is useless. The monitor provides you with instant feedback by showing text and graphic images as you work on or play with your computer.

HSW Web Links

www.howstuffworks.com

How Computer Mice Work
How Digital Television Works
How Hard Disks Work
How LCDs Work
How Modems Work
How Surge Protectors Work
How Televisions Work

Most desktop displays use a cathode ray tube (CRT), and portable computing devices such as laptops and personal digital assistants (PDAs) use a liquid crystal display (LCD).

CRTs

A CRT computer monitor is something like a TV set, but there are two big differences:

- A monitor normally provides much higher resolution than a TV. There are more pixels on a computer screen than on an analog TV screen. The best computer monitors have up to 10 times the resolution of a TV.
- Because of their higher resolution, computer monitors use a special cable to connect the display. This cable separates all the video signals onto separate wires to avoid interference. A normal analog TV set gets all its signals from a composite video signal on a single wire. You can see how the signals sent to a computer monitor are separated by looking at the pinout of a VGA cable, where the numbers have the following meanings:

Pinout	Description
1	Red out
2	Green out
3	Blue out
4	Monitor ID 2 in
5	Ground
6	Red return
7	Green return
8	Blue return
9	Optional +5V output from graphics adapter
10	Sync return
11	Monitor ID 0 in
12	Monitor ID 1 in or data from display
13	Horizontal sync out
14	Vertical sync out (sometimes used as data clock, too)
15	Monitor ID 3 in or data clock

The red, green, and blue guns in the CRT are all controlled separately, and so are the horizontal and vertical sync signals.

A number of factors contribute to the display quality of a CRT, including the following:

- **Maximum resolution**—Resolution refers to the number of individual dots of color, known as pixels, that are visible on a display. Resolution is typically expressed by identifying the number of pixels on the horizontal axis (rows) and the number on the vertical axis (columns), such as 800 x 600 pixels. The monitor's dot pitch, viewable area, and refresh rate all directly affect the maximum resolution a monitor can display.
- **Dot pitch**—Dot pitch is the measure of how much space there is between a display's pixels. When considering dot pitch, remember that smaller is better. Packing the pixels closer together is fundamental to achieving higher resolutions. A display can normally support resolutions that match the physical dot (pixel) size as well as several lesser resolutions. For example, a display with a physical grid of 1,280 rows by 1,024 columns can obviously support a maximum resolution of 1280 x 1024 pixels. It usually also supports lower resolutions such as 1024 x 768, 800 x 600, and 640 x 480.
- **Refresh rate**—In monitors based on CRT technology, the refresh rate is the number of times that the image on the display is drawn each second. If a CRT monitor has a refresh rate of 72 Hz, then it cycles through all the pixels from top to bottom 72 times per second. Refresh rates are very important because they control flicker, and you want the refresh rate to be as high as possible. With too few cycles per second, you will notice a flickering, which can lead to headaches

and eyestrain. The fact that the speed of a monitor's refresh rate depends on the number of rows it has to scan limits the maximum possible resolution. A lot of monitors support multiple refresh rates, usually dependent on the level of resolution that is selected.

- **Display size**—Two measures describe the size of a display: aspect ratio and screen size. Most computer displays today, like most televisions, have an aspect ratio of 4:3. This means that the ratio of the width of the display screen to the height is 4 to 3. The other aspect ratio in common use is 16:9. The display includes a projection surface, commonly referred to as the screen. Screen sizes are normally measured in inches or centimeters from one corner to the corner diagonally across from it. This diagonal measuring system came about because the early television manufacturers wanted to make the screen sizes of their TVs sound more impressive. 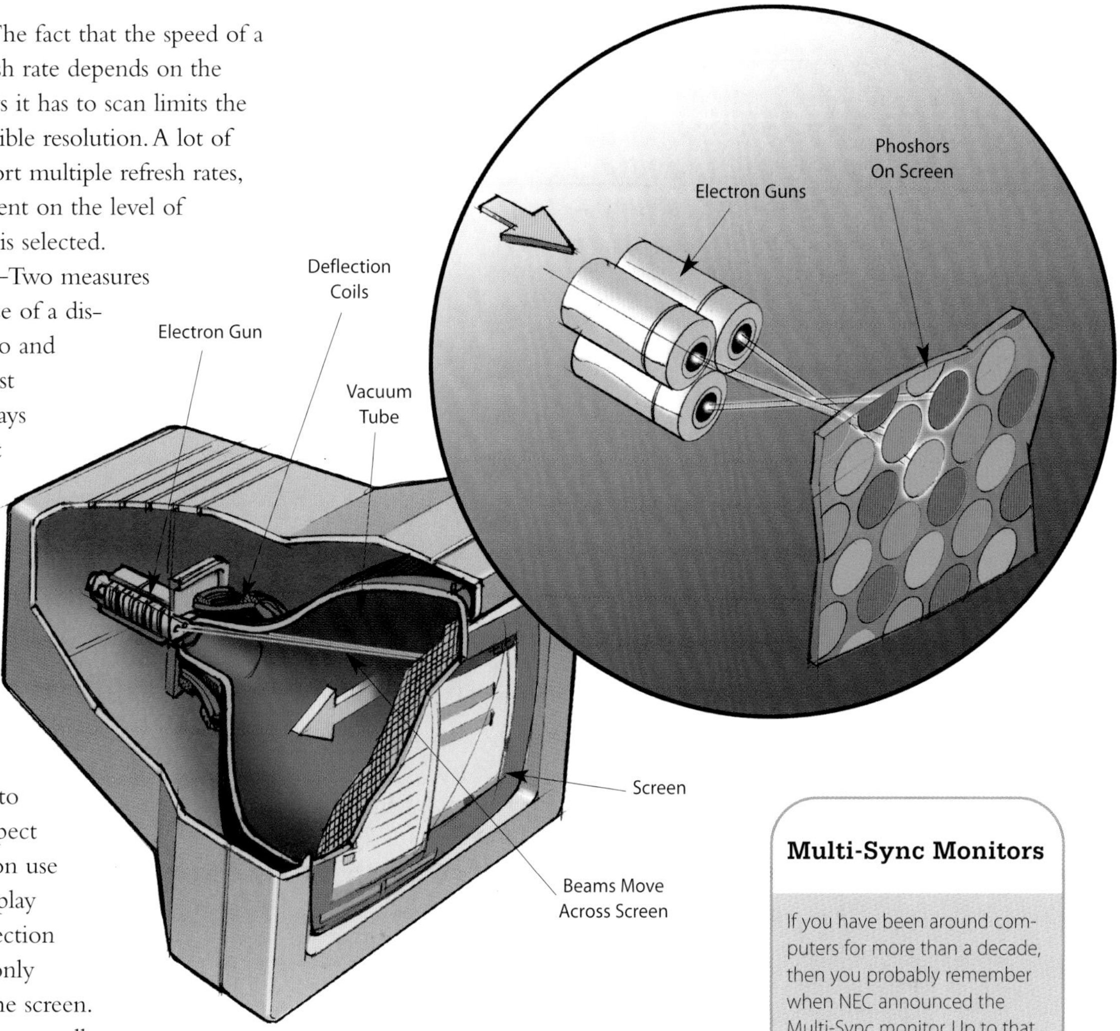
- **Color depth**—Color bit depth refers to the number of bits used to describe the color of a single pixel. The bit depth determines the number of colors that can be displayed at one time. The color depth is more a function of the graphics card than of the monitor. Nearly every monitor sold today can handle 24-bit color using a standard VGA connector.

Other Screen Technologies

CRT technology is the most prevalent system in desktop displays. Because standard CRT technology requires a certain distance between the beam projection device and the screen, monitors that use this type of display technology tend to be bulky. Other technologies make it possible to have much thinner displays, commonly known as flat-panel displays.

LCD technology works by blocking light rather than creating it, and light-emitting diode (LED) and gas plasma work by lighting up display screen positions based on the voltages at different grid intersections. LCDs require far less energy than active technologies such as CRT, LED, and gas plasma, and they are currently the primary technology for notebook and other mobile computers. As flat-panel displays continue to grow in screen size and improve in resolution and affordability, they will gradually replace CRT-based displays because they are so much smaller and lighter.

Multi-Sync Monitors

If you have been around computers for more than a decade, then you probably remember when NEC announced the Multi-Sync monitor. Up to that point, most monitors understood only one frequency, which meant that the monitor operated at a single fixed resolution and refresh rate. You had to match your monitor with a graphics adapter that provided that exact signal, or it wouldn't work.

The introduction of NEC Multi-Sync technology started a trend toward multiscanning monitors. This technology allows a monitor to understand any frequency within a certain bandwidth that is sent to it.

The benefit of a multiscanning monitor is that you can change resolutions and refresh rates without having to purchase and install a new graphics adapter or monitor each time. Because of the obvious advantage of this approach, nearly every monitor you buy today is a multiscanning monitor.

How LASER PRINTERS Work

The laser printer is one of those inventions that "changed everything," by making desktop publishing possible. A laser printer enables anyone to inexpensively create documents—books, brochures, magazines, and so on—that look as good as those created by the largest publishing companies in the world. In a sense, laser printers have given "freedom of the press" to everyone.

HSW Web Links

www.howstuffworks.com

How 3-D Graphics Work
How Centropolis FX Creates Visual Effects
How Lasers Work
How MRI Works
How Offset Printing Works

Did You Know?

Before it can print anything, the printer has to receive and organize the image data. The host computer—your desktop PC, for example—uses a page description language (PDL) to send image information to the printer's on-board computer. PDLs typically describe pages in vector form—that is, as mathematical values of geometric shapes rather than as a series of dots (called a bitmap image).

The printer itself converts this vector image into an array of tiny dots. Using a rotating mirror and a series of lenses, the laser scans these dots onto the drum assembly as short bursts of light. The dots are scanned one horizontal line at a time as the drum revolves vertically. The system is so precise that these dots come together seamlessly as clear, sharp lines. High-end laser printers can resolve thousands of dots per inch.

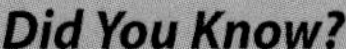

The most important principle at work in a laser printer is static electricity, the same energy that makes clothes in the dryer cling together. Static electricity is an electrical charge built up on an insulated object.

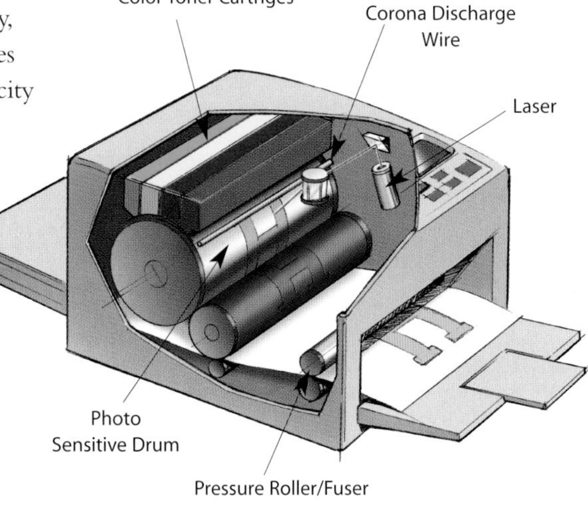

The Drum

A laser printer uses static cling as a sort of temporary glue. The core component of this system is a revolving drum or cylinder. This drum is made out of photoconductive material, which means it conducts electricity when light strikes it.

Initially, the drum is given a total positive charge. Some printers create the charge by using a corona wire, and other printers use charged rollers. As the charged drum revolves, the printer scans a laser beam across the surface. Each point the laser touches loses its charge; that is, the laser light causes the photoconductor to conduct electricity and drains the charge. In this way, the laser draws the page as an electrostatic image on the drum.

After the pattern is set, the printer coats the drum with a fine black powder of positively charged toner. Because it has a positive charge, the toner clings to the discharged areas of the drum, but not to the positively charged "background." This is something like writing on a soda can with glue, and then rolling it over some flour: The flour sticks to only the glue-coated part of the can, so you end up with a message written in powder.

With the powder pattern affixed, the drum rolls over a sheet of paper that is moving along a belt below. Before the paper rolls under the drum, it is given a strong negative charge. This charge is stronger than the negative charge of the electrostatic image, so the paper draws the toner away. Because it is moving at the same speed as the drum, the paper picks up the image pattern exactly. To keep the paper from clinging to the drum, the paper is discharged immediately after grabbing the toner.

The Fuser and Discharge Lamp

Finally, the printer passes the paper through the fuser, a pair of heated rollers that melts the toner, binding it to the paper fibers. The paper then passes on to the output tray, and the drum surface passes the discharge lamp, which is a bright light that exposes the photoreceptor covering. The discharge lamp erases the electrical image before the charged corona wire reapplies the positive charge, starting the process all over again.

Conceptually, a laser printer is very simple. Of course the actual implementation is a major engineering project, as you discover very quickly if you ever have to dislodge a laser printer paper jam.

chapter six

UNWEAVING THE WEB

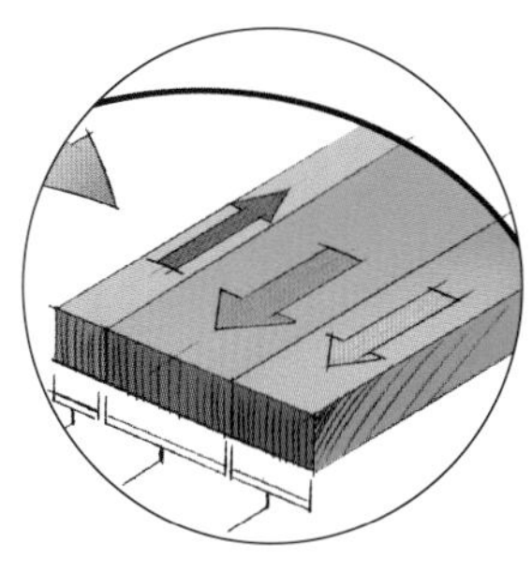

How **WEB PAGES** Work

If you think of the Internet as a landscape, you can think of Web pages as the billions of buildings that cover that land scape. When people are viewing the Internet, they are essentially traveling to different Web pages. People find some of these Web pages by typing a specific address directly into a browser, and others connect to pages by using links.

HSW Web Links

www.howstuffworks.com

- How CGI Scripting Works
- How Domain Name Servers Work
- How Internet Search Engines Work
- How the Internet Works
- How LAN Switches Work
- How Web Pages Work
- How Web Servers and the Internet Work

Publishing Web pages is not all that different from publishing other forms of printed media. First, you write text into a basic word processing program. You then format that text with a language called Hypertext Markup Language (HTML). HTML lets you control how the text and images will be displayed on a page. You then publish Web pages to a Web server. A user opens a Web browser to connect to Web servers that retrieve your page and format it onto the screen. The browser recognizes the HTML tags and displays your Web page accordingly.

Using HTML or HTML Coding

You can create Web pages without HTML, but these pages display only black-and-white text. In order to create pages that have colors, different fonts, headlines, and graphics, you must use HTML.

HTML tags are used to format text for publication to the Web. Although it looks rather intimidating, HTML is surprisingly simple to use and learn.

An HTML tag is a code element that tells the Web browser what to do with a piece of text. Each tag is typed as letters or words between a < (less-than sign) and a > (greater-than sign). For example, to center a piece of text, the author types <center> right before the text that should be centered. At the end of the centered text, the author types the </center> tag. The browser recognizes these tags and centers the text between <center> and </center>.

As shown in the center tag example, most tags come in pairs: one to begin a formatting function and one to end it. For example, tags can be used to apply <b>bold</b>, <i>italics</i>, and <u>underline</u> to words. Other tags are used to put breaks between lines of text or place spaces between paragraphs. HTML tags also allow you to change the color, size and type of the font used on a Web page.

Some more complicated uses of HTML involve tables, which arrange data in rows and columns; links, which let you point a user to another page; and embedded graphics, which add color and life to a page.

You do not have to learn all the HTML tags to create eye-catching Web pages. In most word processing programs, you can simply format a page by using the word processing programs tools, and then select an option such as "Save as HTML," and the program will automatically apply the needed HTML tags to your page.

Publishing a Web Page

After you create a Web page, you need to find a Web server to put it on so that the entire world of Internet users can view it. Many services and institutions are equipped with Web servers, and some of them offer free space for Web authors to publish their pages. Many colleges offer space to students, and many Internet service providers give each subscriber a space for a Web page.

Designers can also place Web pages on the Web through a hosting service. Most of the professional sites on the Web use this type of service. Hosting services provide their clients with disk space for the pages on a reliable machine. Hosting services also give clients the option of owning a specific domain name—a unique URL. Most hosting services register custom domain names for their clients.

The Internet has enabled anyone with a computer and a phone line to cheaply and quickly spread their messages to millions. Not since Johann Gutenberg invented the printing press has a technology so revolutionized the way people distribute written material.

Cool Facts

As of 2001, there are approximately 4 billion Web pages on the Internet.

More than 7 million Web pages are added to the Internet every day.

The United States creates 84% of the Internet's Web pages.

Computer scientist Tim Berners-Lee invented HTML and the World Wide Web in 1989, while at the European Laboratory for Particle Physics (CERN).

How E-COMMERCE Works

If you've ever purchased something on the Web, you have direct, personal experience with e-commerce. Still, you may feel that you don't understand e-commerce at all. But e-commerce is actually not very different from regular commerce.

If you understand commerce, then e-commerce is an easy extension. Commerce is the exchange of goods and services, usually for money. It is the backbone of any economy. We see commerce all around us, in millions of different forms.

When you buy something at a grocery store or a discount store, you are participating in commerce. You are getting goods in exchange for money.

The Roles in Commerce

When you think about commerce, you instinctively recognize several different roles:

- **Buyers**—These are people with money who want to purchase things.
- **Sellers**—These are people who offer goods and services to buyers. Sellers are generally recognized in two different forms: retailers, who sell directly to consumers, and wholesalers or distributors who sell to retailers and other businesses. Sellers—retailers and wholesalers—generally do not produce anything. They buy what they sell from producers.
- **Producers**—These are the people who create the products and services that sellers offer to buyers. A producer is always, by necessity, a seller as well as a producer. The producer sells the products or services to wholesalers, other businesses, retailers, or consumers.

The Elements of Commerce

Imagine that you would like to become a seller—in this case a retailer—and sell something to a customer. Here are some of the things you need to think about:

- If you would like to sell something to a customer, you must have a product or service to offer. A product can be anything from ball bearings to xylophones, and a service can be anything from a back rub to lawn maintenance.
- You must also have a place from which to sell your products. That place can sometimes be very ephemeral. For example, a phone number might be the place. If you are a customer in need of a back rub, you call Judy's Backrubs, Inc. on the telephone to order a back rub. If Judy shows up at your office to give you a backrub, then the phone number is the place where you purchased this service.
- You need to figure out a way to get people to come to your place—whether it is a physical store, a phone number, or a Web site. This process is known as marketing. If no one knows that your place exists, you will never sell anything.
- You need a way to accept orders. At most retail stores, the checkout line handles this. In a mail-order company, the orders come in by mail or phone, or through a Web site, and they are processed by employees of the company.
- You also need a way to accept money. If you are at a grocery store, you know that you can use cash, check, or credit cards to pay for products. Business-to-business transactions, such as a company buying from an office supply store, often use purchase orders. A purchase order is an agreement to pay when you receive an invoice. If, as a seller, you accept purchase orders, it means you may also have to think about the whole area of billing and collections.
- You need a way to deliver the product or service. This is known as fulfillment. At a grocery store, the customer takes the products at the time of purchase. In a mail-order company, the products are mailed.
- Sometimes customers do not like what they buy, so you need a way to accept returns.
- Sometimes a product breaks, so you need a way to honor warranty claims.

HSW Web Links

www.howstuffworks.com

How Affiliate Programs Work
How Banner Ads Work
How CGI Scripting Works
How Domain Name Servers Work
How Web Pages Work

- Many products are so complicated that a company must have customer service and technical support departments to help customers use the products.

You find all these elements in any traditional mail-order company. Whether the company is selling books, consumer products, services, or information in the form of reports and papers, all these elements come into play.

The Elements of E-commerce

In an e-commerce company, you find all the same elements as in regular commerce, but they change slightly so that they fit into a Web model. You must have the following elements to conduct e-commerce:

- A product.
- A place to sell the product; in the case of e-commerce, a Web site displays the products in some way.
- A way to get people to come to your Web site.
- A way to accept orders, normally through an online form of some sort.
- A way to accept money, normally through a merchant account that handles credit card payments. This requires a secure ordering page and a connection to a bank. Or you might use more traditional billing techniques either online or through the mail.
- A fulfillment facility to ship products to customers. You may be able to hire someone else to do this for you. In the case of software and information, fulfillment can occur over the Web through a file download mechanism.
- A way to accept returns.
- A way to handle warrantee claims, if necessary.
- A way to provide customer service (often through email, online forms, online knowledge bases and FAQs, and so on).

Challenges of E-commerce

Several things about e-commerce make it challenging and interesting. There are so many Web sites, and it is so easy to create a new e-commerce Web site on the Internet today, that getting people to look at yours is the biggest problem. Getting them to come back and actually buy from you is the next problem. Imagine that you have opened an e-commerce site, you have chosen your products, you have the products sitting in a warehouse, and you have your Web site up and displaying the products clearly.

...and shipped

...packaged...

ordered is processed...

The first, and possibly the most interesting, challenge you face is figuring out how to get people to visit your site. Generally, you do this by advertising, but advertising tends to be very expensive. After you get a person to visit once, therefore, it becomes important to get the visitor to come back a second time without so much expense. Techniques like newsletters, frequently changing and interesting content, word-of-mouth techniques, and email are all possibilities.

Differentiating yourself from the competition is important in e-commerce. When people come to visit and look at your products on the Internet, they know that other places are likely selling the same products. Price, customer service, and selection are all important to keeping customers on your site and getting them to come back.

Finally, getting people to buy something from your Web site is a challenge. Having

Percentages

At the height of the e-commerce phenomenon, hundreds of large Web sites began selling every imaginable product over the Internet. Sites like furniture.com, hardware.com, and pets.com were born, funded by hundreds of millions of dollars in venture capital. The large majority of these sites later folded.

One reason for many of the failures was an inability to attract visitors efficiently. For example, if a company spends $50 million on advertising and only 100,000 people come to visit, then it is very hard to make a profit.

Another reason has to do with percentages. If 1 million people come to visit, and out of the million only 1 thousand look at products, and out of a thousand only one or two buy anything, it is again difficult to make a profit. The key is to be able to attract lots of people inexpensively and for a good percentage of them to buy something.

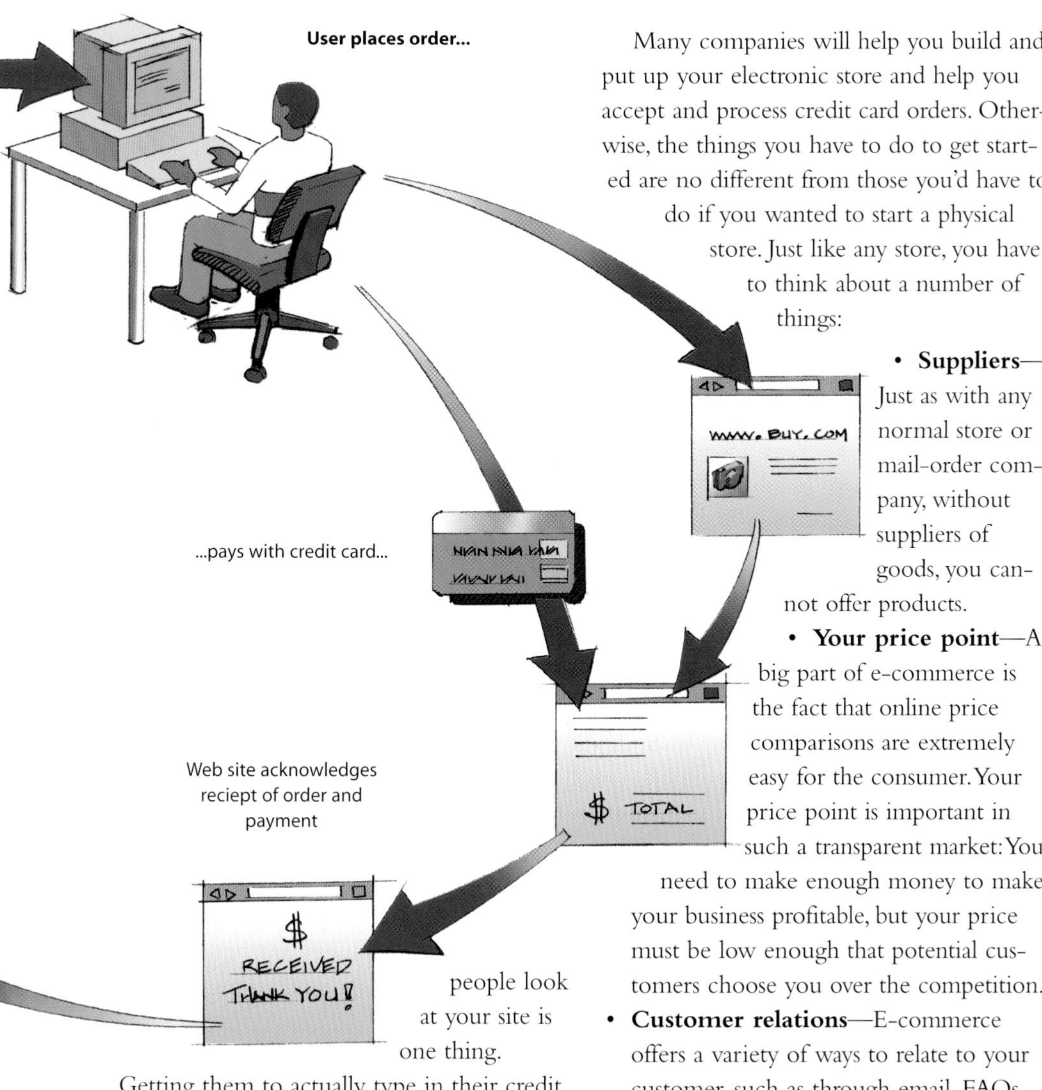

people look at your site is one thing. Getting them to actually type in their credit card numbers is another.

Building an E-commerce Site

You see two different flavors of e-commerce on the Web. There is massive-scale e-commerce through places like Amazon.com and Walmart.com. These Web sites stock hundreds of thousands of products and ship to millions of customers every year. Then there are much more modest efforts, created by individuals and small companies.

If you have ever thought about producing or selling products, e-commerce is one way to get started. The startup costs are low because you do not have to lease a building or hire staff. The things that are easy about e-commerce, especially for small businesses and individuals, include creating the Web site and displaying your products, taking the orders, and accepting payment.

Many companies will help you build and put up your electronic store and help you accept and process credit card orders. Otherwise, the things you have to do to get started are no different from those you'd have to do if you wanted to start a physical store. Just like any store, you have to think about a number of things:

- **Suppliers**—Just as with any normal store or mail-order company, without suppliers of goods, you cannot offer products.
- **Your price point**—A big part of e-commerce is the fact that online price comparisons are extremely easy for the consumer. Your price point is important in such a transparent market: You need to make enough money to make your business profitable, but your price must be low enough that potential customers choose you over the competition.
- **Customer relations**—E-commerce offers a variety of ways to relate to your customer, such as through email, FAQs, knowledge bases, forums, and chat rooms. Integrating these features into your e-commerce offering helps you differentiate yourself from the competition.
- **The back end, including fulfillment, returns, and customer service**—These processes make or break any retail establishment. They define, to a great extent, your relationship with the customer.

E-commerce did not exist until about 1995 or so, and now it is a big part of the economy. Like mail order, it completely changed the relationship between the store and the customer. But the bottom line is that not that much has changed: Any business, whether it's involved in e-commerce or not, needs great products at good prices, and it needs to keep the customer happy.

How INTERNET SEARCH ENGINES Work

Internet search engines are designed to help people find information stored on the millions of Web sites available today. There are differences in the ways various search engines work, but they all perform three basic tasks: They scan the Internet—or select pieces of the Internet—in a process called spidering; they keep an index of the words they find and where they find them, and they allow users to look for words or combinations of words found in that index.

HSW Web Links

www.howstuffworks.com

How Boolean Logic Works
How Bytes and Bits Work
How Domain Name Servers Work
How Web Pages Work
How Web Servers and the Internet Work

Early search engines held an index of a few hundred thousand pages and documents, and they received maybe a couple thousand inquiries each day. Today, a top search engine indexes hundreds of millions of pages, and it responds to tens of millions of queries per day.

Spidering

Before a search engine can tell you where a file or document is, it must find the file or document. To find information on the billions of Web pages that exist, a search engine uses special software robots, called spiders, to build lists of the words found on Web sites. When a spider is looking at pages and building its lists, the process is called spidering or Web crawling.

A spider usually starts by looking at lists of heavily used servers and very popular pages. The spider begins with a popular site, indexes the words on its pages, and follows every link within the site. In this way, the spidering system quickly begins to travel, spreading out across the most widely used portions of the Web.

When a spider looks at an Web page, it usually notes at least two things:

- A list of the words within the page
- Where on the page the words were found

The spider takes note of words that occur in the title, subtitles, meta-tags, and other positions of relative importance so that they come up during a subsequent user search. The spider usually indexes every significant word on a page, leaving out the articles *a*, *an*, and *the*, as well as other common words. This simple optimization saves lots of space in the index.

Other spiders take different approaches. These different approaches usually attempt to make the spider operate faster, allow users to search more efficiently, or both. Some spiders, for example, have a limit on the number of words they index per page.

Building an Index

When spiders have found all the information they can on Web pages, the search engine must store the information in a way that makes it useful. (Note that finding all the information on Web pages is a task that is never actually completed—the constantly changing nature of the Web means that the spiders are always crawling.) There are two key components involved in making the gathered data accessible to users: the information stored with the data and the method by which the information is indexed.

In the simplest case, a search engine could just store the word and the uniform resource locator (URL) where it was found. In reality, this would make for an engine of limited use because there would be no way to tell whether the word was used in an important or a trivial way on the page, whether the word was used once or many times, or whether the page contained links to other pages containing the word.

To make for more useful results, most search engines store more than just the word and URL. The engine might assign a weight to each entry, with increasing values assigned to words if they appear near the top of the document, in subheadings, in links, in the meta-tags, or in the title of the page. Each commercial search engine has a different formula for assigning weight to the words in its index. This is one of the reasons that a search for the same word on different search engines will produce different lists, with the pages presented in different orders.

An index has a single purpose: It allows information to be found as quickly as possible. There are quite a few ways for an index to be built. For example, search engines can be built using a normal Structured Query Language (SQL) database. However, with billions of entries, a normal database often runs into performance barriers. One effective way to build a high-speed index involves hash tables. In hashing, a formula is applied to attach a numerical value to each word. The formula is designed to evenly distribute the entries across a predetermined number of divisions. This numerical distribution is different from the distribution of words across the alphabet, and that is the key to a hash table's effectiveness.

Submitting a Query

Searching through an index involves a user building a query and submitting it through the search engine. The query can be quite simple—a single word at minimum. If you enter a single-word query into a search engine, the search engine finds all the pages in its index that contain the word. It then sorts the pages by rank—usually by some sort of importance measure, as described earlier in this article. For example, the page that uses the word the most might be ranked first.

Building a more complex query requires the use of Boolean operators that allow you to refine and extend the terms of the search. These are the most commonly used Boolean operators:

- **AND**—All the terms joined by AND must appear in the pages or documents. Some search engines substitute the operator + for the word AND.
- **OR**—At least one of the terms joined by OR must appear in the pages or documents.
- **NOT**—The term or terms following NOT must not appear in the pages or documents. Some search engines substitute the operator – for the word NOT.
- **FOLLOWED BY**—One of the terms must be directly followed by the other.
- **NEAR**—One of the terms must be within a specified number of words of the other.
- **Quotation marks ("")**—The words between the quotation marks are treated as a phrase, and that phrase must be found within the document or file.

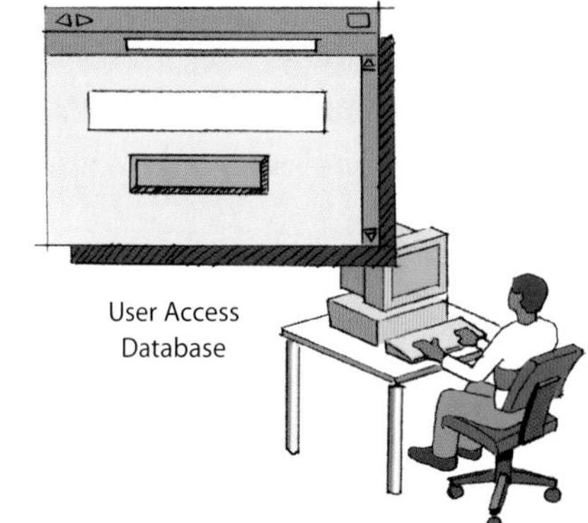

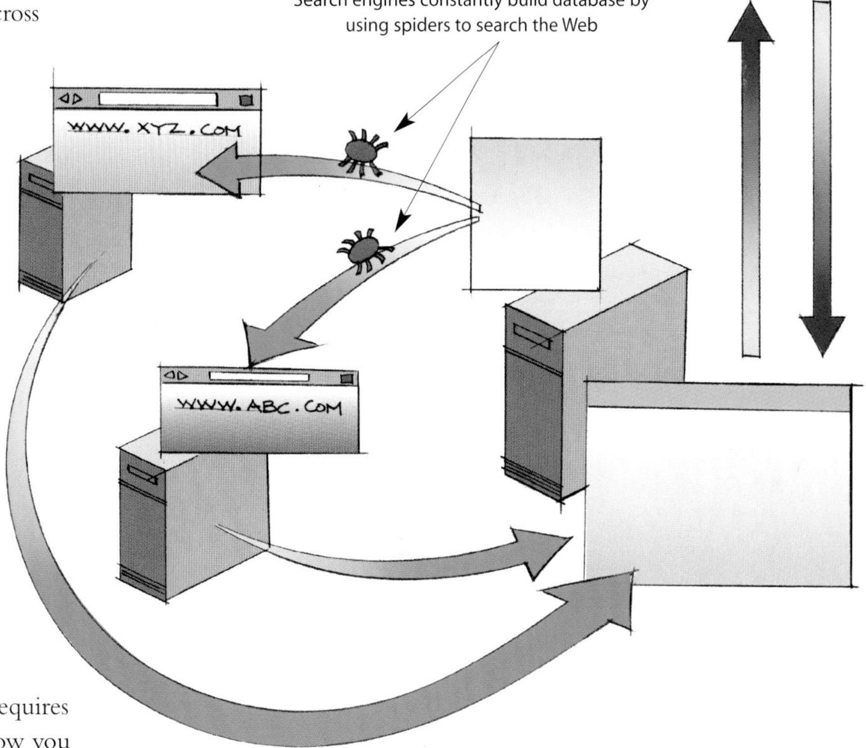

A search engine is a simple system in principle. It is an indexed list of words that point to individual pages on the Web. The three things that make search engines so complex, and so interesting, are the number of words they index, the immense number of pages available on the Web, and the amount of change that occurs on the Web daily. Search engine design and construction are a major engineering endeavor, but the result is a helpful, easy-to-use system that helps people find the sites they need many times every day.

How INTERNET COOKIES Work

Most people don't understand Internet cookies and fear that cookies can somehow invade their privacy, but cookies are actually simple and benign. Cookies provide capabilities that improve the ease of navigating the Web. The designers of almost every major site use cookies because they improve both the user's experience and the site's capability to gather accurate information about the site's visitors.

HSW Web Links

www.howstuffworks.com

How Affiliate Programs Work
How Carnivore Works
How Domain Name Servers Work
How Web Pages Work
How Web Servers and the Internet Work

A cookie is a small piece of text that a Web server can store on a user's hard disk. Cookies allow a Web site to store information on a user's machine and later retrieve it. The pieces of information are stored as name–value pairs. For example, a Web site might generate a unique ID number for each visitor and store the ID number on each user's machine by using a cookie file.

The Contents of a Cookie

If you use Microsoft's Internet Explorer to browse the Web, you can see all the cookies stored on your machine. The most common place for them to reside is in a directory called c:\windows\cookies. When you look in that directory on your machine, you are likely to find hundreds of files. Each file is a text file that contains name–value pairs, and there is one file for each Web site that has placed cookies on your machine.

You can see in the directory that each of these files is a simple, normal text file. You can see which Web site placed the file on your machine by looking at the file name and the information inside the file. You can open each file by clicking on it. Imagine visiting the site howstuffworks.com. This site places a cookie on your machine, and the cookie file might contain the following information:

User 1047243564352587353
www.howstuffworks.com/

There are probably several other values stored in the file after these three: These values are housekeeping information for the browser. The server has stored a single name–value pair on your machine: The name is User, and the value is 1047243564352587353. The first time you visit howstuffworks.com, the site assigns you this unique ID value and stores it on your machine. A site can store one or more name–value pairs.

Data Movement and Cookies

A Web site stores data in a cookie file so that it can later retrieve that information. A Web site can receive only the data it has stored on your machine. It cannot look at any other cookie, nor can it look at anything else on your machine.

The cookie data moves in the following manner:

1) If you type the URL of a Web site into your browser, your browser sends a request to the Web site for the designated page.
2) The browser looks on your machine for a cookie file that the site might have previously stored on your machine. If it finds one, it sends the name–value pairs to the server along with the page request.
3) The Web server receives the request for a page. If name–value pairs are received, the server can use them.

Why the Fury Around Cookies?

You may be wondering why there has been such uproar in the media about cookies and Internet privacy. There are two things that have caused the strong reaction around cookies:

- **On-site profiling:** Using cookies, a Web site can see where you go on their site, what you click on, what you buy, and so forth. They can then develop demographic information about you and use it or sell it to other companies that have certain products or services that fit your profile.
- **Cross-site profiling:** Some companies, particularly ad banner providers, have developed cookies that can track what you do across multiple sites. They can potentially even see the search strings that you type into search engines. Because they can gather so much information about you from multiple sites, such cookies can form very rich profiles.

4) If no name–value pairs are received, the server knows that you have not visited before. The server creates a new ID for you. It sends the cookie information back to your browser with the page, and your browser stores the cookie values on your hard disk.
5) The Web server can change name–value pairs or add new pairs whenever you visit the site and request a page.

There are other pieces of information that the server can send with the name–value pair. One of these is an expiration date. Another is a path, so that the site can associate different cookie values with different parts of the site.

You have control over cookies and how they're used. You can set an option in your browser so that the browser informs you every time a site sends name–value pairs to you. You can then accept or deny the values.

State Information

Cookies evolved because they solve a big problem for the people who implement Web sites. In the broadest sense, a cookie allows a site to store state information on your machine. This information lets a Web site remember what state your browser is in. Web sites use cookies in many different ways. Here are some of the most common examples:

- A site can accurately determine how many readers actually visit the site. Using cookies, a site can determine how many visitors arrive, determine how many are new versus repeat visitors, and so on.
- A site can store user preferences so that the site can look unique for each visitor. For example, the site might say, "Hello, Jane—Thanks for coming back!" If you have registered with the site and told it your first name, the site stores your name with your ID value in the database.
- E-commerce sites can implement things like shopping carts and quick-checkout options.

All the information is stored in the site's database, and a cookie containing your unique ID, is all that is stored on your computer in most cases.

Problems with Cookies

Cookies are not perfect. People share machines, cookies can be erased, and other things can make them not work properly. The main way that sites solve these problems is by allowing users to sign in with an ID and a password. When you sign in, the site can reset the cookie value on your machine, and then the site knows who you are.

Cookies are a very simple mechanism—just little pieces of text that a browser stores on a machine's hard disk. With cookies, Web servers have been able to implement customization and shopping features that have made the Internet a little less static and a lot more interesting.

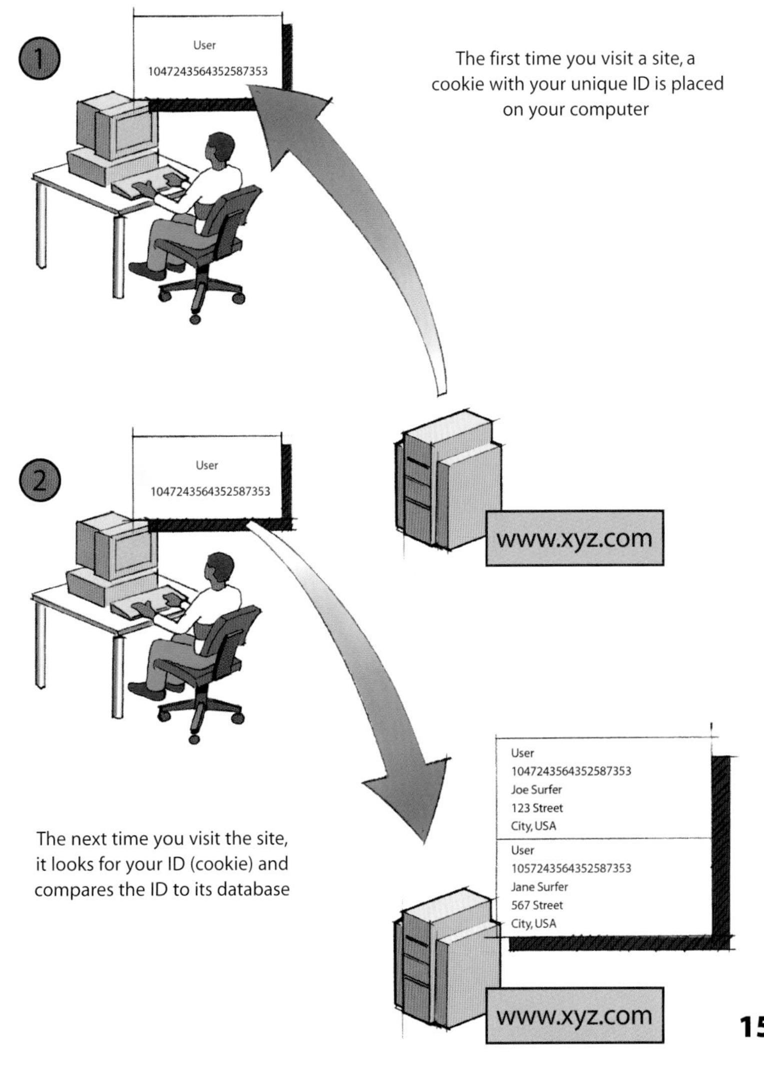

How EMAIL Works

Every day, the citizens of the Internet send each other billions of email messages. You can send business emails, electronic greeting cards, photos of the grandkids, or love letters. If you are online a lot, you may send a dozen or more emails each day without even thinking about it. Email isn't as personal as a handwritten letter or a phone call, but it's obvious that email has become an extremely popular communication tool in a very short time!

HSW Web Links

www.howstuffworks.com

How Carnivore Works
How Domain Name Servers Work
How Internet Odors Will Work
How Web Pages Work
How Web Servers and the Internet Work

The systems that send email from your desktop to a friend halfway around the world are amazing. Despite what you might think, email is an incredibly simple system at its core.

Email Messages

An engineer named Ray Tomlinson sent the first email message in 1971. Before that point, you could only send messages to other users sharing a single machine. Tomlinson's breakthrough was the ability to send messages to other machines on the Internet, using the @ sign to designate the receiving machine.

Starting with Tomlinson's message, an email message has been nothing more than a simple text message sent to a recipient. Email messages tend to be short pieces of text, although the ability to add attachments now makes many email messages quite long.

Understanding Email Clients

To look at email messages, you use some sort of email client. Many people use well-known standalone clients such as Microsoft Outlook, Outlook Express, Eudora, or Pegasus. People who subscribe to free email services such as Hotmail or Yahoo! Mail use a browser-based email client that appears in a Web page. No matter which type of client you are using, you know that an email client generally does four things:

- It shows you a list of all the messages in your mailbox by displaying the message headers (sender, subject, and date).
- It lets you select a message header and read the body of the email message.
- It lets you create and send new messages.
- Most email clients also let you add attachments to messages you send and save the attachments from messages you receive.

A Simple Email Server

If you have an email client on your computer, you are ready to send and receive email. All you need is an email server for the client to connect to. Imagining what the simplest possible email server would look like gives a basic understanding of the process. (See also "How Internet Servers Work," page 170.)

The simplest possible email server might have these features:

- It would have a list of email accounts, with one account for each person who can receive email on the server. Susie Jones's account name might be sjones, John Smith's might be jsmith, and so on.
- It would have a text file for each account in the list. So the server would have a text file in its directory named sjones.txt, another named jsmith.txt, and so on.
- It would allow someone to compose a text message ("Susie, Can we have lunch Monday? John") in an email client, and indicate that the message should go to sjones.
- When the person pressed the Send button, the email client would connect to the email server and pass to the server the name of the recipient (sjones), the name of the sender (jsmith), and the body of the message.
- The server would format those pieces of information and append them to the bottom of the sjones.txt file. The entry in the file might look like this:

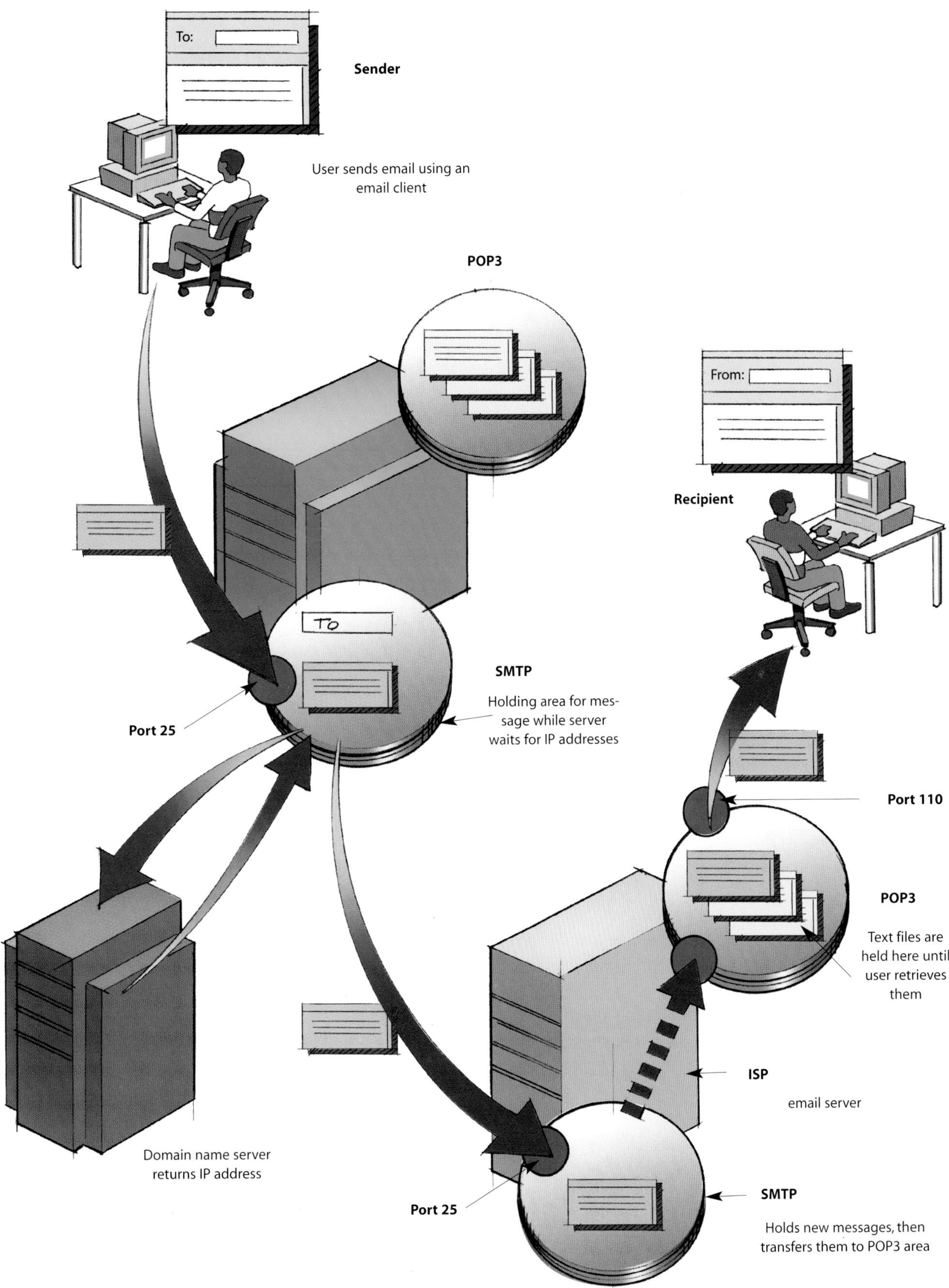

To:
Sender
User sends email using an email client
POP3
From:
Recipient
SMTP
Holding area for message while server waits for IP addresses
Port 25
Port 110
POP3
Text files are held here until user retrieves them
ISP
email server
Domain name server returns IP address
Port 25
SMTP
Holds new messages, then transfers them to POP3 area

From: jsmith
To: sjones
Susie,
Can we have lunch Monday?
John

The server might save in the file several other pieces of information, such as the time and date of receipt and a subject line, but overall you can see that this is a simple process.

As people send mail to sjones, the server would simply append those messages to the bottom of the text file in the order in which they arrive. The text file would accumulate a series of 5 or 10 messages, and eventually Susie would log in to read them. When she wanted to look at her email, her email client would connect to the server machine. In the simplest possible system, the server would do the following:

1) Ask the server to send a copy of the sjones.txt file.
2) Save the sjones.txt file on Susie's local machine.
3) Ask the server to erase and reset the sjones.txt file.
4) Parse the file into the separate messages (using From: as the separator).
5) Show Susie all the message headers in a list.

When Susie double-clicks on a message header, the email client finds that message in the text file and shows her its body.

You have to admit that this is a very simple system. Little text messages are appended to a text file, and the text file is opened to display the messages. Surprisingly, the real email system that you use every day is not much more complicated than this.

Most email systems today consist of two different servers running on a server machine. One is called the SMTP (Simple Mail Transfer Protocol) server, and it handles incoming and outgoing email transport. The other is a POP3 (Post Office Protocol—3) server, and it handles the delivery of messages to a client (takes the mail from the server and delievers it to your PC). (See "How Internet Servers Work," page 170, for details on ports.)

The SMTP Server

Whenever you send a piece of email, your email client interacts with the SMTP server to do the sending. The SMTP server on your host may have conversations with other SMTP servers to actually deliver the email.

Let's assume that Susie wants to send a piece of email. Her email ID is sjones and she has an account on howstuffworks.com. She wants to send an email to the address jsmith@cleartoys.com. She is using the standalone email client Outlook Express.

When Susie sets up her account at howstuffworks, she tells Outlook Express the name of her mail server:

mail.howstuffworks.com

When she composes a message and clicks *send*, here's what happens:

1) Outlook Express connects to the SMTP server at mail.howstuffworks.com, using port 25.
2) Outlook Express has a conversation with the SMTP server. The conversation is an extremely simple set of text commands and responses. Outlook Express tells the SMTP server the address of the sender and the address of the recipient, as well as the body of the message.
3) The SMTP server breaks the To: address (for example, jsmith@cleartoys.com) into two parts: the recipient name (jsmith) and the domain name (cleartoys.com). Because the recipient is at another domain, the SMTP server for howstuffworks.com needs to communicate with cleartoys.com.
4) The SMTP server has a conversation with a domain name server. It says, "Can you give me the IP address of the SMTP server for cleartoys.com?" The domain name server replies with one or more IP addresses for the SMTP server(s) that cleartoys operates. (See "How Domain Name Servers Work," page 172, for details.)
5) The SMTP server at howstuffworks.com connects with the SMTP server at

cleartoys.com using port 25. It has the same simple text conversation that Susie's email client had with the SMTP server for howstuffworks, and it gives the message to the cleartoys server. The cleartoys server recognizes that the domain name for jsmith is at cleartoys, so it hands the message to the cleartoys SMTP server. The cleartoys SMTP server puts the message in jsmith's mailbox.

If, for some reason, the SMTP server at howstuffworks cannot connect with the SMTP server at cleartoys, the message goes into a queue. The SMTP servers on most machines uses a program called sendmail to do the actual sending, so this queue is called the sendmail queue. Sendmail periodically tries to re-send the messages in its queue. For example, it might retry every 15 minutes. After five days, most sendmail configurations give up and return the mail to the sender, undelivered.

The POP3 Server

In the simplest implementations of POP3, the POP3 server really does maintain a collection of text files, one for each email account.

When you check your email, your email client connects to the POP3 server by using port 110. To log in to the POP3 server, you need to enter a valid account name and a password. Then the POP3 server opens your text file and allows you to access it. Like the SMTP server, the POP3 server understands a very simple set of text commands. Here are the most common commands:

- **USER**—Enter the user ID.
- **PASS**—Enter the password.
- **QUIT**—Quit the POP3 server.
- **LIST**—List the message headers and the size of each message.
- **RETR**—Retrieve a message number.
- **DELE**—Delete a message.

Your email client connects to the POP3 server and issues a series of commands to bring copies of your email messages to your local machine. It then deletes the messages from the server, unless you've told it not to.

You can see that the POP3 server simply acts as an interface between the email client and the text file that contains your messages. And you can also see that the POP3 server is extremely simple. You can connect to it with telnet at port 110 and issue the commands yourself if you would like.

Email Attachments

Most email clients allow you to add attachments to email messages you send and also let you save attachments from messages that you receive. Attachments might include word processing documents, spreadsheets, sound files, snapshots, and pieces of software. Because email messages can contain only text information and because attachments are not always just text, a problem needs to be solved.

In the early days of email you solved this problem yourself, by using a program called uuencode. The uuencode program assumes that the file contains binary information. It extracts 3 bytes from the binary file and converts them to four ASCII text characters. What uuencode produces, therefore, is an encoded version of the original binary file that contains only text characters. In the early days of email you would run uuencode yourself and paste the uuencoded file into an email message. The recipient would then save the uuencoded portion of the message to a file and run uudecode on it to translate it back to binary. Modern email clients are doing exactly the same thing, but they run uuencode and uudecode for you automatically.

Email has become an essential communication tool in today's world. When an email virus overruns a company's email system, it can slow the company down tremendously. It is amazing to see how simple this system really is when you look behind the scenes!

Email Viruses

In the late 1990s, email viruses became a major nuisance for users and companies.

An email virus is usually spread via email attachments. The attachment is typically either an EXE or a VBA (Visual Basic for Applications) program. Email viruses take advantage of two facts.

The first fact is that Microsoft has spent a lot of time building "programmer interfaces" into its major applications (like Word and Outlook). These interfaces allow programmers to write programs that manipulate the applications. A programmer can easily write a program that tells Outlook to send an email to every name in the email address book. Someone writing an email virus uses these capabilities to create a program that can resend an infected email message to dozens of people each time the message is opened.

The second fact is that it takes just a sentence or two to convince a person to click an attachment. Something like "I love you! Please double-click on the attachment!" will do it.

It turns out that with just a few lines of software code and a couple of good sentences, you can bring the email systems of hundreds of major companies to their knees. They get totally clogged with millions of email messages.

How DSLs Work

A standard telephone installation in the U.S. consists of a pair of copper wires that the phone company installs in your home. A pair of copper wires has plenty of bandwidth for carrying data in addition to voice conversations. Voice signals use only a fraction of the available capacity on the copper pair. A digital subscriber line (DSL) exploits this remaining capacity to carry information on the wire without disturbing the line's ability to carry conversations.

HSW Web Links

www.howstuffworks.com

How Cable Modems Work
How Ethernet Works
How Home Networking Works
How Modems Work
How Virtual Private Networks Work

To understand DSL, you first need to know a couple things about a normal telephone line. Standard phone service limits the frequencies that the switches, telephones, and other equipment can carry. Human voices, speaking in normal conversational tones, can be carried in a frequency range of 400 to 3,400 Hertz, or cycles per second. The wires themselves have the potential to handle frequencies up to several million hertz in most cases. Modern equipment that sends digital, rather than analog, data can safely use much more of the telephone line's capacity, and DSL does just that.

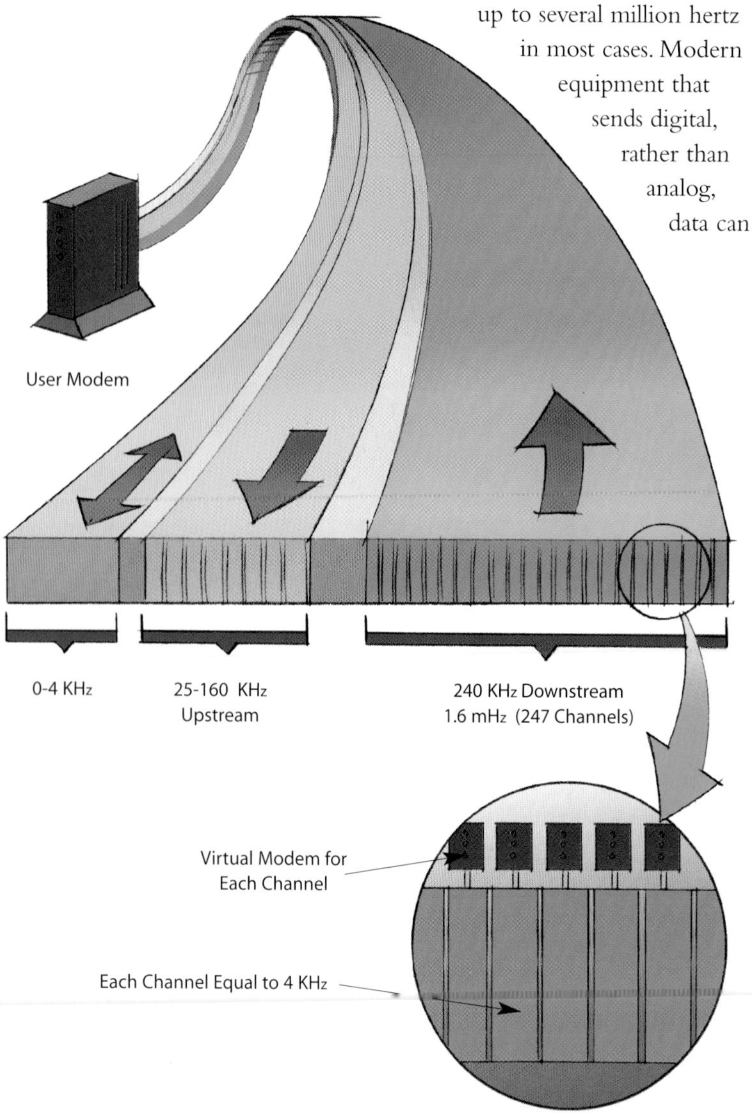

Types of DSL

Most home and small business DSL users are connected to an asymmetrical DSL (ADSL) line. It is called *asymmetric* because the download speed is greater than the upload speed; ADSL works this way because most Internet users look at, or download, much more information than they send, or upload. There are other types of DSL, such as very-high-bit-rate DSL (VDSL), symmetric DSL (SDSL), and rate-adaptive DSL (RADSL). Each type has certain advantages and disadvantages.

DSL and Distance

DSL is a distance-sensitive technology: As the connection's length increases, the signal quality decreases and the connection speed goes down. ADSL service can be run a maximum of 18,000 feet (5,460 m), though for speed and quality of service reasons, many ADSL providers place an even lower limit on the distances. At the extremes of the distance limits, ADSL customers may see speeds far below the promised maximums, and customers nearer the central office or DSL termination point have the potential for seeing very high speeds in the future.

You might wonder why, if distance is a limitation for DSL, it's not also a limitation for voice telephone calls. The answer lies in small amplifiers, called loading coils, that the telephone company uses to boost voice signals. These loading coils are in-compatible with DSL signals, so if there is a voice coil in the loop between your telephone and the telephone company's central office, you cannot receive DSL service. Other factors might disqualify you from receiving DSL:

- **The presence of bridge taps**—These are extensions, between you and the central office, that service other customers.
- **Fiber-optic cables**—ADSL signals can't pass through the conversion from analog to digital and back to analog that occurs if a portion of your telephone circuit comes through fiber-optic cables.
- **Distance**—Even if you know where your central office is (don't be surprised if you don't—the telephone companies don't advertise their locations), looking at a map is no indication of the distance a signal must travel between your house and the office. The wire may follow a very convoluted path between the two points.

VDSL

Very high bit-rate DSL (VDSL) provides an incredible amount of bandwidth, with speeds as high as 52 Mbps downstream (to your home) and 16 Mbps upstream (from your home). Although much faster than ADSL, VDSL's amazing performance comes at a price: It can only operate over the copper line for a short distance, about 4,000 feet (1,200 m).

The key to VDSL is that the telephone companies are replacing many of their main feeds with fiber-optic cable. In fact, many phone companies are planning Fiber to the Curb (FTTC), which means that they will replace all existing copper lines right up to the point where your phone line branches off at your house. At the least, most companies expect to implement Fiber to the Neighborhood (FTTN). Instead of installing fiber-optic cable along each street, FTTN has fiber going to the main junction box for a particular neighborhood.

By placing a VDSL transceiver in your home and a VDSL gateway in the junction box, the distance limitation is neatly overcome. The gateway takes care of the analog-digital-analog conversion problem that disables ADSL over fiber-optic lines. It converts the data received from the transceiver into pulses of light that can be transmitted over the fiber-optic system to the central office, where the data is routed to the appropriate network to reach its final destination. When data is sent back to your computer, the VDSL gateway converts the signal from the fiber-optic cable and sends it to the transceiver. All of this happens millions of times each second!

DSL Equipment

ADSL uses two pieces of equipment: one on the customer end and one at the Internet service provider (ISP), telephone company, or other provider of DSL services:

- **Transceiver**—At the customer's location, there is a DSL transceiver, which may also provide other services.
- **DSL access multiplexer (DSLAM)**—The DSL service provider has a DSLAM to receive customer connections.

Most residential customers call their DSL transceiver a "DSL modem." The engineers at the telephone company or ISP call it an ATU-R. Regardless of what it's called, the transceiver is the point where data from the user's computer or network is connected to the DSL line. The transceiver can connect to a customer's equipment in several ways, though most residential installation uses universal serial bus (USB) or 10BaseT Ethernet connections. Whereas most of the ADSL transceivers sold by ISPs and telephone companies are simply transceivers, the devices used by businesses may combine network routers, network switches, or other networking equipment in the same box.

The DSLAM at the access provider is the equipment that really allows DSL to happen. A DSLAM takes connections from many customers and aggregates them onto a single, high-capacity connection to the Internet. DSLAMs are generally flexible and able to support multiple types of DSL in a single central office. The DSLAM may provide additional functions, including routing or dynamic Internet Protocol address assignment for customers.

The DSLAM provides one of the main differences between user service through ADSL and through cable modems. Because cable modem users generally share a network loop that runs through a neighborhood, more users means lowered performance in many instances. ADSL provides a dedicated connection from each user back to the DSLAM, meaning that users do not see a performance decrease as new users are added—until the total number of users begins to saturate the DSLAM's single, high-speed connection to the Internet. At that point, an upgrade by the service provider can provide additional performance for all the users connected to the DSLAM.

DSL connectivity has completely changed the relationship that many people have with the Internet. Their connections go from slow and intermittent to fast and constant.

How CABLE MODEMS Work

Television brings news, entertainment, and educational programs into millions of homes. Many people get their TV signal from cable television because cable TV provides a clearer picture and more channels than traditional through-an-antenna TV. As it turns out, the existence of a dedicated coaxial cable running into your home makes lots of other things possible. In many communities you can get Internet access and sometimes telephone service over this same cable.

HSW Web Links

www.howstuffworks.com

How Cable Television Works
How DSL Works
How Fiber Optics Work
How Virtual Private Networks Work
How Web Servers and the Internet Work

In a normal analog cable TV system, each television signal is given a 6-megahertz (MHz, or millions of cycles per second) channel on the cable. The coaxial cable used to carry cable television can carry hundreds of channels. (For information, see "How Cable Television Works, page 196.)"

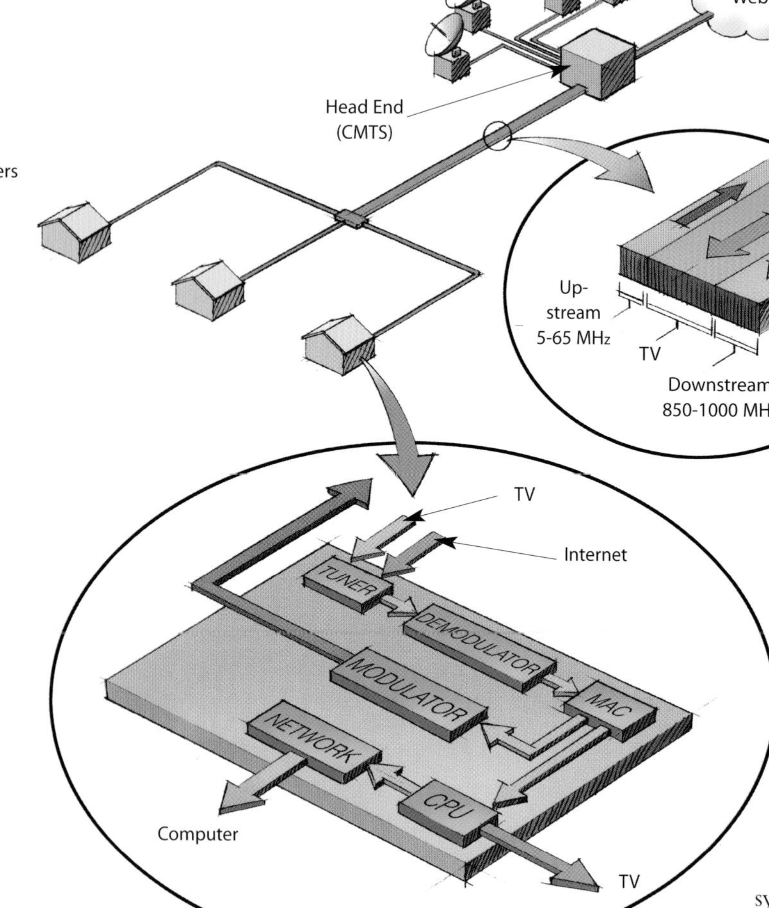

Cable Modem

Internet Access over Cable TV

When a cable company offers Internet access over the cable TV cable, the cable TV company takes one (or more) of the available 6-MHz channels and uses it to send Internet data instead of a television show. On the cable, the data looks just like a TV channel, so Internet downstream data takes up the same amount of cable space as any single channel of programming. Upstream data—information sent from an individual back to the Internet—often requires even less of the cable's bandwidth, just 2 MHz, since most people download far more information than they upload.

A 6-MHz cable TV channel can transmit data at a rate of 30 to 40 megabits per second (Mbps). However, this bandwidth is often shared by up to 1,000 subscribers in a neighborhood. Depending on the number of subscribers, users may see far better—or much worse—performance than is available with standard dial-up modems. The cable company must ensure that a 6-MHz channel is not overloaded by the people using it.

When a channel does get swamped, the cable company simply allocates a new channel on the system to handle another group of users. A cable system might have dozens of channels on the cable, handling thousands of subscribers.

Putting both upstream and downstream data on the cable television system requires two types of equipment: a cable modem on the customer's end and a cable-modem termination system (CMTS) at the cable provider's end. Using these two types of equipment, all the computer networking, security, and management of Internet access over cable television is put into place.

Parts of a Cable Modem

Cable modems can be either internal or external to the computer. In some cases, the cable modem can be part of a set-top cable box, requiring that only a keyboard and mouse be added to the box for standalone Internet access. Regardless of their outward appearance, all cable modems contain certain key components:

- A tuner
- A demodulator
- A modulator
- A media access control (MAC) device and a microprocessor

The tuner in a cable modem is the same as the tuner in a TV. The tuner's job is to separate one channel from all the channels that are coming in. In the case of a cable modem, the tuner extracts a 6-MHz data channel from all the other TV channels on the cable. The tuner sends the data signal to the demodulator.

The signal that the tuner sends to the demodulator looks more like a TV signal than like Internet data. The demodulator's job is to take the signal from the tuner and turn it into the digital data that the computer can understand. This is the same thing that a normal modem does with the signal it receives from a telephone line. The most common technique used today for transmitting data (both on regular modems and cable modems) is called quadrature amplitude modulation (QAM). QAM is the technique that has allowed telephone modems to advance from 300 bits per second (bps) back in the 1970s to 56,000 bps today. A cable modem handles 30 or 40 Mbps, but the technique it uses is similar to what a telephone modem does.

After demodulation, the cable modem also has to sort through the data and discard all the irrelevant packets. Because hundreds of users share a single channel, most of the packets are irrelevant because they are intended for someone else who is using the same channel.

Some cable systems send upstream data—the information you are sending to the Internet to request pages or upload data—over a phone line. True cable modems send the upstream data through the cable. The modulator converts the upstream data into a TV-like signal. Because many people share the cable, something has to be done to keep their upstream data packets from colliding. Dedicated time slots are one common technique.

The microprocessor controls the other parts of the modem, routes data in and out, and communicates with the CMTS to keep things running smoothly.

The CMTS

The CMTS takes the traffic coming in from a group of customers on a single channel and routes it to an Internet service provider (ISP) for connection to the Internet. The downstream information flows to all connected users, just like in an Ethernet network. On the upstream side, information is sent from the user to the CMTS. The narrower upstream bandwidth is divided into slices of time, measured in milliseconds, in which users can transmit one "burst" of data at a time to the Internet. The division by time works well for the very short commands, queries, and addresses that form the bulk of most users' traffic back to the Internet.

A CMTS can enable as many as 1,000 users to connect to the Internet through a single 6-MHz channel. Because a single channel is capable of 30 Mbps to 40 Mbps of total throughput, users may see far better performance than is available with standard dial-up modems.

Cable modems give people who already receive cable TV an easy, fast, and relatively inexpensive way to connect to the Internet. Cable modems are also a great example of reuse—because the cable company has already spent the money to run the cable into your house, and because Internet access is popular, it makes a lot of sense to run multiple services over the same cable.

How FIBER OPTICS Work

You hear about fiber-optic cables whenever people talk about the telephone system, the cable TV system, or the Internet. Fiber-optic lines are strands of extremely clear glass as thin as a human hair that carries digital information over long distances. They are used in many ways, including in medical imaging and in mechanical engineering inspection.

HSW Web Links

www.howstuffworks.com

- How Cable Television Works
- How DSL Works
- How Lasers Work
- How Light Works
- How Routers Work
- How Telephones Work
- How Web Servers and the Internet Work

Over the past 20 years or so, fiber-optic lines have taken over and transformed the long-distance telephone industry. Optical fibers are also a huge part of making the Internet available around the world. When fiber-optic lines replace copper cables in a phone or cable TV system, they dramatically lower costs. Fiber-optic lines cost less, weigh less, take up less space, and use less energy than copper wires.

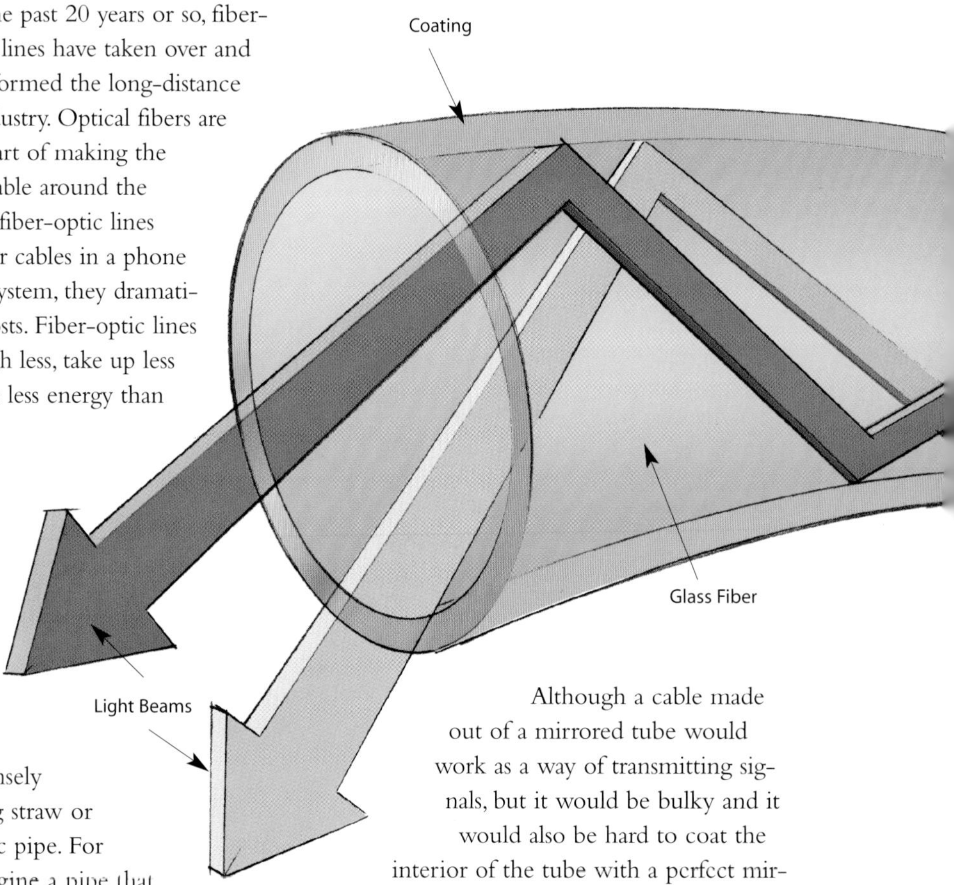

Physics of Total Reflection

When light passes from a medium with one index of refraction (m_1) to another medium with a lower index of refraction (m_2), it bends or refracts away from an imaginary line perpendicular to the surface (normal line). As the angle of the beam through m_1 becomes greater with respect to the normal line, the refracted light through m_2 bends further away from the line. At one particular angle (critical angle), the refracted light will not go into m_2, but instead will travel along the surface between the two media (sin [critical angle] = n_2/n_1 where n_1 and n_2 are the indices of refraction [n_1 is less than n_2]). If the beam through m_1 is greater than the critical angle, then the refracted beam will be reflected entirely back into m_1 (total internal reflection), even though m_2 may be transparent! In physics, the critical angle is described with respect to the normal line.

Fiber-Optic Cable

To understand how a fiber-optic cable works, imagine an immensely long drinking straw or flexible plastic pipe. For example, imagine a pipe that is several miles long. Now imagine that the inside surface of the pipe has been coated with a perfect mirror. Now imagine that you are looking into one end of the pipe. Several miles away, at the other end of the pipe, a friend turns on a flashlight and shines it into the pipe. Because the interior of the pipe is a perfect mirror, the flashlight's light reflects off the sides of the pipe (even though the pipe may curve and twist), and you will see the light at your end. If your friend were to turn the flashlight on and off in a Morse-code fashion, your friend could communicate with you through the pipe. That is essentially how a fiber-optic cable works.

Although a cable made out of a mirrored tube would work as a way of transmitting signals, but it would be bulky and it would also be hard to coat the interior of the tube with a perfect mirror. A real fiber-optic cable is therefore made out of a very thin strand of glass, about the diameter of a human hair. The glass is incredibly pure so that even though it is several miles long, light can still make it through. The glass strand is then coated with two layers of plastic.

By coating the glass in plastic, you get the equivalent of a mirror around the glass strand. This mirror creates total internal reflection, just like a perfect mirror coating on the inside of a tube would. You can experience this sort of reflection with a flashlight and a window in a dark room. If you direct the flashlight through the window at a 90-degree angle, it passes straight through the glass. However, if you shine the flashlight at a very shallow angle (nearly parallel to the glass), the glass will act as a mirror, and you will see the beam reflect off the window and hit the wall inside the room. Light traveling through the fiber bounces at shallow angles like this and stays completely within the fiber.

To send telephone conversations through a fiber-optic cable, analog voice signals are translated into digital signals. (See How Analog and Digital Recording Works," page 212, for details.) A laser or light-emitting diode (LED) at one end of the pipe switches on and off to send each bit. A modern fiber system with a single laser can transmit billions of bits per second—the laser can turn on and off several billion times per second. The newest systems use multiple lasers with different colors to fit multiple signals into the same fiber.

Modern fiber-optic cables can carry a signal quite a distance—perhaps 60 miles (100 km). On a long-distance line, there is an equipment hut every 40 to 60 miles. The hut contains equipment that picks up and retransmits the signal down the next segment at full strength.

How Optical Fibers Are Made

Optical fibers are made of extremely pure optical glass. We think of a glass window as being transparent, but the thicker the glass gets, the less transparent it becomes because of impurities in the glass. The glass in an optical fiber has far fewer impurities than windowpane glass, so even a miles-thick window made of optical-fiber glass would still be clear.

To make an optical fiber, you start by making a cylinder of optically pure glass called a preform. Then you draw the fibers out by using gravity and a tower. The glass for the preform is made through a process called modified chemical vapor deposition (MCVD). In MCVD, oxygen is bubbled through solutions of silicon chloride ($SiCl_4$), germanium chloride ($GeCl_4$), and other chemicals. The gas vapors flow to the inside of a synthetic silica or quartz tube in a special lathe. As the lathe turns, a torch moves up and down the outside of the tube. The extreme heat from the torch causes two things to happen:

- The silicon and germanium react with oxygen, forming silicon dioxide (SiO_2) and germanium dioxide (GeO_2).
- The silicon dioxide and germanium dioxide deposit on the inside of the tube and fuse together to form glass.

The lathe turns continuously to make an even coating of this glass.

After the preform blank has been tested, it is loaded into a fiber-drawing tower. A furnace that reaches 3452°F to 3992°F (1900°C to 2200°C) heats one end of the preform until a molten glob falls, with the help of gravity. As the glob drops, it cools and forms a thread. The furnace operator threads the strand through a series of coating cups onto a spool. The spool mechanism pulls the fiber from the heated preform blank and precisely controls the process by using a laser micrometer.

Fiber-optic cables have helped make long-distance calls incredibly inexpensive, and they allow trillions of bytes of data to flow freely on the Internet every day.

Advantages of Fiber Optics

Some of the advantages fiber optic systems have over conventional metal or copper wire are:

Less expensive — Several miles (or kilometers) of optical cable can be made more cheaply than equivalent sizes of copper wire.

Thinner — Optical fibers can be drawn to smaller diameters than copper wire.

Less signal degradation — The loss of signal in optical fiber is less than in copper wire.

Low power — Because signals in optical fibers degrade less, lower power transmitters can be used instead of high voltage electrical transmitters for copper wires.

Digital signals — Optical fibers are ideally suited for carrying digital information, which is especially useful in computer networks.

Non-flammable — Because no electricity is passed through optical fibers, there is no fire hazard.

How **FIREWALLS** Work

A firewall is like an automated guard. It sits between the Internet and a private network. It looks at everything trying to pass into the private network and makes sure it is okay. The firewall blocks anything that is not okay. A firewall helps protect a network from people with malicious intent.

HSW Web Links

www.howstuffworks.com

How Home Networking Works
How LAN Switches Work
How Network Address Translation Works
How Routers Work
How Web Servers and the Internet Work

If you have been using the Internet for any length of time, and especially if you work at a large company and browse the Web while you are at work, you have probably heard the term *firewall* used. For example, you often hear people in companies say things like, "I can't use that site because they won't let it through the firewall."

What Firewalls Do

Unscrupulous people use many creative ways, such as the following, to access or abuse unprotected computers:

- Remote login
- Application backdoors
- SMTP session hijacking
- Operating system bugs
- Denial-of-service attacks
- Email bombs
- Macros
- Viruses
- Spam
- Redirect bombs
- Source routing

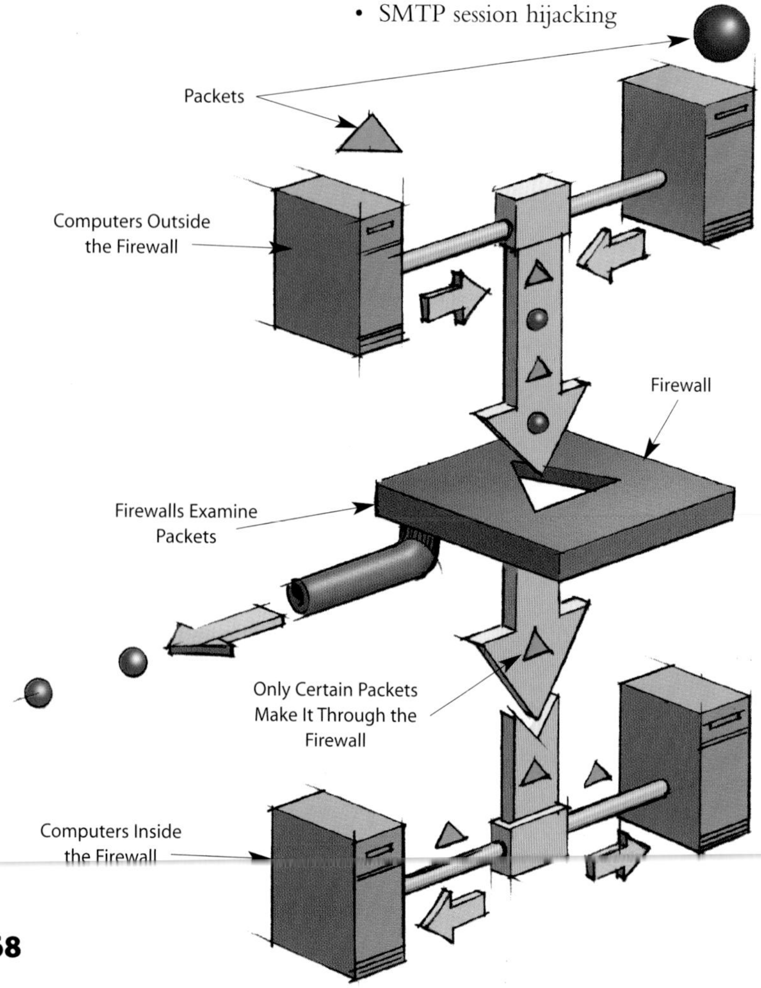

A correctly configured firewall can filter out many of these threats.

For example, say that you work for a company that has 500 employees. The company has hundreds of computers that all have network cards connecting them together. In addition, the company has one or more connections to the Internet through something like T1 or T3 lines. Without a firewall in place, all those hundreds of computers would be directly accessible to anyone on the Internet. A person who knows what he or she is doing can probe those computers, try to make File Transfer Protocol (FTP) connections to them, try to make telnet connections to them, and so on. If one employee makes a mistake and leaves a security hole, hackers can get to the machine and exploit the hole, and then they might be able to wreak havoc on the company's entire system.

With a firewall in place, the landscape is much different. A firewall is a program or a hardware device that filters the information coming through the Internet connection into a private network or computer system. If a firewall's filters flag an incoming packet of information as inappropriate, the packet is not allowed through.

Packets
Computers Outside the Firewall
Firewall
Firewalls Examine Packets
Only Certain Packets Make It Through the Firewall
Computers Inside the Firewall

Firewall Filtering Methods

Firewalls use three different methods, individually and in combination with one another, to control traffic flowing in and out of a network:

- **Packet filtering**—The firewall analyzes packets (small chunks of data) against a set of filters. Packets that make it through the filters are sent to the requesting system, and all others are discarded.
- **Proxy service**—The firewall retrieves information from the Internet and then sends it to the requesting system and vice versa.
- **Stateful inspection**—Stateful inspection is a newer method, in which the firewall doesn't examine the contents of each packet, but instead compares certain key parts of the packet to a database of trusted information.Information traveling from inside the firewall to the outside is monitored for specific defining characteristics, and then incoming information is compared to these characteristics. If the comparison yields a reasonable match, the information is allowed through. Otherwise, it is discarded.

Firewall Customization

Firewalls are customizable. This means that you can add or remove filters based on several conditions. Some of the typical things that filters look at include the following:

- **IP addresses**—A filter might block all traffic to and from a particular IP address. For example, if a certain IP address outside the company is reading too many files from a server, the firewall can block all traffic to or from that IP address.
- **Domain names**—A company might block all access to certain domain names, or it might allow access only to specific domain names.
- **Protocols**—A filter might block a specific protocol, or it might allow it on only certain machines inside the firewall. These are some common protocols:
  - **IP (Internet Protocol)**—The main delivery system for information over the Internet
  - **TCP (Transmission Control Protocol)**—Used to break apart and rebuild information that travels over the Internet
  - **HTTP (Hypertext Transfer Protocol)**—Used for Web pages
  - **FTP (File Transfer Protocol)**—Used to download and upload files
  - **UDP (User Datagram Protocol)**—Used for information that requires no response, such as streaming audio and video
  - **ICMP (Internet Control Messaging Protocol)**—Used by a router to exchange information with other routers
  - **SMTP (Simple Mail Transport Protocol)**—Used to send text-based information (email)
  - **SNMP (Simple Network Management Protocol)**—Used to collect system information from a remote computer
  - **Telnet**—Used to perform commands on a remote computer
- **Ports**—A server machine makes its services available to the Internet by using numbered ports, one for each service that is available on the server. For example, if a server machine is running a Web (HTTP) server and an FTP server, the Web server would typically be available on port 80, and the FTP server would be available on port 21. A company might block port 21 access on all machines except for one inside the company.
- **Specific words and phrases**—The firewall will sniff (that is, search through) each packet of information for an exact match of the text listed in the filter. For example, you could instruct the firewall to block any packet with the word *X-rated* in it. The key here is that it has to be an exact match. So a filter for *X-rated* would not catch *X rated* (no hyphen). But you can include as many words, phrases, and variations of them as you need.

Firewalls are a vital part of a company's security system. As more and more families connect to the Internet with high-speed connections, firewalls are becoming important in homes as well as in businesses. Firewalls help you control very specifically who does what on your network.

Controlling Flow

At HowStuffWorks, one of the most common things we do with the firewall is to control machines that are overwhelming the server. This seems to happen about once a week, so at least once a week we adjust the firewall to shut down a machine that is attempting a port scan or Denial of Service attack on our server.

For example, a student might create a new search engine at a university. If the search engine is configured incorrectly, it might try to open hundreds of connections to the HowStuffWorks servers all at once in an attempt to download all of the pages from HowStuffWorks very quickly. This is a Denial of Service attack—it prevents other users from requesting pages because the HowStuffWorks servers are overwhelmed with too many simultaneous requests. A simple tweak in the firewall configuration will block that new search engine from making any future requests. Many new firewalls can detect Denial of Service attacks and auto-apply blocking.

How **INTERNET SERVERS** Work

The Internet has revolutionized many parts of our lives. Email has changed the way we communicate, and Web sites have changed the way we look up information and buy things. Internet servers make the Internet possible, and all servers use the same basic mechanisms built into the Internet. By understanding these mechanisms, you can see what the Internet is doing behind the scenes to bring so many different things to your desktop.

HSW Web Links

www.howstuffworks.com

How Domain Name Servers Work
How Fiber Optics Work
How LAN Switches Work
How Network Address Translation Works
How Virtual Private Networks Work

The Internet is a collection of millions of machines that are all connected to each other in one way or another. All the machines on the Internet are either servers or clients. The machines that provide services to other machines are servers, and the machines that are used to connect to those services are clients. Web servers, email servers, File Transfer Protocol (FTP) servers, and other types of servers serve the needs of Internet users all over the world.

When you connect to the site www.howstuffworks.com, you are a user sitting at a client machine. You are accessing the HowStuffWorks Web server. The server machine finds the page you requested and sends it to you.

A server machine may provide one or more services on the Internet. For example, a server machine might have software running on it that allows it to act as a Web server and an email server. Clients that come to a server machine do so with a specific intent, so clients direct their requests to a specific software server running on the server machine. For example, if you are running a Web browser on your machine, it will want to talk to the Web server—not the email server—on the server machine.

IP Addresses and Domain Names

To keep all the machines on the Internet straight, a unique address, called an Internet Protocol (IP) address, is assigned to each machine. IP addresses are 32-bit dotted-decimal numbers. A typical IP address looks like this:

192.168.0.12

The four numbers in an IP address are called octets, because they can have values between 0 and 255, which is 2^8 possibilities per octet.

A server has a static IP address that does not change very often. A home machine that is dialing up through a modem, on the other hand, typically has an IP address assigned by the Internet service provider (ISP) every time you dial in. That IP address is unique for your session—it may be different the next time you dial in. With this system, an ISP needs only one IP address for each active modem it supports on its end, rather than one for each customer.

As far as the Internet's machines are concerned, an IP address is all you need to talk to a server. For example, in your browser you can type a URL such as http://192.168.0.12, and you will arrive at another machine's Web server. On some servers, the IP address alone is not sufficient, but on most large servers it is.

Because most people have trouble remembering the strings of numbers that make up IP addresses, and because IP addresses sometimes need to change, all servers on the Internet also have human-readable names, called domain names. For example, www.howstuffworks.com is a permanent, human-readable domain name. It is easier for most of us to remember www.howstuffworks.com than it is to remember something like 192.168.0.12. (To find out more, see "How Domain Name Servers Work," page 172.)

Ports

A server machine makes its services available by using numbered ports—one for each service that is available on the server. For example, if a server machine is running a Web server and an FTP server, the Web server would typically be available on port 80, and the FTP server would be available on port 21. Clients connect to a service at a

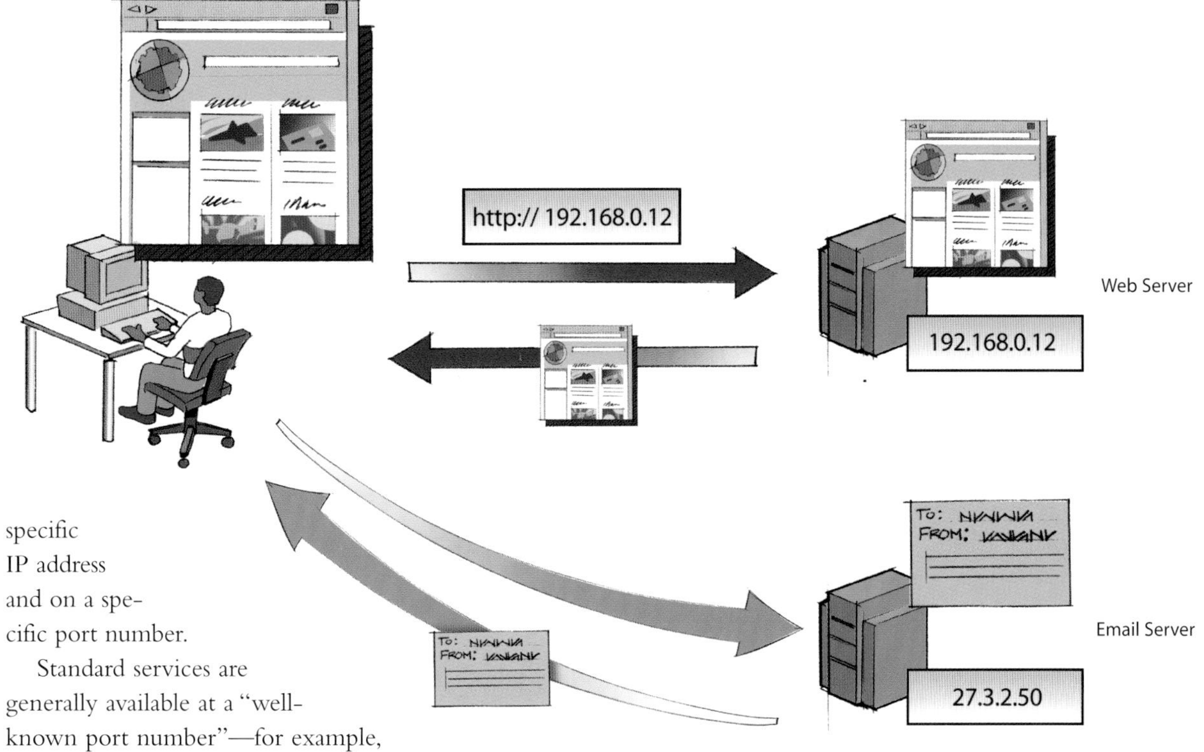

specific IP address and on a specific port number.

Standard services are generally available at a "well-known port number"—for example, port 80 is the well-known port for the Web server. The Web server doesn't have be on port 80, however. If you were to set up your own machine and load Web server software on it, you could put the Web server on port 918 or any other unused port. Then, if your machine were known as xxx.yyy.com, someone on the Internet could connect to your server with the URL http://xxx.yyy.com:918. The :918 explicitly specifies the port number, and it would have to be included for someone to reach your server. When no port is specified, the browser simply assumes that the server is using the well-known port 80.

Protocols

When a client has connected to a service on a particular port, it accesses the service by using a specific protocol. The protocol is the predefined way that someone who wants to use a service talks with that service. The someone could be a person, but more often it is a computer program such as a Web browser. Protocols are often text that simply describes how the client and server will have their conversation. Perhaps the simplest protocol is the daytime protocol. If you connect to port 13 on a machine that supports a daytime server, the server will send you its impression of the current date and time and then close the connection. The protocol is, "If you connect to me, I will send you the date and time and then disconnect." Most protocols are more involved, of course.

Every Web server on the Internet conforms to Hypertext Transfer Protocol (HTTP). The most basic form of the protocol understood by an HTTP server involves just one command: GET. If you connect to a server that understands HTTP and tell it "GET *filename*," the server will respond by sending you the contents of the named file and then disconnecting.

In the original version of HTTP, all you would have sent was the actual filename, such as web-server.htm. The protocol was later modified to handle the sending of the complete URL. This has allowed companies to host virtual domains, where many domains live on a single machine and use one IP address for all the domains they host.

Powerful Web servers and the other servers handle huge volumes of information in milliseconds. Internet servers are extremely important in modern life—without them, there would be no Internet.

Big Servers

If you have a simple site that gets a couple hundred visitors a day, you can pay someone $10 a month to host your Web site.

However, if you have a site that is being visited by hundreds of thousands of visitors a day, or if you want the site to be up all the time or the content to have dynamic components, it can take impressive amounts of hardware to keep the whole thing running.

For example, you might have several machines up front that are handling page requests from users. A load-balancing switch feeds these machines, and then the load-balancer needs a backup in case it has a problem. These machines rely on a database to feed them information, and the database machine is often quite impressive (terabytes of data, half a dozen processors, and so on), with everything redundant to avoid any downtime. A big Web site can consume millions of dollars in hardware and require a small army to keep it running.

How DOMAIN NAME SERVERS Work

If you spend any time on the Internet sending email or browsing the Web, then you use domain name servers without even realizing it. Domain name servers are an incredibly important but very hidden part of the Internet. The Domain name system (DNS) forms one of the largest and most active distributed databases on the planet.

HSW Web Links

www.howstuffworks.com

How Email Works
How LAN Switches Work
How Network Address Translation Works
How Web Pages Work
How Web Servers and the Internet Work

All the Time

One of the most amazing things about the domain name system is the sheer number of requests it handles every day in a totally invisible way. For example, whenever you type in any URL or click on any link in a Web page, whenever you send any email message or download any file, it can generate several queries. In the worst case, there is a request from the root name server, the COM name server, and the site's name server for anything you do.

Let's say in a typical day you look at 50 Web pages, send 15 emails and download a couple files. That could easily generate 100 to 200 name server requests. Multiply that by the tens of millions of users on the Internet each day and you can have billions of name server requests in a single day.

When you use the Web or send an email message, you use a domain name to do it. The uniform resource locator (URL) www.howstuffworks.com, for example, contains the domain name howstuffworks.com. So does the email address brain@howstuffworks.com.

URLs and Addresses

Human-readable names such as howstuffworks.com are easy for human beings to remember, but machines have a hard time with them. All the machines on the Internet use numeric designations called Internet Protocol (IP) addresses to refer to one another. For example, the machine that people refer to as www.howstuffworks.com has an IP address like 192.168.0.12. Every time you use a domain name, the Internet's domain name servers must translate the human-readable domain name into the machine-readable IP address. During a day of browsing and emailing, you might access domain name servers hundreds of times!

Domain name servers translate domain names to IP addresses. That sounds fairly simple, and it would be simple except for five things:

- Billions of IP addresses and domain names are currently in use.
- Many billions of requests are made to domain name servers every day. You alone can easily make 100 or more DNS requests each day, and there are hundreds of millions of people and machines using the Internet.
- Domain names and IP addresses change daily.
- New domain names are created daily.
- Millions of people do the work to change and add domain names and IP addresses every day.

The huge, worldwide scale of the DNS database, the massive traffic that hits it, and the number of changes that occur every day make DNS an interesting challenge!

Parts of a Domain Name

Human beings just are not that good at remembering strings of numbers. We are good at remembering words, however, and that is where domain names come in. You probably have hundreds of domain names stored in your head. Here are some examples of domain names:

- **www.howstuffworks.com**—A typical commercial-site name
- **www.mit.edu**—A popular .edu name **encarta.msn.com**—A Web server that does not start with www
- **www.bbc.co.uk**—A name that uses four parts rather than three, with a .uk ending
- **ftp.microsoft.com**—An FTP server name rather than a Web server name

The .com, .edu, and .uk portions of these sample domain names refer to the top-level, or first-level, domain.

Within every top-level domain is a huge list of second-level domains. For example, these are some of the millions of second-level domains in the com first-level domain:

- howstuffworks
- yahoo
- microsoft

Every name in the com top-level domain must be unique.

In the case of bbc.co.uk, bbc is a third-level domain. Up to 127 levels are possible, although more than 4 or 5 is rare.

The leftmost part of the address, like www, is the host name. It specifies the name of a specific machine (with a specific IP address) in a domain. A given domain can, potentially, contain millions of host names, as long as they are all unique within that domain.

Name Servers

Name servers accept requests from programs and other name servers to convert domain names into IP addresses. When a request comes in, the name server can do one of four things with it:

- It can answer the request with an IP address because it already knows the IP address for the requested domain.
- It can contact another name server and try to find the IP address for the name requested. It may have to do this multiple times.
- It can say, "I don't know the IP address for the domain you requested, but here's the IP address for a name server that knows more than I do."
- It can return an error message because the requested domain name is invalid or does not exist.

Let's say that you type the URL www.howstuffworks.com into your browser. The browser contacts a name server to get the IP address. A name server would start its search for an IP address by contacting one of the root name servers. The root servers know the IP addresses for all the name servers that handle the top-level domains. Your name server would ask the root for www.howstuffworks.com, and the root would say, "I don't know the IP address for www.howstuffworks.com, but here's the IP address for the com name server."

Your name server then sends a query to the com name server, asking it if it knows the IP address for www.howstuffworks.com. The name server for the com domain knows the IP addresses for the name servers handling the howstuffworks.com domain, so it returns those.

Your name server then contacts the name server for howstuffworks.com and asks whether it knows the IP address for www.howstuffworks.com. It does, so it returns the IP address to your name server, which returns it to the browser. Your browser can then contact the server for www.howstuffworks.com to get a Web page.

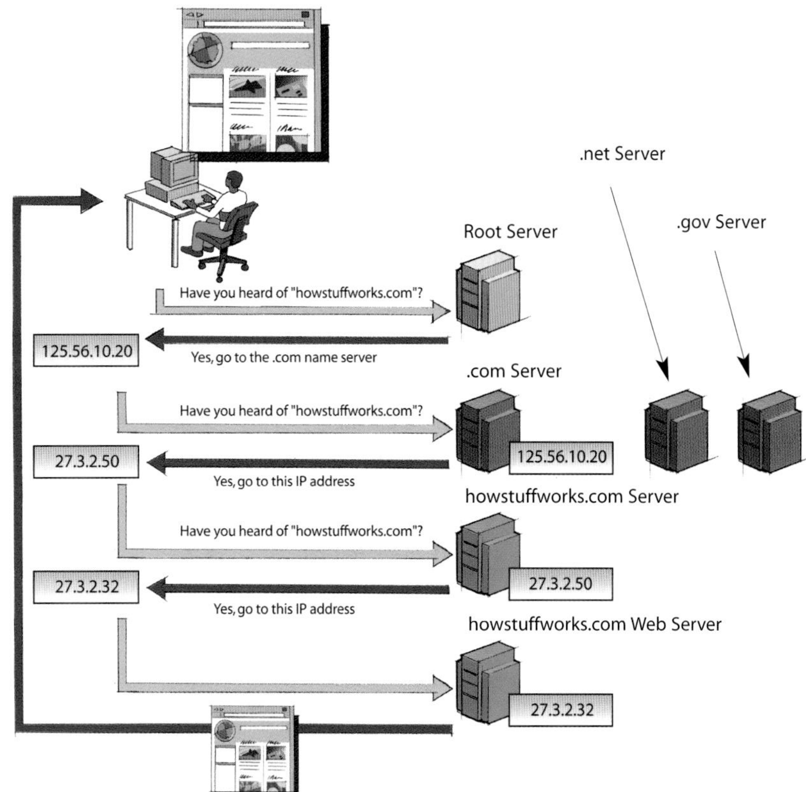

Redundancy and Caching

One of the keys to making DNS work is redundancy. There are multiple name servers at every level, so if one fails, there are others to handle the requests. There are, for example, three different machines running name servers for howstuffworks.com requests. All three would have to fail before there would be a problem.

The other key is caching. After a name server resolves a request, it caches the IP address it receives. After the name server has made a request to a root server for any com domain, it knows the IP address for a name server handling the com domain, so it does not have to bother the root servers for that information again in the future. Name servers can do this for every request, and this caching helps to keep the DNS from bogging down.

Even though it is totally invisible, the DNS handles billions of requests every day and is essential to the Internet's smooth functioning. The fact that this database works so well and so invisibly day in and day out is a testimony to the design.

How the INTERNET INFRASTRUCTURE Works

The Internet is a global collection of networks, both big and small. These networks connect together in many different ways to form the single entity that we know as the Internet. In fact, the very name comes from this idea of interconnected networks. Since its beginning in 1969, the Internet has grown from four host computer systems to tens of millions of them, so no one owns the Internet, per se. This doesn't mean that it is not monitored and maintained in different ways, however. The Internet Society, a nonprofit group established in 1992, oversees the formation of the policies and protocols that define how we use and interact with the Internet.

HSW Web Links

www.howstuffworks.com

How CGI Scripting Works
How Domain Name Servers Work
How LAN Switches Work
How Web Servers and the Internet Work

When the Internet was in its infancy, it consisted of a small number of computers hooked together with modems and telephone lines. Each computer had a unique numeric ID known as an Internet Protocol (IP) address. People could make connections to other machines only by providing the IP address of the computer they wanted to establish a link with. For example, a typical IP address might be 216.27.22.162. This was fine when there were only a few hosts, but it became unwieldy as more and more systems came online.

The first solution to the problem was a simple text file maintained by the Network Information Center that mapped names to IP addresses. Soon this text file became so large that it was too cumbersome to manage. In 1983 the University of Wisconsin created the Domain Name System, which maps text names to IP addresses automatically. This way you only need to remember www.howstuffworks.com, for example, instead of the site's IP address. (See "How Domain Name Servers Work," page 172, for details.)

The Hierarchy of Networks

Every computer that is connected to the Internet—even the one in your home—is part of a network. For example, you may use a modem and dial a local number to connect to an Internet service provider (ISP). When you connect to your ISP, you become part of its network. The ISP may then connect to a larger network and become part of its network. This larger network may have a connection to a network access point (NAP), which is a junction point where dedicated connections to the various networks in a particular area meet. Each NAP is connected to the other NAPs by a backbone, which is a high-capacity connection.

Most large communications companies have their own dedicated backbones connecting between various regions. In each region, the company has a point of presence (POP). The POP is a place for local users to access the company's network, often through a local phone number or dedicated line. Although there is no overall controlling network, there are several high-level networks connecting to each other through the NAPs.

Here's an example. Imagine that Company A is a large ISP. In each major city, Company A has a POP. The POP in each city is a rack full of modems that the ISP's customers dial in to. Company A leases fiber-optic lines from the phone company to connect the POPs together. Imagine that Company B is a corporate ISP. Company B builds large buildings in major cities, and corporations locate their Internet server machines in these buildings. Company B is such a large company that it runs its own fiber-optic lines between its buildings so that they are all interconnected.

In this arrangement, all of Company A's customers can talk to each other, and all of Company B's customers can talk to each other, but there is no way for Company A's customers and Company B's customers to intercommunicate. Therefore, Company A and Company B both agree to connect to NAPs in various cities, and traffic between the two companies flows between the networks at the NAPs. In the real Internet, dozens of large ISPs interconnect at NAPs in various cities, and trillions of bytes of data flow between the individual networks at these points.

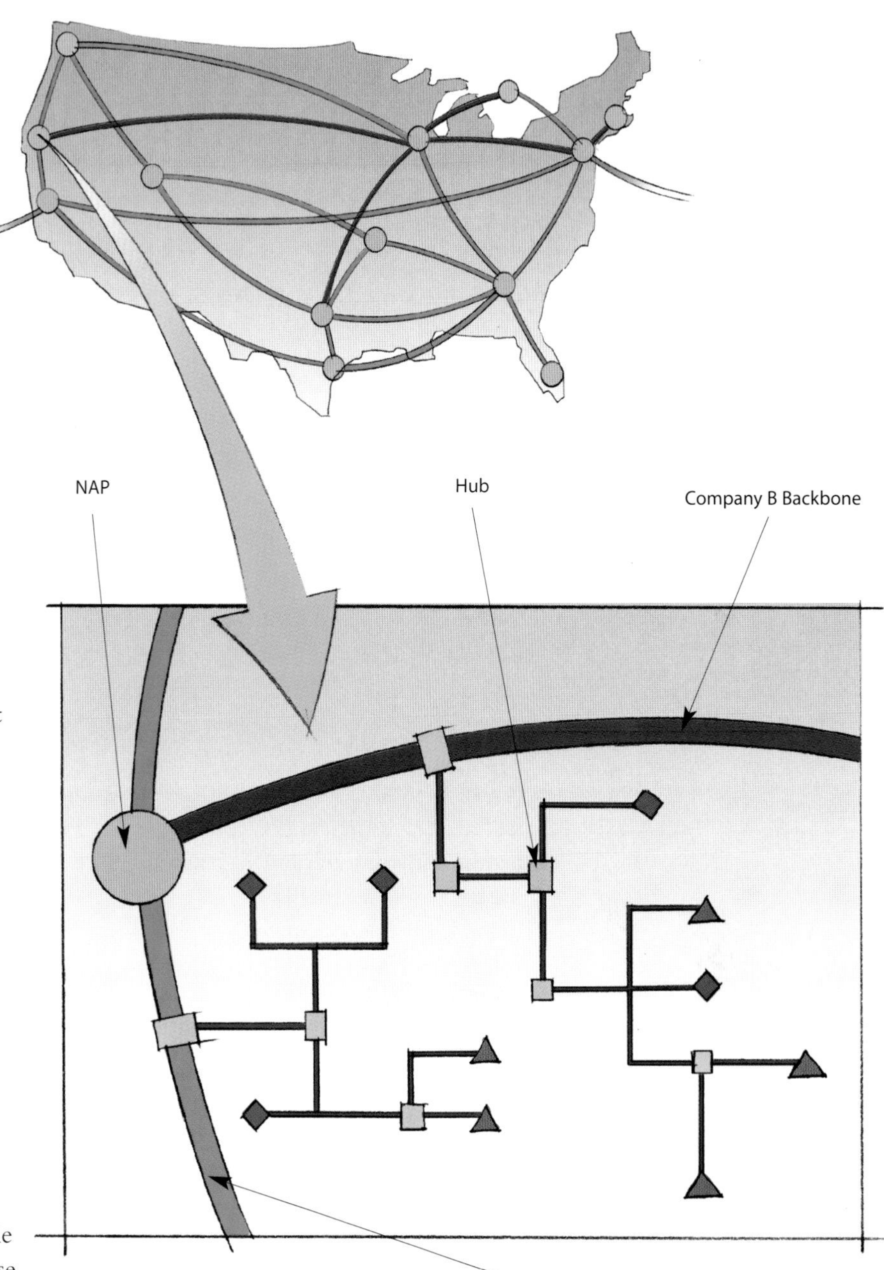

Backbones

All the connected networks rely on NAPs, backbones, and routers to talk to each other. The routers determine where to send information from one computer to another. A message can leave one computer and travel halfway across the world through several different networks and arrive at another computer in a fraction of a second! (For even more information, you will want to read "How Routers Work," page 176.)

The National Science Foundation (NSF) created the first high-speed backbone in 1987. Called NSFNET, this backbone was a T1 line that connected 170 smaller networks together and operated at 1.544 million bits per second (Mbps). IBM, MCI, and Merit worked with NSF to create the backbone and developed a T3 (45-Mbps) backbone the following year.

Backbones are typically fiber-optic trunk lines. The trunk line has multiple fiber-optic cables, combined to increase the capacity. Fiber-optic cables are designated with the letters OC (for optical carrier), as in OC-3, OC-12, and OC-48. An OC-3 line is capable of transmitting 155 Mbps, and an OC-48 line can transmit 2,488 Mbps (2.488 Gbps). Compare that to a typical 56-Kbps modem, and you see just how fast a modern backbone is.

Today many companies operate their own high-capacity backbones, and all of them interconnect at various NAPs around the world. In this way, everyone on the Internet, no matter where they are and what company they use, is able to talk to everyone else on the planet. The entire Internet is a gigantic, sprawling agreement between companies to intercommunicate freely.

The Internet started with just a few computers and a couple of phone lines connecting them together. Today it has grown to a point where thousands of companies cooperate to connect millions of computers into a single, worldwide network. The design of the Internet infrastructure has allowed both an incredible rate and an incredible volume of expansion in a short period of time.

How **ROUTERS** Work

When you send email to a friend on the other side of the country, how does the message know to end up on your friend's computer, rather than on one of the millions of other computers in the world? Much of the work to get a message from one computer to another is done by routers; routers are the crucial devices that let messages flow between, rather than just within, networks. Routers are specialized computers that send messages speeding to their destinations along thousands of pathways.

HSW Web Links

www.howstuffworks.com

How Ethernet Works
How LAN Switches Work
How Network Address Translation Works
How Virtual Private Networks Work
How Web Servers and the Internet Work

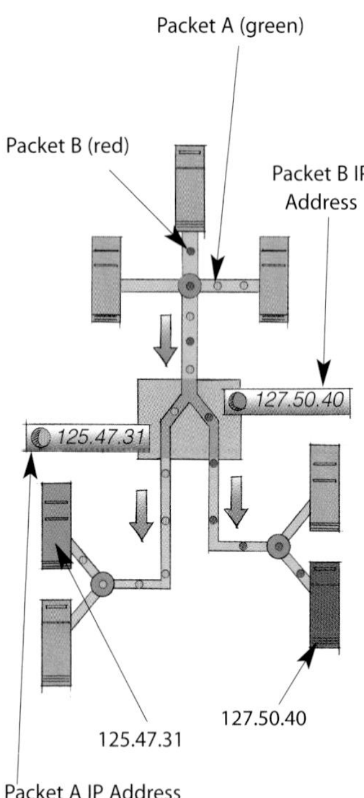

The Internet is a packet-switched network. Internet data—whether in the form of a Web page, a downloaded file, or an e-mail message—travels over the Internet as a collection of packets.

Routing Packets

The data in a message or file is broken up into packages that are each about 1,500 bytes long. Each of these packages gets a wrapper that includes information on the sender's address, the receiver's address, the package's position in the entire message, and how the receiving computer can be sure that the package arrived intact. Each data package, called a packet, is then sent to its destination via the best available route—a route that might be taken by all the other packets in the message or by none of the packets in the message.

The routers that make up the main part of the Internet can reconfigure the paths that packets take because they look at the information surrounding the data packet and they tell each other about line conditions. Routers inform each other about things such as delays in receiving and sending data and traffic on various pieces of the network. Not all routers do so many jobs, however. Therefore, routers have different sizes.

A router connects two or more networks together. The router decides two things:

- Whether the packet needs to be routed at all or whether it needs to stay on its network
- Which path a routed packet should take on its next hop through the Internet

One of the crucial tasks for any router is knowing when a packet of information stays on its local network. For this, the router uses a subnet mask. The subnet mask looks like an IP address and usually reads 255.255.255.0. This tells the router that all messages with the sender and receiver having an address sharing the first three groups of numbers are on the same network and shouldn't be sent out to another network. For example, a packet sent from machine number 192.169.0.24 to 192.169.0.53 would not be routed because both machines are on the same network.

Making Connections

In order to handle all the users of a large network, millions and millions of traffic packets must be sent at the same time. Even with the computing power available in a very large router, how does a router know which of the many possibilities for outbound connection a particular packet should take? The router takes a number of steps:

1) It scans the destination address and matches that IP address against rules in the configuration table. The rules might say that packets in a particular group of addresses go in a specific direction.
2) It checks the performance of the primary connection in that direction against another set of rules. If the performance of the connection is good enough, the packet is sent, and the next packet is handled. If the connection is not performing up to expected parameters, then an alternate is chosen and checked.
3) Finally, a connection is found with the best performance at a given moment, and the packet is sent on its way.

A big router can handle this process millions of times per second.

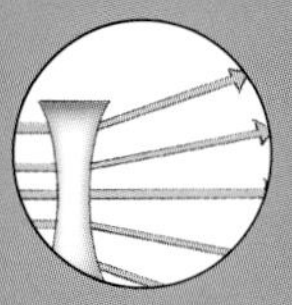

chapter seven

PICTURE PERFECT

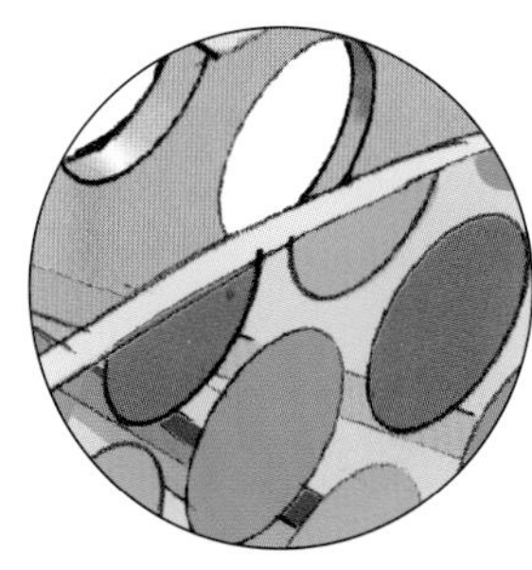

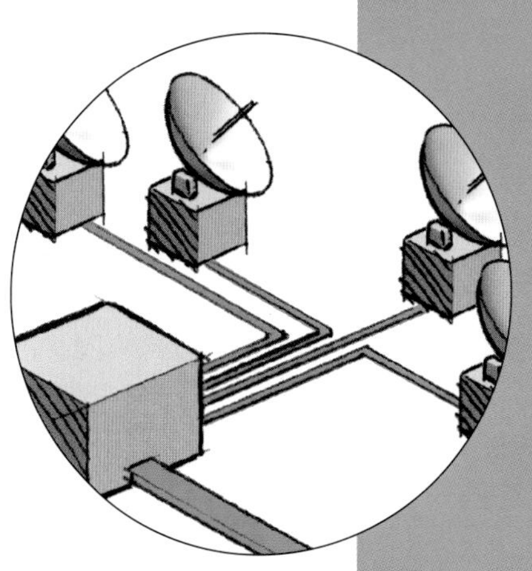

How **CAMERAS** Work

Photography is certainly one of the most important inventions in history. It has completely changed the way we think about and look at the world. Thanks to photography, we can see all sorts of things without actually being anywhere near them! The device that makes this possible—the camera—is remarkably simple.

HSW Web Links

www.howstuffworks.com

How Autofocus Systems Work
How Camcorders Work
How Digital Cameras Work
How DVDs and DVD Players Work
How Photographic Film Works

A camera records a picture by combining three basic processes: one optical process, one chemical process, and one mechanical process.

The Optical Process: The Lens

The optical part of a camera focuses beams of light so that they form a real image that looks just like the image in front of the camera. The component that makes this possible is the lens.

A lens is a piece of glass or plastic that is curved so that it refracts light. Refraction is actually a fairly simple process. As light travels from the one medium to another, it changes speed. Light travels more quickly through air than it does through glass, so a lens slows it down. When lightwaves enter a piece of glass at an angle, one part of the wave reaches the glass before another, and so it starts slowing down first. This is something like pushing a shopping cart from pavement to grass, at an angle. The right wheel hits the grass first, and so it slows down while the left wheel is still on the pavement. Because the left wheel is for the moment moving more quickly than the right wheel, the shopping cart turns to the right as it moves onto the grass.

The effect on light is similar to the effect of moving the shopping cart from pavement to grass: As light enters the glass at an angle, it bends in one direction. It bends again when it exits the glass, because parts of the lightwave enter the air and speed up before other parts of the wave.

In a converging—or convex—lens, each side of the glass curves out. This means rays of light passing through bend toward the center of the lens on entry and exit.

This effectively reverses the path of light from an object. A light source—say, a candle—emits light in all directions. The rays of light all start at the same point, the candle's flame, and then are constantly diverging. A converging lens takes those rays and redirects them so that they are all converging back to one point.

The exact position of the real image depends on the lens structure and the distance between the lens and the object. You can observe this phenomenon with a simple experiment. Light a candle in the dark, and hold a magnifying glass between it and the wall. You will see an upside-down image of the candle on the wall that appears somewhat blurry. To focus the image, you move the magnifying glass closer to or farther away from the candle.

To focus a camera on an object, you move the lens in and out. This changes the distance between the lens and the film, so the real image of the object is lined up exactly with the film surface. When you take the picture, the image is exposed to the film, which records the light information.

The Chemical Process: The Film

The active component in photographic film is a collection of tiny light-sensitive grains of silver halide. These grains are spread out in a chemical suspension on a strip of plastic. When exposed to light, the grains undergo a chemical reaction, recording the pattern of light and dark that make up the real image.

A roll of film is developed by being exposed to other chemicals, which react with the light-sensitive grains. In black-and-white film, the developer chemicals convert the grains that were exposed to light into opaque silver and wash away the remaining silver halide, leaving the transparent plastic film base. This produces a negative, in which lighter areas in the original scene appear darker and darker areas appear lighter. This negative is then converted into a positive image in printing.

Color film has three different layers of light-sensitive materials that respond to the three colors red, green, and blue. When the film is developed, these layers are exposed to chemicals that dye the layers of film to create a color negative. When you overlay the color information from all three layers, you get a full-color picture.

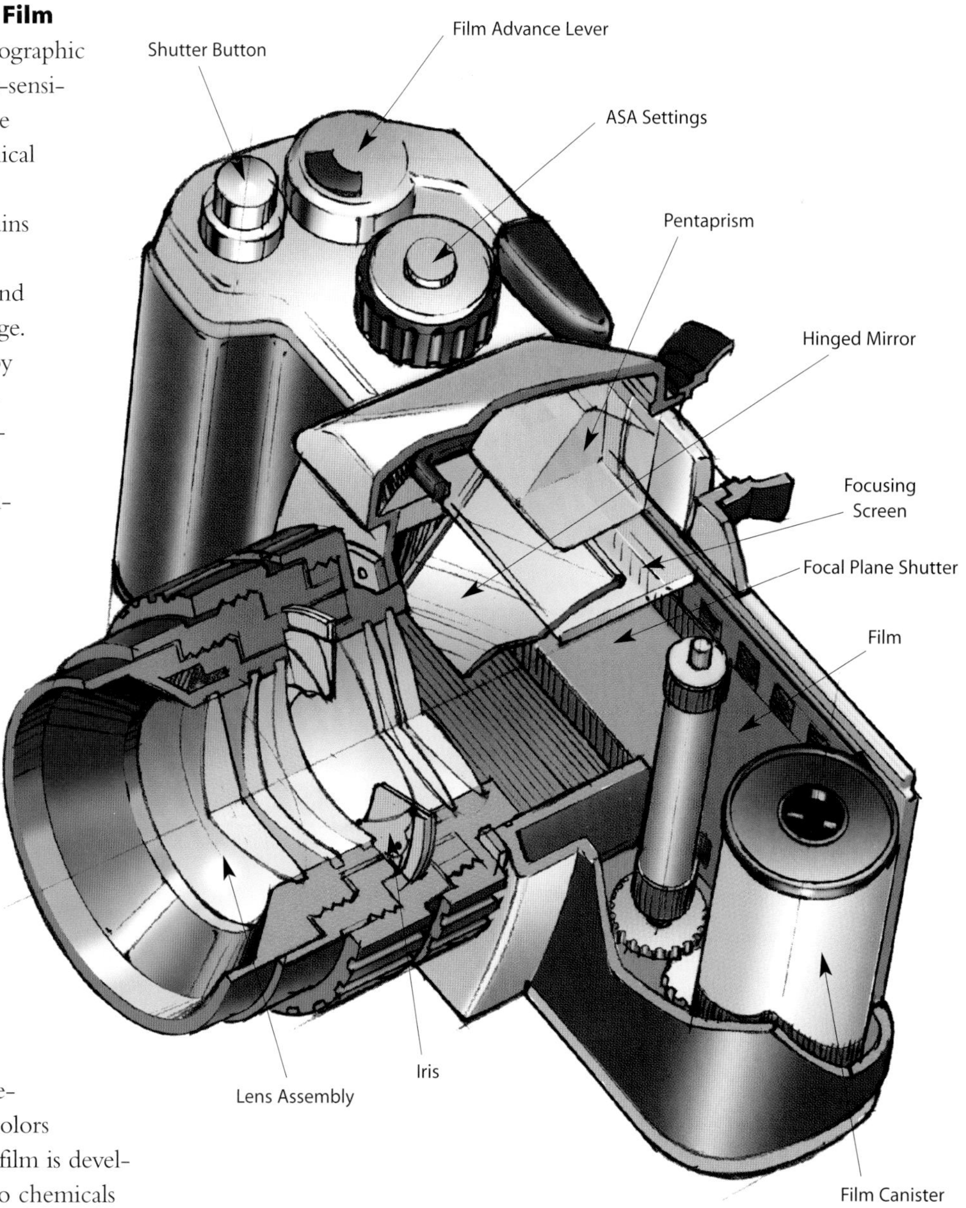

The Mechanical Process: The Camera

To keep the light-sensitive grains from being exposed, film must be kept in complete darkness until you take pictures. To get a clear, balanced picture, you need to carefully control how much light is exposed at the moment that you open the shutter. If you let too much light hit the film, too many grains will react, and the picture will appear washed out. If you don't let enough light hit the camera, too few grains will react, and the picture will be too dark. The camera has a precise mechanical or electronic system to control this exposure level.

The camera has two mechanisms for controlling the exposure. You can adjust the size of the aperture—the opening in the lens that light passes through—with a simple iris diaphragm. This mechanism consists of several overlapping metal plates that can fold in on each other or expand out. Essentially, the iris in a camera works the same way as the iris in your eye—it opens or closes in a circle, to shrink or expand the diameter of the lens. When the lens is smaller, it captures less light, and when it is larger, it captures more light.

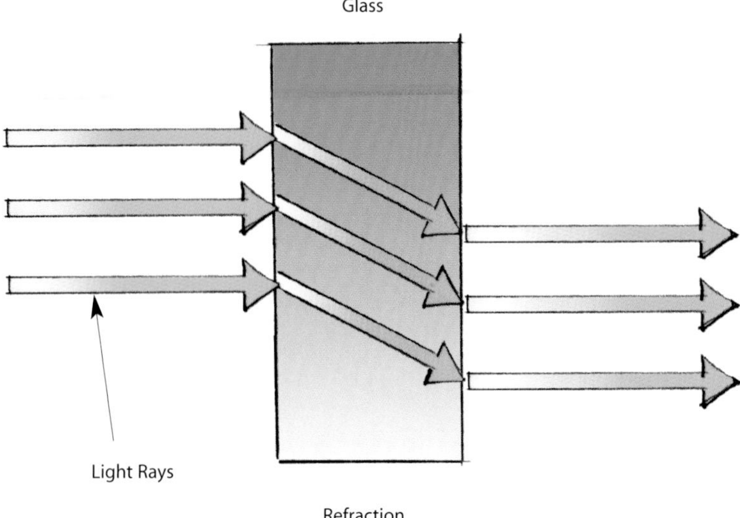

Did You Know...

A camera lens is actually several lenses combined into one unit. A single converging lens could form a real image on the film, but it would not be a very good image.

One of the most significant warping factors is that different colors of light bend differently when moving through a lens. This chromatic aberration produces an image in which the colors are not lined up correctly. Cameras compensate for this by using several lenses made of different materials. The lenses each handle colors differently, and when you combine them in a certain way, they correctly realign the warped colors.

The other variable for controlling the exposure is shutter speed. The shutter in a standard single-lens reflex (SLR) camera is very simple—it basically consists of two "curtains" between the lens and the film. Before you take a picture, the first curtain is closed, so the film won't be exposed to light. When you take the picture, the curtain slides open, so the film is exposed. After a certain amount of time, the second curtain slides in from the other side, to stop the exposure.

You can vary the exposure by varying the amount of time the shutter is open. To get a clear, properly exposed image, photographers balance shutter speed and aperture size.

Most cameras have a built-in light meter to help you get the proper balance. The main component of the light meter is a panel of semiconductor light sensors that are sensitive to light energy. The sensors express this light energy as electrical energy. The light meter system interprets this information based on the film and shutter speeds.

The mechanical system in a camera is very intricate because it has to balance so many elements precisely. It does this with a complex combination of gears and springs that move in unison to reset the camera for each shot. In this way, a camera is a lot like a watch or pendulum clock—but modern electronic cameras use a microprocessor to do the same thing.

There are two types of consumer film cameras on the market—SLR and viewfinder (VF) cameras. The main difference between them is how they let the photographer see the scene. In a VF camera, the viewfinder is a simple window through the body of the camera. You don't see the real image formed by the camera lens, but you get a rough idea of what is in view.

In an SLR camera, you see the real image that the film will see. If you take the lens off an SLR camera and look inside, you'll see how this works. The camera has a slanted mirror positioned between the shutter and the lens. The mirror aims the light up at a translucent glass layer. A prism above the glass layer aims the image at the viewfinder window. You see the image hitting the translucent glass layer. This layer is exactly the same distance from the lens as the film, so the real image appears just as it will be recorded.

When you take a picture with an SLR camera, the camera quickly switches the mirror out of the way so that the image is directed at the exposed film. This is why the viewfinder blacks out when you take a picture.

In an SLR camera, the mirror and prism are set up so that they present the real image exactly as it will appear on the film. The advantage of this design is that you can adjust the focus and compose the scene, so you get exactly the picture you want. For this reason, professional photographers typically use SLR cameras.

There have been many different camera systems over the years, but they all combine the same basic elements: a lens system to create the real image, a light-sensitive film to record the real image, and a mechanical or an electronic system to control how the real image is exposed to the film. Conceptually, that's all there is to photography!

How **AUTOFOCUS CAMERAS** Work

Most cameras now have some sort of autofocus system to make it easier to take a good picture. Video cameras, digital cameras, single-lens reflex (SLR) cameras, and point-and-shoot cameras all use some kind of automatic system to focus the lens.

There are two common autofocus systems: Active and Passive. Active autofocus systems emit energy—normally ultrasound or infrared light—that bounces off an object. Passive systems use light already reflected off the object to focus the camera.

Active Autofocus

One example of an active autofocus system is sound navigation ranging (sonar) autofocus. The system emits ultra-high-frequency sounds and then listens for the echo—the sound waves reflected by an object. Because sound travels at a constant speed, the autofocus system can figure out the distance to an object based on how long it took the sound wave to reflect and return. The autofocus processor then activates a small motor that adjusts the camera's lens so it focuses on objects at that distance.

This system has limitations. For example, if you try taking a picture through a window, the sound waves will bounce off the glass, and the motor will focus the lens incorrectly. For this reason, sonar is not used in many modern cameras.

Today, active autofocus cameras typically use an infrared beam instead of sound waves. On any camera that uses an infrared system, you can see both the infrared emitter and the receiver on the front of the camera, normally near the viewfinder.

Infrared sensing can have problems, such as the following:

- A source of infrared light from an open flame (such as birthday cake candles) can confuse the infrared sensor.
- A dark-colored subject may absorb the outbound infrared beam.
- The infrared beam can bounce off of something in front of the subject—such as glass or bars—rather than making it to the subject, causing the focus to be in the wrong spot.

One advantage of an active autofocus system is that it works in the dark, making flash photography easy.

Passive Autofocus

Passive autofocus systems determine the distance to the subject by analyzing the image itself. They are considered passive because they don't emit any energy of their own.

One common passive autofocus system uses a charge-coupled device (CCD), typically a single strip of 100 or 200 light-sensitive diodes. The CCD actually looks at the scene through the lens so that it will see the image as the film will see it.

In this system, light from the scene hits the CCD strip, and the microprocessor looks at the values from each diode. When the image is in focus, there are very sharp areas of contrast from diode to diode. When the image is out of focus, the light is blurred, and the values change gradually. The system searches for the point where there is a maximum contrast between adjacent diodes. This is the point where the image is focused most precisely.

This sort of system must see contrasts in light in order to work correctly. If you try to take a picture of a blank wall or a large object of uniform color (such as the sky), the camera cannot compare adjacent diodes, so it cannot focus.

If you want to take pictures very quickly or you don't want to bother with focusing a picture yourself, an autofocus system is extremely useful. This technology has certainly made it easier for inexperienced photographers to take good pictures, which has made the art of photography accessible to a much wider range of people.

HSW Web Links

www.howstuffworks.com

How Camcorders Work
How Digital Cameras Work
How DVDs and DVD Players Work
How Photographic Film Works

How DIGITAL CAMERAS Work

A digital camera, a camcorder, and a webcam all work about the same way: They turn light into electrical signals and record it. In a digital camera, the signals are recorded in flash memory or on a disk. In a camcorder, the signals are recorded on videotape at 30 frames per second. In a webcam, they are recorded in a file for a Web page.

www.howstuffworks.com

How Camcorders Work
How DVDs and DVD Players Work
How Offset Printing Works
How PDAs Work
How Serial Ports Work

At its most basic level, a digital camera is a device that measures light bouncing off objects and converts this information into a digital file in some standard format. Most digital cameras produce JPEG or TIFF files by taking the following steps:

1) Focus the light from the scene onto a sensor.
2) Convert the measured light on the sensor into an electrical charge.
3) Convert this analog information into a digital form.
4) Save the digital information in the proper file format.

Rear Housing

Viewfinder Opening

Display Windows

Light Capture

A digital camera focuses light the same way a film camera does—with a series of lenses. But whereas a film camera uses chemically treated celluloid to capture an image, a digital camera focuses light onto an electronic image sensor instead.

The standard sensor technology for most digital cameras is a charge-coupled device (CCD). The CCD is a collection of tiny light-sensitive diodes (called photosites), which convert photons (that is, light) into electrons. Each photosite is sensitive to light—the brighter the light that hits a single photosite, the greater the electrical charge that accumulates at that site. The number of photosites on a CCD determines its maximum resolution.

The next step is to read the value (that is, the accumulated charge) of each cell in the image. In a CCD, the charge is actually transported across the chip and read at one corner of the array. An analog-to-digital converter (ADC) turns each pixel's value into a digital value.

The ADC is a sophisticated piece of equipment, but the basic concept is very simple. Think of each photosite on the CCD as a bucket. Now think of the photons of light as raindrops. As the raindrops fall into the bucket, water accumulates (in reality, electrical charge accumulates). Some buckets have more water than others, representing brighter and darker sections of the image. The ADC measures the depth of the water in each bucket one by one. Then it records this information as a binary value. Even the simplest digital image contains thousands of buckets, and the best digital cameras have millions of photosites on the CCD.

Color Capture

A photosite measures just the total intensity of the light striking its surface. It cannot distinguish different colors of light from one

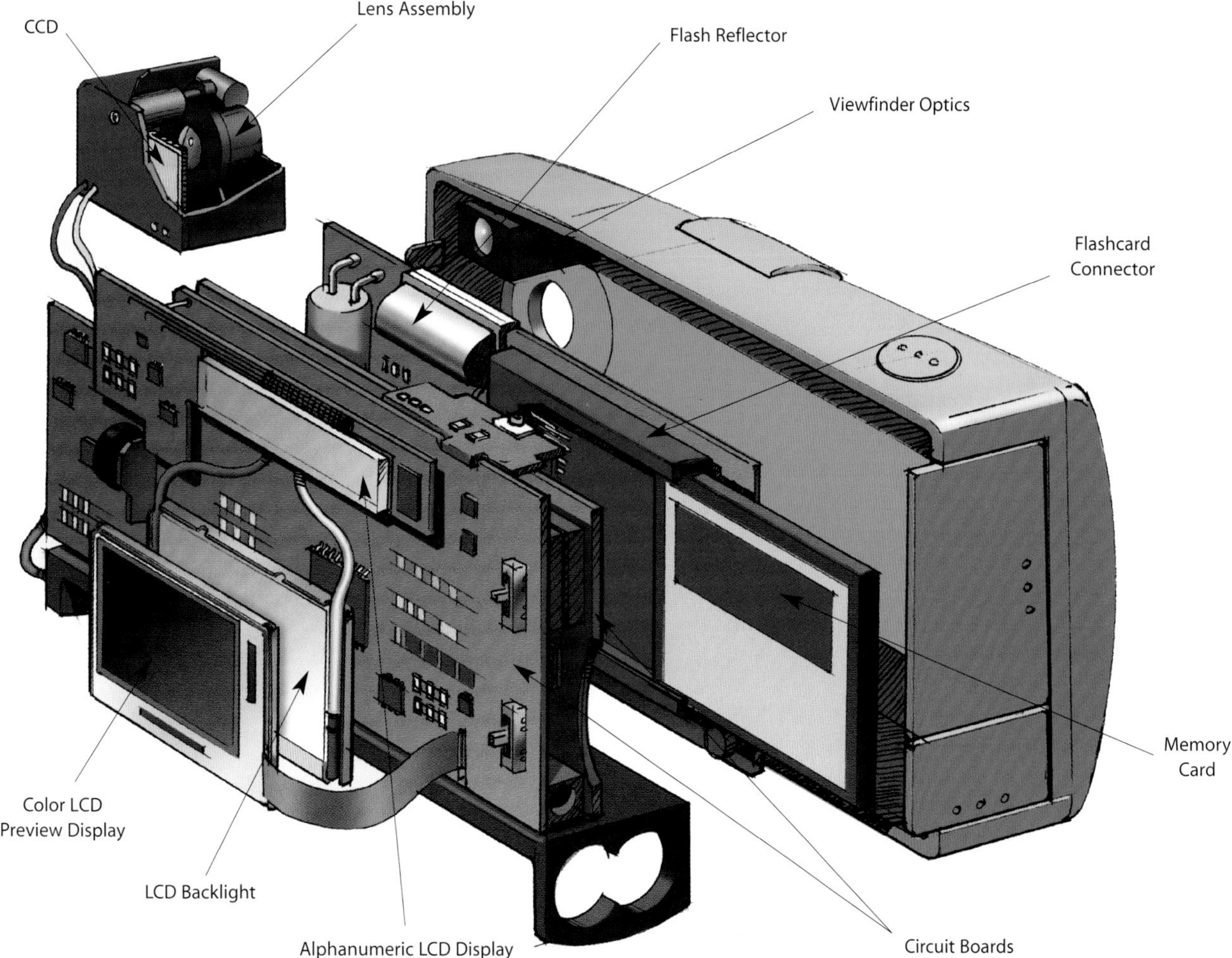

another. To get a full-color image, digital cameras use special filters that separate light into the three colors red, green, and blue. By combining these colors in different proportions, you can create all the colors in the visible spectrum.

The highest-quality cameras use three separate CCDs, each with a different filter over it. A beam splitter sends the same light information to all three sensors so that each sensor gets an identical look at the image. When you overlay the red, green, and blue images, you get a full-color picture. The advantage of this method is that the camera records each of the three colors at each pixel location. Unfortunately, cameras that use this method tend to be bulky, power-hungry, and expensive.

A second method for capturing color is to rotate a series of red, blue, and green filters in front of a single sensor. The sensor records three separate images in rapid succession. This method also provides information on all three colors at each pixel location. But because the three images aren't taken at precisely the same moment, both the camera and the target of the photo must remain stationary for all three readings. This isn't practical for candid photography or handheld cameras.

A more economical and practical way to record the three primary colors from a single image is to permanently place a filter over each individual photosite. By breaking up the sensor into a variety of red, blue, and green pixels, it is possible to get enough information in the general vicinity of each sensor to make very accurate guesses about the true color at that location. This process of looking at the other pixels in the neighborhood of a sensor and making an educated guess is called interpolation.

The most common pattern of permanent filters is the bayer filter pattern. This pattern alternates a row of red and green filters with a row of blue and green filters. You may be surprised to find that there are so many green pixels. In fact, there are as many green pixels as there are blue and red pixels combined. This is because the human eye is not equally sensitive to all three colors. It's necessary to include more information from the green pixels in order to create an image that the human eye will perceive as a true color.

The advantages of using a bayer filter are that only one sensor is required and all the color information (red, green, and blue) is recorded at the same moment. That means the camera can be smaller, cheaper, and more useful in a wider variety of situations. In other words, it makes it possible to create an affordable handheld digital camera. The raw output from a sensor with a bayer filter is a mosaic of red, green, and blue pixels of different intensities.

Output, Storage, and Compression

Most digital cameras on the market have a liquid crystal display (LCD) screen so that you can view your picture right away. This is one of the great advantages of a digital camera: You get immediate feedback on what you capture. When the image leaves the CCD sensor (by way of the ADC and a microprocessor), it is ready to be viewed on the LCD.

Of course, you also want to be able to load the picture into your computer or send it directly to a printer. There are several ways to store images in a camera and then transfer them to a computer.

Early generations of digital cameras had fixed storage inside the camera. To get the pictures out, they needed to be hooked up directly to a computer by cables so that the images could be transferred. Although most of today's cameras are capable of connecting to a serial, parallel, SCSI, and/or USB port, they usually also provide you with some sort of removable storage device. Most portable memory devices use solid-state technolgy—memory storage with no moving parts—and you can buy a memory reader to attach to your computer. Some cameras use standard floppy disks or writable CDs instead.

Zoom

Most digital cameras have a traditional optical zoom lens, just like film cameras, but they may also have digital zoom technology.

An optical zoom magnifies an image by changing the focal length of the lens. The advantage of an optical zoom is that it focuses the magnified image on the entire surface of the CCD.

A digital zoom, on the other hand, uses only a small portion of the photosites on the CCD. Then it uses interpolation techniques to add detail to the photo. Although it may look like you are shooting a picture with twice the magnification, you can get the same results by shooting the photo without a zoom and blowing up the picture by using computer software.

Because digital cameras record light values in a digital form, they create images you can manipulate with computers, send over phone lines, or post on a Web site. They are so useful that they have already become standard equipment for newspaper and magazine employees, freelance photographers, and anybody else who deals with photography daily. The resolution and storage capacities of digital cameras are rapidly increasing, and prices are falling. Eventually digital cameras will equal the quality and price of film cameras.

Did You Know?

Like a digital camera, a camcorder captures light with a CCD. The main difference between a digital camera and a camcorder is that a camcorder is constantly capturing images, which it encodes in an analog video signal. This signal is composed of scan lines—the lines a television electron beam paints on the screen. A camcorder records this information on tape with a built-in VCR. A digital camcorder converts the analog video signal to a digital form before encoding it to tape.

A webcam is essentially a digital camera that takes pictures regularly say every 10 seconds—and transmits them directly to a computer. You can program a computer to upload these images to a Web page, as they are transmitted. Depending on the frequency of the pictures, Web viewers will see the series of digital pictures as a moving image.

How **DVDs & DVD PLAYERS** Work

Digital videodiscs (DVDs) are to movies what compact discs (CDs) are to music, and DVDs are to videotape what CDs are to cassette tape. A DVD is a digital format for storing movies with incredible picture and sound quality. A DVD can also store huge quantities of digital data and software for use on a computer.

A DVD is very similar to a CD, but it has a much larger data capacity. (See "How CDs Work," page 210, for more information.) A standard single-layer DVD holds about seven times more data than a CD. This huge capacity means that a DVD has enough room to store a full-length, MPEG 2–encoded movie, as well as a lot of other information.

A DVD movie typically includes the following:

- Up to 133 minutes of high-resolution video, in letterbox or pan-and-scan format, with 720 dots of horizontal resolution. The video compression ratio is typically 40:1, using MPEG-2 compression.
- A soundtrack presented in up to eight languages, using 5.1 channel Dolby digital surround sound.
- Subtitles in up to 32 languages.

DVD vs. CD Storage

DVDs can store more data than CDs for a few reasons:

- **Higher-density data storage**—A single-sided, single-layer DVD can store about seven times more data than a CD. A large part of this increase is a result of the pits and tracks being smaller on DVDs. There is room for about 4.5 times as many pits on a DVD as on a CD.
- **Less overhead and more area**—On a CD, a lot of extra information is encoded to allow for error correction. The error-correction scheme that a CD uses is inefficient compared to the method used on DVDs. The DVD format doesn't waste as much space on error correction, which enables it to store much more real information.
- **Multilayer storage**—A DVD can have up to four layers, two on each side. The laser that reads the disc can focus, for example, on the second layer through the first layer. The capacity of a DVD doesn't double when you add a second layer to the disc. On a two-layer disc, the pits have to be a little longer, on both layers, than when a single layer is used. This helps to avoid interference between the layers.

A DVD also has the potential to store almost eight hours of CD-quality music per side.

Data Storage on a DVD

DVDs are of the same diameter and thickness as CDs, and they are made using some of the same materials and manufacturing methods as CDs. As on a CD, the data on a DVD is encoded in the form of small pits or bumps, arranged in a spiral track on a mirrored disc.

A DVD is composed of several layers of plastic about 1.2 mm thick. DVDs can have one or two layers per side and can be single-sided or double-sided. In a multilayer DVD, the laser needs to be able to focus separately on either the first or second layer. To allow this sort of two-layer focusing, the first layer uses a semitransparent coating of gold, and the second layer uses an opaque aluminum mirror, like a CD does. The laser can focus on the gold layer, or it can shine through the gold and focus on the aluminum layer.

Each layer of the DVD is made just like a CD, with a piece of injection-molded plastic impressed with billions of tiny bumps. A layer of aluminum or gold is sprayed onto the bumps to create the reflective coating. After all the layers are made, each one is

HSW Web Links

www.howstuffworks.com

How Camcorders Work
How Digital Cameras Work
How MP3 Players Work
How PDAs Work
How VCRs Work

coated with lacquer, they are squeezed together, and the disc is cured under infrared light. For single-sided discs, the label is silk-screened onto the nonreadable side. Double-sided discs are printed only on the nonreadable area near the hole in the middle.

Each layer of a DVD has a spiral track of data. On single-layer DVDs, the track always circles from the inside of the disc to the outside. The data track is incredibly small. The elongated bumps that make up the track are each 320 nanometers (nm, or billionths of a meter) wide, a minimum of 400 nm long, and 120 nm high. Just 740 nm separate one track from the next.

DVD Video Format

Even though a DVD's storage capacity is huge, the uncompressed video data of a full-length movie would never fit on a DVD. In order to fit a movie on a DVD, you need video compression. A group called the Moving Picture Experts Group (MPEG) establishes the standards for compressing moving pictures.

A movie is usually filmed at a rate of 24 frames per second. The MPEG encoder that creates the compressed movie file analyzes each frame and decides how to encode it. Each frame can be encoded in one of three ways:

- **As an intraframe**—An intraframe contains the complete image data for that frame. This method of encoding provides the least compression.
- **As a predicted frame**—A predicted frame contains just enough information to tell the DVD player how to display the frame based on the most recently displayed intraframe or predicted frame. This means that the frame contains only the data that relates to how the picture has changed from the previous frame.
- **As a bidirectional frame**—In order to display a bidirectional frame, the player must have the information from the surrounding intraframe or predicted frames. Using data from the closest surrounding frames, the player uses interpolation, something like averaging, to calculate the position and color of each pixel.

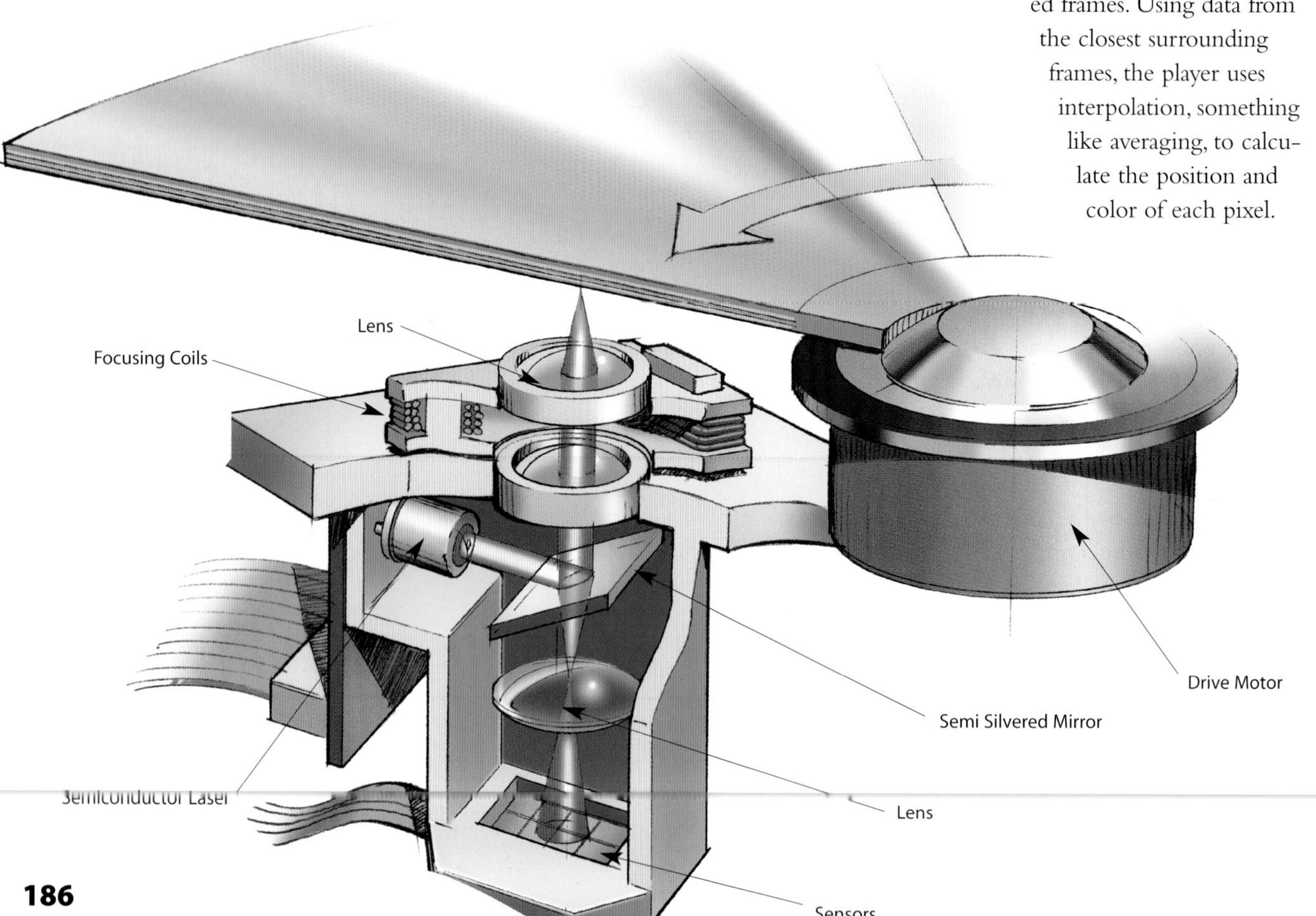

Depending on the type of scene being converted, the encoder will decide which types of frames to use. If a newscast were being converted, a lot more predicted frames could be used because most of the scene is unaltered from one frame to the next. On the other hand, in a very fast action scene—in which things changed very quickly from one frame to the next—more intraframes would have to be encoded. The newscast would compress to a much smaller size than the action sequence.

DVD Players

A DVD player is very similar to a CD player. It has a laser assembly that shines the laser beam onto the surface of the disc to read the pattern of bumps. The DVD player decodes the MPEG 2–encoded movie, turning it into a standard composite video signal. The player also decodes the audio stream and sends it to a Dolby decoder, where it is amplified and sent to the speakers.

The DVD player has the job of finding and reading the data stored as bumps on the DVD. Considering how small the bumps are, the DVD player has to be an exceptionally precise piece of equipment. The drive consists of three fundamental components:

- **A drive motor to spin the disc**—The drive motor is precisely controlled to rotate between 200 and 500 RPM, depending on which track is being read.
- **A laser and a lens system to focus on the bumps and read them**—The light from this laser has a smaller wavelength (640 nm) than the light from the laser in a CD player (780 nm), which allows the DVD laser to focus on the smaller DVD pits.
- **A tracking mechanism**—This mechanism can move the laser assembly so that the laser's beam can follow the spiral track. The tracking system has to be able to move the laser at micron resolutions.

The DVD player's laser can focus either on the semitransparent reflective material behind the closest layer or, in the case of a double-layer disc, through this layer and onto the reflective material behind the inner layer. The laser beam passes through the polycarbonate layer, bounces off the reflective layer behind it, and hits an opto electronic device that detects changes in light. The bumps reflect light differently than the "lands" (the flat areas of the disc), and the optoelectronic sensor detects that change in reflectivity. The electronics in the drive interpret the changes in reflectivity in order to read the bits that make up the bytes.

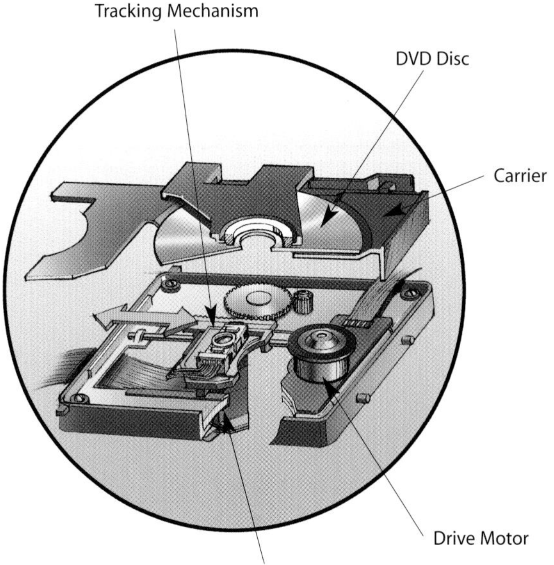

The hardest part of reading a DVD is keeping the laser beam centered on the data track, which is the job of the tracking system. As the DVD is played, the tracking system has to move the laser outward continually. As the laser moves outward from the center of the disc, the bumps move past the laser at an increasing speed. As the laser moves outward, the spindle motor slows down so that the data comes off the disc at a constant rate.

An interesting thing to note is that if a DVD has a second layer, the start of that layer's data track can be at the outside of the disc instead of the inside. This allows the player to transition quickly from one layer to the next, without a delay in data output, because it doesn't have to move the laser back to the center of the disc to read the next layer.

With closer track spacing, smaller bumps and multiple layers, the DVD can store much more data than a CD. And with clever video compression, it has enough capacity to store full-length movies with multiple soundtracks and subtitles. On top of all this, the quality of DVD video is much better than that of VHS videotapes.

Cool Facts

- The first DVD player hit the market in March 1997.
- If an average DVD movie were uncompressed, it would take at least a year to download it over a normal phone line.
- The Sony PlayStation 2 was the first videogame system able to play DVDs.
- DVDs often have special features hidden on them. These "Easter eggs" can be previews of other movies, computer software, or music.
- Some DVDs carry commentary tracks, in which the filmmaker talks about the movie while it is running. This feature is exciting for film buffs. DVDs can also contain extra, previously unreleased scenes. And a DVD is sometimes a director's cut—the film as the director originally intended it rather than the version that played in theaters.
- Software loaded from a DVD can contain more information than software loaded from a CD-ROM. For example, an entire encyclopedia that can fit onto one DVD would require multiple CD-ROMs.

How VCRs Work

The introduction of VCRs is one of a handful of events that stand out as being extremely important to television history. The VCR gave people, for the first time, control of what they could watch on their TVs and when.

HSW Web Links

www.howstuffworks.com

How Camcorders Work
How Cell Phones Work
How DVDs and DVD Players Work
How TV Works
How Video Game Systems Work

VCRs (videocassette recorders) basically work like audiotape recorders: They use small electromagnets called tape heads to record a varying magnetic field on ferromagnetic tape. The varying magnetic signal represents the varying electrical signal used to transmit TV video and audio.

Parts of a VCR

A VCR has a more difficult job than a tape deck because a video signal contains roughly 500 times as much information as a sound signal. An audiotape recorder can record a signal linearly, in a long, straight track on the tape. To record or play a track, a tape player moves the tape past the tape head at a rate of two or three inches per second. To record a video signal in a linear track, the VCR would have to spool many feet of tape past the tape head every second. It would take 50 miles of tape to store a 2-hour movie!

VCRs have an ingenious mechanism that solves this problem—the tape heads themselves move, along with the tape. Two (or four) tape heads ride on a rotating drum. The drum is tilted at an angle in relation to the tape, so that it moves along the tape diagonally. Each tape head records or plays one diagonal band on every pass. This helical scanning system allows the VCR to fit a lot of information on a length of tape, and the tape can move at a reasonable speed past the rotating heads.

A unique part of this design is that the VCR has to wrap the videotape up against the rotating head when it plays or records. The VCR needs to do a few other things as well:

- It needs to read and encode linear audio and control tracks on the same tape.
- It needs to keep the tape moving at exactly the right speed.
- It needs to detect the end of the tape.

If you are recording, the VCR also has to move the tape past the erase head to get rid of previously recorded material. All this requires a lot of precision machinery, in both the VCR and the videocassette.

Tape Movement

When you insert a tape into your VCR, the first thing that happens is that a pin releases and opens the plastic guard on the bottom of the cassette so that the tape is exposed. The VCR also inserts a pin into part of the cassette to disengage a lock on the tape spools. Then two movable arms pull the tape out of the cassette to fit it around the rotating drum as well as all the other heads and rollers that the tape must travel past. A typical VCR pulls more than a foot of tape out of the cassette to accomplish this.

Next, the pinch roller and inertia roller press the tape onto the audio head, the control head, and the erase head. It's like watching a ballet when you see all this take place—there are some pretty amazing mechanical engineers working on this stuff!

The tape is also spooled between a small light and a light sensor. The beginning and end of a videotape are clear so that the VCR knows to stop when the light shines through the tape.

The control track keeps everything operating correctly. It tells the VCR whether the tape was recorded in standard play (SP), long play (LP), or extended play (EP) mode, and it gets the tape heads to line up with the diagonal tracks correctly. It also makes sure the tape is moving at the right speed. Tape can stretch over time, so the VCR may have to speed up to keep the video playing correctly.

There are many different VCR designs, with numerous special features, but they all have the same sort of system that pulls tape through a series of rollers, drums, and sensors. When you consider how much complex machinery is involved, you might be surprised that a VCR ever works—or that you can buy one for less than $100!

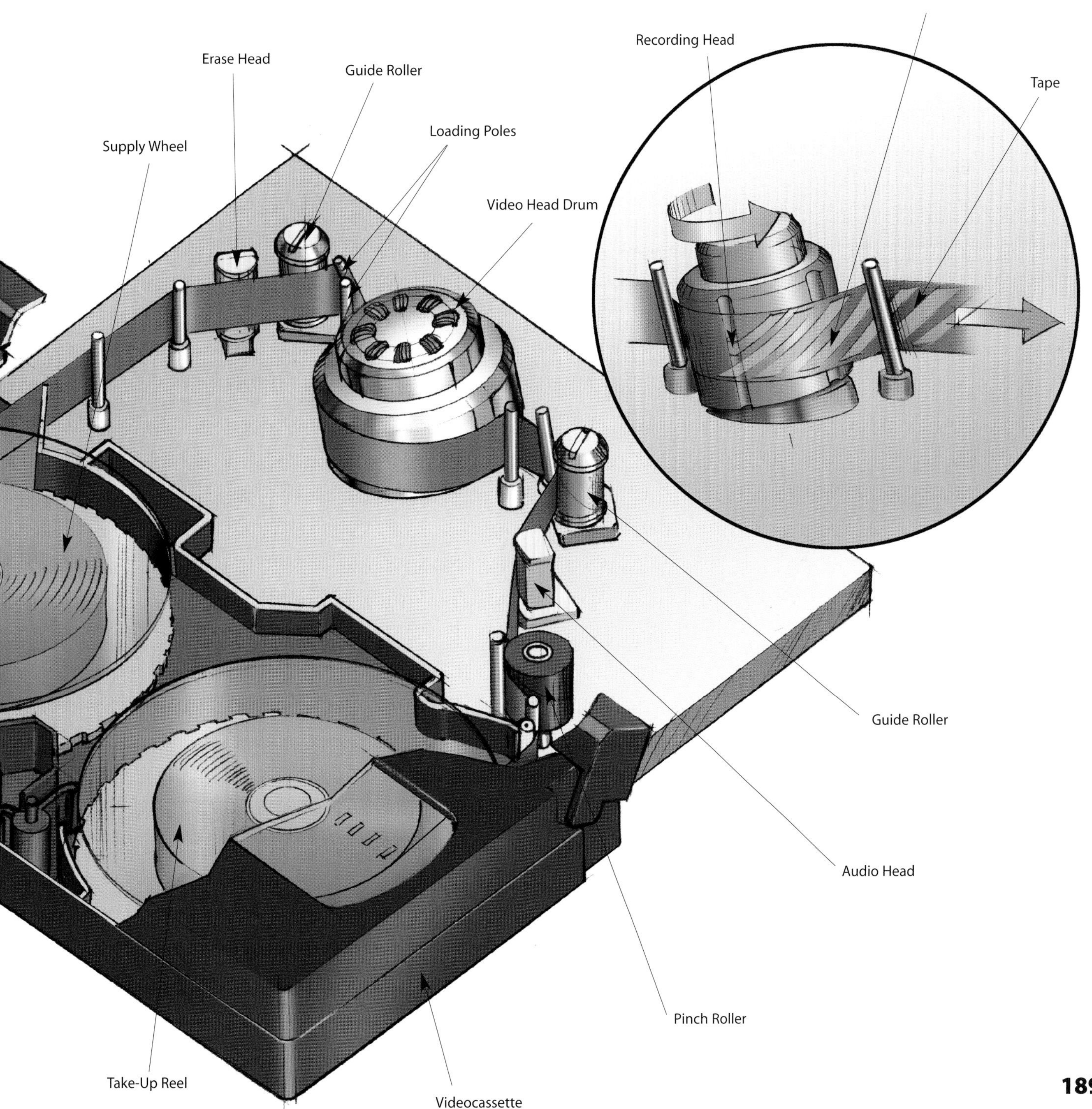

How **TELEVISION** Works

Television is one of the most influential forces of our time. Through a TV set, you are able to receive news, sports, entertainment, information, and commercials. Have you ever wondered about the technology that makes TV possible?

HSW Web Links

www.howstuffworks.com

How Digital TV Works
How Cable TV Works
How HDTV Works
How Projection TV Works
How Racef/x Works

At its heart, what you see on a TV screen is a composite image created from tens of thousands of tiny phosphorescent molecules.

Parts of a TV

Almost all TVs in use today rely on a device known as the cathode ray tube (CRT) to display images. CRTs have several important parts:

- The tube part is a large hollow piece of gas filled with a vacuum.
- The cathode ray part, also known as an electron gun, is able to generate, aim, and steer a very fine beam of electrons at the screen.
- The part of the CRT that we look at is a screen covered with phosphor.

The electron beam hits the phosphor screen to light it up. Because the electron beam is very thin, the beam has to rapidly scan across and down the entire screen to paint an image.

In a CRT, the inside of the screen is coated with phosphor. Phosphor is a material that, when exposed to energy, emits visible light. The energy might be ultraviolet light or a beam of electrons. Any fluorescent color is really a phosphor—fluorescent colors absorb invisible (to humans)

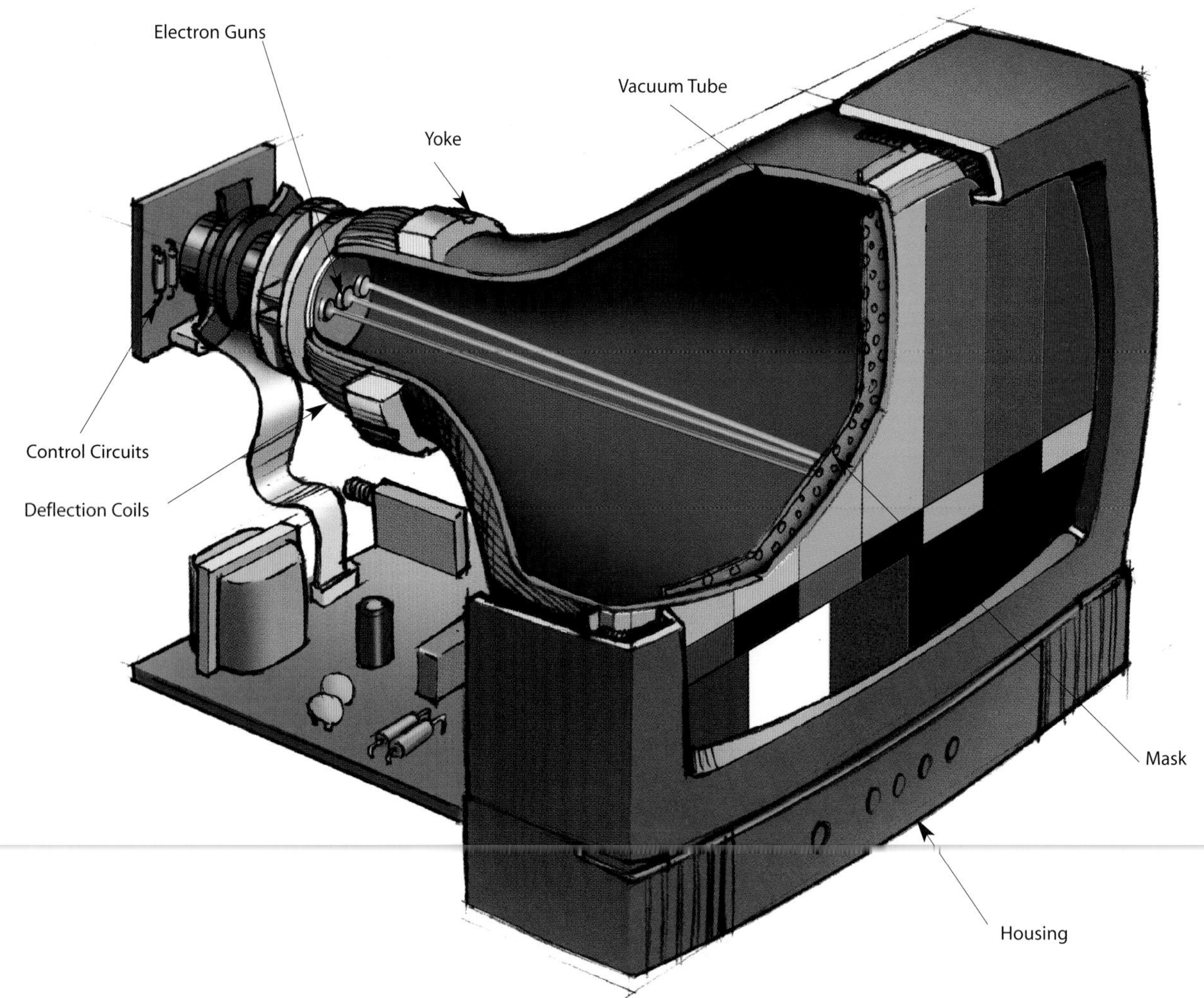

ultraviolet light and emit visible light of a characteristic color.

When the electron beam strikes the phosphor inside a TV screen, it makes the screen glow. In a black-and-white screen, there is one phosphor that glows white when struck. In a color screen there are three phosphors, arranged as dots or stripes, that emit red, green, and blue light. There are also three electron beams to illuminate the three different colors together.

The terms *anode* and *cathode* are used in electronics as synonyms for positive and negative terminals, respectively. In a CRT, the cathode is a heated filament (like the filament in a light bulb). The heated filament exists in a vacuum created inside a glass tube. The ray is a stream of negatively charged electrons that naturally pour off a heated cathode into the vacuum. The anode is positive, so it attracts the electrons pouring off the cathode. In a TV's CRT, a focusing anode focuses the electrons into a tight beam, and then an accelerating anode speeds them up. This tight, high-speed beam of electrons flies through the vacuum in the tube and hits the flat screen at the other end of the tube. This screen, coated with phosphor, glows when it is struck by the beam.

There's not a whole lot to a basic CRT. There is a cathode, a pair (or more) of anodes, the phosphor-coated screen, and a conductive coating inside the tube to soak up the electrons that pile up at the screen end of the tube. The CRT has no way to steer the beam. That's why, if you look inside any TV set, you will find that the tube is wrapped in coils of wire.

The steering coils are simply copper windings. These coils are able to create magnetic fields inside the tube, and the electron beam responds to the fields. One set of coils creates a magnetic field that moves the electron beam vertically, and another set moves the beam horizontally. By controlling the voltages in the coils, you can position the electron beam at any point on the screen.

Black-and-White TV Signals

In a black-and-white TV, the screen is coated with white phosphor, and the electron beam paints an image onto the screen by moving the electron beam across the phosphor a line at a time. To paint the entire screen, electronic circuits inside the TV use the magnetic coils to move the electron beam in a "raster scan" pattern across and down the screen: The beam paints one line across the screen from left to right, and then it quickly flies back to the left side, moves down slightly, and paints another horizontal line, and so on down the screen.

As the beam paints each line, the intensity of the electron beam changes, to create different shades of black, gray, and white across the screen. Because the lines are spaced very closely together, the human brain integrates them into a single image. A TV screen normally has 525 lines visible from top to bottom.

All analog TVs use an interlacing technique when painting the screen. In this technique, the screen is painted 60 times per second, but only half the lines are painted per frame. The beam paints every other line as it moves down the screen—for example, it might paint every odd-numbered line. Then the next time it moves down the screen, it paints the even-numbered lines, alternating back and forth between even-numbered and odd-numbered lines on eachpass. All 525 lines on the screen, in two passes, are painted 30 times every second, for a total of 15,750 lines per second. (Some people can actually hear this frequency as a very high-pitched sound emitted when the TV is on.)

When a TV station wants to broadcast a signal to your TV, or when a VCR wants to display the movie from a videotape, the sig-

Monitors vs. TV

Your computer probably has a "VGA monitor" that looks a lot like a TV but is smaller, has a lot more pixels and has a much crisper display. The CRT and electronics in a monitor are much more precise than is required in a TV because of the higher resolutions. Most computer monitors use a technique called progressive scanning to paint the screen. Progressive scanning, which paints every line on the screen 60 times a second, significantly reduces flicker. In addition, the plug on a VGA monitor is not accepting a composite signal—a VGA plug separates out all of the signals so that the monitor can interpret them more precisely.

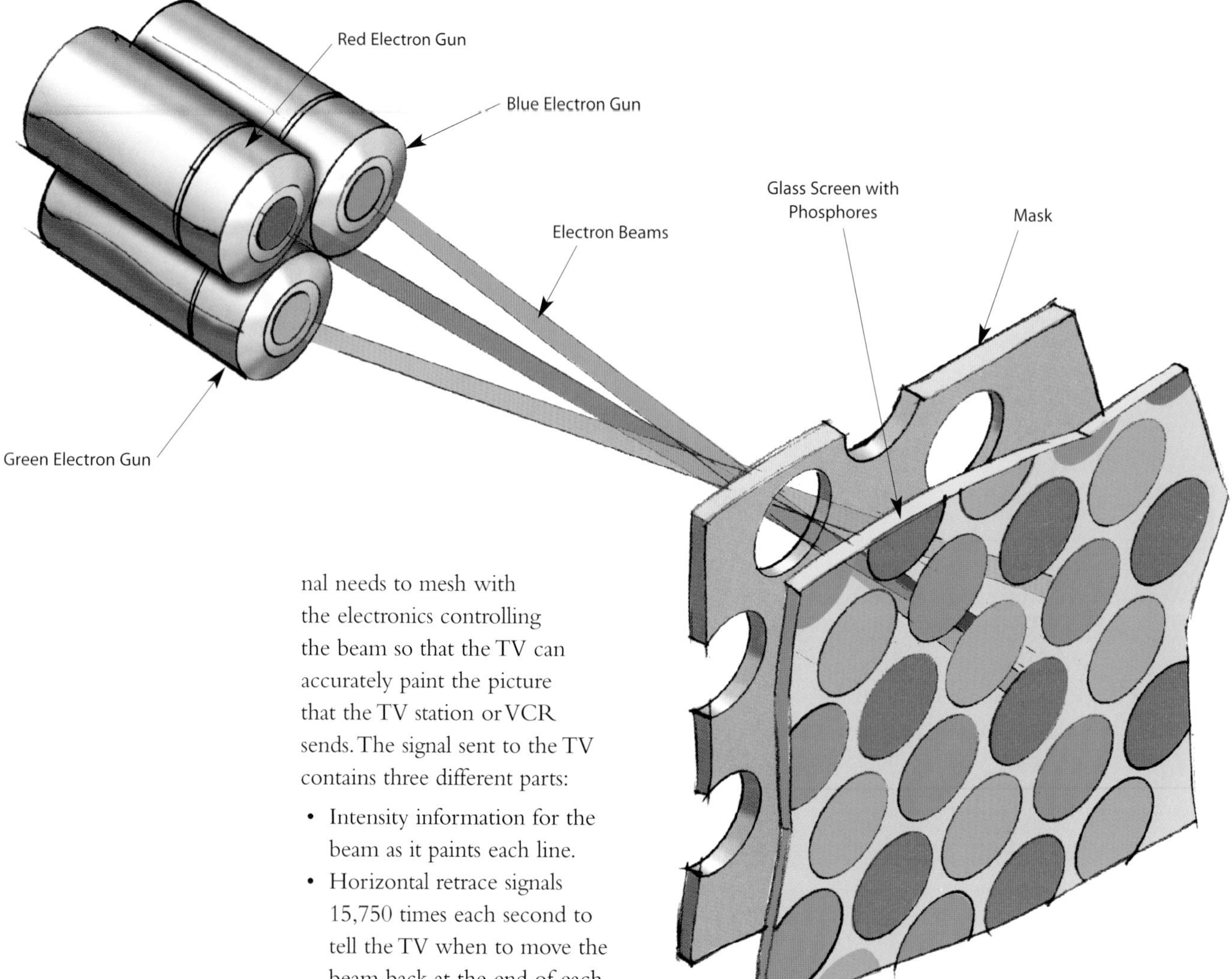

nal needs to mesh with the electronics controlling the beam so that the TV can accurately paint the picture that the TV station or VCR sends. The signal sent to the TV contains three different parts:

- Intensity information for the beam as it paints each line.
- Horizontal retrace signals 15,750 times each second to tell the TV when to move the beam back at the end of each line.
- Vertical retrace signals 60 times per second to move the beam from bottom right to top left.

A signal with all three of these components is called a composite video signal. The horizontal retrace signals are 5-microsecond (ms) pulses at 0 volts. Electronics inside the TV can detect these pulses and trigger the beam's horizontal retrace with them.

The actual signal for the line is a varying wave between 0.5 volts and 2.0 volts, with 0.5 volts representing black and 2.0 volts representing white. This signal drives the intensity circuit for the electron beam.

A vertical retrace pulse is similar to a horizontal retrace pulse, but it is 400 ms to 500 ms long. The vertical retrace pulse is serrated with horizontal retrace pulses in order to keep the horizontal retrace circuit in the TV synchronized.

Color TV

A color TV screen differs from a black-and-white screen in three ways:

- Three electron beams move simultaneously across a color TV screen: the red, green, and blue beams.
- The screen is not coated with a single sheet of phosphor, but with red, green, and blue phosphors arranged in dots or stripes.
- On the inside of the tube, very close to the phosphor coating, is a thin metal screen called a shadow mask. This mask is perforated with very small holes that are aligned with the phosphor dots on the screen.

When a color TV needs to create a red dot, for example, it fires the red beam at the red phosphor. To create a white dot, red, green, and blue beams are fired simultaneously, and the three colors mix together to create white. To create a black dot, all three beams are turned off as they scan past the dot. All other colors on a TV screen are combinations of red, green, and blue.

A color TV signal starts off looking just like a black-and-white signal. An extra chrominance signal is added by superimposing a 3.579545-megahertz (MHz) sine wave onto the standard black-and-white signal. Right after the horizontal sync pulse, eight cycles of a 3.579545-MHz sine wave are added as a color burst.

Following these eight cycles, a phase shift in the chrominance signal indicates the color to display. The amplitude of the signal determines the saturation. A black-and-white TV filters out and ignores the chrominance signal. A color TV picks it out of the signal and decodes it, along with the normal intensity signal, to determine how to modulate the three color beams.

Signal Transmission

There are five different ways to get a signal into a TV set:

- Broadcast programming received through an antenna
- VCR or DVD content run from the VCR or DVD to the antenna terminals
- Cable TV arriving in a set-top box that connects to the antenna terminals
- Large (6- to 12-foot) satellite dish antenna arriving in a set-top box that connects to the antenna terminals
- Small (1- to 2-foot) satellite dish antenna arriving in a set-top box that connects to the antenna terminals

These five types of signals all use standard analog composite video signals. A typical analog TV signal requires 4 MHz of bandwidth. By the time you add sound and other extras, a TV signal requires 6 MHz of bandwidth. Therefore, the Federal Communications Commission (FCC) allocates three bands of frequencies in the radio spectrums chopped into 6-MHz slices to accommodate TV channels:

- 54 MHz to 88 MHz for Channels 2 through 6
- 174 MHz to 216 MHz for Channels 7 through 13
- 470 MHz to 890 MHz for UHF Channels 14 through 83

The composite TV signal described earlier can be broadcast to your house on any available channel. The composite video signal is amplitude modulated into the appropriate frequency, and then the sound is frequency modulated (±25 KHz) as a separate signal. For example, a program transmitted on Channel 2 has its video carrier at 55.25 MHz and its sound carrier at 59.75 MHz. The tuner in your TV, when tuned to Channel 2, extracts the composite video signal and sound signal from the radio waves that transmitted them to the antenna.

VCRs are essentially their own little TV stations. Almost every VCR has a switch on the back that allows you to select Channel 3 or 4. The videotape contains a composite video signal and a separate sound signal. The VCR has a circuit inside that takes the video and sound signals off the tape and turns them into a signal that, to the TV, looks just like the broadcast signal for Channel 3 or 4.

The cable in cable TV transmits a large number of channels on the cable. You select the channel on the box; the box then decodes the right signal and transmits it to the TV on Channel 3 or 4, just like a VCR does.

Large-dish satellite antennas pick off unencoded or encoded signals beamed to earth by satellites. The set-top box receives the signal, decodes it if necessary, and then once again sends it to Channel 3 or 4.

Small-dish satellite systems are digital. The TV programs are encoded in MPEG-2 format and transmitted to earth. The set-top box does a lot of work to decode MPEG-2 and then converts it to a standard analog TV signal and sends it to your TV on Channel 3 or 4.

The analog TV system has been around for more than half a century. During that time it has totally transformed society in ways that its inventors could have never imagined.

Two Amazing Things about the Brain

There are two amazing things about your brain that make television possible. By understanding these two facts, you gain a good bit of insight into why televisions are designed the way they are.

The first principle is this: If you divide a still image into a collection of small colored dots, your brain will reassemble the dots into a meaningful image. The only way we can see that this is actually happening is to blow the dots up so big that our brains can no longer assemble them.

The human brain's second amazing feature is this: If you divide a moving scene into a sequence of still pictures and show the still images in rapid succession, your brain will reassemble the still images into a single moving scene.

Without these two capabilities, TV as we know it would not be possible.

How **DIGITAL TELEVISION** Works

Digital television (DTV) promises to be the next big step in the evolution of TV in the United States, and it will eventually replace the analog television system. In the process, DTV will change several different things we have taken for granted with analog TV, including the aspect ratio, the number of pixels, and the number of channels per station.

HSW Web Links

www.howstuffworks.com

How Camcorders Work
How Computer Monitors Work
How HDTV Works
How Television Works
How Radio Works

To understand digital TV, it is helpful to understand analog TV so that you can see the differences. Analog television uses analog cameras, analog transmission, and an analog TV for reception.

True digital TV is completely digital. It uses digital cameras working at a much higher resolution than analog cameras, digital transmission, and a digital display on a digital television at a very high resolution.

From Analog to Digital TV

Given that analog TV has worked fine for 50 years, and that it works fine with broadcast TV, cable TV, VCRs, satellite dishes, camcorders, and so on, an obvious question is, "What's wrong with analog TV?"

The main problem is resolution. The resolution of the TV controls the crispness and detail in the picture you see and is a function of the number of pixels on the screen. An analog TV set can display 525 horizontal lines of resolution every 1/30 second. That has been fine for years.

Now we have all become conditioned by computer monitors to be comfortable with much better resolution. The lowest-resolution computer monitor today displays 640 x 480 pixels, and this is better resolution than an analog TV.

So the worst computer monitors you can buy have more resolution than the best analog TV sets, and the best computer monitors are able to display up to 10 times more pixels than analog TV sets. There is simply no comparison between a computer monitor and analog TV in terms of detail, crispness, image stability, and color. If you look at a computer monitor all day at work, and then go home and look at a TV set, the TV set can look very fuzzy.

So the drive toward digital TV is fueled by the desire to give TV the same crispness and detail as a computer screen. If you have ever looked at a true digital TV signal displayed on a good digital TV set, you can certainly understand why—the digital version of TV looks fantastic! With 10 times more pixels on the screen, all displayed with digital precision, the picture is incredibly detailed and stable.

Signal, Format, and Aspect Ratio

Digital TV in the United States combines three new and different ideas:

- The signal
- The formats
- The aspect ratio

The FCC gave television broadcasters a new signal frequency to use for digital broadcasts. The digital channel carries a 19.39 megabits-per-second (Mbps) stream of digital data that a digital TV receives and decodes.

Each broadcaster has one digital TV channel, but one channel can carry multiple subchannels if the broadcaster chooses that option. On its digital channel, each broadcaster sends a 19.39-Mbps stream of digital data. Broadcasters have the ability to use this stream in several different ways, such as these:

- A broadcaster can send a single program at 19.39 Mbps.
- A broadcaster can divide the channel into several different streams (for example, four streams of 4.85 Mbps each). These streams are called subchannels. For example, if the digital TV channel is channel 53, then 53.1, 53.2, and 53.3 could be three subchannels on that channel. Each subchannel can carry a different program.
- Broadcasters can create subchannels because digital TV standards allow three different digital formats:

- **480p**—The picture is 704 x 480 pixels, sent at 60 complete frames per second.
- **720p**—The picture is 1280 x 720 pixels, sent at 60 complete frames per second.
- **1080i**—The picture is 1920 x 1080 pixels, sent at 60 interlaced frames per second (that is, 30 complete frames per second).

The *p* and *i* designations stand for *progressive* and *interlaced*. In a progressive format, the full picture updates every 1/60 second. In an interlaced format, half of the picture updates every 1/60 second. The 480p and 480i formats are called the standard-definition (SD) formats, and 480i is roughly equivalent to a normal analog TV picture. When analog TV shows are broadcast on digital TV stations, they are broadcast in 480p or 480i.

The 720p and 1080i formats are high-definition (HD) formats. When you hear about HDTV, this is what is being discussed—a television receiving a digital signal in the 720p or 1080i format.

The HD formats of digital TV have a different aspect ratio than analog TV. An analog TV has a 4:3 aspect ratio, meaning that the screen is four units wide and three units high. For example, a "25-inch diagonal" analog TV is 15 inches high and 20 inches wide. The HD format for digital TV has a 16:9 aspect ratio, much like the format of an in-theater movie.

MPEG-2 and Bit Rates

The idea of sending multiple programs within the 19.39-Mbps stream is unique to digital TV and is made possible by the digital compression system. To compress the image for transmission, broadcasters use MPEG-2 compression, and MPEG-2 allows the broadcaster to select both the screen size and bit rate when encoding the show. A broadcaster can choose a variety of bit rates within any of the three resolutions.

Many variables determine how the picture will look at a given bit rate. Here are two examples:

- A sporting event, where there is lots of movement in the scene, requires the entire 19.39 Mbps at 1080i to get a high-quality image.
- A newscast showing a newscaster's head can use a much lower bit rate. A broadcaster might transmit the newscast at 480p resolution and a 3-Mbps bit rate, leaving 16.39 Mbps of space for other subchannels.

It is possible for broadcasters to send three or four subchannels for a period of time (say, during the day) and then switch to a single high-quality show that consumes the entire 19.39 Mbps (at night, for instance).

Digital TV makes TV images larger and much clearer and moves us one step closer to the point where the computer and the TV will merge into one interactive entertainment device.

And Another Thing...

When you buy a normal, analog TV, you are really buying two parts that have been integrated together: the screen and the tuner. The tuner is able to receive broadcast and cable channels.

With most digital TVs, what you are buying is a DTV monitor. You typically buy the digital tuner separately, and plug it into the DTV monitor through connections on the back of the TV. The connection is similar to the connection between a computer monitor and the computer, and in fact many DTV monitors can be connected directly to a computer's video card.

Seeing The Difference

It is hard to explain the difference between a digital TV signal and an analog signal without giving an actual demonstration, but a comparison can help you understand the idea. These pictures show the difference between an analog and a digital TV image:

You can see that the analog TV picture is much fuzzier than the digital TV image. The significant difference in picture quality is even more obvious when the image is moving. This is the main force that drives the interest in digital TV. Digital TV also offers much better sound than analog TV.

How **CABLE TELEVISION** Works

When cable TV was first introduced, it was a revolutionary idea because the thought of running a wire to nearly every house in the United States seemed impossible. Now we take it completely for granted.

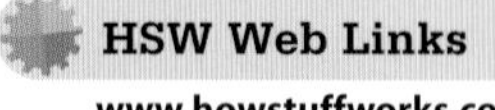

HSW Web Links

www.howstuffworks.com

How Cable Modems Work
How Digital TV Works
How Fiber Optics Work
How HDTV Works
How TV Works

In essence, cable works like this: A satellite broadcasts a digital signal to a dish antenna owned by a cable provider. The cable provider transmits the signal to paying subscribers through a combination of coaxial and fiber-optic cables, rather than by using the airwaves.

Airwaves vs. Cables

The earliest cable systems were basically strategically placed antennas with very long cables connecting them to subscribers' TV sets. Because the signal from the antenna became weaker as it traveled through the long cable, cable providers had to put amplifiers at regular intervals to boost the strength of the signal.

The spread of cable TV stations and cable systems led manufacturers to add a switch to most new TV sets. People could set their TVs to tune to channels based on the Federal Communications Commission (FCC) frequency allocation plan for the airwaves, or they could set them for the plan used by most cable systems.

In both tuning systems, each TV station was given a 6-megahertz (MHz) slice of the radio spectrum, but the two systems differed in important ways. (See "How Television Works," page 190, for details.) The FCC originally devoted parts of the very high frequency (VHF) spectrum to 12 TV channels. The channels weren't put into a single block of frequencies, but were instead broken into two groups to avoid interfering with existing radio services. As TV grew in popularity, the FCC allocated frequencies in the ultrahigh frequency (UHF) portion of the spectrum as well.

Because they used cable instead of antennas, cable TV systems didn't have to worry about existing frequencies. The cable TV/antenna switch tells the TV's tuner whether to tune around the restricted blocks in the FCC broadcast plan or to tune

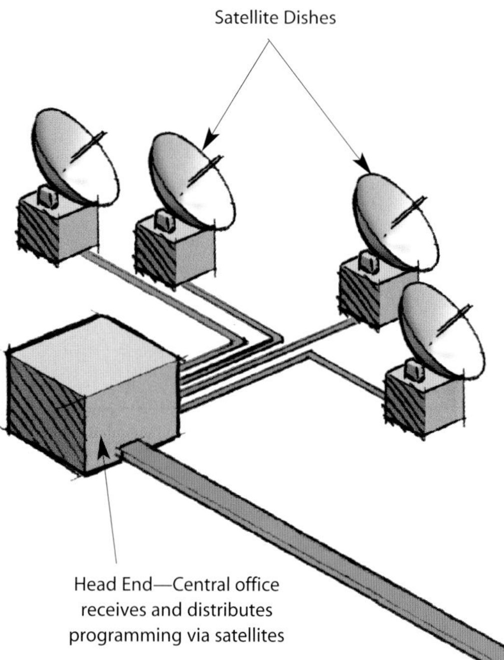

"straight through" for cable reception. In the cable TV position, the switch tells the tuner to start at 88 MHz and go straight up in 6-MHz slices, with no break.

Satellites, Fiber Optics, and Digital

In 1972 a regional cable system in Pennsylvania started the first "pay-per-view" channel, where you could pay to watch individual movies or sporting events. They called the new service Home Box Office (HBO). In 1975 HBO went national by transmitting its signal to cable systems in Florida and Mississippi from a satellite in geosynchronous orbit. Satellite-delivered programming is the basis of the modern cable system.

As the number of program options grew, the bandwidth of cable systems also increased. Early systems operated at 200 MHz, which allowed 33 channels. As technology progressed, the bandwidth increased to 550 MHz, which allowed 91 possible channels. Two additional advances in technology improved features and broadcast quality and made even more channels possible: using fiber-optic cable and moving from analog to digital.

In 1976 cable systems started using fiber-optic cable for the trunk cables that carry signals from the cable TV head-end to neighborhoods. (The head-end is where the cable system receives programming from various sources, assigns the programming to channels, and retransmits it onto cables.) The advantage of fiber-optic cable is that it doesn't suffer the same signal losses as coaxial cable, which eliminates the need for so many amplifiers. The number of amplifiers between the head-end and customer was reduced from the 30 or 40 needed then to the 1 or 2 amplifiers required for most customers now. Decreasing the number of amplifiers made dramatic improvements in signal quality and system reliability.

An analog TV signal consumes an entire 6-MHz TV channel. By using MPEG compression, cable TV systems installed today can transmit up to 10 channels of video in the 6-MHz bandwidth of a single analog channel. Therefore, more than 1,000 channels are now possible! In addition, digital technology allows for error correction to ensure the quality of the received signal.

The move to digital technology also changed one of cable TV's most visible features: the scrambled channel. Scrambling systems insert a signal slightly offset from the channel's frequency to interfere with the picture and then filter the interfering signal out of the mix at the customer's TV. In a digital system, the signal isn't scrambled; rather, it's encrypted. The encrypted signal must be unencrypted with the proper key, or the digital-to-analog converter will not be able to turn the stream of bits into anything usable by the TV's tuner. When a "nonsignal" is received, the cable system substitutes an advertisement or the familiar blue screen.

Prior to cable, most homes in the United States could receive only a few TV stations over the airwaves. Cable TV has brought hundreds of TV channels into most homes. You can now watch channels devoted to anything from cartoons to gardening to shopping. By making hundreds of channels possible, cable TV has changed the entire video landscape.

Did You Know?

CATV systems don't use the same frequencies for stations broadcasting on channels 1 to 6 that those stations use to broadcast over the airwaves. Cable equipment is designed to shield the signals carried on the cable from outside interference, and televisions are designed to accept signals only from the point of connection to the cable or antenna, but interference can still enter the system, especially at connectors. When the interference comes from the same channel that's carried on the cable, there is a problem because of the difference in broadcast speed between the two signals.

Radio signals travel through the air at a speed very close to the speed of light. In a coaxial cable like the one that brings CATV signals to your house, radio signals travel at about two-thirds the speed of light. When the broadcast and cable signals get to the television tuner a fraction of a second apart, you see a double image called "ghosting."

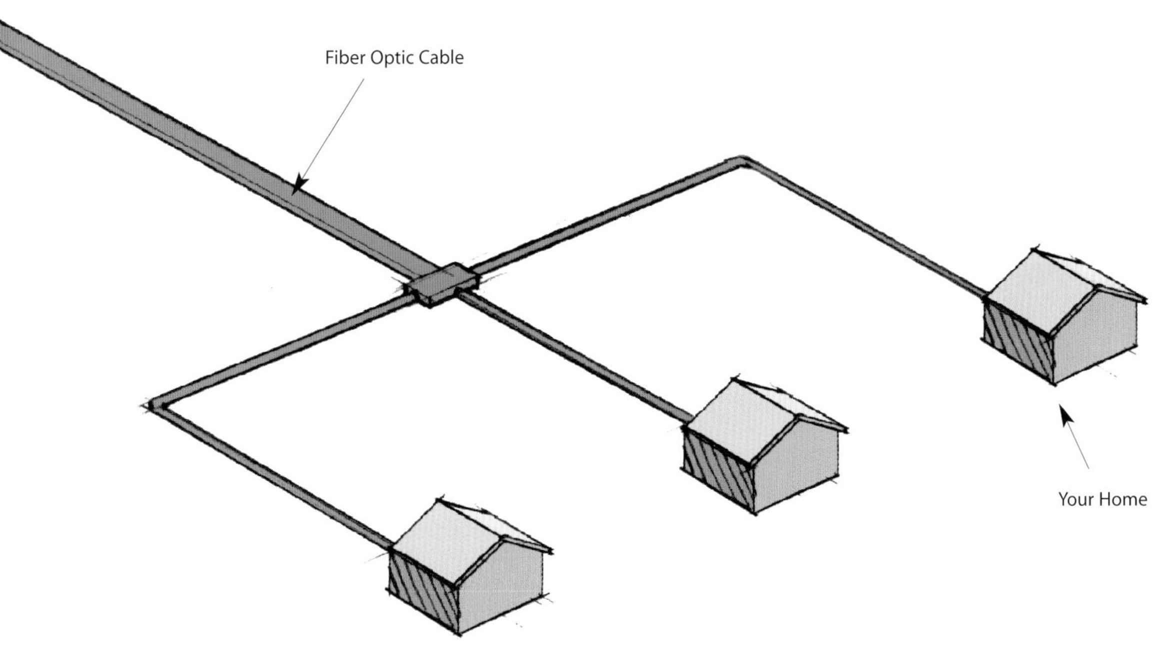

How BLUE–SCREEN SPECIAL EFFECTS Work

How do they make the impossible happen in movies and make it look totally real? For example, in the movie ET, *how did they make it look real when the boys' bicycles began flying? Or in the movie* Star Wars, *how did they make it look real when Luke flew his X-Wing fighter down the trench of the* Death Star *with TIE fighters in close pursuit?*

HSW Web Links

www.howstuffworks.com

How Centropolis FX Creates Visual Effects
How Movie Projectors Work
How Movie Sound Works
How Movie Screens Work
How Video Formatting Works

Many movie illusions are created by using a special-effects technique known as the traveling matte, or the blue-screen technique. This technique allows a director to put actors and scale models in totally unbelievable or imaginary situations—in spaceships, dangling from rope bridges over gorges, flying through the air (à la *Superman*), and so on—and to have it look completely real in the theater. The technique is used so often now that you probably don't even realize it or think anything of the remarkable things you see in movies.

Static Mattes

Mattes have been used in the movie industry practically since the beginning of film to create special effects. This is one of the oldest special-effects techniques used in the industry. By understanding static mattes, it is easy to see how the traveling matte technique evolved.

A very common effect can be created easily by using a double-exposure matte. Let's say that the director would like to create a spooky scene where the actors are walking across a large, flat plain while the sky boils with dark clouds. To create this effect, the camera operator first shoots the actors on the plain. When this shot is created, however, a piece of black paper or tape is used on the lens so that the area of the sky is masked out and left unexposed on the film. The scene is shot normally, but in the camera, the film is exposed on only half of the frame. Then the camera operator rewinds the film in the camera, puts a piece of black paper on the lens to mask out the portion of the film that's already exposed, and then films the clouds of a thunderstorm. The camera operator might film the clouds with a slow film speed, so that when played at normal speed, they look like they are boiling across the sky.

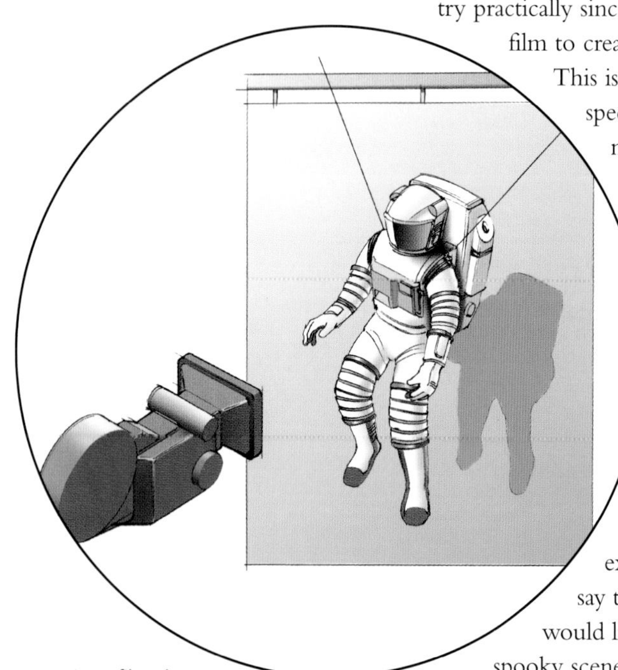

Actor filmed in front of blue screen. In this example, an astronaut is suspended from cables so that she appears to be weightless.

Two variations on this technique are very common:

- The two scenes might be shot separately on two pieces of film and then brought into the special effects department to be combined onto a third piece of film, using a technique called optical compositing.
- Rather than actually film the sky outside, the director might decide to make the sky computer generated.

Traveling Mattes

Another special-effects technique that's commonly used is the traveling-mattes technique. Say that you are a director and you would like to film a scene where the actress is dangling from a rope over a deep river gorge, screaming for the hero to save her. You have several ways to do this:

- If the actress is up for it, you can actually have her dangle from the rope. But most actors and actresses are too valuable to risk in that way.
- You can use a stunt person to stand in for the actress and shoot the scene from far away so people can't tell. Doing this, you will lose the emotional effect of seeing a close-up of the actress's face as she screams.

- You can use blue-screen photography to make it look like the actress is dangling from the rope. Or you might use the blue screen for the close-ups and the stunt person for the long shots to get the best of both worlds.

To use the blue-screen technique, you first film the river gorge on location. This shot is the background plate. You then film the actress dangling from a rope 2 feet off the ground in a studio. Behind the actress in the studio you place a bright blue background screen (hence the name blue screen). You end up with two pieces of film.

In the special–effects department you can easily use special filters to form two mattes from the shot of the actress. One shows the actress's silhouette in black, and the other is the reverse.

These mattes are easy to create because the bright blue background color, when run through a red filter, turns black. By using high-contrast black-and-white film to create the mattes, you can create the silhouettes. So now you have four pieces of film: the two originals and the two mattes. By combining these pieces of film in layers, you can create the final piece of film for the shot. First you combine the background with the actress's silhouette. Then you rewind the film and reexpose it to lay the actress into the "hole" that the matte created.

This is called a traveling matte because the matte is different for each frame of the film. In a static matte, you simply tape black paper over the lens, and that single matte is the same for the entire shot. In a traveling matte shot, you need to create a matte that is exactly the same shape as the actress. In each frame the actress moves, so a new matte is required for each frame. It is possible to create these individual mattes by hand, but that takes a tremendous amount of time. The blue screen behind the actress makes it possible to create all the mattes easily and automatically, using optical or digital techniques.

The technique is also used extensively in science-fiction films such as *Star Wars* and *Star Trek* to make the spacecraft models look real. The models are filmed separately on blue backgrounds and then combined in multiple layers to make the final film. Very complex shots with hundreds of layers have been created.

The blue screen shot is layered or "composited" with a shot of a spacecraft and the earth for the final scene.

In order for a blue-screen shot to look convincing, several things are important:

- The actress (or model, in the case of spaceships) has to have the right level of diffusion to match the background. You have probably seen bad blue-screen techniques used in TV shows, where the foreground actor is very crisp and the background plate is diffused. You immediately know it is fake because of the mismatch.
- The blue screen's color cannot reflect onto the actor or model. If it does, the actor acquires a blue "fringe" around the edges that looks very bad.
- The actor cannot wear anything blue, or there will be a hole in the actor!

With computers, blue-screen shots are especially easy to create because the computer can make the mattes and combine the shots automatically.

The next time you go to a movie, you will understand how they make some of the impossible-looking shots, but you can still be amazed at how real they look!

How MOVIE PROJECTORS & SCREENS Work

If you have ever looked at a strip of movie film in a theater, you know that it is a long collection of still images. A movie projector turns those still images into a fluid, moving picture on a huge screen. You can think of a movie projector as a rapid-fire slide projector—it projects 24 slides per second.

HSW Web Links

www.howstuffworks.com

How Cell Phones Work
How GPS Receivers Work
How the Radio Spectrum Works
How TV Works
How VCRs Work

It takes a great deal of film to make a movie. Most movies are shot on 35mm film stock (which is identical to the film in a 35mm camera, except that it's much longer). You can get 16 frames (that is, individual pictures) on 1 foot (0.3 m) of film. Movie projectors move the film at a speed of 24 frames per second, so it takes 1.5 feet (0.46 m) of film to create every second of a movie. A 120-minute film is more than 2 miles (3.2 km) long!

Early Projectors

Because a feature-length film is so long, distributors divide it into segments that are rolled onto reels. A typical 2-hour movie is usually divided into five or six reels. In the early days, films were shown with two projectors. One projector was threaded with the first reel of the movie and the other projector with the second reel. The projectionist would start the film on the first projector, and when it was 11 seconds from the end of the reel, a small circle flashed briefly in the corner of the screen. This alerted the projectionist to get ready to change to the other projector. Another small circle flashed when 1 second was left, and the projectionist pressed a changeover pedal to start the second projector and stop the first one. While the second reel was rolling, the projectionist removed the first reel on the other projector and threaded the third reel. This swapping continued throughout the movie.

In the 1960s, a device called a platter began to show up in theaters. The platter consists of two to four large discs, about four or five feet in diameter, stacked vertically. A payout assembly on one side of the platter feeds film from one disc to the projector and takes the film back from the projector to spool onto a second disc. Each disc is large enough to hold one large spool of the entire film, which the projectionist assembles by splicing together all the lengths of film from the different reels.

Moving the Film

After a projectionist splices the film and loads it on the feed platter, he or she threads the film through the platter's payout assembly and into the top of the projector. A strip of film has small square holes along each side, called sprocket holes. These holes fit over the teeth of special gear-like wheels in the projector that are called sprockets. The sprockets, driven by an electric motor, pull the film through the projector. Cambers—small spring-loaded rollers—provide tension to keep the film from bunching up or slipping off the sprockets.

To project the movie, the film needs to advance one frame, pause for a fraction of a second while the frame appears on the screen, and then advance to the next frame. In addition, the projector's light must be blocked during the movement so that the image does not smear. This is accomplished using one of two mechanisms.

The first mechanism uses a small lever known as the claw, which is mounted on a bar next to the film's path. The claw is connected to the outer edge of a wheel that acts as the crank. The circular motion of the crank makes the claw lift up and out so that it catches each sprocket hole one at a time. The speed of the sprockets is closely synchronized with the lever action of the claw, to make sure that the claw is consistently advancing the film at a rate of 24 frames per second.

The second mechanism uses another sprocket wheel, mounted just below the aperture gate. This intermittent sprocket rotates just far enough to pull the film down one frame, pauses, and then rotates again.

Automating the Process

Tens of thousands of movie theaters across the United States use projectors like this.

Projectionists have developed many innovative techniques to ensure the show proceeds as it should. Cue tape is one of the more interesting and useful of these. It is a short strip of metal fastened to the edge of the film at a specific location. At the appropriate time, the film passes two electrical contacts and the cue tape completes a circuit between the contacts. This circuit acts like a switch, and it can serve a variety of functions. A cue-tape switch can:

- Dim the house lights
- Turn off the house lights
- Change the lens setting
- Change the sound format
- Change the screen masking (masking is the use of curtains to frame the screen)

Intermittent sockets provide more reliable performance and do not wear out the sprocket holes as quickly as the claw does.

The film is stretched over two bars as it passes in front of the lens. The bars keep the film tight and properly aligned. Depending on the projector's configuration and the sound format used, the film passes through an optical audio decoder that is mounted either before or after the lens. For digital sound, the film travels through a special digital decoder attached to the top of the projector. As the film leaves the projector (or the digital audio decoder), it is carried on a series of rollers back to the platter's payout assembly and spooled to a take-up platter.

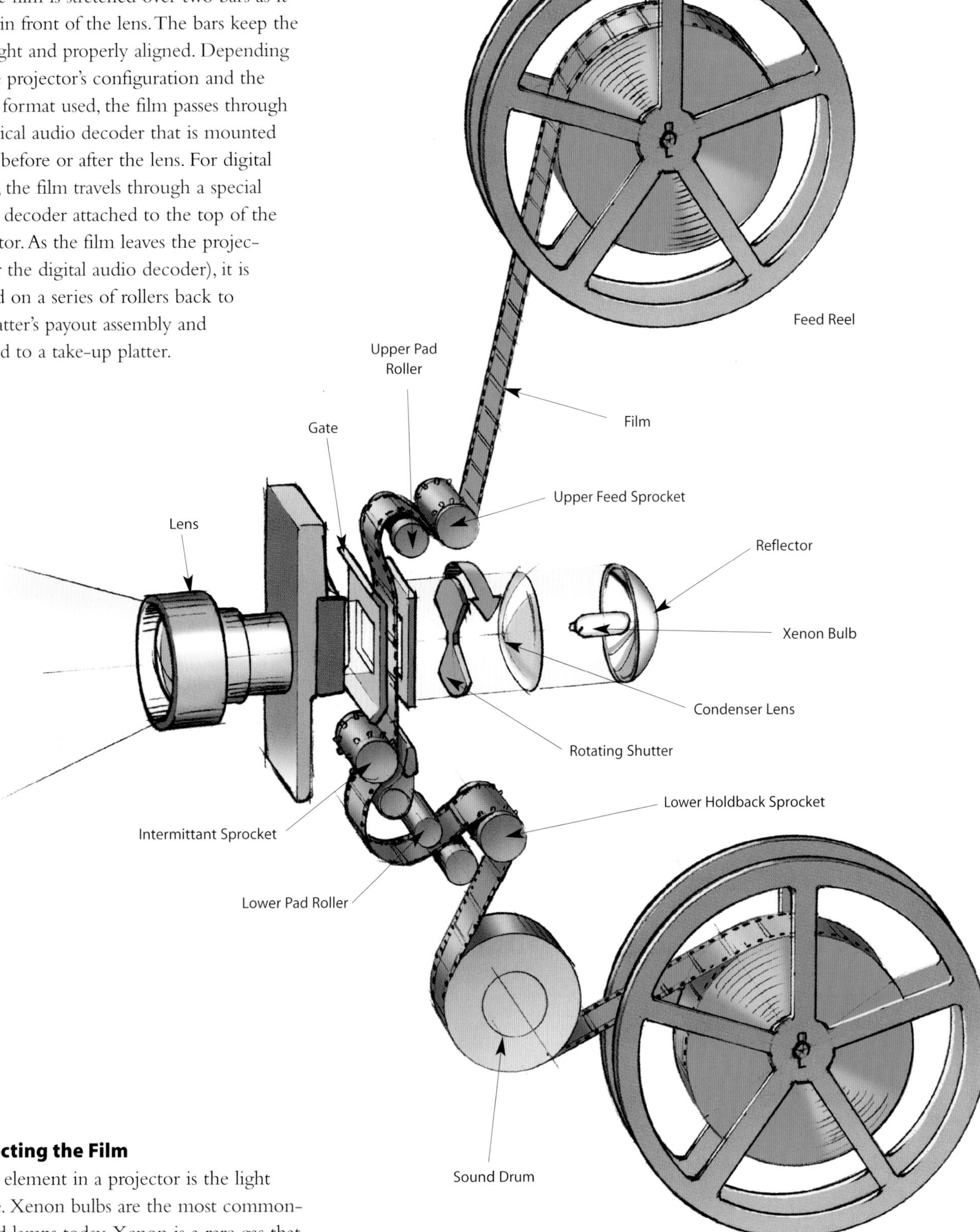

Projecting the Film

A key element in a projector is the light source. Xenon bulbs are the most commonly used lamps today. Xenon is a rare gas that has some properties that make it especially useful in projectors:

- In dense enough quantities, it conducts electricity.
- As a conductor, it glows very brightly.
- It provides bright illumination for a substantial amount of time (2,000 to 6,000 hours).

A xenon bulb is mounted in the center of a parabolic mirror located in the lamp house. The mirror reflects light from the bulb and focuses it on the condenser. The condenser is a pair of lenses used together to further intensify the light and focus it on the main lens assembly. The heat generated by this focused light is incredible—that's why film melts so quickly when a projector stops spooling it.

As the focused light leaves the lamp house and enters the projector, it is intercepted by the shutter, which is a small propeller-like device that rotates 24 times per second. Each blade of the shutter blocks the path of the light as it comes to a certain point in its revolution. This blacking out is synchronized with the advancement of the film so that the light doesn't project during the fraction of a second when the film is moving from one frame to the next and cause flicker. Many projectors use double shutters that rotate in opposite directions. This causes the light to be cut off from both the top and bottom of each frame, further reducing the possibility of flicker.

Before the light gets to the film, it also passes through an aperture gate. The aperture gate is a small removable metal frame that blocks the light from illuminating anything but the part of the film that you want to see on the screen. Aperture gates have a variety of sizes, which correspond to the screen format of a movie.

From the aperture gate, the light passes through the film and into the main lens. The lens is removable and can change depending on the format of the film. The two most common lenses are flat and CinemaScope.

The Screen

A movie screen is usually made of heavy white vinyl and is categorized by the amount of light it reflects. There are four main categories:

- **Matte white**—This type of screen has 15% reflectivity; black appears dark gray, the image is bright. This type of screen provides the best overall contrast.
- **Silver**—This type of screen has 30% reflectivity; black appears medium gray, the image is very bright, and dark colors can seem a little dull. To make a silver screen, a reflective coating is added to the matte white vinyl.
- **Glass bead**—This type of screen has 40% or more reflectivity; black appears light gray, the image is usually too bright.
- **Pearlescent**—Pearlescent is the most commonly used screen for a typical movie theater. As with a silver screen, to make a pearlescent screen, a reflective coating is added to the matte white vinyl.

Most movie screens have tiny perforations in them so the audience will be able to hear sound from speakers placed behind the screen. In a typical theater, you'll find three speakers behind the screen, located at the far left, the center, and the far right. This makes the sound seem more realistic, particularly when someone is talking. The audio is delivered through the appropriate speaker so that a sound seems to come from the person or thing talking or making noise.

A theater can have screens with three different curves:

- **Flat screen**—This type of screen has no curve at all.
- **Horizontal-curve screen**—This type of screen curves toward the audience slightly at each end. Curving the ends of the screen toward the projector equalizes the distance the light travels.
- **Torex screen**—Torex screens take the idea of a horizontal-curve screen to the next step. Not only do they curve in at each end, they also curve in at the top and bottom of the screen, creating a concave surface.

The film, the projector, and the screen work together in a darkened theater to totally immerse the audience in the movie. The process is just as popular today as it was a century ago.

Early Film Projectors

The earliest film projectors actually showed up in the late 1600s, but they presented only still images. Everything changed with the invention of Thomas Edison's kinetoscope in 1891. The kinetoscope used a motor to revolve a strip of film in front of a light source. The light source projected the image from the film on a screen in a booth. As it became obvious that people were willing to pay money for this type of entertainment, many inventors began to design variations of Edison's original device. One such variation, the manually operated kinora, was invented by the Lumiere brothers and enjoyed great success into the 1930s.

The Lumiere brothers, Louis and Auguste, created the astounding cinematographe in 1895. This portable device was a camera, film processing lab, and projector all in one package! The brothers traveled the French countryside shooting films that lasted a few minutes at most. They then processed and projected the film on location! The next year, the vitascope (which was another variation of the kinetoscope) heralded the dawn of a new age of entertainment. The vitascope worked like a basic kinetoscope with an essential difference: The image was projected on a large screen in a room instead of a small one in a booth.

chapter eight

NOW HEAR THIS!

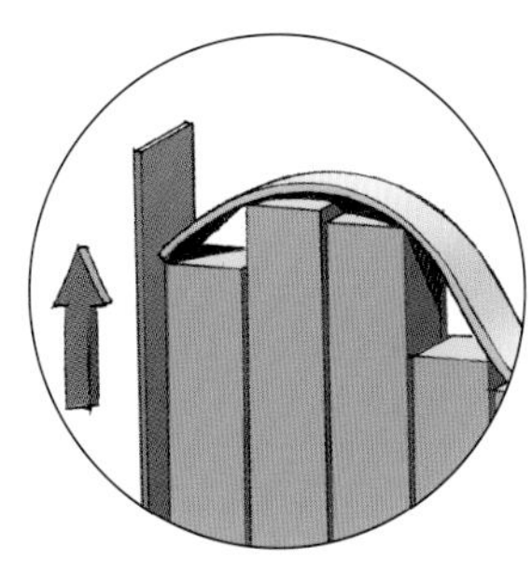

How **CELL PHONES** Work

A cell phone is a miracle if you think about it. Almost anywhere, you can pick up a little box not much bigger than a deck of cards, and in several seconds you can be talking to nearly anyone on the planet. Imagine what Benjamin Franklin or George Washington would have thought about a device like that!

HSW Web Links

www.howstuffworks.com

How Cell Phone Services Work
How Cordless Telephones Work
How Disposable Cell Phones Will Work
How the Radio Spectrum Works
How Telephones Work

One of the most interesting things about a cell phone is that it is really a radio—an extremely sophisticated radio, but a radio nonetheless. There is not a huge difference between a cell phone and the simple three-transistor walkie-talkie that you played with as a kid. The genius of the cellular system is that cities are divided into small cells, with a cell phone tower at the center of each cell. Cells allow extensive frequency reuse across a city, which means that millions of people can use cell phones simultaneously.

You can understand a good bit about cell phones by thinking back to your walkie-talkie. Here are three differences that you have probably noticed between walkie-talkies and cell phones:

- A walkie-talkie can only transmit about 1,000 feet (305 m) to another walkie-talkie, but your cell phone can transmit several miles to a tower. That's because a walkie-talkie has a 0.25-watt transmitter, whereas a cell phone has a 3-watt transmitter.
- Only one person can talk at a time on a walkie-talkie (or a CB radio), and you have to push a button on the walkie-talkie when you talk. On a cell phone both people can talk at the same time and there is no button. That's because two walkie-talkies must use the same radio frequency, but each cell phone uses two radio frequencies—one to carry your voice to the tower and the other to carry the other person's voice from the tower to you.
- If six people in the neighborhood have walkie-talkies, they can all hear each other because all the walkie-talkies share the same single frequency. On a cell phone each conversation is private; each cell phone can understand hundreds or thousands of different frequencies, and each cell uses a distinct pair of frequencies for every individual call.

Battery Type

Cell phones use two main battery technologies:

- **NiMH (Nickel Metal Hydride)**—high capacity battery that provides extra power for extended use
- **Li-ion (Lithium Ion)**—has a lot of power in a lightweight package but usually costs more than NiMH batteries

The Cell System

The key to the cell phone system is the cell. The cell phone system divides a city into a set of cells (often thought of as hexagons on a hexagonal grid). In the center of each cell is a cell phone tower, and the cell might cover an area with a diameter of 2 or 3 miles around the tower. With a walkie-talkie you have to transmit directly to another walkie-talkie, but with a cell phone you transmit to the tower. The tower then connects you to the normal land-based telephone system to route the call. A handoff has to occur when you move from one cell to another. For example, when you are driving you might move from one cell to another every couple minutes, and the system has to handle it smoothly, so you don't even realize it is happening!

Cell phones have low-power transmitters in them. Many cell phones have two signal strengths: 0.6 watts and 3 watts (in comparison, most CB radios transmit at 4 watts). The base station in each cell also transmits at low power. Low-power transmitters have two advantages:

- The transmissions of a base station and the phones within its cell do not make it very far outside that cell. The same frequencies can be reused across the city.
- The power consumption of the cell phone, which is normally battery operated, is relatively low. Low power means small batteries, and this is what has made handheld cellular phones possible.

A typical large city can have hundreds of towers. Towers are expensive, but because so many people are using cell phones, costs per user remain relatively low. Each carrier in each city also runs one central office, called the mobile telephone switching office (MTSO). This office handles all the hand-offs—the phone connections to the normal land-based phone system—and it controls all the towers in the region.

Here is what happens when someone tries to call you on your cell phone:

1) When you first power up the phone, it listens for a control channel, which is a special frequency that the phone and tower use to talk to one another about things like call setup and channel-changing. If the phone cannot find any control channels to listen to, it knows it is out of range and displays a no-service message.
2) The phone also transmits a registration request, and the MTSO keeps track of your phone's location in a database. This way, the MTSO knows which cell you are in when it wants to ring your phone.
3) The MTSO gets the call, and it tries to find you. It looks in its database to see which cell you are in.
4) The MTSO picks a frequency pair that your phone will use in that cell to take the call.
5) The MTSO communicates with your phone over the control channel to tell it what frequencies to use, and when your phone and the tower switch on those frequencies, the call is connected. You are basically talking to another person on a walkie-talkie type of call at this point!
6) As you move toward the edge of the cell, the cell's tower notes that your signal strength is diminishing. Meanwhile, the tower in the cell you are moving toward (which is listening and measuring signal strength on all frequencies, not just its own) can see your phone's signal strength increasing. The two towers coordinate themselves through the MTSO. At some point, your phone gets a signal on a control channel that tells it to change frequencies. This handoff switches your phone to the new cell, and if it works right, you can't even tell it happened. But when it doesn't work right, you lose the call.

Analog and Digital

Advanced Mobile Phone System (AMPS), the first common cell phone system in the United States, uses a range of frequencies between 824 MHz and 894 MHz for analog cell phones. To encourage competition and keep prices low, the U.S. government requires the presence of two carriers in every market, known as A and B carriers. One of the carriers is normally the local exchange carrier (LEC), a fancy way of saying the local phone company.

Carriers A and B are each assigned 832 frequencies: 790 for voice and another 42 for data. A pair of frequencies (the transmit and receive frequencies) is used to create one channel. The frequencies used in analog voice channels are typically 30 KHz wide. This size was chosen as the standard size because it gives voice quality that is comparable to what you get on a wired telephone.

The transmit and receive frequencies of each voice channel are separated by 45 MHz, to keep them from interfering with each other. Each carrier has 395 voice channels as well as 21 control/data channels to use for housekeeping activities such as registration and paging. Each cell uses about one-seventh of the frequencies. That way a hexagonal cell, along with the six cells that surround it, are all using unique frequencies.

Cool Facts

- Most new digital cellular phones have some sort of entertainment programs on them, ranging from simple dice-throwing games to memory and logic puzzles.
- Approximately 20 percent of American teens (more girls than boys) own a cellular phone.
- Cellular phones are more popular in European countries than they are in the United States, with nearly 2/3 of Europeans owning a phone, compared to only about 1/4 of Americans.
- The GSM standard for digital cell phones was established in Europe in the mid-1980's—long before digital cellular phones became commonplace in American culture.
- The technology is now possible to locate a person using a cellular phone down to a range of a few meters, anywhere on the globe.
- 3G (third generation wireless) phones may look more like PDAs, with features such as video-conferencing, advanced personal calendar functions, and multi-player gaming.

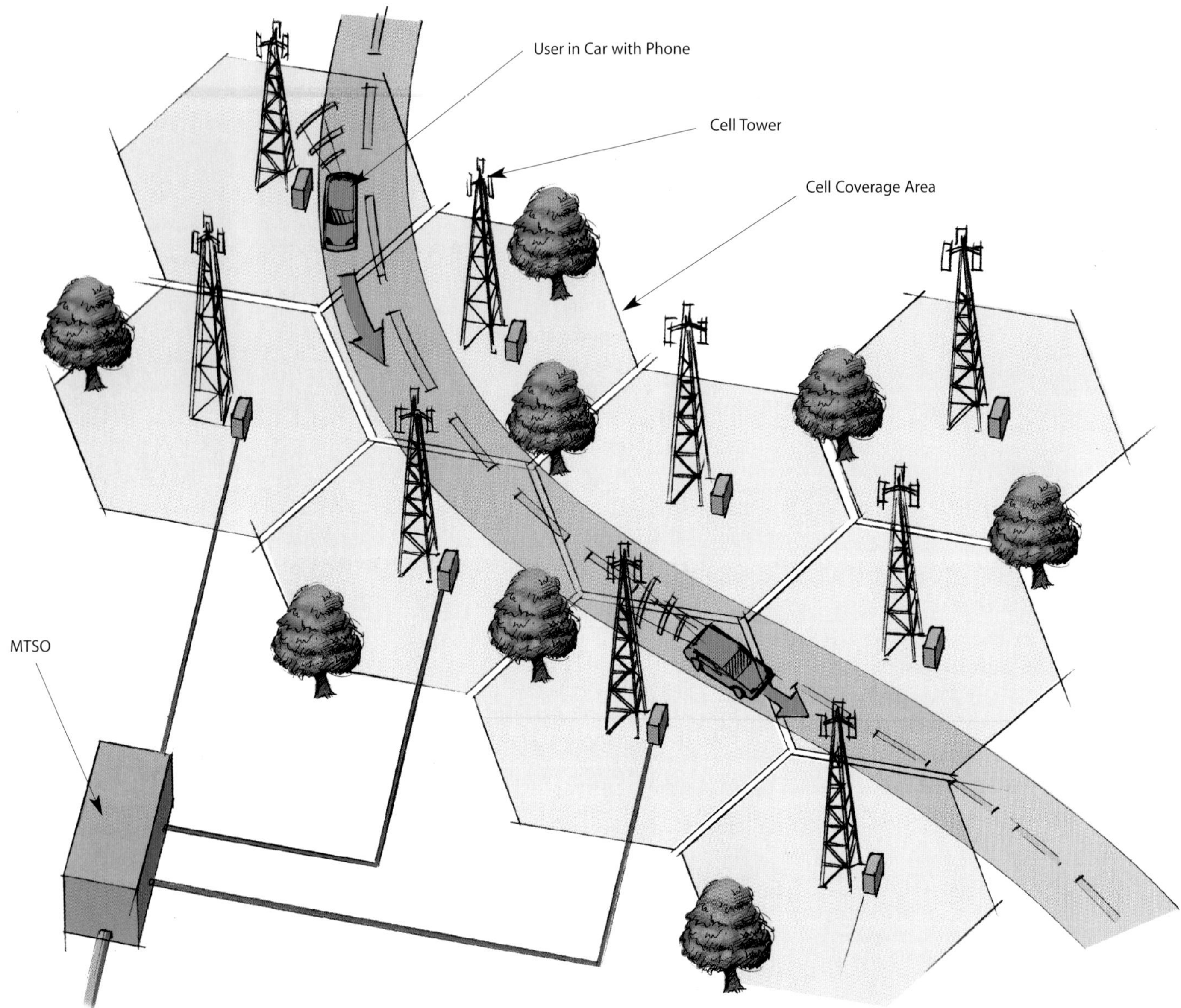

Digital cell phones use the same radio technology as analog phones, but they use it in a different way. Digital phones convert voice into binary information (1s and 0s) and then compress it. This compression allows between 3 and 10 digital cell phone calls to occupy the space of a single analog cell phone call.

Cellular Access Technologies

Cell phone networks use three common technologies for transmitting information:

- **Frequency-Division Multiple Access (FDMA)**—FDMA puts each call on a separate frequency.
- **Time-Division Multiple Access (TDMA)**—TDMA assigns each call a certain portion of time on a designated frequency.
- **Code-Division Multiple Access (CDMA)**—CDMA gives a unique code to each call and spreads it over the available frequencies.

These technologies sound rather intimidating, but you can get a good sense of how they work just by breaking down the name of each one. The first word tells you what the access method is and the second word, *division*, lets you know that it splits calls based on that access method. The last part of each name is *multiple access*, which simply means that more than one user *(multiple)* can use *(access)* each cell at the same time.

FDMA separates the spectrum into distinct voice channels by splitting it into uniform chunks of bandwidth. To better understand FDMA, think of radio stations. Each station sends its signal at a different frequency within the available band. FDMA is used mainly for analog transmission. Although it is certainly capable of carrying digital information, FDMA is not considered to be an efficient method for digital transmission.

In TDMA, a 30-KHz-wide analog channel is chopped into 6.7-millisecond time slices, and each 6.7-millisecond slice is split into three time slots. Each conversation gets the radio for one-third of the time. This is possible because voice data that has been converted to digital information is compressed so that it takes up significantly less transmission space than analog data. TDMA is the access technology for the global system for mobile communications (GSM). GSM operates in the 900-MHz and 1800-MHz bands in Europe and Asia and in the 1900-MHz (sometimes referred to as 1.9-GHz) band in the United States. It is used in digital cellular and PCS-based systems.

CDMA spreads data out over the entire bandwidth it has available, by using spread spectrum technology. Let's say that there are 40 frequencies available in a cell. Each phone will transmit on all those frequencies, switching from one frequency to another dozens of times each second. Each phone uses a different random number sequence to decide which frequency it will use when. At the receiver, that same unique code is used to recover the signal. Because a CDMA system needs to put an accurate time stamp on each piece of a signal, it uses the global positioning system (GPS) to get this information. (See page 91 for more information on GPS.)

Parts of a Cell Phone

On a scale of complexity per cubic inch, cell phones are some of the most intricate devices people play with on a daily basis. Modern digital cell phones can process millions of calculations per second in order to compress and decompress the voice stream.

If you ever take apart a cell phone, you will find that it contains just a few individual parts:

- An amazing circuit board that contains the brains of the phone
- An antenna
- A liquid crystal display (LCD)
- A keyboard
- A microphone
- A speaker
- A battery

There are also several computer chips inside a cell phone.

- The analog-to-digital and digital-to-analog conversion chips translate the outgoing audio signal from analog to digital and the incoming signal from digital back to analog.
- The digital signal processor (DSP) is a highly customized processor designed to perform signal manipulation calculations at high speed.
- The microprocessor handles all the housekeeping chores for the keyboard and display, deals with command and control signaling with the cell tower, and coordinates the rest of the functions on the board. The read-only memory (ROM) and flash memory chips provide storage for the phone's operating system and customizable features, such as the phone directory. The radio frequency (RF) and power section handle power management and recharging, and it also deals with the hundreds of FM channels. Finally, the RF amplifiers handle signals into and out of the antenna.

Millions of people in the United States and around the world use cell phones. As a business tool, as an added sense of security during road travel, or for simple convenience—whatever the reason, these handy devices have made an incredible impact on our daily lives.

Cell Phones and CBs

A good way to understand the sophistication of a cell phone is to compare it to a CB radio or a walkie-talkie:

- **Simplex vs. Duplex**—Both walkie-talkies and CB radios are simplex devices. That is, two people communicating on a CB radio use the same frequency, so only one person can talk at a time. A cell phone is a duplex device. That means that you use one frequency for talking and a second, separate frequency for listening. Both people on the call can talk at once.
- **Channels**—A walkie-talkie typically has one channel, and a CB radio has 40 channels. A typical cell phone can communicate on 1,664 channels or more!
- **Range**—A walkie-talkie can transmit about 1 mile using a 0.25 watt transmitter. A CB radio, because it has much higher power, can transmit about 5 miles using a 5 watt transmitter. Cell phones operate within cells, and they can switch cells as they move around.

How **TELEPHONES** Work

Although you probably take it completely for granted, the telephone is one of the most amazing devices ever created. If you want to talk to someone, all you have to do is pick up the phone and dial a few numbers. Instantly, you are connected to that person, and you can have a complete, two-way conversation. And because the telephone network extends worldwide, you can reach almost anyone on earth with it. When you compare that to the state of the world just 100 years ago, when it might have taken up to several weeks to get a one-way written message to someone, you realize just how incredible the telephone is!

HSW Web Links

www.howstuffworks.com

How Cell Phones Work
How Cordless Telephones Work
How Electromagnets Work
How Home Networking Works
How Modems Work

Despite its usefulness, a telephone is one of the simplest devices you have in your house. One of the reasons it is so simple is because the telephone connection to your house has not changed in nearly a century. If you had a phone from the 1920s and the proper converter, you could connect it to the wall jack in your house and it would work fine.

Parts of a Phone

A basic telephone contains three simple parts:

- **A hook switch**—The hook switch connects and disconnects the phone from the network when you lift and replace the handset.

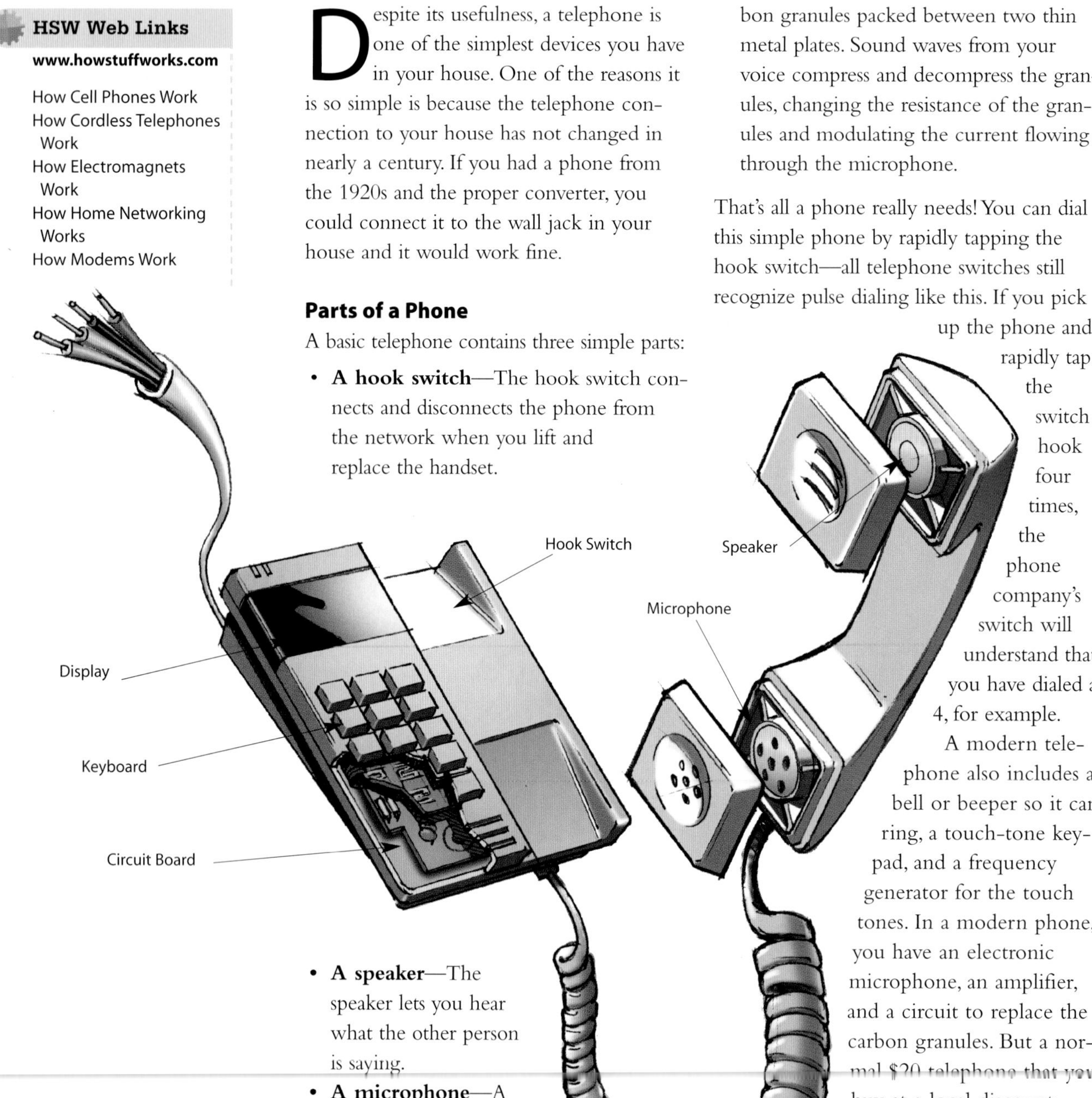

- **A speaker**—The speaker lets you hear what the other person is saying.
- **A microphone**—A telephone microphone can be as simple as carbon granules packed between two thin metal plates. Sound waves from your voice compress and decompress the granules, changing the resistance of the granules and modulating the current flowing through the microphone.

That's all a phone really needs! You can dial this simple phone by rapidly tapping the hook switch—all telephone switches still recognize pulse dialing like this. If you pick up the phone and rapidly tap the switch hook four times, the phone company's switch will understand that you have dialed a 4, for example.

A modern telephone also includes a bell or beeper so it can ring, a touch-tone keypad, and a frequency generator for the touch tones. In a modern phone, you have an electronic microphone, an amplifier, and a circuit to replace the carbon granules. But a normal $20 telephone that you buy at a local discount store remains a very simple device.

The Telephone Network

A pair of copper wires (one red and one green) run from a box at the road to a box (often called an entrance bridge) at your house. From there, the wires connect to each phone jack in the house. If your house has two phone lines (such as a residence line and a home office or computer line), then two separate pairs of copper wire run from the road to your house. The second pair is usually colored yellow and black to distinguish these wires from the other line's red and green wires. Buried in the ground along the road is a thick cable, packed with 100 or more copper pairs. Depending on where you're located, this thick cable will run directly to the phone company's switch in your area or to a box about the size of a refrigerator that acts as a digital concentrator somewhere in your neighborhood.

The Parts of a Telephone Network

The connection between your house and the phone company is even simpler than the telephone. In fact, you can easily create your own intercom system by using two telephones, a pair of wires, a 9-volt battery, and a 300-ohm resistor that costs about $1.

Your connection to the phone company consists of two copper wires—one green and one red. The green wire is common, and the red wire supplies your phone with DC voltage at about 30 milliamps. Think about the simple carbon granule microphone in early phones: All it's doing is modulating that current, letting more or less current through, depending on how the sound waves compress and relax the granules. The speaker at the other end "plays" that modulated signal.

The easiest way to wire up a private intercom is to connect the two phones to each other with two long pieces of wire. Cut one of the wires, strip the ends, and hook in the battery and resistor in series. What you have created is a current loop between the two phones, with the battery providing the current. When two people pick up the phones together, they can talk to each other just fine. This sort of arrangement will work at lengths up to several miles apart.

The only thing your little intercom-type network cannot do is ring the phone to tell the person at the other end to pick up. The ring signal is a 90-volt AC wave at 20 Hertz (Hz).

Phone Calls

If you think about the days of the manual switchboard, it is easy to understand how the larger phone system works. In those days, a pair of copper wires (called a loop) ran from every house to a central office in the middle of town. The switchboard operator sat in front of a board, with a jack for every pair of wires entering the office. Above each jack was a small light. A large battery supplied current through a resistor to each wire pair. When someone picked up the handset on a telephone, the hook switch would complete the circuit and let current flow through wires between the house and the office. The light bulb above that person's jack on the switchboard would glow, and the operator would connect her headset into that jack and ask who the person would like to talk to.

The operator would then send a ring signal to the receiving party and wait for the party to pick up the phone. The ring signal is a 20-Hz jolt of electricity at about 80 volts. (In the old phones on which you turned a crank, the crank connected to a generator that generated the ring signal on the wire.) When the receiving party picked up, the operator would connect the two people together by connecting their two loops with the battery. It was that simple!

In a modern phone system, the operator has been replaced by an electronic switch. When you pick up the phone, the switch senses the completion of your loop and it plays a dial tone sound so you know the switch and your phone are working. The dial tone sound is simply a combination of a 350-Hz tone and a 440-Hz tone. When you dial the number on your phone's touch-tone keypad, the switch interprets the different dialing sounds as numbers.

The telephone has totally changed the world. With the phone network, you can communicate with nearly anyone, or at least his or her voicemail, instantly.

Dialing Sounds

The different dialing sounds are made of pairs of tones. The following combinations of tones are what you hear when you press a particular button on the telephone key pad:

1—697 Hz and 1209 Hz
2—697 Hz and 1336 Hz
3—697 Hz and 1477 Hz
4—770 Hz and 1209 Hz
5—770 Hz and 1336 Hz
6—770 Hz and 1477 Hz
7—852 Hz and 1209 Hz
8—852 Hz and 1336 Hz
9—852 Hz and 1477 Hz
*—941 Hz and 1209 Hz
0—941 Hz and 1336 Hz
#—941 Hz and 1477 Hz

The dial tone sound is simply a combination of a 350 hertz tone and a 440 hertz tone, and a busy signal is made up of a 480 hertz and a 620 hertz tone, with a cycle of 1/2 second on and 1/2 second off.

How **CDs** Work

Compact discs—commonly called CDs—are everywhere these days. Whether they are used to hold music, data, or computer software, they have become the standard medium for distributing large quantities of information in an inexpensive, reliable package. CDs are so easy and inexpensive to produce that many companies send out millions of them every year in the form of junk mail.

HSW Web Links

www.howstuffworks.com

How Analog and Digital Recording Work
How DVDs and DVD Players Work
How IDE Controllers Work
How MP3 Files Work

MP3 Files on CDs

A typical 4-minute song on a normal music CD consumes about 40 megabytes of space, and a music CD can hold about 700 megabytes of music. Fifteen to 20 songs fit on a CD.

The traditional way to save MP3 files on a CD has been to uncompress the MP3 file. This turns a 3-megabyte MP3 file into a 40-megabyte CD track. By writing 15 or 20 of these tracks onto a CD-R, you create a CD that you can play anywhere. Many newer CD players can now read MP3 files straight from data CDs. You write the MP3 files onto the CD directly rather than uncompressing them. This allows a single CD to hold several hundred songs.

A CD can store huge amounts of digital information on a very small surface that is incredibly inexpensive to manufacture. The design that makes this possible is simple: The CD surface is a mirror covered with billions of tiny bumps that are arranged in a long, tightly wound spiral. The CD player reads the bumps with a very precise laser and interprets the information as bits of data.

Parts of a CD

The spiral of bumps on a CD starts in the center of the CD. This means that you can create a partially filled CD that contains just a few songs. Or you may have even seen CD business cards or baseball cards that contain smaller amounts of information.

CD tracks are so small that they have to be measured in millionths of a meter, called microns. The CD track is approximately 0.5 microns wide, with 1.6 microns separating one track from the next. The elongated bumps are each 0.5 microns wide, a minimum of 0.83 microns long, and 125 nanometers high. (A nanometer is a billionth of a meter.) If you could somehow unwind this extremely thin track from one CD like it was a piece of string, it would extend for several miles!

The bumps and tracks on a CD need to be very small so that the CD can store lots of data. A CD holds up to 74 minutes of music. There are 44,100 samples per second and 2 bytes per sample, and a stereo recording requires 2 tracks. So the total amount of digital data that must be stored on a CD can be figured like this:

44,100 samples/channel/second
x 2 bytes/sample x 2 channels
x 74 minutes x 60 seconds/minute
= 783,216,000 bytes

Most of the mass of a CD is an injection-molded piece of clear polycarbonate plastic that is about 1.2 millimeters thick. During manufacturing, this plastic is impressed with the microscopic bumps that make up the long, spiral track. A thin, reflective aluminum layer is then coated on the top of the disc, covering the bumps. Then a thin protective acrylic layer is sprayed over the aluminum. The CD label is printed on this layer of acrylic.

The CD Player

The tricky part of CD technology is reading all the tiny bumps correctly, in the right order, at the right speed. To do all this, the CD player has to be exceptionally precise when it focuses the laser on the track of bumps.

When you play a CD, the laser beam passes through the CD's polycarbonate layer, reflects off the aluminum layer, and hits an optoelectronic device that detects changes in light. The bumps reflect light differently than the flat parts of the aluminum layer, which are called lands. The optoelectronic sensor detects this change in reflectivity, and the electronics in the CD player's drive interpret the changes as data bits.

The most difficult job of the CD player is to keep the laser beam centered on the data track. This centering is the job of the tracking system. The tracking system, as it plays the CD, has to continually move the laser outward. As the laser moves outward from the center of the disc, the bumps move past the laser faster. As the laser moves out-

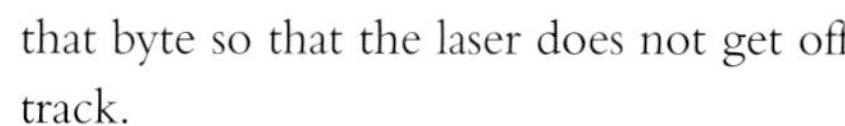

ward, the spindle motor must slow the speed of the CD. That way, the bumps travel past the laser at a constant speed, and the data coming off the disc has a constant rate.

CD Encoding

In addition to storing recorded data, the bumps on a CD include information that helps the player do its job. This format encoding addresses the main problems a player might encounter while reading a CD. The following is a fairly complete list of standard CD format information:

- Because the laser is tracking the spiral of data by using the bumps, there cannot be extended gaps where there are no bumps in the data track. To solve this problem, data is encoded using eight-fourteen modulation (EFM). In EFM, 8-bit bytes are converted to 14 bits, some of which are guaranteed to be 1s. Even if the byte contains all 0s, EFM will force some 1s into that byte so that the laser does not get off track.
- If the laser needs to be able to move between songs, the CD must have data encoded into the music that tells the drive where it is on the disc. This subcode data tells the absolute and relative position of the laser in the track, as well as song titles and other information.
- Because the laser may misread a bump, the CD needs error-correcting codes. Extra data bits are added that allow the drive to detect single-bit errors and correct them.
- A scratch or a speck on the CD might cause the player to misread a whole packet of bytes; this is called a burst error. Interleaving the data on the disc solves this problem. With interleaving, data is stored nonsequentially around the disc's circumference. The drive reads data one revolution at a time, and it uninterleaves the data in order to play it.

Bumps

If a few bytes are misread in music, the worst thing that can happen is that the sound might be a little fuzzy during playback. When computer data is stored on a CD, however, any data error is catastrophic. Therefore, additional error correction codes are used when storing data on a CD-ROM.

Writeable CDs

You can write anything you want onto a CD-R — a recordable CD. In a CD-R you have a plastic substrate, a dye layer, and a reflective gold layer. On a new CD-R disk the entire surface of the disk is reflective — the laser can shine through the dye and reflect off the gold layer. When you write data to a CD-R, the writing laser (which is much more powerful than the reading laser) heats up the dye layer and changes its transparency. This is a permanent change and both CD and CD-R drives can read the modified dye as a bump later on.

Because CDs can store a huge amount of digital data, they have an incredible number of applications. In addition to music, they can store software applications, images, video, and all kinds of text and numeric data. And because they are made out of simple plastic and aluminum, they are extremely inexpensive to produce. All these factors have made CDs a revolutionary force in the modern world. Thanks to CDs, companies can send volumes of information in slim, lightweight envelopes, and it's easier than ever to back up computer files. It's absolutely amazing what you can do with a simple piece of plastic!

How **ANALOG & DIGITAL RECORDING** Work

CDs were first introduced in the early 1980s, and their single purpose was to hold music in digital format. Now we use CDs for data, too, because digital music is really just binary data. At its most fundamental, the great sound that you hear from a CD is simply the conversion of analog waves to digital data and back to analog again.

HSW Web Links

www.howstuffworks.com

How Cell Phones Work
How CDs Work
How Floppy Disk Drives Work
How MP3 Players Work
How Speakers Work

What A CD Can Hold

A CD's sampling rate and precision together produces a lot of data. On a CD, the digital numbers produced by the ADC are stored as bytes, and it takes 2 bytes to represent 65,536 gradations. There are two sound streams being recorded (one for each of the speakers on a stereo system), and a CD can store up to 74 minutes of music. So a CD can hold 783,216,000 bytes.

That is a lot of bytes! To store that many bytes onto a cheap piece of plastic tough enough to survive the abuse most people put a CD through is no small task, especially when you consider that the first CDs came out two decades ago.

Thomas Edison created the first device for recording and playing back sounds in 1877. His approach used a very simple mechanism to store an analog sound wave. In Edison's original phonograph, a diaphragm directly controlled a needle, and the needle scratched an analog signal onto a tin cylinder.

You spoke into Edison's device while rotating a tin cylinder. Your voice vibrated a thin diaphragm, which vibrated a needle, and those vibrations impressed themselves onto the tin. To play the sound back, the needle moved over the groove that was scratched during recording. During playback, the vibrations pressed into the tin caused the needle to vibrate, which in turn caused the diaphragm to vibrate and play the sound.

Emil Berliner improved this system in 1887 by producing the gramophone, which is also a purely mechanical device that uses a needle and diaphragm. The gramophone's major improvement over Edison's device was that it used flat records with spiral grooves, and that made mass production of the records easy because it's a lot easier to mass produce a flat disk than a cylinder. An album turntable works nearly the same way as the gramophone, but a turntable's needle reads signals that are amplified electronically rather than through direct vibration of a mechanical diaphragm.

Analog Waves

What did the needles in Edison's phonograph and Berliner's gramophone scratch onto the tin? They scratched analog waves representing the vibrations that sound creates. The vibrations are very quick: The diaphragm is vibrating back and forth thousands of times every second.

The storage and playback of an analog wave can be very simple, as in the case of scratching the wave onto tin. The problem with the simple approach is that the fidelity is not very good. For example, when you use Edison's phonograph, there is a lot of scratchy noise stored with the intended signal, and the signal is distorted in several different ways. And if you play a phonograph record repeatedly, eventually the needle will wear out the groove.

Digital Data

With a CD or any other digital recording technology there are two goals:

- **High fidelity**—There should be very great similarity between the original signal and the reproduced signal.
- **Perfect reproduction**—The recording should sound the same every time you play it, no matter how many times you play it.

To accomplish these two goals, digital recording converts analog sound waves into a stream of numbers and records the numbers instead of the wave. A device called an analog-to-digital converter (ADC) does the conversion.

To play back the music, a digital-to-analog converter (DAC) converts the stream of numbers back to an analog wave. The analog wave that the DAC produces is amplified and fed to the speakers to produce the sound.

The DAC will produce the same analog wave every time, as long as the numbers are not corrupted. The DAC will also produce an analog wave that is very similar to the original analog wave if the ADC sampled at a high rate and produced accurate numbers.

To get better sound quality, you must increase the number of samples.

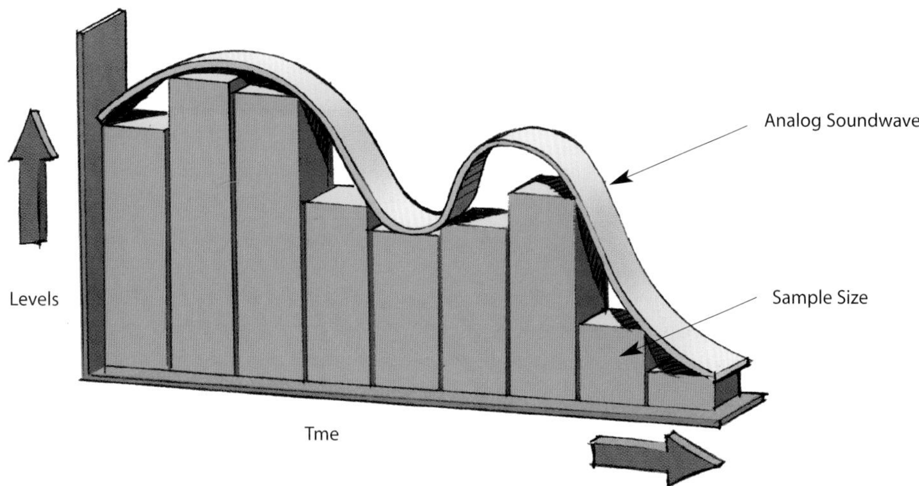

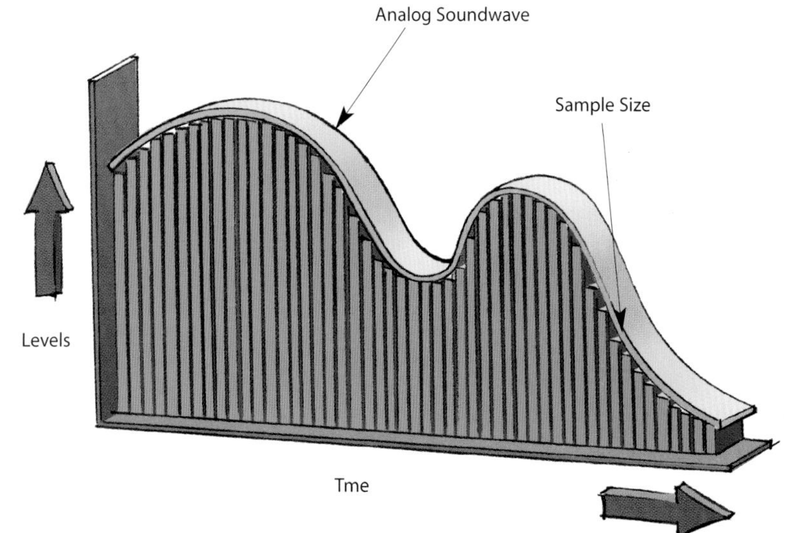

Analog-to-Digital Conversion

You can understand why CDs have such high fidelity if you understand the analog-to-digital conversion process. Let's say you have a sound wave, and you sample it with an ADC. When you do this, you have control over two variables:

- **Sampling rate**—The rate controls how many samples are taken per second.
- **Sampling precision**—The precision controls how many different gradations (or *quantization levels*) are possible when taking the sample.

For example, let's assume that the sampling rate is 1,000 samples per second and the precision is 16 different levels. Every 1/1000 second the ADC looks at the sound wave and picks a number between 0 and 15 that most closely represents the part of the wave that's being examined. These numbers are a digital representation of the original wave. Then the DAC re-creates the wave from these numbers.

With 1,000 samples per second and 16 quantization levels, sound quality from the DAC will be very poor. To get better sound quality, you can increase the number of samples and the number of levels. In the case of CD sound, fidelity is an important goal, so the sampling rate is 44,100 samples per second and the number of gradations is 65,536. At this level, the output of the DAC so closely matches the original wave form that the sound is essentially perfect to most human ears.

The switch from analog to digital music storage starts with CDs. The interesting side-effect was MP3 files, which simply compress CD tracks so that they can be sent over the Internet and stored on MP3 players. The same sort of digitization process brought movies from video tape to DVDs.

How **MP3s** Work

The MP3 movement is one of the most amazing phenomena that the music industry has ever seen. Unlike other movements—for example, the introduction of the cassette tape or the CD—the MP3 movement started not with the music industry itself, but with a huge audience of music lovers on the Internet. The MP3 format for digital music has had, and will continue to have, a huge impact on how people collect, listen to, and distribute music.

HSW Web Links

www.howstuffworks.com

How Analog and Digital Recording Work
How CDs Work
How DVDs work
How File Compression Works

When people talk about "an MP3," they mean a song encoded in the MP3 audio file format. The MP3 format is a compression system for music that reduces the number of bytes in a song without hurting the quality of the song's sound. Remarkably, this highly compressed MP3 music still sounds almost as good as it did in CD format. To see how this works, let's consider how music is encoded on a CD.

Sound Compression

Music on a CD is in digital form. The analog signal of the original music is sampled and represented by bytes—strings of 1s and 0s. In a music CD, music is sampled 44,100 times per second, the samples are each 2 bytes (16 bits) long, and separate samples are taken for both the left and right speakers in a stereo system.

If you add this up, you see that a 3-minute song takes up about 32 million bytes, or 32 MB. This is a very big file: Using a 56-kilobytes-per-second (Kbps) modem, it would take approximately 2 hours to download just one song. But if you compress the music into the MP3 format, the file size is only about one-tenth of the original size—perhaps 3 MB. At that size, the data moves across the Internet, even on a slow modem, very easily, and it does not take up very much hard disk space.

The MP3 format is able to shrink a sound file by using various compression algorithms. The most interesting compression scheme that MP3 uses is called perceptual noise shaping. This process is built around the way human beings hear sound. For example, there are certain sounds that the human ear cannot hear, and there are certain sounds that the human ear hears

Did You Know?

MPEG is an acronym for Moving Picture Experts Group. This group has developed compression systems for video data. For example, DVD movies, HDTV broadcasts, and DSS satellite systems use MPEG compression to fit video and movie data into smaller spaces. The MPEG compression system includes a subsystem to compress sound, called MPEG audio Layer-3, which we know by its abbreviation, MP3.

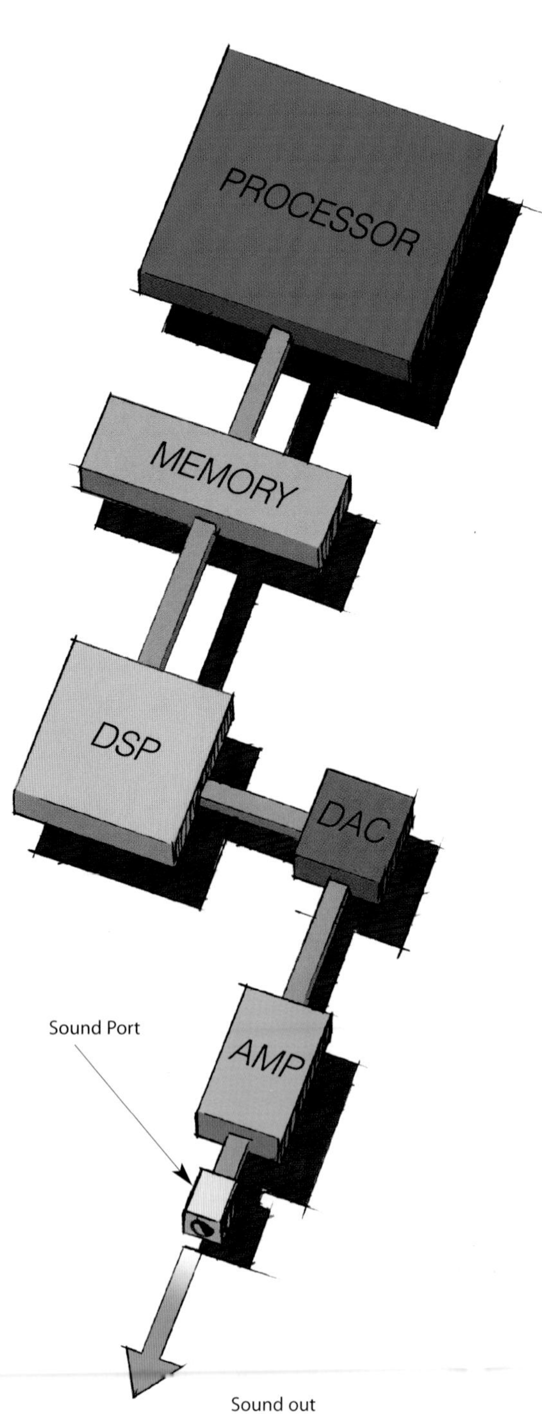

better than it does other sounds. And if there are two sounds playing simultaneously, we hear the louder one but cannot hear the softer one. So if you eliminate all the parts of the song that the human ear does not hear very well, you can reduce the file size quite a bit without changing the sound very much.

MP3s are described as having near-CD quality, meaning that a song doesn't sound exactly the same as on the original CD, but it is close. It is something like FM radio, which does not have the bandwidth to sound as good as a CD, but it sounds fine to most people for casual listening.

After you download an MP3 file, there are several different things you can do with it:

- You can use software to play it through your computer's speakers.
- You can download it to a portable MP3 player and listen to it.
- You can decompress it back to the normal CD file size and write it to an audio CD with a writable CD-ROM drive. It will not sound any better than the MP3 file, but you can listen to the song in any CD player after you load it onto a CD.

MP3 Players

A sound format that works only on a computer is fairly limiting. To let you take MP3s anywhere, electronics manufacturers have developed portable MP3 players. These units are something like portable cassette or CD players, but they use solid-state memory to hold the music instead of a physical medium such as a tape or a CD. All the players currently on the market include a software application that lets you transfer MP3 files into the player. Most players also include utilities for copying music from CDs or Web sites and the ability to create custom play lists.

The MP3 player is a great example of the invention process. All the components in a typical MP3 player are existing technologies; by combining these components in a new way, and writing some code to control it all, manufacturers have created an entirely new line of consumer products!

The job of the MP3 player is pretty straightforward. When you play a song, the player must do several things:

1) Pull the song from memory byte by byte
2) Decompress the MP3 encoding
3) Run the decompressed bytes through a digital-to-analog converter
4) Amplify the analog signal so that you can hear it

An MP3 player also has a controller that keeps track of all the songs in memory and handles the keypad and display.

An MP3 player plugs into a computer's universal serial bus (USB) port or parallel port to transfer data. The MP3 files are saved in the player's memory. Most players use some form of solid-state memory such as flash memory, but some use small hard disks. The advantage of solid-state memory is its small size and the fact that there are no moving parts, which means better reliability and no skips in the music.

The microprocessor is the brains of the player. It monitors user input through the playback controls, displays information about the current song on the LCD panel, and sends directions to the digital signal processor (DSP) chip that tells it exactly how to process the audio.

The DSP pulls the song data from memory, decompresses it, applies any special effects needed, and streams it to the amplifier. The DSP runs a decompression algorithm that undoes the compression of the MP3 file, and then a digital-to-analog converter turns the bytes back into waves. The amplifier boosts the strength of the signal and sends it to the audio port, where you connect a pair of headphones.

MP3 technology compresses a song into a smaller size, making it easier to transfer over the Internet and store to a disc. Portable MP3 players are small computers that take the files from the Internet or your hard disk, store them in memory, and decompress them when you want to listen to songs.

Did You Know?

The first MP3 players started with 32 megabytes or 64 megabytes of Flash memory. Since a typical MP3 file is about 3 megabytes, Flash-based players can therefore hold 10 to 20 songs.

MP3 players evolved on two fronts so that they could hold more songs. One style incorporates a small hard disk into the MP3 player. A typical hard disk like this can hold several gigabytes, meaning the player can hold a thousand songs or more. Another style lets the user save MP3 files on data CDs. Since a data CD can hold 650 megabytes, its possible to store several hundred songs per CD and carry multiple CDs. The CD-based players are larger, but have the advantage of inexpensive CD-R disks for storage.

How **TAPE RECORDERS** Work

Magnetic recording is a backbone technology of the electronic age. It is used to record sound, video, and data in a variety of different devices. A basic cassette tape recorder is the simplest example of magnetic recording technology. It takes an electrical audio signal and stores it as a magnetic signal, by using two components: the recorder and the tape.

HSW Web Links

www.howstuffworks.com

How Camcorders Work
How DVDs Work
How Floppy Disk Drives Work
How MP3s Work
How VCRs Work

Tape History

Audiotapes have gone through several format changes over the years.

- The original format was not tape at all, but thin steel wire. Valdemar Poulsen invented the wire recorder in 1900.
- German engineers perfected the first tape recorders using oxide tapes in the 1930s. Tapes originally appeared in a reel-to-reel format.
- The cassette—a plastic case that housed audiotape—was patented in 1964. It eventually beat out 8-track and reel-to-reel tapes to become the dominant tape format in the audio industry.

The storage medium that tape recorders today use is audiotape—a length of thin plastic base material coated with ferric oxide (Fe_2O_3) powder. The ferric oxide is mixed with a binder, which adheres it to the plastic, and a dry lubricant, which minimizes wear and tear on the recorder. Ferric oxide is simply an oxide of iron, just like common rust. It is used in tapes because it is ferromagnetic—that is, if you expose it to a magnetic field it will stay magnetized. This is exactly what a tape recorder does: It runs the tape by the magnetic field created by a small electromagnet.

Electromagnets and Tape Recorders

An electromagnet is a length of wire coiled around a piece of magnetic metal, such as iron. Electricity flowing through the wire creates a magnetic field, which magnetizes the metal. Because it is activated by electricity, you can turn the magnetic field on and off. You can also reverse the polar (north–south) orientation of the magnetic field by simply reversing the direction in which the electrical current is flowing. (See "How Electromagnets Work," page 90, for more information.)

An audio signal is an alternating current—it constantly reverses direction, just like the pressure fluctuations that make up a sound wave do. The recorder's electromagnet is activated by this audio signal. As the current reverses direction, so does the electromagnet's magnetic field. The recorder spools tape by the electromagnetic head at a constant speed, recording the sound's signal as a long pattern of varying magnetic fields.

When you play back the tape on the recorder, the process is essentially reversed. The deck spools the tape, with its varying magnetic fields, past an electromagnet. The core of the electromagnet picks up the changing magnetic field, which generates an electrical current in the electromagnet. The current reverses itself as the magnetic field reverses itself, creating an alternating current just like the one that originally charged the recording electromagnet. This signal then travels onto the amplifier and the speakers, where it is translated into sound.

In essence, this is all that's involved in recording sound on tape and then playing it back. Of course, to record and play the sound signals correctly and efficiently, these basic components have to be arranged in a very precise system.

Parts of a Tape Recorder

A normal cassette player uses two electromagnets that, together, are about as wide as one-half of the tape's width. The two heads record the two channels of a stereo signal simultaneously in two tracks on one-half of the tape surface. You can record two more audio tracks on the tape surface by reversing the cassette in the recorder deck and recording on the other half of the tape.

The tape is spooled past the electromagnets by a series of sprockets and wheels. Two motorized sprockets engage the spools inside the cassette. These sprockets spin one of the spools to take up the tape during recording, playback, fast-forward, and reverse.

Most recorders have two tape heads. The record/playback head contains the two electromagnets that record and the play audio signals. The other one, the erase head, essentially wipes the tape clean of any existing signals before recording and improves the recording quality. Not every recorder has an erase head.

The capstan and the pinch roller spool the tape past the electromagnets. The capstan

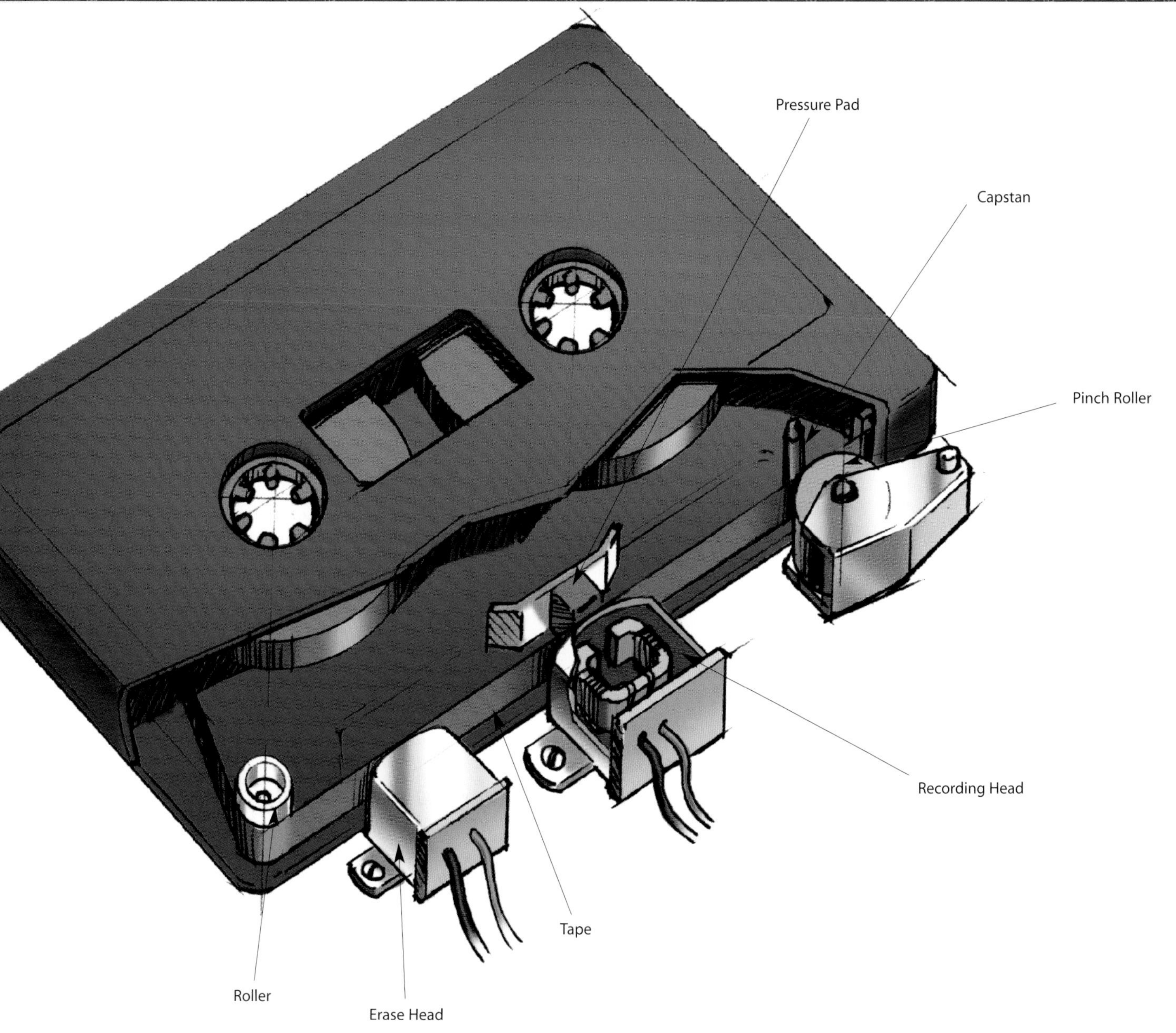

revolves at a very precise rate, pulling the tape across the head at exactly the right speed—typically 1.875 inches (4.76 cm) per second. The roller simply applies pressure, holding the tape tight against the capstan.

A tape recorder is a fascinating combination of precise mechanical machinery built around a simple process—using electromagnets to convert electrical signals to magnetic patterns and back again. Amazingly, this very simple design lets you record and play rich, high-quality sound, over and over again. The same basic process is used on VCR tapes and floppy disks as well, so magnetic recording is a very versatile technology!

Types of Tape

There are four types of tape in common use today:

- **Type 0**—The original ferric oxide tape. Very rarely seen these days.
- **Type 1**—Standard ferric oxide tape. Also referred to as "Normal Bias."
- **Type 2**—"Chrome" or CrO2 tape. The ferric oxide particles are mixed with chromium dioxide.
- **Type 4** —"Metal" tape. Metallic particles rather than metal oxide particles are used in the tape.

Sound quality improves as you go from one type to the next, with metal tapes having the best sound quality. A normal tape deck cannot record onto a metal tape - the deck must have a setting for metal tapes in order to record onto them. Any tape player can play a metal tape, however.

How SPEAKERS Work

In any sound system, the quality of the sound ultimately depends on the speakers. The speaker is the component that takes the electric signal stored on things like CDs and tapes and turns it back into actual sound that humans can hear.

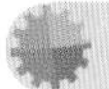

HSW Web Links

www.howstuffworks.com

How CDs Work
How Electromagnets Work
How Movie Sound Works
How Mufflers Work
How Tape Recorders Work

To understand how speakers work, you first need to understand how sound works.

Vibrations and Sound

Inside your ear is a very thin piece of skin called the eardrum. When your eardrum vibrates—usually because of rapid fluctuations in air pressure—your brain interprets the vibrations as sound.

When something vibrates, it moves the air particles around it. Those air particles in turn move the air particles around them, carrying the pulse of the vibration through the air as more and more particles are pushed farther from the source of the vibration.

To see how this works, think about a simple vibrating object—a bell. When you ring a bell, the metal vibrates—that is, it flexes in and out—rapidly. When it flexes out on one side, it pushes out on the surrounding air particles on that side. These air particles then collide with the particles in front of them, which collide with the particles in front of them, and so on. When the bell flexes away, it pulls in on these surrounding air particles, creating a drop in pressure that pulls in on more surrounding air particles. This creates another drop in pressure that pulls in particles that are even farther out, and so on. This decreasing of pressure is called rarefaction.

When this wave of air pressure fluctuation from the bell reaches your ear, it vibrates the eardrum. The human brain interprets different sounds because of variations in two things:

- **Sound wave frequency**—A higher wave frequency simply means that the air pressure fluctuates faster, which sounds to the human ear like a higher pitch. When there are fewer fluctuations in a period of time, the wave frequency is lower and so is the pitch.
- **Wave amplitude**—The wave's amplitude determines how loud the sound is. Sound waves with greater amplitudes move eardrums more, and you register this sensation as a higher volume.

A microphone works something like human ears. It has a diaphragm that is vibrated by sound waves. The vibration in a microphone is encoded on a tape or CD as an electrical signal.

Parts of a Speaker

A speaker takes the electrical signal from an amplifier and translates it back into physical vibrations to create sound waves. When everything is working as it should, the speaker produces nearly the same vibrations that the microphone originally recorded and encoded on a tape or CD.

Traditional speakers have one or more drivers. A driver produces sound waves by rapidly vibrating a flexible cone, or diaphragm. The cone, which is usually made of paper, plastic, or metal, is attached on the wide end to the surround. The surround is a rim of flexible material that allows the cone to move and is attached to the driver's metal frame, called the basket. The narrow end of the cone is connected to the voice coil, and the coil is attached to the basket by the spider, a ring of flexible material that holds the coil in position but allows it to move back and forth freely.

Chunks of the Frequency Range

To produce quality sound more effectively, loudspeakers typically break the entire range into smaller chunks that are handled by specialized drivers. Quality loudspeakers will typically have three drivers: a woofer, a tweeter, and a midrange driver, all included in one enclosure.

In a three-driver speaker, the crossover breaks the audio signal into three different chunks—low frequency, high frequency, and sometimes midrange frequencies.

The most common system—a passive crossover—uses inductors and capacitors, circuitry components that only become good conductors under certain conditions. A crossover capacitor will conduct the current very well when the frequency exceeds a certain level, but will conduct poorly when the frequency is below that level. A crossover inductor acts in the reverse manner—it is only a good conductor when the frequency is below a certain level. The crossover system feeds the electric signal through a series of inductors and capacitors so that each driver mainly receives a certain frequency range.

The voice coil is a basic electromagnet. An electromagnet is a coil of wire, usually wrapped around a piece of magnetic metal, such as iron. (See "How Electromagnets Work," page 90, for more information.) Running electrical current through the wire creates a magnetic field around the coil and magnetizes the metal it is wrapped around. The field acts just like the magnetic field around a permanent magnet: It has a polar orientation—a north end and a south end—and it attracts iron objects. But unlike a permanent magnet, in an electromagnet you can alter the intensity and orientation of the poles. If you reverse the flow of the current, the north and south ends of the electromagnet switch.

A stereo signal works just like a voice coil—it constantly reverses the flow of electricity. If you've ever hooked up a stereo system, then you know that there are two output wires for each speaker—typically a black one and a red one.

Essentially, the amplifier is constantly switching the electrical signal, fluctuating between a positive charge and a negative charge on the red wire. Since electrons always flow in the same direction between positively charged particles and negatively charged particles, the current going through the speaker moves one way and then reverses and flows the other way. This alternating current causes the polar orientation of the electromagnet to reverse itself many times each second.

The electromagnet is positioned in a constant magnetic field created by a permanent magnet. These two magnets—the electromagnet and the permanent magnet—interact with each other as any two magnets do. The positive end of the electromagnet is attracted to the negative pole of the permanent magnetic field, and the negative pole of the electromagnet is repelled by the permanent magnet's negative pole. When the electromagnet's polar orientation switches, so does the direction of repulsion and attraction. In this way, the alternating current constantly reverses the magnet forces between the voice coil and the permanent magnet. This pushes the coil back and forth rapidly, like a piston.

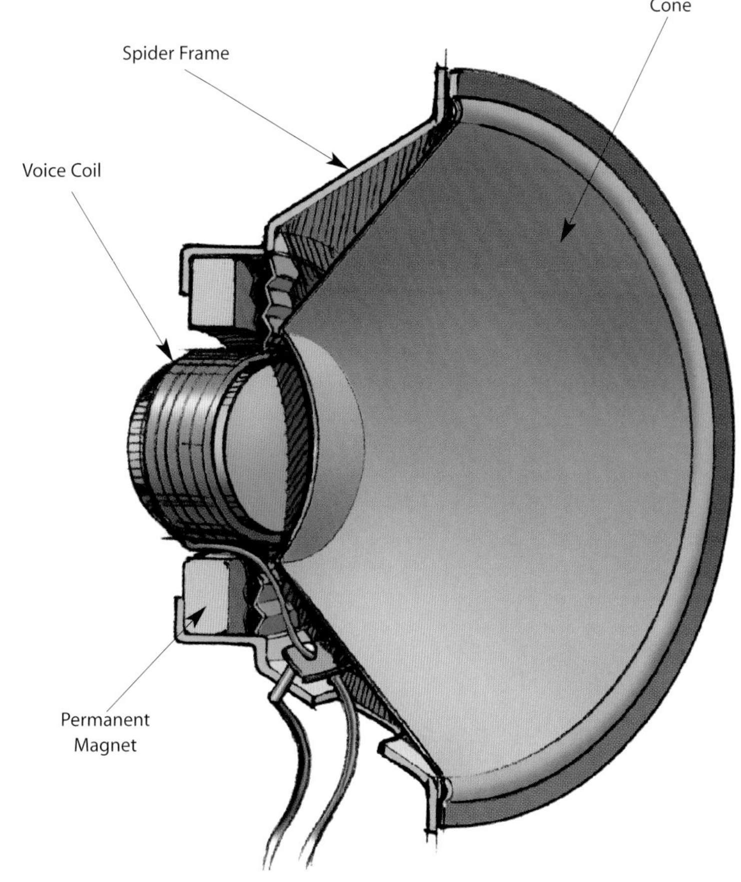

When the coil moves, it pushes and pulls on the speaker cone. This vibrates the air in front of the speaker, creating sound waves. The electrical audio signal can also be interpreted as a wave. The frequency and amplitude of this wave, which represents the original sound wave, dictates the rate and distance that the voice coil moves. This, in turn, determines the frequency and amplitude of the sound waves produced by the diaphragm. When everything is working right, you hear almost exactly the same sound the microphone recorded!

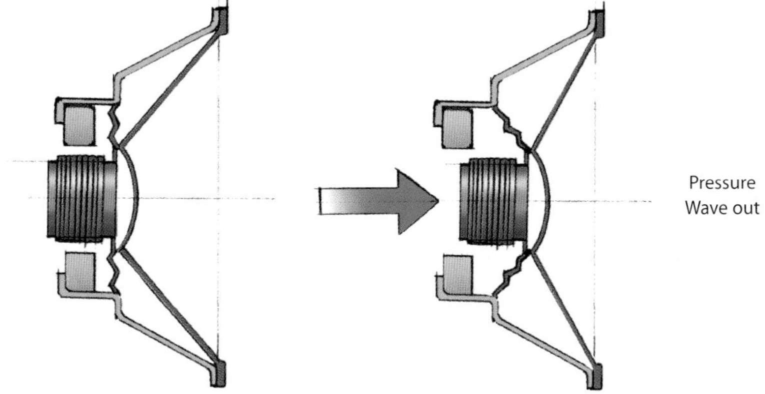

How **RADIO** Works

Radio waves transmit music, conversations, pictures, and data invisibly through the air, often over millions of miles. This happens every day in thousands of different ways!

HSW Web Links

www.howstuffworks.com

How Cell Phones Work
How Cordless Telephones Work
How Ham Radio Works
How Radio Scanners Work
How Radio-Controlled Toys Work
How Television Works

Make a Radio Transmitter

This experiment shows how simple a radio can be:

1) Find a fresh 9-volt battery, an AM radio, and a coin.
2) Tune the AM radio to an area of the dial where you hear static.
3) Now hold the battery near the antenna and quickly tap the two terminals of the battery with the coin (so that you connect them together for an instant).
4) You will hear a crackle in the radio that is caused by the connection and disconnection of the coin.

Your battery/coin combination is a radio transmitter! It's not transmitting anything useful (just static) and it will not transmit very far (just a few inches) because it's not optimized for distance. But if you use the device to tap out Morse code, you can communicate over several inches with this device!

Many things you see and use every day depend on radio waves, such as AM and FM radio broadcasts, cordless phones, garage door openers, wireless networks, radio-controlled toys, television broadcasts, cell phones, GPS receivers, ham radios, satellite communications devices, police radios, and wireless clocks.

All radios today use continuous sine waves to transmit information such as audio, video, and other data. Each radio signal uses a different sine wave frequency, and that is how they are all separated and can be picked up properly by different devices.

Any radio setup has two parts:

- The transmitter
- The receiver

The transmitter takes some sort of message (it could be the sound of someone's voice, pictures for a TV set, data for a radio modem, or something else), encodes it onto a sine wave, and transmits it with radio waves. The receiver receives the radio waves and decodes the message from the radio sine wave.

Transmitting Signals

You can get an idea of how a radio transmitter works by starting with a battery and a piece of wire. A battery sends electricity (a stream of electrons) through a wire if you connect the wire between the two terminals of the battery. The moving electrons create a magnetic field surrounding the wire, and that field is strong enough to affect a compass. In a radio antenna, the same basic process converts a sine wave from the transmitter into a varying electromagnetic wave that a receiver can pick up.

One characteristic of a sine wave is its frequency, or the number of times the wave oscillates up and down per second. When you listen to an AM radio broadcast, your radio is tuning in to a sine wave with a frequency around 1,000,000 cycles per second (cycles per second is also known as Hertz, or Hz). For example, 680 AM is 680,000 cycles per second.

Modulation

If you have a sine wave and a transmitter that is transmitting the sine wave into space with an antenna, you have a radio station. But in order to broadcast sound, the sine wave must contain some information. You need to modulate the wave in some way to encode information on it. There are three common ways to modulate a sine wave:

- **Pulse modulation (PM)**—In PM, you simply turn the sine wave on and off. This is an easy way to send Morse code. PM is not that common, but one good example of it is the radio system that sends signals to radio-controlled clocks in the United States. One PM transmitter is able to cover the entire United States!
- **Amplitude modulation (AM)**—Both AM radio stations and the picture part of an analog TV signal use amplitude modulation to encode information. In AM,

AM

the amplitude of the sine wave (that is, its peak–to–peak voltage) changes. So, for example, the sine wave produced by a person's voice is overlaid onto the transmitter's sine wave to vary its amplitude.

- **Frequency modulation (FM)**—FM radio stations and hundreds of other wireless technologies (including the sound portion of a TV signal, cordless phones, and cell phones) use FM. In FM, the transmitter's sine wave frequency changes very slightly based on the information signal.

Receiving Signals

When you tune your car's AM radio to a station—for example, 680 AM—the transmitter's sine wave is transmitting at 680,000 Hz (or cycles per second, where the sine wave repeats 680,000 times per second). The radio DJ's voice is modulated onto that carrier wave by varying the amplitude of the transmitter's sine wave. An amplifier amplifies the signal to something like 50,000 watts for a large AM station. Then the antenna sends the radio waves out into space.

So how does your car's AM radio—a receiver—receive the 680,000-Hz signal that the transmitter sent and extract the information (the DJ's voice) from it? Here are the steps:

1) Unless you are sitting right beside the transmitter, your radio receiver needs an antenna to help it pick the transmitter's radio waves out of the air. An AM antenna is simply a wire or a metal stick that increases the amount of metal the transmitter's waves can interact with.
2) Your radio receiver needs a tuner. The antenna receives thousands of sine waves, so the tuner needs to separate one desired sine wave from the thousands of radio signals that the antenna receives. In this case, the tuner is tuned to receive the 680,000-Hz signal. Tuners work by using a principle called resonance; tuners resonate at, and amplify, one particular frequency and ignore all the other frequencies in the air. It is easy to create a resonator with a capacitor and an inductor.
3) The tuner causes the radio to receive just one sine wave frequency (in this case, 680,000 Hz). The radio has to extract the DJ's voice out of that sine wave. A part of the radio called a detector or demodulator does this. In the case of an AM radio, the detector is made with an electronic component called a diode, which allows current to flow through in one direction but not the other, so it clips off one side of the wave.
4) The radio next amplifies the clipped signal and sends it to the speakers or a set of headphones. The amplifier is made of one or more transistors, and more transistors means more amplification and therefore more power to the speakers. What you hear coming out the speakers is the DJ's voice!

FM radio works basically the same way as AM radio—the antenna, tuner, and amplifier are largely the same for both AM and FM. The main difference between AM and FM is that the detector is different: In FM, the detector turns frequency changes into sound.

Antennas

An antenna in a radio transmitter launches the radio waves into space. A receiver picks up as much of the transmitter's power as possible and supplies it to the tuner. For satellites that are

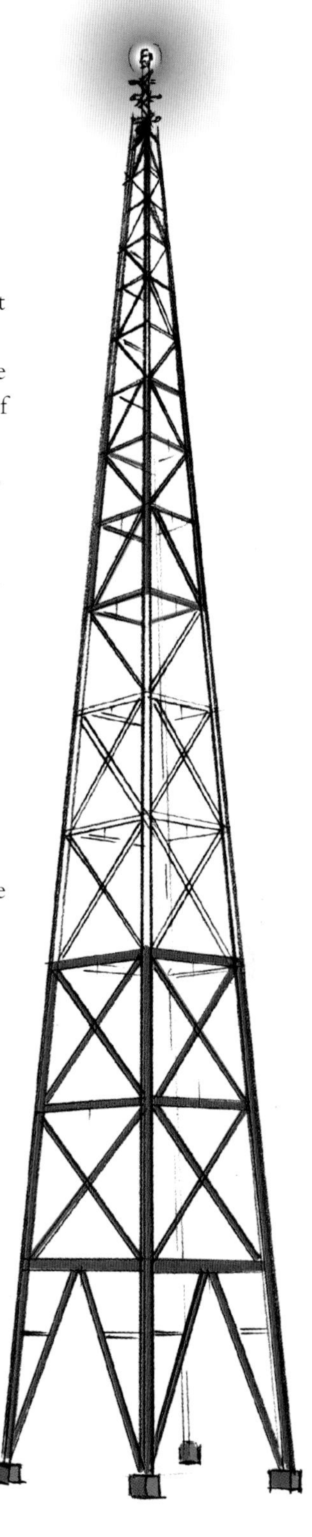

FM

EPIRB Distress Radio

Emergency Position Indicating Radio Beacon (EPIRB) is meant to help rescuers locate you in an emergency situation, and these radios have saved many lives since their creation in the 1970s. Boaters and pilots are the main users of EPIRBs.

A modern EPIRB is a sophisticated device that contains:

- A 5-watt radio transmitter operating at 406 MHz (see "How the Radio Spectrum Works," on the next page, for details on frequencies).
- A 0.25-watt radio transmitter operating at 121.5 MHz.
- A GPS receiver.

Once activated, both of the radios start transmitting. 24,000 or so miles up in space, a GOES weather satellite in a geosynchronous orbit can detect the 406 MHz signal.

Embedded in the signal is a unique serial number and, if the unit is equipped with a GPS receiver, the exact location of the radio. If the EPIRB is properly registered, the serial number lets the Coast Guard know who owns the EPIRB. Rescuers in planes or boats can home in on the EPIRB using either the 406 MHz or 121.5 MHz signal.

Older EPIRBs did not contain the GPS, so the GOES satellite received only a serial number. To locate the EPIRB, another set of satellites (like the TIROS-N satellite) orbiting the planet in a low polar orbit could pick up the signal as it passed overhead. This would give a rough fix on the location, but it takes several hours for a satellite to come into range.

millions of miles away, NASA uses huge dish antennas that are up to 200 feet (60 m) in diameter!

The size of an optimum radio antenna is related to the frequency of the signal that the antenna is trying to transmit or receive. This relationship has to do with the speed of light (186,000 miles per second, or 300,000 km per second) and the distance electrons can travel as a result.

Let's say that you are trying to build a radio tower for radio station 680 AM. It is transmitting a sine wave with a frequency of 680,000 Hz. In one cycle of the sine wave, the transmitter is going to move electrons in the antenna in one direction and then switch and pull them back. In other words, the electrons will change direction two times during one cycle of the sine wave. If the transmitter is running at 680,000 Hz, this means that every cycle completes in 1/680,000 second, or 0.00000147 second per cycle. One-quarter of that is 0.0000003675 second. At the speed of light, electrons can travel 0.0684 miles (0.11 km) in 0.0000003675 seconds. That means the optimal antenna size for the transmitter at 680,000 Hz is about 361 feet (110 m). So you can see that AM radio stations need very tall towers. For a cell phone working at 900,000,000 Hz (900 MHz), on the other hand, the optimum antenna size is about 3 inches (8.3 cm). This is why cell phones can have such short antennas.

AM stations are so strong in cities that it doesn't really matter if your car's receiving antenna is the optimal length. But if you want to set up an AM radio antenna very far away from the transmitter, matching the antenna's length to the frequency can dramatically improve reception.

Radio waves now touch many parts of our lives. In your home, your AM/FM radio, TV, cordless phone, cell phone, pager, GPS, baby monitor, and garage door opener all depend on radio waves to do their jobs. Space travel and airline flights would be nearly impossible without radio. It's amazing how much of an impact this technology has had on the world!

How the **RADIO SPECTRUM** Works

You have often heard about AM and FM radio, VHF and UHF television, and citizens band radio and shortwave radio. Have you ever wondered what all those different names for radio really mean?

A radio wave is an electromagnetic wave propagated by an antenna. Radio waves have different frequencies, and by tuning a radio receiver to a specific frequency, you can pick up a specific signal. In the United States, the Federal Communications Commission (FCC) decides who is able to use what frequencies for what purposes, and it issues licenses to stations for specific frequencies.

Similarly, AM radio is confined to a band from 535 kilohertz (KHz) to 1,700 KHz (that is, 535,000 to 1,700,000 cycles per second). So an AM radio station that says, "This is AM 680 WPTF!" means that the radio station is broadcasting an amplitude modulation (AM) radio signal at 680 KHz and has the FCC-assigned call letters WPTF.

Common frequency bands include the following:

HSW Web Links

www.howstuffworks.com

How Cell Phones Work
How Cordless Telephones Work
How GPS Receivers Work
How Ham Radio Works
How Radio Works

Frequency Bands

When you turn to a radio station and the announcer says, "You are listening to 106.1 FM—WRDU, The Home of Rock 'N' Roll!" the announcer means that you are listening to a radio station broadcasting frequency modulation (FM) radio signal at a frequency of 106.1 megahertz, with the FCC-assigned call letters WRDU. Megahertz (or MHz) means "millions of cycles per second," so "106.1 MHz" means that the transmitter at the radio station is oscillating at a frequency of 106,100,000 cycles per second. If you are in range of the station's signal, your FM radio can tune in to that specific frequency and give you clear reception of that station. All FM radio stations transmit in a band of frequencies between 88 MHz and 108 MHz. In the United States, this band of the radio spectrum is used only for FM radio broadcasts.

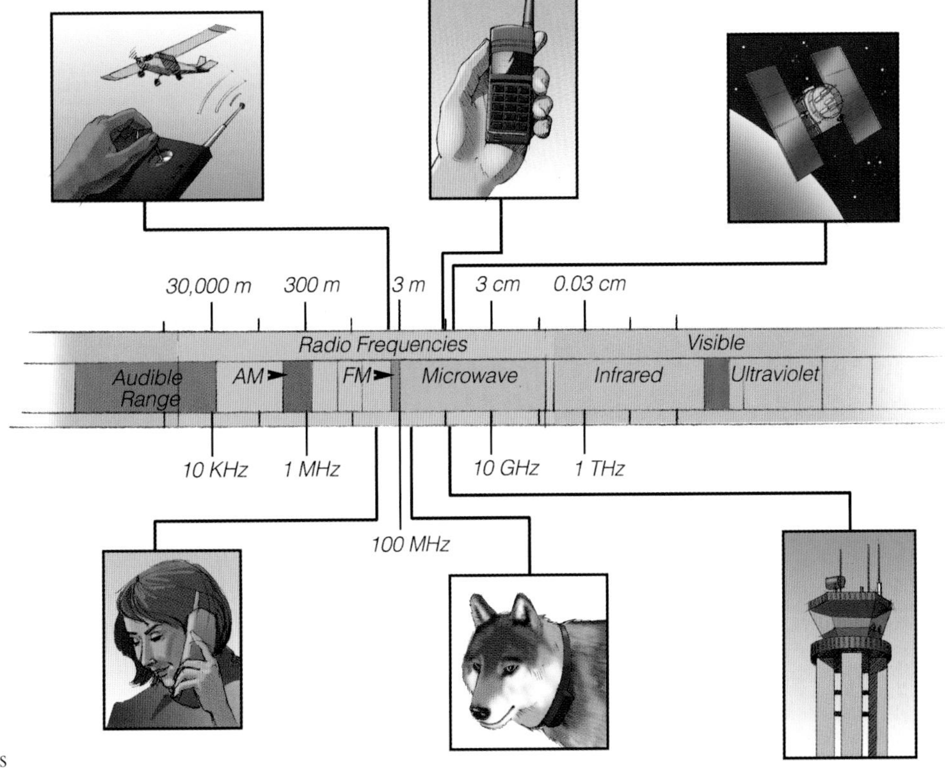

- **AM radio:** 535 KHz to 1.7 MHz
- **Shortwave radio:** 5.9 MHz to 26.1 MHz
- **Citizens band (CB) radio:** 26.96 MHz to 27.41 MHz
- **Television stations:** 54 MHz to 88 MHz for channels 2-6

- **FM radio:** 88 MHz to 108 MHz
- **Television stations:** 174 MHz to 220 MHz for channels 7-13

What is funny is that every wireless technology you can imagine has its own little band, too. There are hundreds of them! Here are some examples:

- **Garage door openers and alarm systems:** around 40 MHz
- **Standard cordless phones:** from 40 MHz to 50 MHz
- **Baby monitors:** 49 MHz
- **Radio-controlled airplanes:** around 72 MHz
- **Radio-controlled cars:** around 75 MHz
- **Wildlife tracking collars:** 215 MHz to 220 MHz
- **Cell phones:** 824 MHz to 849 MHz
- **900-MHz cordless phones:** around 900 MHz (obviously)
- **Air traffic control radar:** 960 MHz to 1,215 MHz
- **Global positioning systems:** 1,227 MHz to 1,575 MHz
- **Deep space radio communications:** 2,290 MHz to 2,300 MHz

Why is AM radio in a band at 535 KHz, to 1.7 MHz and FM radio is in a band at 88 MHz to 108 MHz? It is all completely arbitrary, and a lot of it has to do with history. For example, AM radio has been around a lot longer than FM radio. The first radio broadcasts occurred in 1906 or so, and frequency allocation for AM radio occurred during the 1920s. (The predecessor to the FCC was established by Congress in 1927.) In the 1920s radio and electronic capabilities were fairly limited, so people used relatively low frequencies for AM radio. Television stations were nearly nonexistent until 1946 or so, when the FCC allocated commercial broadcast bands for TV. By 1949 1 million people owned TV sets, and by 1951 there were 10 million TVs in the United States. FM radio was invented in order to make high-fidelity (and static-free) broadcast of music possible. The first FM radio station went on the air in 1939, but FM did not become really popular until the 1960s.

Better Reception at Night

Radio waves naturally travel in straight lines, so you would naturally expect (because of the curvature of the earth) that no radio station would transmit farther than 30 or 40 miles. And that is exactly the case for ground-based (as opposed to satellite) TV transmissions. The curvature of the earth prevents ground-based transmissions from going much farther than 40 miles for TV.

Certain radio stations, however, especially in the short-wave and AM bands, can travel much farther. Short-wave can circle the globe, and AM stations transmit hundreds of miles at night. The reason that is possible is because there is a layer of the atmosphere called the ionosphere. It is called the ionosphere because, when the sun's rays hit this layer, many of the atoms lose electrons and turn into ions. As it turns out, the ionosphere reflects certain frequencies of radio waves. So the waves bounce between the ground and the ionosphere and make their way around the planet. The composition of the ionosphere is different between night and day because of the presence or absence of the sun, and it turns out the reflection characteristics are better at night.

Scanners

Most radios that you see in your everyday life are single-purpose radios. For example, an AM radio can listen to any AM radio station in the frequency band from 535 KHz to 1.7 MHz, but it can't pick up anything else. An FM radio can listen to any FM radio station in the band from 88 MHz to 108 MHz and nothing else. A CB radio can listen to the 40 channels devoted to CB radio and nothing else. Scanners, however, are radio receivers that have extremely wide frequency ranges so that you can use them to listen to all kinds of radio signals. Typically scanners are used to tune in to police, fire, and emergency radio in the local area (so scanners are often called police scanners), but you can use a scanner to listen to all kinds of conversations. Typically you do one of the following:

- You can set up a scanner to scan (that is, switch between) a whole range of frequencies and then stop scanning when it detects a signal on any of the frequencies it is scanning. If you are interested in learning what the police are doing, you will scan the police radio frequencies in your local area. When a patrol car calls in to report a problem, the scanner will stop on that frequency and let you hear the conversation.
- You can set a scanner to a specific frequency and listen to that channel. You could listen to the chatter between the control tower and airplanes at the local airport by listening to the specific frequency used at the airport.

Because a scanner can receive a huge range of frequencies, you can set it to receive nearly anything on the air. In order to use a scanner, you need to have good frequency tables so you know where the action is.

Because radio airwaves are chopped into distinct bands, thousands of different technologies can share the frequency spectrum without interfering with each other. Your garage door opener, baby monitor, cordless phone, and cell phone can all work at the same time because of this frequency allocation.

chapter nine

AROUND THE HOUSE

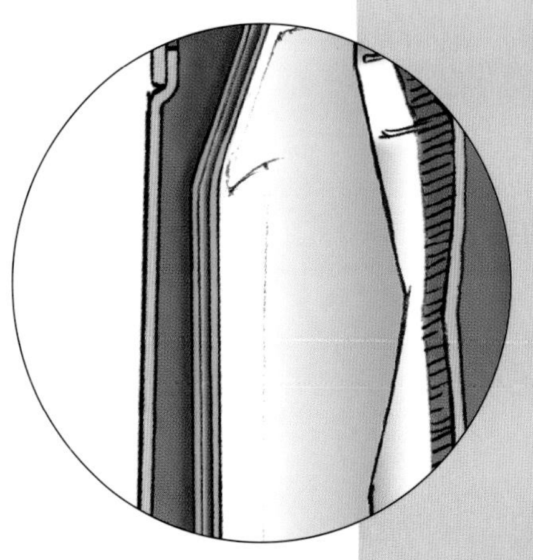

How CHRISTMAS LIGHTS Work

As November and December roll around, you see strands of Christmas lights everywhere—on Christmas trees, houses, shrubs, bushes, and even occasionally cars. Some strands stay lit constantly, and some flash in interesting patterns that range from chase sequences to a gentle dimming effect. Although they are very simple, today's Christmas lights are interesting because they have saved the country millions of megawatts of power!

HSW Web Links

www.howstuffworks.com

How Christmas Trees Work
How Christmas Works
How Electricity Works
How Lights Work

If you were to go back in time 30 or 40 years and look at how people decorated their houses and trees with lights, you would find that most people used small 120-volt incandescent bulbs. Each bulb was a 5- or 10-watt bulb like the bulb you find in a night light. You can still find strands of these bulbs today, but they aren't very common anymore for three reasons:

- They consume a lot of power. If you have a strand of 50 5-watt bulbs, the strand consumes 250 watts. Consider that most people use two or three strands to decorate a tree and 5 or 10 strands to decorate a house, and you are talking about a lot of power!
- Because the bulbs consume so much power, they generate a lot of heat. When used indoors, three strands at 250 watts per strand are generating as much heat as a 750-watt space heater. The heat from the individual bulbs can also melt things.
- They are expensive. You can buy a 10-pack of miniature bulbs for about a dollar. The large bulbs might cost 5 to 10 times more.

The one advantage of old-style bulb strands is that one bulb's failure has absolutely no impact on the rest of the bulbs. That's because a 120-volt bulb system places the bulbs in parallel. You can have 2 or 200 bulbs in a strand that is wired in parallel. The only limit is the amount of current that the two wires can carry.

Mini-lights

The 1970s saw a revolution in decorative lighting. Mini-lights were introduced and they now dominate the Christmas lights market. A mini-light is a small 2.5-volt incandescent bulb that is not much different from an incandescent flashlight bulb.

How can you plug these 2.5-volt mini-lights into a 120-volt outlet? The key to using these small, low-voltage bulbs with normal house current is to connect them in series. If you multiply 2.5 volts by 48, you get 120 volts, and originally, that's how many bulbs each strand had. A typical strand today has 2 more bulbs, so that there are 50 lights in the strand—a nice round number. Adding the 2 extra bulbs dims the set imperceptibly, so it doesn't make much difference in the look of the strand.

You can now see why mini-light strands are so sensitive to the removal of one bulb: It breaks the circuit, so none of the bulbs can light. When mini-lights were first introduced, if any bulb burned out, the entire strand would darken. Today, one or more bulbs can burn out and the strand will stay lit, but if you pop one of the bulbs out of its socket, the whole strand will go dark. The difference in behavior occurs because

the new bulbs contain an internal shunt.

If you look closely at a bulb you can see the shunt wire wrapped around the two posts inside the bulb. The shunt contains a coating that gives it fairly high resistance, until the filament fails. The heat caused by current flowing through the shunt burns off the coating and reduces the shunt's resistance.

The big advantages of mini-light strands are the low wattage (about 25 watts per 50-bulb strand) and the low cost (the bulbs, sockets, and wire are all much less expensive than a 120-volt parallel system). The big disadvantage is the problem of loose bulbs. Unless there is a shunt in the socket, a loose bulb will cause the whole 50-bulb strand to fail. It's not hard to have a loose bulb because the sockets are pretty flimsy. A tester can help find loose bulbs faster; you point the tester at each bulb, and it tells you which bulb is loose.

Blinking Lights

Two different techniques—one crude and the other sophisticated—are used to create blinking lights.

The crude method involves the installation of a special blinker bulb at any position in the strand. A typical blinker bulb contains a bimetallic strip. The current runs from the strip to the post to light the filament. When the filament gets hot, it causes the strip to bend, breaking the current and extinguishing the bulb. As the strip cools it bends back, reconnects the post, and relights the filament so that the cycle repeats. Whenever this blinker bulb is not lit, the rest of the strand is not getting power, so the entire strand blinks in unison. Obviously, these bulbs don't have a shunt (if they did, the rest of the strand would not blink), so when the blinker bulb burns out, the rest of the strand will not light until the blinker bulb is replaced.

The most sophisticated light sets today have 16-function controllers that can run the lights in all sorts of interesting patterns. This type of system typically has a controller box that drives four separate strands of mini-bulbs that are braided together. If you take a controller box apart, you will find that it is rather simple. It contains an integrated circuit and four transistors, or triacs—one to drive each strand. The integrated circuit simply turns on a triac to light one of the four strands. By sequencing the triacs appropriately, you can create all kinds of effects.

Christmas lights are not only a fun decoration; they are another great example of how technological improvements impact our lives in all sorts of ways. Today's lights cost a lot less than their older counterparts and they are more energy efficient.

Christmas Light Testers

You can pick up a Christmas light tester for about $4 at a discount store and they work pretty well. You plug a strand of non-working lights into a wall socket and point the left end of the tester at a bulb. If the bulb is getting power (even if it is not lit), then the LED on the tester will light when you push the button. If the bulb is not getting power — because a bulb upstream is loose — then the LED on the tester will not light.

These testers are handy for other things as well. For example, you can point the tester at any wall outlet and it will tell you whether the outlet has power or not. It has a range or 3 or 4 inches, so you can sometimes use the tester to find where wires are running through the walls as well. These testers detect the alternating electromagnetic waves given off by any wire carrying power. Current running through a wire generates a field that can affect a compass. In the case of household power, the current is alternating at 50 (European standard) or 60 (U.S. standard) oscillations per second. These oscillations set up a field near the wire that is fairly strong. The bulb tester picks up this field and a simple amplifier amplifies it. The amplified signal is either strong enough to light the LED in the tester or it is not. The level of amplification is set so that only a strong electromagnetic field very near a wire will light the LED. All other sources of electromagnetic radiation are not strong enough to light the light.

How **REFRIGERATORS** Work

Prior to refrigeration, the only way to preserve meat was to salt it, and cold or iced beverages in the summer were a real luxury. Today, without refrigeration, we'd be throwing out our leftovers instead of saving them for another meal. Although many of us take it completely for granted, the refrigerator is one of those miracles of modern living that totally changes life.

HSW Web Links

www.howstuffworks.com

How Air Conditioners Work
How Food Preservation Works
How Ice Rinks Work
How Thermometers Work
How Tire Pressure Gauges Work
How Vacuum Flasks Work

The basic idea behind a refrigerator is very simple: A refrigerator uses the evaporation of a liquid to absorb heat from a food storage area. This is the same process that cools your skin as you sweat: As the water in sweat evaporates, it absorbs heat.

Principles of Evaporation

A few basic principles can help you understand how evaporation works in a refrigerator:

- When you heat a liquid to a specific temperature, it will boil and evaporate molecules as a gas. Different liquids boil at different temperatures, and all liquids have lower boiling points if you lower the pressure.
- The evaporation process uses a lot of energy. The energy comes from heat in the surrounding environment. Evaporation cools the surrounding environment.
- When a gas is compressed, its temperature and pressure rise. If a gas is compressed enough and allowed to dissipate the heat of pressurization, it turns into a liquid. When a gas expands (that is, the volume it occupies increases), its temperature and pressure fall.

A refrigerator uses these three principles to move heat from one place to another. In a refrigerator, heat moves from the food storage area to the radiator on the back of the refrigerator. The working fluid, known as the refrigerant, absorbs and moves the heat. A refrigerator cools air by evaporating and condensing the refrigerant in a continuous cycle.

Parts of a Refrigerator

A refrigerator system has five basic components:

- A compressor
- A serpentine, or coiled, set of heat-exchanging pipes outside the refrigerator
- An expansion valve
- A serpentine, or coiled, set of heat-exchanging pipes inside the refrigerator
- A refrigerant

The refrigerant is the liquid that evaporates inside the refrigerator to create the cold temperatures. Many industrial installations use pure ammonia as the refrigerant. Pure ammonia evaporates at -27°F (-32°C).

The compressor powers the entire process and also produces the humming noise you hear when your refrigerator is running. It compresses the refrigerant gas, which raises the refrigerant's pressure and temperature. The heat-exchanging coils outside the refrigerator allow the refrigerant to dissipate the heat of pressurization—that is, they cool the refrigerant.

As the refrigerant cools, it condenses into liquid form. Then it flows through the expansion valve, a small passageway between a high-pressure zone and a low-pressure

Why Do We Need Refrigerators?

The basic idea behind refrigeration is to slow down the activity of bacteria (which all food contains) so that it takes longer for the bacteria to spoil the food. For example, bacteria will spoil milk in 2 or 3 hours if the milk is left out on the kitchen counter at room temperature. However, by reducing the temperature of the milk by storing it in the refrigerator, the milk will stay fresh for 1 or 2 weeks. The cold temperature inside the refrigerator decreases the activity of the bacteria that much! By freezing the milk you can stop the bacteria altogether and the milk can last for many months (until effects such as freezer burn begin to spoil the milk in non-bacterial ways).

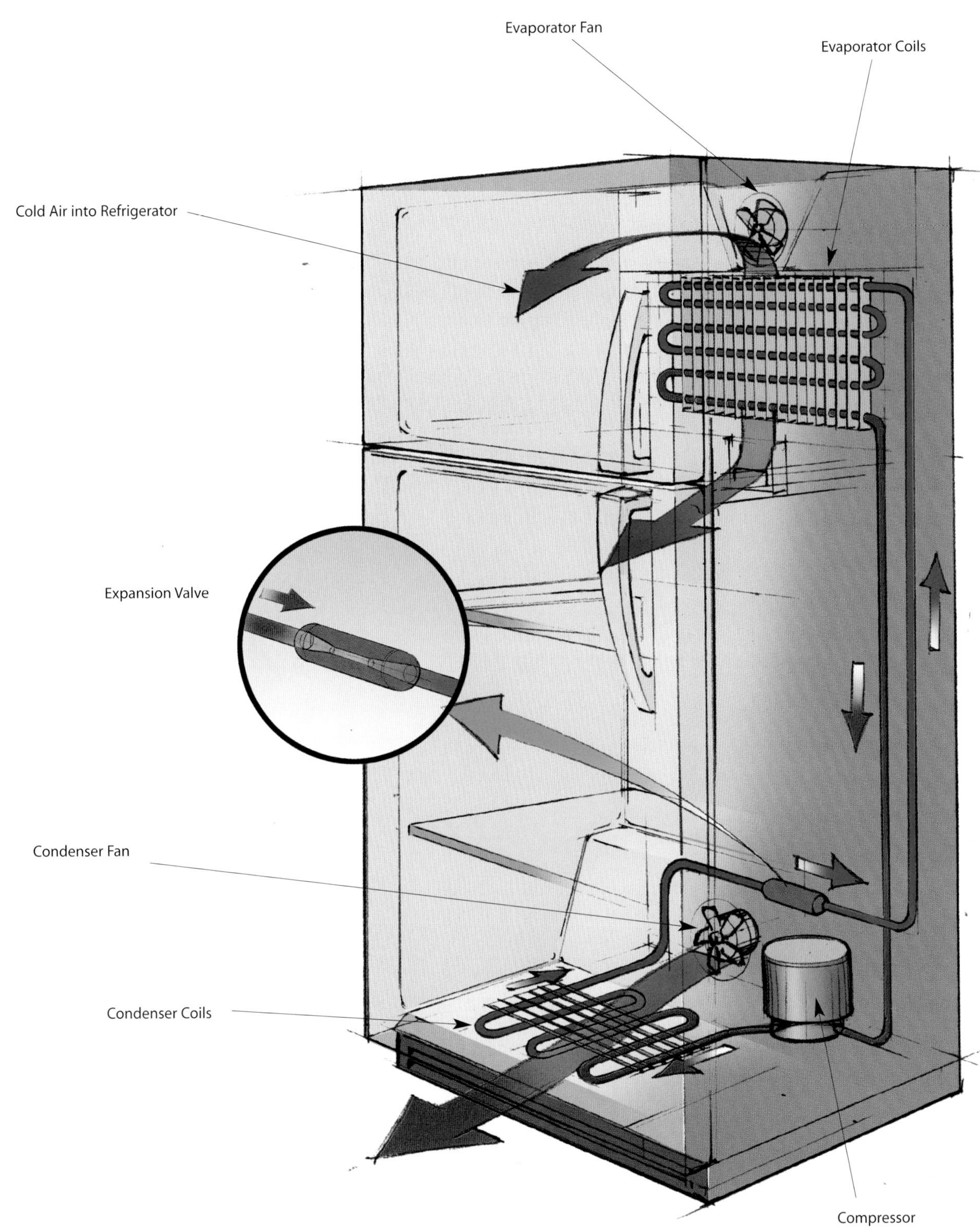

zone. In the low-pressure zone—the heat-exchanging coils inside the refrigerator—the refrigerant evaporates. As it evaporates, the refrigerant absorbs heat from the area around it—the space inside your refrigerator.

A simple cycle has done incredible things for your food supply. It allows you to refrigerate food so that it lasts a week or two before spoiling—or freeze it so that it lasts for months.

How **AIR CONDITIONERS** Work

An air conditioner is basically the same thing as a refrigerator. It cools air, using the same process of evaporating and condensing a refrigerant. But an air conditioner cools something as big as a house, whereas a refrigerator cools only a small, insulated box. That means the equipment for these two devices looks totally different.

HSW Web Links

www.howstuffworks.com

How Clothes Dryers Work
How Food Preservation Works
How Hair Dryers Work
How Ice Rinks Work
How Refrigerators Work

Like a refrigerator, an air conditioner absorbs heat from inside and releases heat outside. There are several types of air-conditioners: window units, split-system air conditioners, and chilled-water systems.

Window Units

A window air conditioner unit is a complete air conditioner in a small space. If you examine a window unit, you will find these parts:

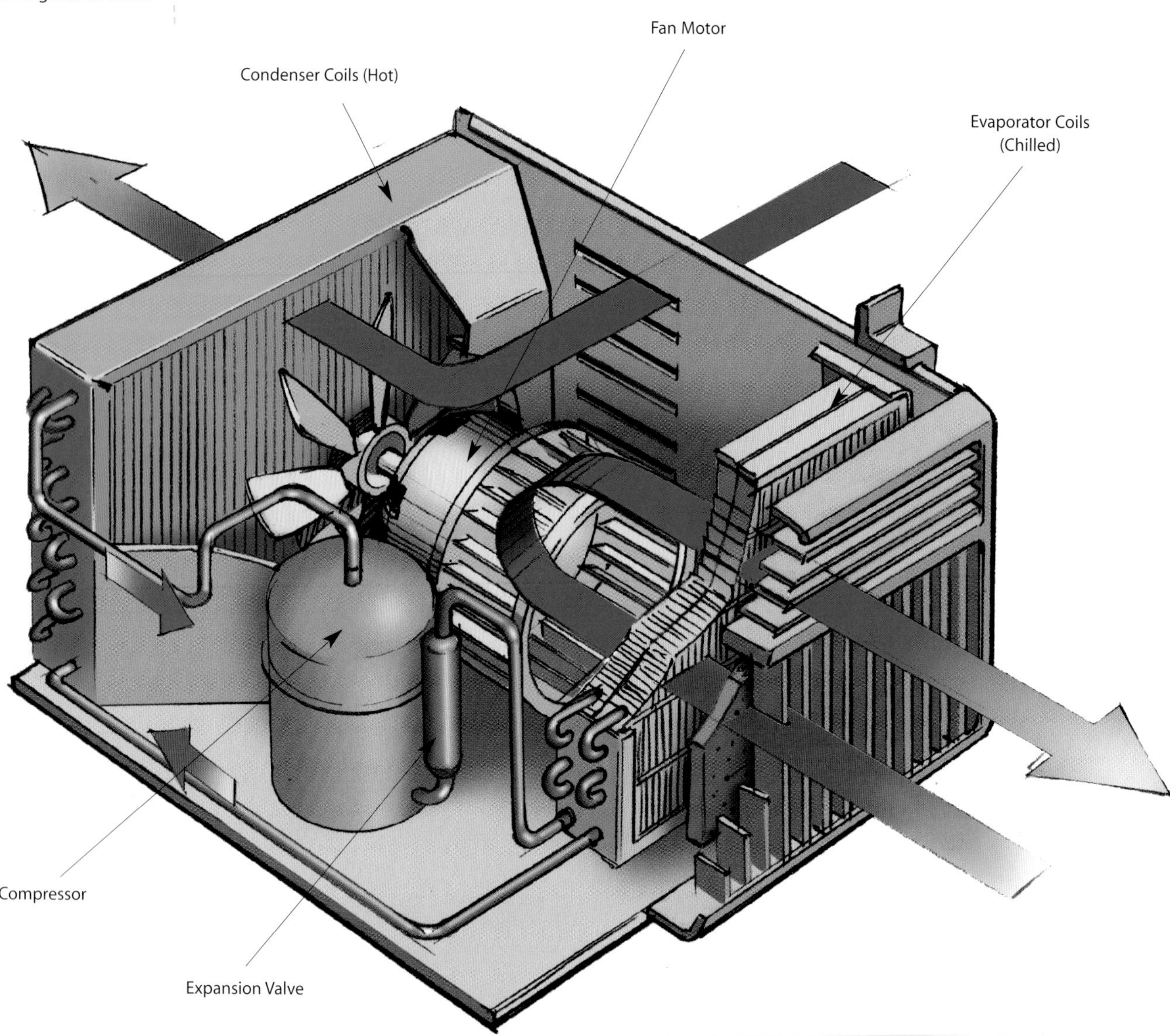

- A compressor
- An expansion valve
- A hot coil (on the outside)
- A chilled coil (on the inside)
- Two fans
- A control unit

This is just like a refrigerator—the room with the window unit acts as the cooled "box." The fans blow air over the coils to improve their ability to dissipate heat (to the outside air) and cold (to the room being cooled). (See "How Refrigerators Work," page 228, for more information.)

Split-System Air Conditioners

A split-system air conditioner splits the hot side and the cold side of the system into two separate units. The cold side, made up of the expansion valve and the cold coil, is generally placed in a furnace or some other air handler. The air handler blows air through the coil and routes the air throughout the building by using a series of ducts. The hot side, known as the condensing unit, is outside the building.

The condensing unit consists of a long spiral coil shaped like a cylinder. It has a fan inside to blow air through the coil, along with a weather-resistant compressor and some control logic. This makes the system relatively low cost, and it moves the noisy parts outside the building to keep the inside of the building quiet.

There is no difference between a split-system and a window air conditioner except for the fact that in a split system the hot and cold sides are split apart and the capacity is greater (making the coils and compressor larger).

In warehouses, malls, large department stores, and so on, the condensing unit is typically a massive structure on the roof. Alternatively, you might have many smaller units on the roof, each attached inside to a small air handler that cools a specific zone in the building.

Chilled-Water Systems

In large buildings, particularly multistory buildings, the split-system approach begins to have problems. Either the pipe running between the condenser and the air handler is too long to work, or the amount of ductwork and the length of ducts becomes a problem. Such a large building might need a chilled-water system.

In a chilled-water system, the entire air conditioner is on the roof or behind the building. It cools water to 40°F (4.4°C) or 45°F (7.2°C) and pipes it through the building. The pipes are connected to small air handlers throughout the building that cool the air. There is no practical limit to the length of a chilled water pipe if it is well insulated.

Air conditioners are really very simple devices, but they have had a profound impact on the world. If you live in a hot area, air conditioning significantly improves your comfort, especially in the summer.

Heat Index

During an average day your body burns about 2,000 calories (when you are exercising heavily it burns a lot more). During waking hours that means you are burning 2 calories or so a minute. These 2 calories have the ability to raise the temperature of 1 kilogram of water 2°C. If you weigh 50 kilograms (110 pounds), your body temperature rises 1/25th °C (1/12th °F) every minute.

Your body needs a way to dump that excess heat. If it doesn't, then body temperature rises into the danger zone in a matter of half an hour. Up to about 80°F (24°C) it's easy to dump excess heat simply through radiation (this is why air temperature "feels" comfortable up to about 80°F). Above 80°F your body does not have enough surface area to radiate the heat away fast enough, so your body turns on your sweat glands to make evaporative cooling possible.

Evaporative cooling works great if the air is dry. In high humidity, however, sweat doesn't work very well—the sweat cannot evaporate because the air is already saturated with humidity. In high temperature/high humidity environments your body can get into a dangerous situation where it cannot radiate or evaporate the heat away. The heat index is designed to make you aware of these dangerous situations.

The heat index takes the day's temperature and humidity into account and calculates what the temperature would be if the air were at 25% humidity or so (very dry). On this scale, high humidity can be excruciatingly hot because your body has no way to eliminate excess heat. For example, 100°F with 100% humidity is the equivalent to 195°F—nearly the boiling point of water!

How THERMOMETERS Work

In most houses you can find several different thermometers. There might be a thermometer in the window that tells you the temperature outside and a thermometer in the medicine cabinet to check a person's body temperature. There might also be thermometers on the thermostat, water heater, oven, and refrigerator. So it seems that knowing the temperature is pretty important!

HSW Web Links

www.howstuffworks.com

How Air Conditioners Work
How Microcontrollers Work
How Pop-Up Turkey Timers Work
How Refrigerators Work

Did You Know?

Do you know who invented the thermometer? The answer is not as simple as you might think. Galileo Galilei (1564–1642) is given credit for inventing the thermometer, but there is some debate about this fact. Some believe the thermometer is the invention of the Italian physician Santorio Santorio (1561–1636). Santorio was the first to apply a numeric scale to the thermoscope, which later evolved to become the thermometer. Then there is German physicist Daniel Fahrenheit (1686–1736), who invented the modern alcohol and mercury thermometers, which are closer to the thermometers that we use today than are the earlier measurement devices.

If you look around your house, you will find lots of different devices whose sole purpose is to detect or measure changes in temperature:

- The thermometer in the backyard tells how hot or cold it is outside.
- Meat and candy thermometers in the kitchen measure food temperatures.
- The thermometer for the furnace tells it when to turn on and off.
- Oven thermometers and refrigerator thermometers maintain a set temperature.
- A medical thermometer accurately measures a small range of body temperatures.

Thermometers use two common temperature scales: Fahrenheit and Celsius. Daniel Fahrenheit decided that the freezing and boiling points of water would be separated by 180°, and he pegged freezing water at 32°. So he made a thermometer, stuck it in freezing water, and marked the level of the mercury on the glass as 32°. Then he stuck the same thermometer in boiling water and marked it 212°. He then put 180 evenly spaced marks between those two points. Anders Celsius arbitrarily decided that the freezing and boiling points of water would be separated by 100°, and he pegged the freezing point of water at 100° and the boiling point at 0°. (His scale was later inverted, so the boiling point of water became 100° and the freezing point became 0°.) So both of the temperature scales we commonly use are completely arbitrary! You could come up with your own scale if you want.

Thermometers fall into three categories: bulb thermometers, bimetallic strip thermometers, or electronic thermometers.

Bulb Thermometers

The bulb thermometer is the common glass thermometer that's been around for ages. It contains some type of fluid, usually mercury. Bulb thermometers rely on the simple principle that a liquid changes its volume relative to its temperature. Liquids contract when they are cold and expand when they are warm.

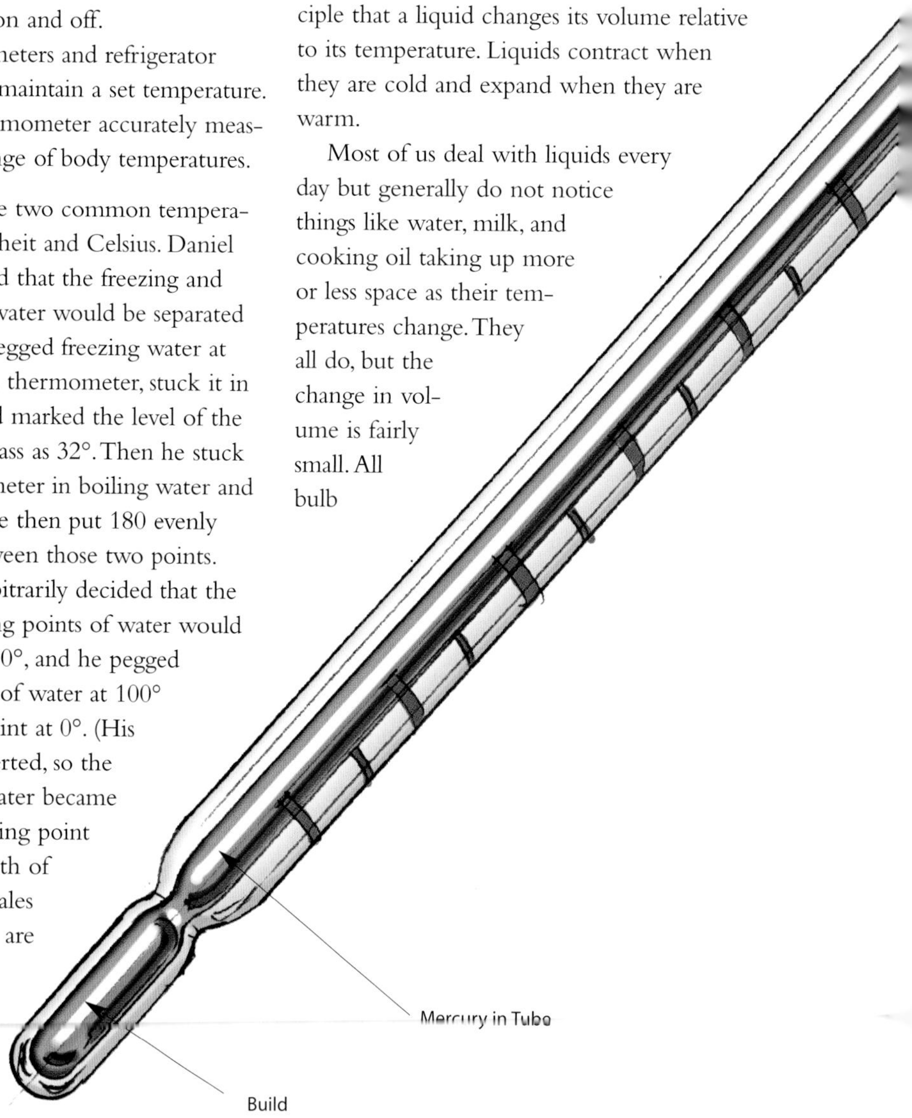

Most of us deal with liquids every day but generally do not notice things like water, milk, and cooking oil taking up more or less space as their temperatures change. They all do, but the change in volume is fairly small. All bulb

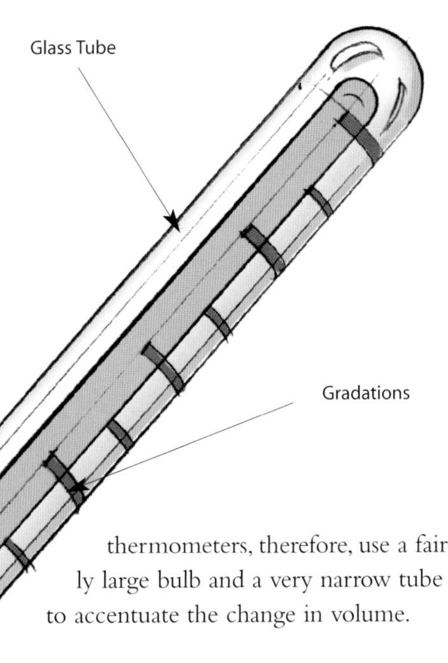

thermometers, therefore, use a fairly large bulb and a very narrow tube to accentuate the change in volume.

Eardrum Thermometers

The eardrum is an extremely accurate point to measure body temperature from because it is recessed inside the head. The problem with the eardrum is that it is so fragile. You don't want to be touching the eardrum with a thermometer. This makes the detection of the eardrum's temperature a remote sensing problem. Granted it is not very remote — just a centimeter or so. But remote nonetheless! It turns out the remote sensing of an object's temperature can be done using its infrared radiation. This technique is a very good way to detect the temperature of a person's eardrum.

All of the objects around you are radiating infrared energy right now. Human beings don't have any sensors that can detect subtle differences in infrared, but our skin can detect objects radiating lots of infrared energy. When you warm yourself by standing close to a fire, the "warmth" is infrared energy that you are absorbing. The idea behind the temperature sensor in the ear thermometer is to create a device that is sensitive to very subtle changes in infrared emission. One common sensor is the thermopile, which can be accurate to the tenth of a degree. The thermopile sees the eardrum and measures its infrared emissions. The emission is converted into a temperature and displayed on an LCD.

Bimetallic Strip Thermometers

Bulb thermometers are good for measuring temperature accurately, but they do not do a good job of turning things on and off in response to changes in temperature. The bimetallic strip thermometer, because it is made of metal, is good at controlling temperatures.

The principle behind a bimetallic strip thermometer relies on the fact that different metals expand at different rates as they warm up. By bonding two different metals together, you create a piece that bends as the temperature changes. Because it is made of metal, a bimetallic strip thermometer is a good controller that can withstand fairly high temperatures. This sort of controller is often found in ovens.

In thermometers, long bimetallic strips are often coiled into spirals. This is the typical layout of a backyard dial thermometer. Coiling a very long strip increases its sensitivity to small temperature changes. In a furnace thermostat, this coiling technique is used and a mercury switch is attached to the coil to turn the furnace on and off.

Electronic Thermometers

It is now common to measure temperature with electronics. The most common sensor is a thermoresistor (or thermistor). This device changes its resistance with changes in temperature. A computer or another circuit measures the resistance and converts it to a temperature, either to display it or to make decisions about turning something on or off.

Galileo Thermometers

The Galileo thermometer is a simple, fairly accurate thermometer mostly used as decoration. It consists of a glass tube that is filled with water and several floating bubbles. The bubbles are glass spheres filled with a colored liquid mixture (usually alcohol). The idea is that each bubble differs very slightly in density (the ratio of mass to volume) from the others, and the density of all of them is very close to the density of the surrounding water. As the temperature changes, the water surrounding the bubbles expands or contracts, changing its density. At any given density, some of the bubbles will float and others will sink. Each bubble carries a little metal tag that indicates a temperature, and the bottom-most bubble of the floating bubbles indicates the current temperature.

In your house you probably have 10 different thermometers hidden in places like your oven, refrigerator, thermostat, and coffee machine. Temperature is important to our comfort and to chemistry, and thermometers are the way we sense and control it.

How MICROWAVE OVENS Work

Microwave ovens have completely changed the typical kitchen. Using radio waves, microwaves can fix foods like potatoes in minutes rather than an hour.

HSW Web Links

www.howstuffworks.com

How Gas Lanterns Work
How Military Pain Beams Will Work
How Pendulum Clocks Work
How Radar Works
How Telephones Work
How Vacuum Flasks Work

The microwave heating process is different from the conventional heating process because you are exciting atoms in a microwave oven rather than conducting heat.

Traditional Cooking

You often hear that microwave ovens cook food "from the inside out." You can understand what that means by thinking about an example. Let's say you want to bake a cake in a conventional oven. Normally you would bake a cake at 350°F or so. But let's say you accidentally set the oven at 600° instead of 350° In this situation, the outside of the cake will burn before the inside even gets warm. In a conventional oven, the heat has to migrate (by conduction) from the outside of the food toward the middle. You also have dry, hot air on the outside of the food evaporating moisture. So the outside of the food can be crispy and brown (for example, bread forms a crust) while the inside is moist.

Microwave Cooking

In microwave cooking, radio waves penetrate the food and excite water and fat molecules evenly throughout the food. Heat does not have to migrate toward the interior by conduction. There is heat everywhere all at once because the molecules are all excited together. There are limits—radio waves penetrate unevenly in thick pieces of food (they don't make it all the way to the middle), and hot spots can be caused by wave interference.

In a microwave oven, the air in the oven is at room temperature, so there is no way to form a crust on the food you're cooking. That is why some microwavable foods come with a little cardboard/foil sleeve. You put the food in the sleeve and then microwave it. The sleeve reacts to microwave energy by becoming very hot, and this exterior heat makes the crust become crispy as it would in a conventional oven.

Microwave ovens are popular because they cook food incredibly quickly. A microwave oven is also extremely efficient in its use of electricity because it heats only the food—not the food's container or the air inside the microwave or anything else.

Superheating

Superheating can occur when you heat water in a microwave oven, especially in smooth glass containers. The temperature of the water rises well above the boiling point, but the water does not boil because the water heats uniformly and there are no imperfections in the glass to start the process (known as nucleation points). When you take the water out, it can boil explosively when you add something like coffee or sugar to it.

How TOASTERS Work

Millions of people have toast with their breakfast every morning, making toasters one of the most popular home appliances ever created. At the same time, toasters are incredibly simple and ingenious.

The idea behind any toaster is simple: Infrared radiation heats a piece of bread. When you put bread in the toaster, the coils glow red and feel hot in the process of producing infrared radiation. The radiation gently dries and chars the surface of the bread.

Infrared Radiation

The most common way for a toaster to create infrared radiation is to use nichrome wire. Nichrome is an alloy of nickel and chromium. The wire wraps back and forth across a sheet of mica, which is a good insulator and does not burn or melt. The nichrome wire has two nice features:

- It has a fairly high electrical resistance compared to something like copper wire—it has enough resistance to get quite hot.
- It does not oxidize when heated. Iron wire would rust very quickly at the temperatures in a toaster. Nichrome wire lasts for years.

Parts of a Toaster

The simplest toaster would have two mica sheets wrapped in nichrome wire, and they would be spaced to form a slot about 1 inch (2.5 cm) wide. The nichrome wires would connect directly to a plug. With this device, this is how you'd make toast:

1) Drop a piece of bread into the slot.
2) Plug in the toaster and watch the bread.
3) Unplug the toaster when the bread gets dark enough.
4) Tip the toaster upside-down to get the toast out.

Most people don't have this sort of patience, and they don't like crumbs all over the counter. So a toaster normally has two other features:

- A spring-loaded tray that pops the toast out and keeps you from having to turn the toaster upside-down
- A timer that turns off the toaster automatically and at the same time releases the tray so the toast pops up

When you push the toaster handle down, a few things have to happen:

- Some sort of mechanism needs to hold the handle down to keep the toast inside the toaster for a period of time.
- Power needs to be applied to the nichrome wires.
- A timer needs to release the holder at the proper time so that the toast pops up.

A plastic bar and a piece of metal are attached to the handle. As the handle is pressed down, the plastic bar presses into a pair of contacts on the circuit board to apply power to the nichrome wires and the elctromagnet, and the piece of metal is attracted to an electromagnet to hold the toast down. When enough time has passed, the timer turns off the electromagnet and allows the toast to pop up.

Some toasters use electronic timers, and others use a bimetallic strip to turn off the electromagnet. As the strip heats up (due to rising temperatures inside the toaster), the strip bends and eventually trips a switch that kills the power to the electromagnet.

Now that you know how your toaster works, watch—the next time you toast a piece of bread, you will look at the process completely differently!

HSW Web Links

www.howstuffworks.com

How Air Conditioners Work
How Clothes Dryers Work
How Hair Dryers Work
How Smoke Detectors Work

How TOILETS Work

The toilet. The commode. The john. No matter what you call it, it is inevitable that this book discuss this device because every home has at least one. More importantly, we must discuss the toilet because it is a technological marvel—a really cool water-handling system!

HSW Web Links

www.howstuffworks.com

How House Construction Works
How Oscillating Sprinklers Work
How Sewer Systems and Septic Systems Work
How Water Heaters Work

It's unfortunate that there are so many bad connotations of the toilet in people's minds. There are the things we do with a toilet, the germs we associate with it, the images of public restrooms, the fact that we have to clean it; all these details leave the toilet somewhat tainted. But if you can put those things aside and look at it as a machine, a toilet is fascinating.

Every toilet has three systems that work together:

- The bowl siphon
- The flush mechanism
- The refill mechanism

Let's look at each part separately to reveal the secrets of the toilet!

The Bowl Siphon

Let's say that you somehow disconnect the toilet tank, and all you have in your bathroom is the bowl. You still have a functional toilet. Even without any moving parts, the bowl can perform all the essential tasks of a toilet. The crucial mechanism that is molded into the bowl is called the bowl siphon.

To understand how the siphon works, try the following experiments. First, pour a cup of water into the bowl. You will notice that very little happens. No matter how many cups of water you pour in individually, the level of the water in the bowl never rises. Each cup of water added immediately spills over the edge of the siphon tube and drains away.

Next, pour a 2-gallon (8-l) bucket of water into the bowl. Pouring in this amount of water causes the bowl to flush. Almost all the water is sucked out of the bowl, and you hear the recognizable flush sound. Here's what happens: If enough water is poured into the bowl fast enough, it will fill the siphon tube. When the tube fills up, the rest is automatic. The siphon sucks the water out of the bowl and down the sewer pipe. As soon as the bowl empties, air enters the siphon tube, producing that distinctive gurgling sound and stopping the siphoning process.

So even if the water to your bathroom is cut off, you can still flush the toilet: All you need is a bucket containing a couple gallons of water.

The Flush Mechanism

The purpose of the tank is to act like a bucket of water. You have to get enough water into the bowl fast enough to activate the siphon. The tank holds several gallons of water. When you flush, all the water in the tank is dumped into the bowl in about 3 seconds—it is the equivalent of pouring in a bucket of water.

A chain is attached to the handle on the side of the tank. When you push on the handle, it pulls the chain, which is connected to the flush valve. The chain lifts the flush valve, which then floats out of the way, revealing a drain hole that is approximately 3 inches (7-cm) in diameter. Uncovering this hole allows the water to enter the bowl. In most toilets, the bowl has been molded so that the water enters the rim, and some of it drains out through holes in the rim. A good portion of the water flows down to a larger

Making Porcelain

Most toilet bowls are made of porcelain, which is a hard, fine-grained ceramic that usually consists of kaolin, quartz, and feldspar. In other words, porcelain is pottery made of a special white form of clay. It is molded in a liquid form, dried as greenware, painted with glaze and then fired in a kiln. A toilet bowl is generally molded in two halves, which are attached together in the greenware state.

hole at the bottom of the bowl, known as the siphon jet. It releases most of the water directly into the siphon tube. Because all the water in the bowl enters the tank in about 3 seconds, it is enough to fill and activate the siphon, and all the water and waste in the bowl are sucked out.

The Refill Mechanism

After the tank has emptied, the flush valve replaces itself on the bottom of the tank, covering the drain hole so the tank can refill. The refill mechanism fills the tank with enough water to start the whole process again.

The refill mechanism has a valve that turns the water on and off. This valve turns the water on when the filler float (or ball float) falls. The float falls when the water level in the tank drops. The refill valve sends water in two directions. Some of the water flows down the refill tube and starts refilling the tank. The rest goes through the bowl refill tube and down the over-flow tube into the bowl. This refills the bowl slowly.

As the water level in the tank rises, so does the float. Eventually the float rises far enough to turn the valve off. It's possible for the tank to overflow if the float becomes detached or the filler valve jams. The overflow tube prevents that from happening, directing the extra water into the bowl instead of onto the floor.

Toilets are amazing devices that perform a thankless task. They've played a huge role in eliminating diseases such as cholera, and they have also ended trips to the outhouse. Can you imagine the world without them?

Did You Know?

In July 2000, archeologists unearthed a 2,000-year-old toilet, which is believed to be the oldest known toilet. This ancient toilet wasn't just a hole in the ground, as you might imagine. It is actually very similar to the toilets we use today. Found in the tomb of a king of the Western Han Dynasty, this royal commode had running water, a stone seat, and an armrest.

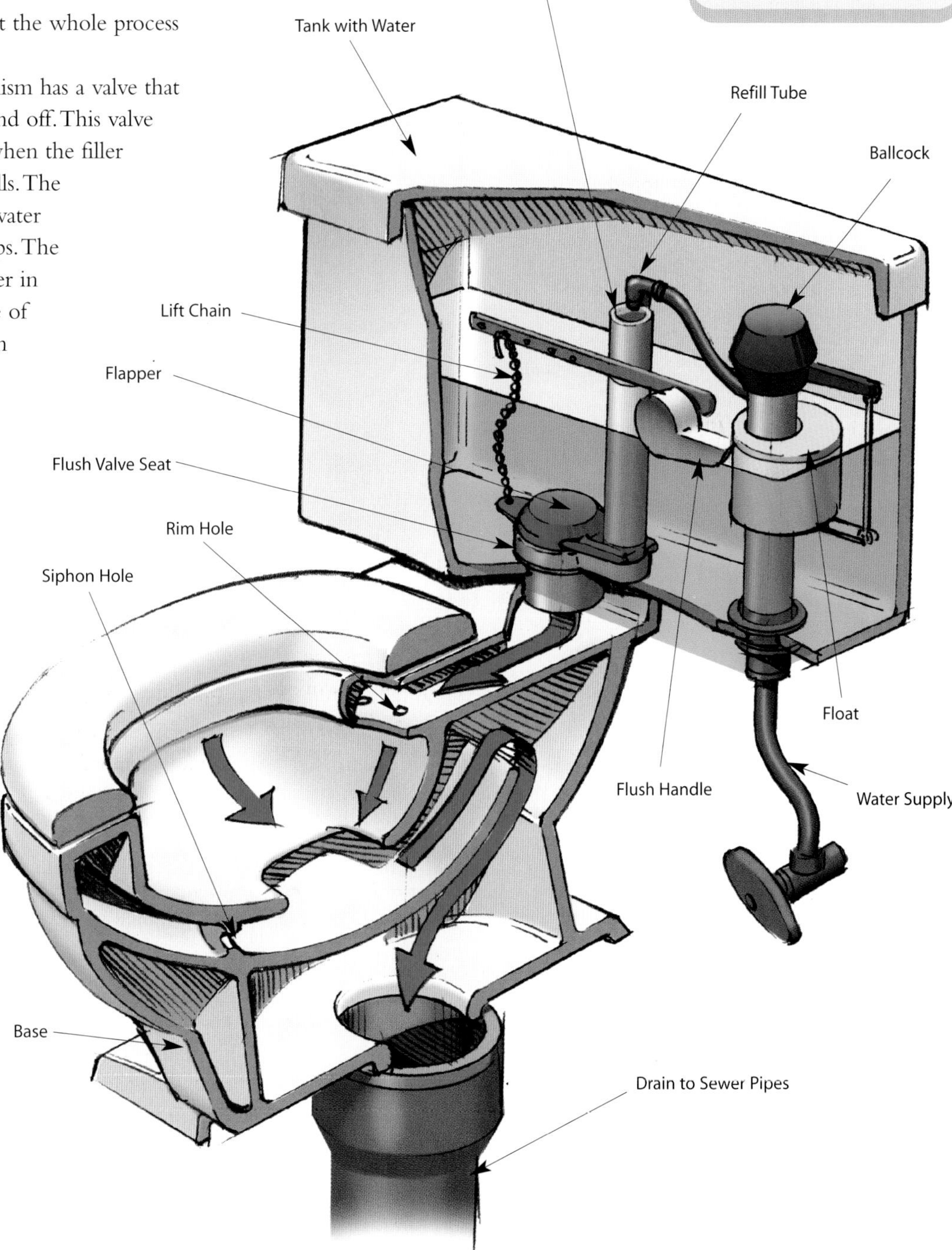

How BURGLAR ALARMS Work

Security systems come in a huge variety of forms, from the incredibly complex ones in movies like Mission: Impossible *to very simple ones that involve a couple switches and a buzzer. You find them in offices, cars, and homes all over the world.*

HSW Web Links

www.howstuffworks.com

How Anti-Shoplifting Devices Work
How Light Works
How Radar Works
How Relays Work
How Ultrasound Work

Generally, security systems fall into two categories: circuit alarms that guard the perimeter and motion detectors that guard large areas.

Circuit Alarms

The most basic burglar alarm is a simple electric circuit that is a lot like the circuit in a flashlight. To open or close a circuit in a flashlight, you simply throw a switch. When you do, the batteries connect to the bulb and the light turns on.

In a burglar alarm, the switch is thrown by the act of intrusion—opening a door or window, for example. The light is replaced by a beeper, buzzer, or horn of some sort. These sorts of alarms fall into two categories:

- **Closed-circuit system**—In this system, the electric circuit is closed when the door is shut. As long as the door is closed, electricity can flow from one end of the circuit to the other. But if somebody opens the door, electricity can't flow, and this triggers an alarm.
- **Open-circuit system**—In this system, opening a door or window closes the circuit, so electricity begins to flow. In this system, the alarm is triggered when the circuit is completed.

There are a number of ways to build an open circuit into an entryway. A magnetic sensor is very simple, inexpensive, and effective. A magnetic sensor in an open circuit consists of a few simple components. For the most basic design, you need

- A battery powering a circuit
- A buzzer
- A metal switch built into a doorframe
- A magnet embedded in the door, lined up with the switch

When the door is closed, the magnet pulls the metal switch open, so the circuit is broken. When you move the magnet, by opening the door, a spring snaps the switch back into place. This completes the circuit and powers the buzzer. You can also build this sort of system into a window. If an intruder pushes a window open, the magnet slides out of line with the switch, and the buzzer is activated. With just a battery and buzzer, this design makes for a fairly flawed security system. After all, the burglar only needs to close the door or cut the wire to the switch to turn off the buzzer.

Most modern burglar alarms incorporate another piece into the circuit—the control box. The control box is hooked up to one or more alarm circuits, and it also has its own power supply. It monitors the circuits and sounds the alarm when they are closed or opened (depending on the design). When the alarm is triggered, the control box won't cut it off until somebody enters a security code at a connected keypad. For added security, the control box is usually hidden in an out-of-the-way spot so the intruder can't find it.

Sound the Alarm

When a security system detects an intruder, it can do a number of things. It might

- Sound a siren or another loud alarm noise
- Flash outdoor lights
- Use an auto-dialer to call the police or security company

The siren and lights serve three functions: They alert occupants and neighbors that someone has broken into the house, drive the intruder away, and signal to police which house has been broken into. The auto-dialer will either call the police and play a recorded distress message or call the security company and tell it which systems have been triggered.

Using these basic concepts, you can create all sorts of alarm systems; just imagine what a burglar might do to break into a house, and then turn that action into the circuit switch. If an intruder might break through a window, for example, you could make the glass itself a circuit. The easiest way to do this is to run a current through a thin line of foil wire affixed to the surface of the glass. If a burglar breaks the glass, the circuit is broken, triggering the alarm. Floor mats are another simple option. A basic floor mat uses an open-circuit design with two metal strips spaced apart. When somebody steps on the mat, the pressure pushes the two metal strips together, completing a circuit.

All these circuit systems are best for guarding the perimeter of a house or business—the points where an intruder would enter the building.

Motion Detectors

Switch-activated alarms aren't very good at detecting intruders inside a building because the intruder's actions are highly unpredictable. You don't know where the intruder will go or what he or she will touch, so a specific trigger isn't very effective. To detect an intruder who is already in the house, you need a motion detector that can cover a wide area.

One common type of motion detector uses radar. A box sends out bursts of microwave radio energy (or ultrasonic sound waves) and waits for the reflected energy to bounce back. If there is nothing happening in the room, the radio energy will bounce back in the same pattern in which it went out. But if somebody enters the area, the reflection pattern is disturbed and the sensor sends an alarm signal to the control box.

Passive infrared (PIR) motion detectors are also common. These sensors "see" the infrared energy emitted by an intruder's body heat. When an intruder walks into the field of view of the detector, the sensor detects a sharp increase in infrared energy. Of course, there will always be gradual fluctuation of heat energy in an area, so PIR detectors are designed to trigger the alarm only when infrared energy levels change very rapidly.

A number of burglar alarm systems can be combined in a house to offer complete coverage. In a typical security system, the control box does not sound the alarm immediately when the motion detectors are triggered. There is a short delay, to give the homeowner time to enter a security code that turns the system off. If the security code is not entered, the control box will activate various alarms to signal an intrusion to residents, neighbors, and the police.

Motion Sensing Lights

The "motion sensing" feature on most lights uses PIR (Passive InfraRed) detectors or pyroelectric sensors. In order to make a sensor that can detect a human being, you need to make the sensor sensitive to the temperature of a human being. Humans, having a skin temperature of 93°F or so, radiate infrared energy with a wavelength between 9 and 10 micrometers. Therefore the sensors are typically sensitive in the range of 8 to 12 micrometers.

The devices themselves are simple electronic components not unlike a photosensor. The infrared light bumps electrons off a substrate, and these electrons can be detected and amplified into a signal.

If you have a motion sensor light, you've probably noticed that your light is sensitive to motion, but not to a person who is standing still. That's because the electronics package attached to the sensor is looking for a fairly rapid change in the amount of infrared energy it is seeing. When a person walks by, the amount of infrared energy in the field of view changes rapidly and is easily detected. You do not want the sensor detecting slower changes, like the sidewalk cooling off at night.

Your motion sensing light has a wide field of view because of the lens you can see covering the sensor. Infrared energy is a form of light, so you can focus and bend it with plastic lenses. But it is not like there is a 2-D array of sensors in there. There is may be a single sensor, or in some cases, two sensors inside looking for changes in infrared energy.

How **SMOKE DETECTORS** Work

Smoke detectors are one of those amazing inventions that, because of mass production, have quite a low purchase price. You can get a smoke detector for as little at $5. But despite their low cost, smoke detectors do their job very well: They save thousands of lives each year!

HSW Web Links

www.howstuffworks.com

How Batteries Work
How Burglar Alarms Work
How Hair Dryers Work
How Nicotine Works

All smoke detectors consist of two basic parts: a component that senses the presence of smoke particles in the air and an electronic horn that wakes people up. Two popular particle-detection systems are in widespread use: photoelectric and ionization systems.

Photoelectric Detectors

Photoelectric smoke detectors use a system similar to the photo beam detectors that chime a bell when you walk through a store entrance. The chime system detects motion by using two basic components: a source of focused light (such as a laser beam) and a photo beam detector. The photo beam detector is a device that can "see" light. When you get in the way of the light source, the detector notices a drop in light and activates the chime. You could make a smoke detector with this basic system, but it wouldn't be very efficient. It would take a lot of smoke to block the light. To make the system more sensitive, the light and beam detector have to be configured differently.

In a photoelectric detector, the beam is aimed so that it misses the detector—it is aimed perpendicular to the detector. In this system, the detector triggers the alarm when it senses light, not when the light is interrupted. When there is no smoke, the light shines straight ahead and doesn't hit the detector. But when smoke particles enter the smoke detector chamber, the light beam bounces off them. Some of the scattered light hits the detector, which then triggers the electronic horn. This system works best at detecting fairly heavy smoke, like the smoke from a smoldering mattress.

Ionization Detectors

Ionization smoke detectors use an ionization chamber and a source of ionizing radiation

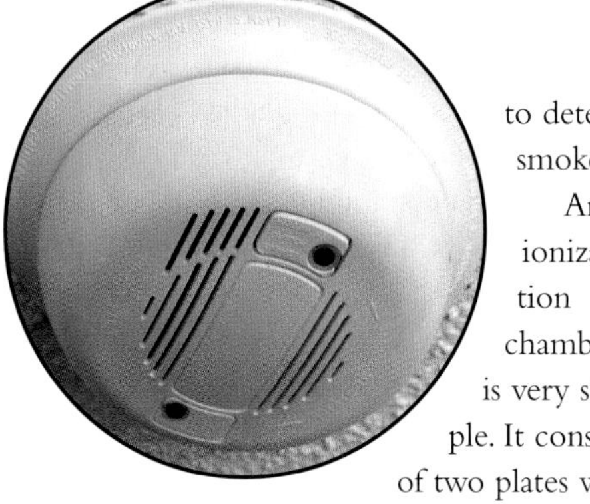

to detect smoke. An ionization chamber is very simple. It consists of two plates with a voltage across them (one that is positively charged and one that is negatively charged), positioned near a radioactive source of ionizing radiation, such as a small amount of Americium-241. This element releases alpha particles—particles of nuclear radiation—into the air.

The alpha particles ionize the oxygen and nitrogen atoms in the air. Basically, this means they knock an electron off each atom. When you knock an electron off an atom, you end up with a free electron, which has a negative charge, and an atom missing one electron, which has a positive charge.

Because opposite charges attract, the negative electron becomes attracted to the plate with a positive voltage, and the positive atom becomes attracted to the plate with a negative voltage. The electrons and ions moving toward the plates create a small electrical current between the two plates.

When smoke enters the ionization chamber, the smoke particles attach to the ions and neutralize them. This reduces the current flowing between the two plates. The electronics in the smoke detector sense this drop in current and set off the alarm.

It's really extraordinary that something this simple and inexpensive can save so many lives each year. But a smoke detector only works if its battery does, so be sure to replace those batteries!

Did You Know?

The words *nuclear radiation* set off an alarm in people's minds. But in the case of smoke detectors, there's not much to be concerned about. The amount of radiation involved is extremely small, and it is predominantly alpha radiation. Alpha radiation can't even penetrate a sheet of paper, and in the detector, it is blocked by several centimeters of air.

The Americium in the smoke detector would pose a risk if you inhaled it. For this reason, it's a very bad idea to open up and mess around with the Americium in a smoke detector.

How **THREE-WAY SWITCHES** Work

In your home you probably have at least one light that is controlled by two separate switches. Whenever you flip either one of the switches, the light changes its state: If the light is on the switch turns it off, and if the light is off the switch turns it on. The switches that make this possible are three-way switches.

Let's start by looking at a normal light circuit. Power runs from the fuse box on two wires. In the United States the two wires are normally black and white. The black one is hot, and the white one is neutral. Power in the form of 120-volt AC flows from the hot wire, through a bulb, and back on the white wire. To turn the light on and off, a switch on the black wire can connect or disconnect it from the bulb. The switch in this circuit is a normal two-way switch—it can be either on or off.

Parts of a Three-Way Switch

The normal two-way on/off switch works great until you want to control a light from two separate places. For example, you might want to be able to turn a light on from both the top and the bottom of the stairs. To create lights that operate with two separate switches like this, an electrician uses two special pieces of equipment in the circuit:

- Three-way switches
- A traveler wire (wiring that has an extra red insulated wire along with the black and white wires within the sheath)

In a normal switch, the two terminals are either connected or disconnected. When they are connected, the switch is on, and when they're disconnected the switch is off. In a three-way switch, the top terminal connects to one or the other of the bottom two terminals. By using a red wire running between the switches, you can see that flipping either switch causes the light to turn on or off.

Three-Way Switch Wiring

There are several different ways to wire three-way switches to a light. For example, the power from the fuse box could come in at the light fixture, and there could be two switches in series from that. Or power could enter at the fixture and then two switches could be arranged in parallel from the light. Or the power could enter a switch and cross over to the other switch and the light.

If you are trying to understand how a set of switches in your house is wired, using an ohmmeter or a continuity detector is the only way to reverse-engineer what the electrician has done (make sure you turn the power off at the fuse panel before doing anything with electrical wiring). If you know the basic idea behind three-way switches and three-way wiring, it's really easy to figure out how the switches are wired.

HSW Web Links

www.howstuffworks.com

How LAN Switches Work
How Relays Work
How Smoke Detectors Work
How Solar Yard Lights Work

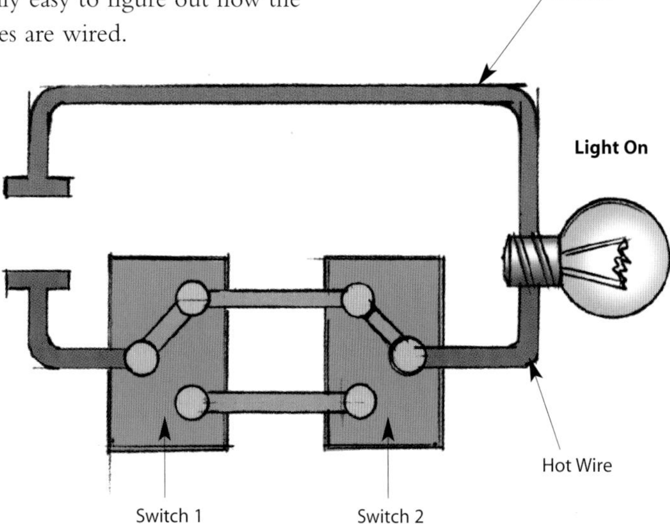

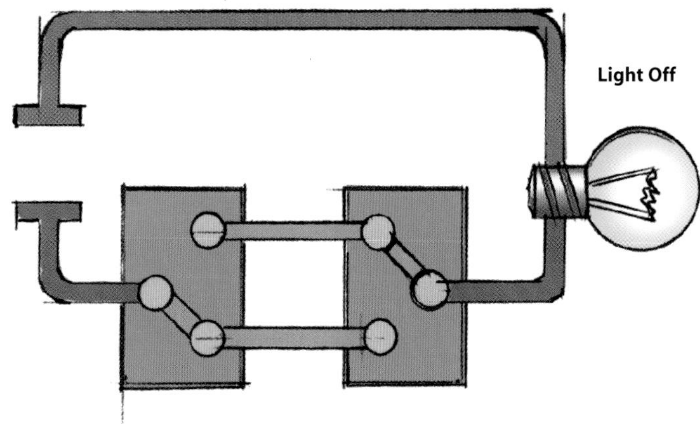

How **VACUUM FLASKS** Work

A vacuum flask, or thermos, is a simple but amazing device that works better than you would ever expect it to. If you pour in something hot, it remains hot all day. If you pour in something cold, it can remain cold for a day or two. The key to the capability is the incredible insulation provided by a vacuum.

HSW Web Links

www.howstuffworks.com

How Car Cooling Systems Work
How Clothes Dryers Work
How Refrigerators Work
How Water Heaters Work

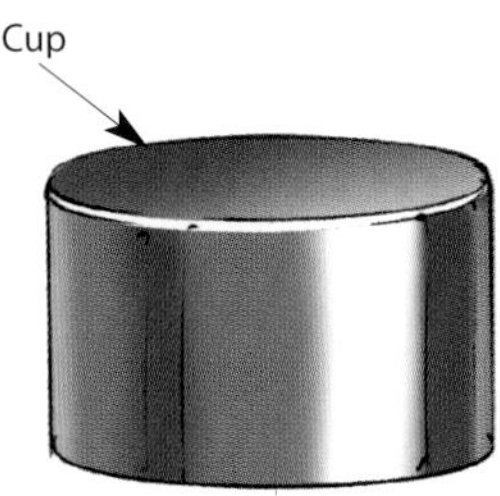

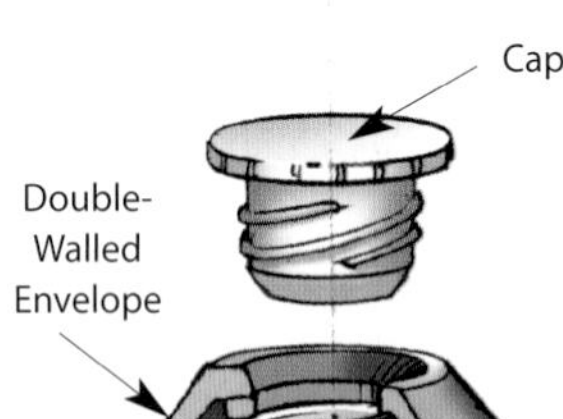

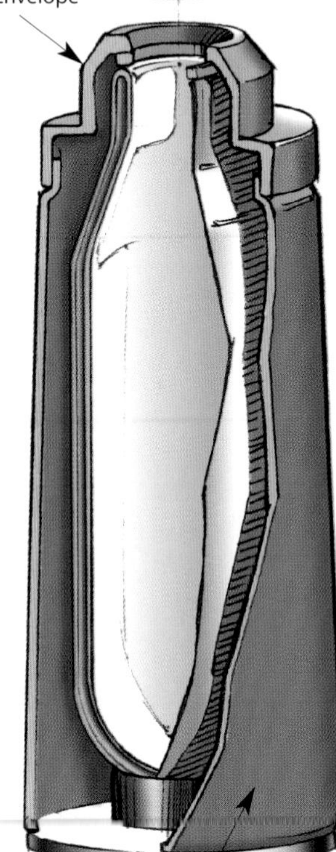

A vacuum flask limits heat transfer through the flask walls. That lets the fluid stay at a nearly constant temperature for a long time.

A vacuum flask is essentially an insulated jar in which the insulation is in the form of a vacuum rather than a thick blanket.

Heat Transfer

Let's say you set a glass of ice water on the kitchen table. You know what will happen. The glass of ice water will warm up to room temperature. Similarly, if you have a bowl of hot soup on the table, it will cool down to room temperature. It is a thermodynamic fact of life that any two objects with different temperatures together in a closed system will resolve themselves to the same temperature by heat transfer. So a room and a hot bowl of soup resolve themselves to the same temperature by the heat transfer process—the room gets slightly warmer and the bowl of soup gets a lot colder.

If you want to keep a bowl of soup hot as long as possible—that is, if you want to slow down the natural heat transfer process as much as you can—you have to slow down the three processes that cause heat transfer:

- **Conduction**—An example of conduction would be to take a metal bar and heat one end of it. The other end will get warm and then hot through conduction.
- **Radiation**—Objects give off infrared radiation, which is a form of light. Your eyes are unable to see infrared, but your skin can feel it. An example of infrared radiation is the heat you feel radiating from the bricks in a fireplace after the fire has gone out.
- **Convection**—When a liquid or gas gets hot, it rises. This is the basis of convection. If you have a hot bowl of soup on the table, it heats a layer of air surrounding the bowl. That layer then rises because it is hotter than the surrounding air. Cold air fills in the space left by the rising hot air, and the cycle repeats. It is possible to speed up convection—that is why you blow on hot soup to cool it down. You can also slow down convection by using insulation.

To keep things hot or cold for long periods of time, you want to reduce all three of these heat transfer processes as much as possible.

Vacuum Insulation

One way to build an insulated container is to wrap a jar in foam. Foam acts as an insulator because of two principles. First, the plastic and air in the foam are not very good heat conductors, so heat and cold do not readily pass through the foam. Second, the foam largely eliminates convection inside the foam, so heat transfer through foam is pretty minor.

But there is an even better insulator than foam: a vacuum. A vacuum is a lack of atoms, and without atoms, you eliminate conduction and convection completely. A perfect vacuum contains zero atoms. It is nearly impossible to create a perfect vacuum, but you can get close.

A vacuum flask has a glass or steel envelope holding a vacuum. The glass is mirrored on the inside of the flask to reduce infrared radiation. The combination of a vacuum and the mirroring greatly reduces heat transfer by convection, conduction, and radiation.

So why do hot things in a thermos ever cool down? A thermos has two paths for heat transfer. The big one is the cap. The other is the glass, which provides a conduction path at the top of the flask where the inner and outer walls meet. Heat transfer through these paths is small, but it is not zero.

How **BALLPOINT PENS** Work

In this electronic age of voicemail, email, and cell phones, there is still no substitute for pen and paper. Even as you browse the Web, you probably have a pen within easy reach to jot down notes, scribble phone numbers, and doodle. Modern ballpoint pens are very inexpensive—you might have a cup on your desk that contains a dozen or two different pens that have wandered in from who knows where.

The key to a ballpoint pen, of course, is the ball. The genius behind the ball is the fact that it acts as a cap to keep the ink from drying out. At the same time, the ball is the device that dispenses the ink onto the paper in a very even line. The ball rotates freely and rolls out the ink as it is continuously fed from the ink reservoir (usually a narrow plastic tube filled with ink).

The Ball

A ballpoint pen's ball is held between the ink reservoir and the paper by a socket. As the pen moves across the paper, the ball turns and gravity forces the ink down the reservoir and onto the ball, where it is transferred onto the paper.

Because the tip of a normal ballpoint pen is so tiny, it is hard to visualize how the ball and socket actually work. One way to understand it is to look at a bottle of roll-on antiperspirant, which uses the same technology on a much larger scale. Like a ballpoint pen, a typical container of roll-on wants to keep air out of the liquid antiperspirant while at the same time making it easy to apply.

If you open up a container of antiperspirant, you will see that it is extremely simple—the ball is exposed on the backside so it can pick up the liquid antiperspirant, and it is loose enough in its socket that it can roll freely. A ballpoint pen works exactly the same way. The tiny ball is held in a socket, and the back of the ball is exposed so that it can pick up ink from the reservoir. The ball fits into the socket with just enough space to move freely.

The Ink

Ink is a fluid or paste that can be a number of colors. To make ink, pigment or dye is dissolved in a liquid called the vehicle.

The ink vehicle can be either plant based (such as linseed oil), which dries by penetration and oxidation, or solvent based (such as kerosene), which dries through evaporation.

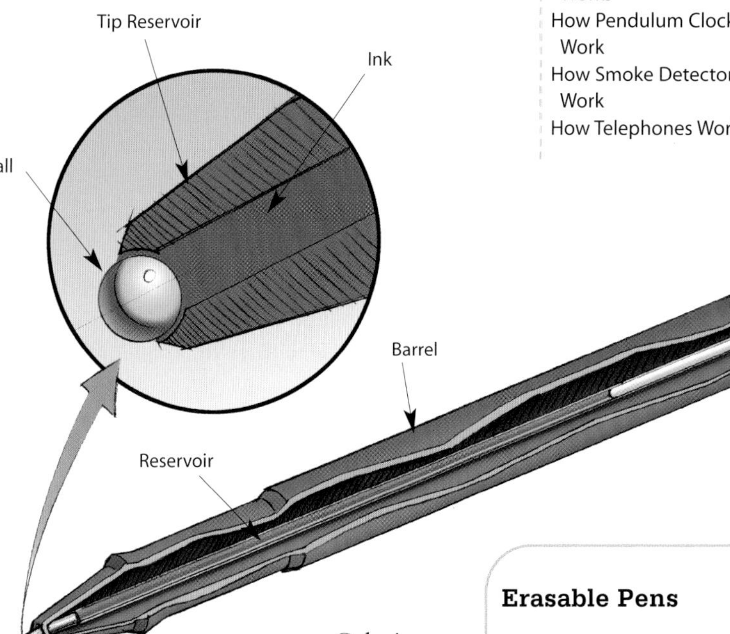

Coloring ingredients can be pigments, which are fine, solid particles suspended in the vehicle, or dyes made from chemicals that dissolve in the vehicle. Additives are used to stabilize the mixture or give it different properties.

Black ink is often made with carbon as the pigment. Colored pigments are inorganic compounds of chromium (yellow, green, and orange), molybdenum (orange), cadmium (red and yellow), and iron (blue).

Ballpoint pen ink is very thick and quick drying. It is thick so that it doesn't spill out of the reservoir, but it is thin enough so that it responds to gravity. That is why a normal ballpoint pen cannot write upside-down—it needs gravity to pull the ink onto the ball.

HSW Web Links

www.howstuffworks.com

How Gas Lanterns Work
How Magna Doodle Works
How Pendulum Clocks Work
How Smoke Detectors Work
How Telephones Work

Erasable Pens

Erasable pens were tremendously popular when they were introduced in the early 1980s. They combine the readability of ink with the erasability of a pencil. While the pens are still manufactured, they aren't as popular as they once were.

What makes erasable ballpoint pens so different from traditional ballpoint pens is the ink—instead of being made from oils and dyes, the ink in an erasable pen is made of liquid rubber cement. As you write, the ballpoint rolls on the paper and dispenses the rubber cement ink. You can rub off the rubber cement for several hours, but after it hardens, you cannot erase it.

How SOLAR YARD LIGHTS Work

If you have a yard and have ever thought about lighting it at night, then you have probably heard about solar yard lights. You don't have to run any wiring for them. In about 15 seconds, you can install a light in any location that gets direct sunlight.

HSW Web Links

www.howstuffworks.com

How Batteries Work
How Solar Cells Work
How Solar Eclipses Work
How Solar Sails Will Work
How the Sun Works

Solar yard lights are interesting because they are miniature solar-powered systems. They have the same properties as an orbiting satellite or a solar-powered home, but on a much smaller scale. They generate and store their own power during the day and then release it at night from a battery. This is just like a satellite that stores solar energy while it is on the sunny side of the planet and then uses that energy on the dark side.

Parts of a Solar Yard Light

A typical solar yard light consists of the following components:

- A plastic case
- A solar cell on top
- A single AA nickel-cadmium (NiCad) battery
- A small controller board
- A light-emitting diode (LED) light source
- A photoresistor to detect darkness

A single solar cell produces a maximum of 0.45 volts and a varying amount of current, depending on the size of the cell and the amount of light striking it. In a typical yard light, therefore, you need four cells wired in series. (See "How Batteries Work," page 26, for a discussion of series wiring.) In a typical yard light, the four cells will produce 1.8 volts and a maximum of about 100 milliamps in full, bright sunlight.

The solar cells are wired directly to the battery through a diode. The battery is a standard AA NiCad battery, which produces about 1.2 volts and can store a maximum of approximately 700 milliamp-hours. The battery reaches maximum charge during the day, except on short winter days or on days that are heavily overcast.

A solar yard light uses several different technologies to produce light: LEDs, fluorescent light, incandescent light, or a combination of these. Because LEDs are inexpensive, rugged, and don't use much power, they are very common.

The controller board accepts power from the solar cell and battery, as well as input from the photoresistor. It has a 3-transitor circuit that turns on the LED when the photoresistor indicates darkness. At night the solar cells stop producing power and the photoresistor turns on the LED.

An LED draws about 45 milliamps, with the battery producing about 1.2 volts (0.05 watts). It produces about half the light that a candle would. The NiCad battery, when fully charged, can operate the LED for about 15 hours.

Half of a candle's light is not very much, and if you have ever purchased one of these yard lights, you know that it really is not enough to provide illumination. You use them more for marking a trail—they are bright enough to see, but not bright enough to illuminate the ground to any great degree.

Alternatives and Cost

Some more expensive lights may offer a combination of LED and small halogen flashlight bulb. The LED is on all the time, and the light bulb turns on for a minute or two when a motion sensor detects movement.

Solar yard lights are expensive right now because of the solar cells and, to a lesser degree, the NiCad battery. Solar cells are much less expensive than they were 10 or 20 years ago, but they are still fairly pricey. However, if you are intrigued by solar power and want to use it to run something at your house, solar yard lights are an easy and fun project.

Batteries for Photovoltaic Cell Systems

What kind of batteries are used in PV systems? Although several different kinds are commonly used, the one characteristic that they should all have in common is that they are deep cycle batteries. Unlike your car battery, which is a shallow cycle battery, deep cycle batteries can discharge more of their stored energy while still maintaining long life. Car batteries discharge a large current for a very short time—to start your car—and are then immediately recharged as you drive. PV batteries generally have to discharge a smaller current for a longer period (such as all night), while being charged during the day.

chapter ten

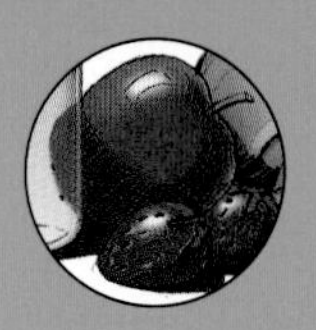

IT'S ALL ABOUT YOU

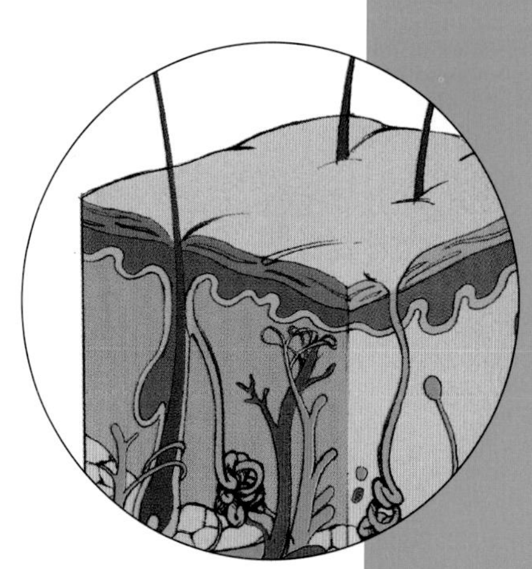

How CHOCOLATE Works

Chocolate bars, chocolate fudge, chocolate cake, chocolate chip cookies, chocolate ice cream, chocolate milk, chocolate cereal, hot chocolate, chocolate sauce... There is something special about this amazing substance—so special that the average person in the United States eats 10 pounds of chocolate every year! The U.S. love affair with chocolate accounts for about half of the world's consumption.

HSW Web Links

www.howstuffworks.com

How Bread Works
How Fats Work
How Food Preservation Works
How Food Works
How Low-Fat Baking Works

Chocolate starts with the cacao tree (*Theobroma cacao*). This evergreen tree grows in moist, shady areas in equatorial regions such as Central and South America, West Africa, and Indonesia. The tree produces seed pods that may contain up to 50 white beans each. The beans are harvested in May and late October, fermented for about a week, dried in the sun (where they turn brown), packed into bags, and then shipped to chocolate makers.

Processing Cacao Beans

A chocolate maker roasts the beans to bring out the flavor. Beans from different regions and different trees have different qualities and flavors, so they are often blended to produce a distinctive mix.

After the beans are roasted and blended, they are ground, and the ground beans form chocolate liquor. This is pure, unsweetened chocolate. (The word *liquor* has nothing to do with alcohol—if you look in the dictionary, the first definition for liquor is "a liquid or juice.")

All seeds contain some fat. For example, corn oil comes from the fat in corn seeds. The fascinating thing about cocoa beans is that they are half fat. This is why the ground beans form a liquid. Similarly, when you grind up peanuts to make real peanut butter, you see that real peanut butter is a thick liquid. The difference between peanut oil and cocoa oil is that peanut oil is liquid at room temperature, whereas cocoa oil is a solid up to about 90°F (32°C). So if you can let chocolate liquor cool, it will solidify.

Pure, unsweetened chocolate liquor is pretty bitter, but it is possible to acquire a taste for it. You can do two different things with chocolate liquor. You can pour it into a mold and let it solidify—this is unsweetened chocolate, often called baking chocolate. Or you can press it in a hydraulic press to squeeze out the fat. The squeezing process leaves a dry cake of the ground cocoa bean solids. The fat that is squeezed out is called cocoa butter, and it is useful in everything from tanning products to white chocolate. If you grind up the cocoa bean cake, you have cocoa powder, which you can buy in the grocery store as pure, unsweetened cocoa powder or as sweetened powder for chocolate milk and hot cocoa.

Types of Chocolate

Baking chocolate	Pure cocoa liquor (ground cocoa beans) with nothing added.
Cocoa powder	Cocoa bean solids. Cocoa liquor pressed to remove the cocoa butter.
Cocoa butter	The fat from a cocoa bean. It is similar to peanut oil (fat from peanuts) or corn oil (fat from corn), but cocoa butter is a solid at room temperature.
Semisweet chocolate	Pure cocoa liquor with extra cocoa butter and some sugar.
Milk chocolate	Pure cocoa liquor with extra cocoa butter, sugar, and milk solids. There is more milk than chocolate liquor in milk chocolate.
White chocolate	Cocoa butter plus sugar and milk. There are no cocoa bean solids in white chocolate.

Making Chocolate Candy

So far, we have taken the seeds of a tree, roasted them, and ground them up. Now we can begin the process of making the chocolate we eat.

A chocolate maker must do three things to make chocolate candy:

- **Add ingredients**—The chocolate that we eat contains sugar, other flavors (such as vanilla) and often milk (in milk chocolate). The chocolate maker adds these ingredients according to a recipe.
- **Conch**—A special conching machine is used to massage the chocolate in order to blend the ingredients together and smooth the chocolate out. For fine chocolates, conching can take as long as 3 days!
- **Temper**—Tempering is a carefully controlled heating process. The chocolate is slowly heated, and then slowly cooled, allowing the cocoa butter molecules to solidify. Without tempering, the chocolate does not harden properly and the cocoa butter separates out—just like cream separates from milk.

Storing Chocolate

Whether it is white chocolate, baking chocolate, milk chocolate, or some kind of chocolate confection, proper storage is key. Since it can easily absorb flavors from food or other products situated nearby, chocolate should be tightly wrapped and stored away from pungent odors. The ideal temperature for storage is somewhere between 65°F to 68°F, with no more than 50% to 55% relative humidity. If stored properly, you can expect milk and white chocolate to be good for up to as many as 6 months and other types of chocolate can have an even longer shelf life.

There's nothing quite like opening a much-anticipated box on Valentine's Day only to find discolored, slightly gray candy. There are two things that could be the culprit—sugar bloom or fat bloom. Sugar bloom is normally caused by surface moisture—the moisture causes sugar in the chocolate to dissolve. Once the moisture dries, sugar crystals remain on the surface. If this process is repeated, the surface can become sticky and even more discolored. Sugar bloom is most often the result of overly humid storage, but can happen when chocolate has been stored at a relatively cool temperature and is then moved too quickly into much warmer surroundings. When this happens, the chocolate will sweat.

Fat bloom is similar to sugar bloom, except that it is fat or cocoa butter that is separating from the chocolate and depositing itself on the outside of the candy. The most common cause of fat bloom is temperature changes and too-warm temperatures.

The art of chocolate making involves roasting and blending cocoa beans, adding ingredients, conching, and tempering. How the steps in this process are handled control the quality, taste, and texture of the chocolate produced, and they are often closely guarded secrets!

Cool Facts

- Among the Aztecs, the cocoa bean was used as currency, and its liquid was used as a drink to promote wisdom and power.
- Chocolate was first made available in solid form in the 17th century.
- There are 5 mg to 10 mg of caffeine in 1 ounce of bittersweet chocolate and about 10 mg in one 6-oz. cup of hot cocoa.
- The Walter Baker Company of Massachusetts first manufactured chocolate in the United States in 1765.
- Chocolate naturally contains a chemical known as phenalethylamine (PEA), which simulates an amorous feeling in the human body.
- The Swiss lead the world in the amount of chocolate each person eats annually—an astounding 22 pounds!
- Chocolate syrup, not blood, covered Janet Leigh in the famous *Psycho* shower scene.

How FOOD Works

It is safe to say that one thing you'll do today is eat some food, because food is pretty important to all animals. If you don't eat, you can have all sorts of problems, including hunger, weakness, and starvation. Food is essential to life.

HSW Web Links

www.howstuffworks.com

How Bread Works
How Chocolate Works
How Fats Work
How Food Preservation Works
How Your Kidneys Work

Think about some of the things you have eaten today: maybe cereal, bread, milk, juice, ham, cheese, an apple, and potatoes. All these foods (and any other normal food that you can think of) contain seven basic components:

- Carbohydrates (both simple and complex)
- Proteins
- Fats
- Vitamins
- Minerals
- Water
- Fiber

Let's look at each of these basic components to understand what they really do and why they are so important to your body.

Carbohydrates

You have probably heard of carbohydrates and complex carbohydrates. Carbohydrates provide your body with its basic fuel. Your body uses carbohydrates like a car engine uses gasoline.

The simplest carbohydrate is glucose. Glucose is also called blood sugar or dextrose. Glucose flows in the bloodstream so that it is available to every cell in your body. Your cells absorb glucose and convert it into energy to drive the cells. Specifically, cells perform a set of chemical reactions on glucose to create ATP (adenosine triphosphate), which has a phosphate bond that powers most of the machinery in any human cell.

If you drink a solution of water and glucose, the glucose passes directly from your digestive system into the bloodstream, and it does this very quickly.

The word *carbohydrate* comes from the fact that glucose is made up of carbon (*carbo*) and water (*hydrate*). The chemical formula for glucose is $C_6H_{12}O_6$. Glucose is a simple sugar, meaning that to your tongue it tastes sweet. You hear about other simple sugars all

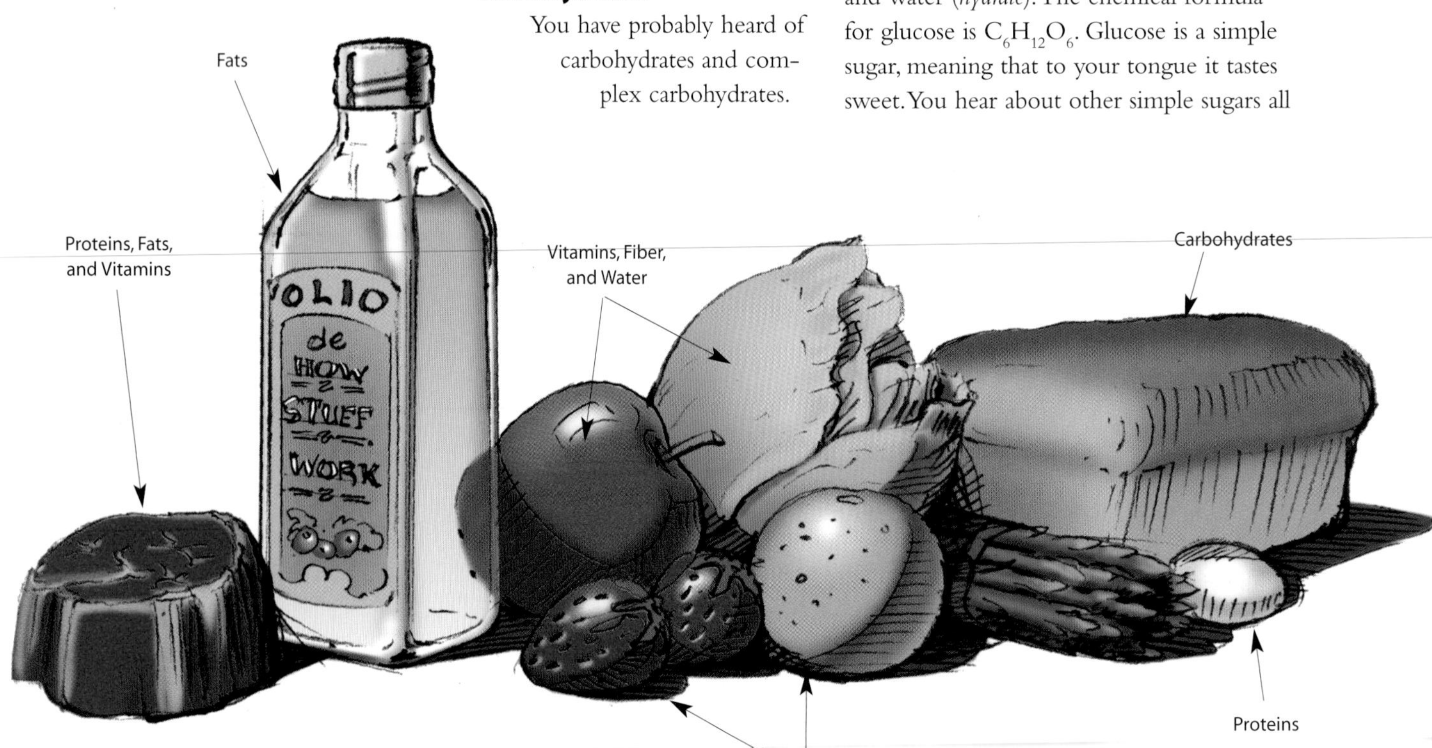

the time. Fructose is the main sugar in fruits, and it has the same chemical formula as glucose ($C_6H_{12}O_6$), but the atoms are arranged slightly differently. Your liver converts fructose to glucose. Sucrose, also known as white sugar or table sugar, is made of one glucose molecule bonded with one fructose molecule. Lactose, the sugar found in milk, is made of one glucose and one galactose molecule bonded together. Galactose, like fructose, has the same chemical components as glucose, but the atoms are arranged differently. The liver also converts galactose to glucose.

Glucose, fructose, and galactose are monosaccharides and are the only carbohydrates that can be absorbed into the bloodstream through the intestinal lining. Lactose, sucrose, and maltose are disaccharides (they contain two monosaccharides) and are easily converted to their monosaccharide bases by enzymes in most people's digestive tracts. You may know someone, however, who is lactose intolerant. This means that the person is not producing the enzyme that breaks lactose apart. Instead of being absorbed in the bloodstream, the lactose stays in the intestines, where it feeds bacteria and causes lots of gas and discomfort.

Monosaccharides and disaccharides are called simple carbohydrates. They are also sugars: They all taste sweet, and they all digest quickly and enter the bloodstream quickly.

Complex carbohydrates are commonly known as starches. A complex carbohydrate is made up of chains of glucose molecules. Plants use starches to store energy. Most grains (such as wheat, corn, oats, and rice) and things like potatoes and plantains are high in starch. Your digestive system breaks a complex carbohydrate (or starch) back down into its component glucose molecules so that the glucose can enter your bloodstream.

It takes a lot longer for your body to break down a starch than a simple sugar. If you drink a can of soda that's full of sugar, glucose will enter your bloodstream at a rate of something like 30 calories per minute. A complex carbohydrate is digested more slowly, so glucose enters the bloodstream at a rate of only about 2 calories per minute.

Simple carbohydrates are harder on your body than complex carbohydrates because they rapidly increase the glucose level in the blood. This causes your body to secrete extra insulin very quickly, which then decreases blood sugar levels. This imbalance causes your body to release extra adrenaline, which causes you to be nervous and irritable. This shows that the foods you eat and the way you eat them can affect your mood and your temperament. Foods affect the levels of different hormones in your bloodstream over time.

Proteins

A protein is a chain of amino acids. An amino acid is a small molecule that acts as the building block of a cell. Carbohydrates provide cells with energy, and amino acids provide cells with the building material they need to grow and maintain their structure. Your body is about 20% protein by weight, and it is about 60% water. Most of the rest of your body is composed of minerals (for example, calcium in your bones). Amino acids are molecules that all contain an amino group (NH_2) and a carboxyl group (COOH), which is acidic. In between there are a variety of molecular structures, and these structures give each amino acid its unique properties. The human body is constructed of 20 different amino acids (there are perhaps 100 different amino acids in nature).

In your body there are two different types of amino acids: essential and nonessential. Nonessential amino acids are amino acids that your body can create out of other chemicals found in your body. Essential amino acids cannot be created, and therefore the only way to get them is through food.

Protein in the diets comes from both animal and vegetable sources. Most animal sources (such as meat, milk, and eggs) provide "complete protein," meaning that they contain all the essential amino acids. Vegetable sources are usually low on or missing certain essential amino acids.

However, different vegetable sources are deficient in different amino acids, so you can

Food Preservation

The basic idea behind all forms of food preservation is either: 1) to slow down the activity of disease-causing bacteria, or 2) to kill the bacteria altogether. In certain cases a preservation technique may also destroy enzymes naturally found in a food that causes it to spoil or discolor quickly. An enzyme is a special protein that acts as a catalyst for a chemical reaction, and enzymes are fairly fragile. By increasing the temperature of food to 150°F or so, enzymes are destroyed.

A food that is sterile contains no bacteria. Unless sterilized and sealed, all food contains bacteria. For example, bacteria naturally living in milk will spoil the milk in 2 or 3 hours if the milk is left out on the kitchen counter at room temperature. By putting the milk in the refrigerator you do not eliminate the bacteria already there, but you slow down the bacteria enough that the milk will stay fresh for 1 or 2 weeks.

Different ways we have to preserve food include:

- Refrigeration and freezing
- Canning
- Irradiation
- Dehydration
- Freeze drying
- Salting
- Pickling
- Pasteurizing
- Fermentation
- Carbonation
- Cheese-making
- Chemical preservation

get all the essential amino acids by combining vegetables throughout the course of the day. Some vegetable sources contain quite a bit of protein—things like beans, nuts, and soybeans are all high in protein. By combining them you can get complete coverage of all essential amino acids.

The digestive system breaks down all proteins into their individual amino acids or short amino acid chains so that they can enter the bloodstream. Cells then use the amino acids as building blocks. A 150-pound person needs at least 54 grams of protein per day. A can of tuna contains about 30 grams of protein. A glass of milk contains about 8 grams of protein. A slice of bread might contain 2 or 3 grams of protein. You can see that it is not that hard to meet the recommendation for protein with a normal diet.

Fats

You commonly hear about two kinds of fats: saturated and unsaturated. Saturated fats are normally solid at room temperature, and unsaturated fats are liquid at room temperature.

Vegetable oils are examples of unsaturated fats, and lard, shortening, and the animal fat you see in raw meat are saturated fats. Most fats contain a mixture of both types of fat.

Unsaturated fats are currently thought to be more healthy than saturated fats, and monounsaturated fats (as found in olive oil and peanut oil) are thought to be healthier than polyunsaturated fats (such as sunflower and corn oil).

Fats that you eat enter the digestive system and meet with an enzyme called lipase. Lipase breaks the fat into its two parts: glycerol and fatty acids. These components are then reassembled into triglycerides for transport in the bloodstream. Muscle cells and fat (adipose) cells absorb the triglycerides, either to store them or to burn them as fuel.

You need to eat fat for several reasons:

- Certain vitamins are fat soluble, so the only way to absorb them into your body is by eating fat.
- In the same way that there are essential amino acids, there are essential fatty acids (for example, linoleic acid is used to build cell membranes). You must obtain these fatty acids from food you eat because your body has no way to make them.
- Fat is a good source of energy. Fat contains twice as many calories per gram as carbohydrates or proteins. Your body can burn fat as fuel when necessary.

Vitamins

Vitamins are smallish molecules that your body needs to keep itself running properly. They do not provide energy or function as building units, but they are very important because they help regulate metabolic processes. The body can produce its own vitamin D, but generally vitamins must be provided in food. The human body needs 13 different vitamins:

- Vitamin A (fat soluble; retinol, comes from beta-carotene in plants—when you eat beta-carotene an enzyme in the stomach turns it into vitamin A)
- Vitamin B1 (water soluble; thiamine)
- Vitamin B2 (water soluble; riboflavin)
- Vitamin B3 (water soluble; niacin)
- Vitamin B6 (water soluble; pyridoxine)
- Vitamin B12 (water soluble; cyanocobalamin)
- Folic acid (water soluble)
- Vitamin C (water soluble, ascorbic acid)
- Vitamin D (fat soluble; calciferol)
- Vitamin E (fat soluble; tocopherol)
- Vitamin K (fat soluble; menaquinone)
- Pantothenic acid (water soluble)
- Biotin (water soluble)

In most cases, a lack of a vitamin causes severe problems because without the vitamin your body can't carry out necessary functions completely. For example, a lack of vitamin C causes scurvy because the body cannot manufacture collagen without vitamin C, and a lack of vitamin A causes night blindness because an important chemical reaction that governs night vision cannot proceed in the eyes. A diet of fresh, natural food usually provides all the vitamins you need. Processing tends to destroy vitamins, so many processed foods are "fortified" with manufactured vitamins.

Minerals

Minerals are elements your body must have to create specific necessary molecules. For example, your body needs calcium to create and maintain teeth and bones. Your body needs iron because red blood cells do not work unless they contain iron atoms. Your body needs sodium and potassium in order for nerve cells to fire properly. Food provides these minerals. If they are lacking in your diet, then various problems and diseases arise because the body lacks certain necessary molecules.

Water

As mentioned previously, your body is about 60% water. A person at rest loses about 40 ounces (1.8 l) of water per day. Water leaves your body in urine, in your breath when you exhale, by evaporation through your skin, and through a number of other routes. Obviously, if you are working and sweating hard, then you can lose extra water—your body can produce up to 1 liter (33.81 ounces) of sweat per hour in extreme cases.

Because you are losing water all the time, you must replace it. Obviously, you need to take in at least 40 ounces (1.8 l) per day in the form of moist foods and liquids. In hot weather and when exercising, your body may need twice that amount. Many foods, especially fruits, contain a surprising amount of water. Pure water and other beverages provide the rest.

Fiber

Fiber is the broad name given to the things you eat that your body cannot digest. You eat three different kinds of fiber on a regular basis:

- Hemicellulose is found in the hulls of different grains, such as wheat. Bran is an example of hemicellulose.
- Cellulose is the structural component of plants that gives a vegetable its shape. It is a complex carbohydrate—a chain of glucose molecules. Some animals and insects can digest cellulose. Both cows and termites have no problem with it because bacteria in their digestive systems secrete enzymes that break down cellulose into glucose. Human beings have neither the enzymes nor these beneficial bacteria, so cellulose is fiber for people.
- Pectin is found most often in fruits, and it is soluble in water but nondigestible. Pectin is often called water-soluble fiber and forms a gel.

When you eat fiber, it simply passes straight through your body, untouched by the digestive system.

You've been eating all your life, but you probably haven't thought much about the components of food. As you can see, there are many complex things going on every time you finish a meal. The process of turning food into life is absolutely fascinating!

Starvation

A normal person who is eating three meals per day and snacking between meals gets almost all of his or her energy from the glucose that carbohydrates provide. But what happens if you stop eating? Your body goes through several phases in its attempt to keep you alive in the absence of food.

The first line of defense against starvation is the liver, which stores glucose by converting it to glycogen. It holds perhaps a 12-hour supply of glucose in its glycogen. When you finish digesting all the carbohydrates that you last ate, the liver starts converting its stored glycogen back into glucose and releases it to maintain glucose in the blood. Lipolysis also starts breaking down fat in the fat cells and releasing fatty acids into the bloodstream. Muscle cells are able to easily burn fatty acids, so they stop burning glucose and start burning the fatty acids. This reduces the glucose demand so that nerve cells get the glucose.

When the liver runs out of glycogen, the liver converts to a process called gluconeogenesis, which turns amino acids into glucose.

The liver then begins producing ketone bodies from fatty acids that are made available in the blood by lipolysis. Brain and nerve cells convert over from being pure consumers of glucose to being partial consumers of ketone bodies for energy.

How DIETING Works

You have probably heard of the Atkins Diet, the Zone Diet, and a number of other diets—you may have tried one of these diets yourself. In fact, on any given day, a huge portion of the U.S. population is "on a diet" and "counting calories" in some way. Advertisements for diets and diet products on television and in magazines surround you—you even hear them on the radio.

HSW Web Links

www.howstuffworks.com

How Calories Work
How Fat Cells Work
How Fats Work
How Food Works
How Your Kidneys Work

Have you ever wondered why, for so many people, weight gain seems to be a fact of life? It's because the human body is way too efficient. It doesn't take much energy to maintain the human body at rest; and when exercising, the body is amazingly frugal about turning food into motion.

Burning Calories

At rest (when you're not moving around), the human body burns only about 12 calories per pound (26 calories per kilogram) of body weight per day. That means that if you weigh 150 pounds (68 kg), your body uses only about:

150 X 12 = 1,800 calories per day

Those 1,800 calories are used to do everything you need to stay alive, including keeping your heart beating and lungs breathing, your internal organs operating properly, your brain running, and your body warm.

In motion, the human body also uses energy efficiently. A person running a marathon (26 miles, or 42 km) burns only about 2,600 calories—only about 100 calories per mile (about 62 calories per km). That level of efficiency is the main reason it's so easy to gain weight: It is incredibly easy to eat 1,800 calories without even realizing it.

Taking In Calories

The 1,800 calories that a typical person needs per day is not that many. If you go to your neighborhood hamburger restaurant for lunch and order a combo meal, you will get a sandwich, a large order of fries, and a large soda. This meal contains calories along these lines:

- 700 calories in the sandwich
- 550 calories in the fries
- 300 calories in the drink

This one meal provides 1,550 calories. If you get a bowl of ice cream for dessert, you'll get 600 more calories. At this one quick meal, you're consuming almost 2,200 calories—400 more than you need in a normal day. Add breakfast and dinner to that, and it is possible to consume 4,000, 5,000, or even 6,000 calories in a day and not even realize it. It is incredibly easy to find and consume lots of calories in today's culture.

Your body is extremely efficient at capturing and storing excess calories. Whenever your body has extra calories on hand, it converts them to fat and saves them for a rainy day. It only takes 3,500 excess calories to create 1 pound of new fat on your body. If you're taking in just 500 extra calories per day, you're gaining 1 pound of fat per week.

Failing at Diets

The reason most diets ultimately fail is that they are not sustainable. A person gains weight because he or she consumes more calories per day than are needed. A diet creates a temporary deficit. When the person ends the diet, he or she resumes normal eating and regains the weight.

Let's say you weigh 150 pounds, and therefore you burn about 1,800 calories per day. You burn perhaps 200 more calories per day just living your life—walking up and down steps, carrying in the groceries, and so on. So you burn about 2,000 calories per day. Let's imagine that, on average, you consume 2,050 calories per day. On a daily basis, your body is taking in, and therefore storing, 50 more calories than it needs. If

Dieting Myths

There are dozens of weight-loss myths that help to derail people. Here are some of the most common, so you can avoid them:

The myth that some kinds of calories are different from others. A calorie is a calorie. If you consume 4,000 calories by eating 1,000 grams of white sugar or 4,000 calories by eating 444 grams of fat, your body doesn't care. It is still 4,000 calories.

The myth that you can eat as much as you want if it is low fat. A product can have 0 grams of fat but can still have lots of calories. Many fat-free foods replace the fat with sugar and contain just as many or more calories as a fat-containing product.

The myth that any passive device—weird acupressure rings and bracelets or soaps or whatever—can help. There is no way to burn calories but to burn them.

that 50-extra-calories-per-day trend continues, over the course of a year you'll gain 5 pounds. This is the pattern for a big portion of the U.S. population. If you overconsume by just a few calories per day, you will, over time, gain noticeable weight. And considering that one small cookie contains 50 calories, overconsuming is incredibly easy.

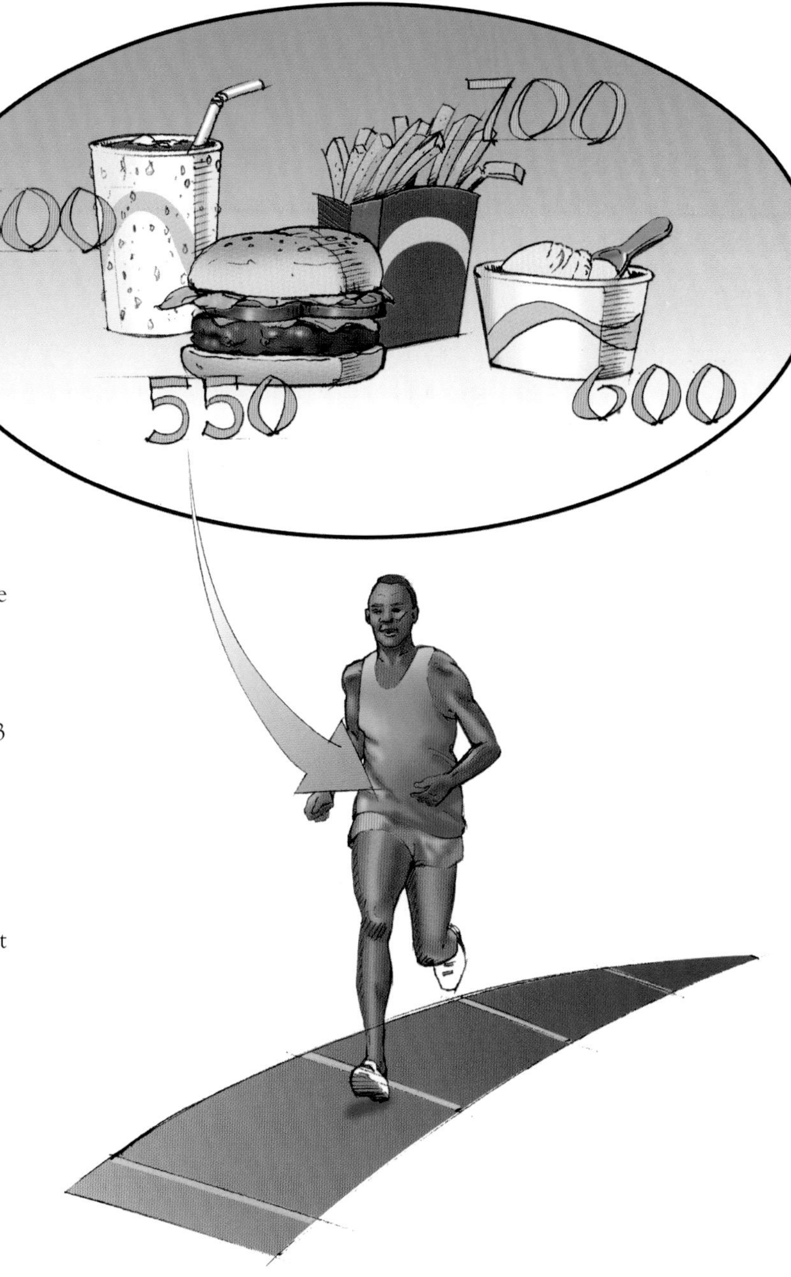

Say you go on a diet—the amazing Palm Beach Miracle Diet—on which you consume nothing but 2 cups of brown rice and a can ofVienna sausages, along with all the onions you care to eat, every day. You are consuming only 1,000 calories per day. You also start jogging 2 miles per day. So, on a typical day, you are consuming 1,200 calories less than you need. Over the course of 3 days, you consume 3,600 calories less than you need, so you lose 1 pound of weight. You stay on this diet for 2 months and lose 20 pounds.

The day you go off this diet, the first thing you're probably going to do is eat a lot more than you did before you went on the diet—after all, you've been eating nothing but rice, Vienna sausages, and onions for 2 months. Then, you may settle into your original eating pattern. Eventually, all the weight will come back.

This is why diets don't work for most people. You may lose weight, but when you go off the diet, you regain the weight. What you need instead is a sustainable diet—a food consumption and exerciseplan—that lets you live a normal life and eat normal foods in a normal way. And it's good to keep in mind that it is best to lose about 1 or 2 pounds a week.

Fad Diet Pitfalls

Hundreds of people each day can't resist the lure of fad diets. "Lose 10 pounds in 10 days." "Eat these fat burning foods in the morning and have whatever you want at night." "Lose weight while you sleep." The statement "It's too good to be true," should be considered when you hear promises like these. If a diet suggests that you can lose 10 pounds in 10 days, that is an average loss of 1 pound per day. Studies have shown that it is not safe for most individuals to lose more than 2 pounds per week. Several diets that promise rapid weight loss do so based on an extremely low-calorie and unhealthy plan. Side effects including nausea, diarrhea, constipation, and fatigue have been associated with extremely low-calorie diets. Although most of these side effects are temporary, they can persist long enough to cause even more serious side effects such as dehydration or gallstones.

How Much Sugar?

Have you ever wondered just how much sugar you are consuming each day from beverages alone? Here's an easy way to figure it out:

Buy a packet of "unsweetened soft drink mix"—that is, the kind you add sugar to when you are making it. You will be instructed to add one cup of sugar and enough water to make two quarts (64 ounces). A cup of sugar contains 48 teaspoons of sugar. Therefore, a 16-ounce serving of one of these beverages contains 12 teaspoons of sugar!

A typical carbonated soft drink will have 200 calories in a 16-ounce serving. All of those calories come from sugar, and sugar contains 16 calories per teaspoon. By this measurement, a 16-ounce serving contains 12.5 teaspoons of sugar.

So go down to the kitchen and get out a 16-ounce glass, a teaspoon and some sugar. Measure 12 teaspoons of sugar into the glass—it's an amazing amount! Then multiply that by however many sodas you typically consume in a day.

Building a Sustainable Diet

Building a sustainable diet and exercise plan is the key to maintaining a consistent weight. This is not easy for many people.

The first step is to start counting the calories that you consume in a day so that you become conscious of two things:

- Exactly how many calories you eat on a normal day
- Where each calorie comes from

In the United States, any food that you buy in the grocery store is required by the U.S. Food & Drug Administration to have a nutritional label displaying that food's calorie content. Any chain restaurant will also supply you with nutritional information, usually both at the store and on the Web.

The second step is to figure out how many calories you need in a day. You can use the 12 calories per pound rule. Next, pick your "ideal weight"—the weight that you would like to maintain. Then calculate how many calories per day you can consume to maintain that weight.

The third step is to compare the two numbers—calories you need versus calories you currently take in. You may be startled by the difference between the numbers. That discrepancy is where the extra pounds come from.

The fourth step is to figure out how to bring the two numbers in line. You'll soon realize that 1,600 or 2,000 calories per day is not that many. You have to watch and count everything you eat and drink every day, and you have to stick to your daily limit.

The fifth step might be to add exercise to the mix. Burning 250 calories per day through exercise means that you can consume an extra 250 calories per day and still be within your allowed number of calories.

Here are some strategies to reduce the number calories you take in:

- Be conscious of every calorie you consume, and keep a daily journal. Write down everything you eat and drink.
- Eliminate all calories that come in through drinking.
- Eliminate white sugar.
- Eliminate fried foods.

Another strategy is to try to replace high-calorie foods (such as cookies) with low-calorie foods (such as bananas). Twelve little cookies contain 600 calories; so do six bananas. Most people can eat a dozen little cookies without even thinking about it. Most people would explode trying to eat six bananas. That's the difference between high-density and low-density foods.

Fitting in Exercise

Exercise is a great weight-control tool because it increases the number of calories you burn in a day. One way to make the most of exercise is to integrate it into your daily routine:

- Find some type of exercise that you enjoy. Do it every day for 30 minutes or more. It might be walking, riding a bike, or working out in a gym during your lunch hour.
- Fit micro-exercises into your daily life. Instead of taking the elevator, take the stairs. Park farther away from stores when you go shopping.
- Keep a hand weight at your desk and use it three or four times during the day, as you think or talk on the phone.
- Find an exercise partner.
- Try to exercise daily. It is easiest to remember to do something if you do it every day.

It has become so easy to consume calories in today's society that it is incredibly easy to gain weight. All that the body can do with the excess calories is turn them into fat. To lose weight or maintain a consistent weight, you have to make sure that the number of calories you take in every day is equal to the number of calories your body burns.

How **ASPIRIN** Works

If you have a sore muscle, you can take aspirin to make it feel better for several hours. But how does the aspirin find the muscle, and what is it doing to block the pain? By understanding how aspirin works, you can learn a good bit about how your body works.

As long ago as the fifth century BC, Hippocrates wrote about a bitter powder extracted from willow bark that could ease pain and reduce fevers. Scientists later discovered that the important part of willow bark is a chemical known as salicin. The human body turns salicin into salicylic acid.

The problem with salicylic acid is that it is hard on the stomach. By putting salicylic acid through a couple chemical reactions that cover up one of the acidic parts with an acetyl group, you can convert it to acetylsalicylic acid (ASA). ASA is not nearly as hard on the stomach as salicylic acid, and it works just as well on fever, aches, and pains. The Bayer company coined the term *aspirin* for ASA, and that's what we've called it ever since.

Producing Pain

Let's say you hit your finger with a hammer. The part of your finger that is damaged has nerve endings in it. The damaged tissue in your finger releases chemicals that make the pain nerve endings register pain even better. Some of these chemicals are prostaglandins, and working cells in the damaged tissues make prostoglandins by using an enzyme called cyclooxygenase 2 (COX-2).

Because of the prostaglandins, the nerve endings that are involved send a strong signal to your brain, where your mind registers it as pain. The prostaglandins seem to contribute just a portion of the total signal that means pain, but this portion is an important one. In addition, prostaglandins also cause the finger to swell up to bathe the tissues in fluid from your blood that will help it to heal. The pain serves a purpose here—it reminds you that your finger is damaged and that you need to be careful with it. The problem is that, sometimes, things hurt without a good reason. Headaches and arthritis are two good examples of pain that serves no useful purpose.

Finding and Fighting Pain

Aspirin helps pain problems by stopping cells from making prostaglandins. COX-2 is a protein that creates prostaglandins, and tissue that has been damaged in some way makes lots of COX-2. Aspirin sticks to COX-2 and prevents it from doing its job. By taking aspirin, you don't stop the problem that's causing the pain. You simply lower the volume on the pain signals getting to your brain.

When you take aspirin, it dissolves in your stomach and enters the bloodstream, so it goes through your entire body. Although it is everywhere, it only works where there are prostaglandins being made—it doesn't stick to anything else.

The problem is that your body actually needs prostaglandins in certain places. Another enzyme, called COX-1, makes a prostaglandin that seems to keep your stomach lining nice and thick. And aspirin keeps COX-1 from working as well. For some people aspirin can thin the stomach lining, which means digestive juices can irritate the stomach. Some types of prostaglandins cause platelets in your blood to stick together to form a blood clot. By inhibiting prostaglandin production, aspirin slows down clot production. Therefore, many adults now take aspirin regularly to prevent clots that can cause heart attacks.

HSW Web Links

www.howstuffworks.com

How Allergies Work
How Anesthesia Works
How Sunburns and Sun Tans Work
How Your Heart Works
How Your Immune System Works

Did You Know?

As with almost all chemicals, your body has ways of getting rid of aspirin. Your liver, stomach, and other organs change aspirin to salicylic acid! The liver slowly changes this chemical a bit more, by sticking other chemicals onto the salicylic acid so that your kidneys can filter it out of your blood and send it out in your urine. This whole process takes about 4 to 6 hours, so you need to take another pill every 4 hours or so to keep the pain-killing effect going.

How **SWEAT** Works

You are about to do something that makes you really nervous—maybe go for a job interview, go on a first date, or stand up at your wedding—and you notice that your palms and underarms are sweating. You've just completed an aerobic workout and your whole body is drenched in sweat. How can such different activities have such a similar effect on your body? What is sweat and why do we make it?

HSW Web Links

www.howstuffworks.com

How Cells Work
How Congestive Heart Failure Works
How Exercise Works
How Sunburn and Sun Tans Work
How Your Lungs Work

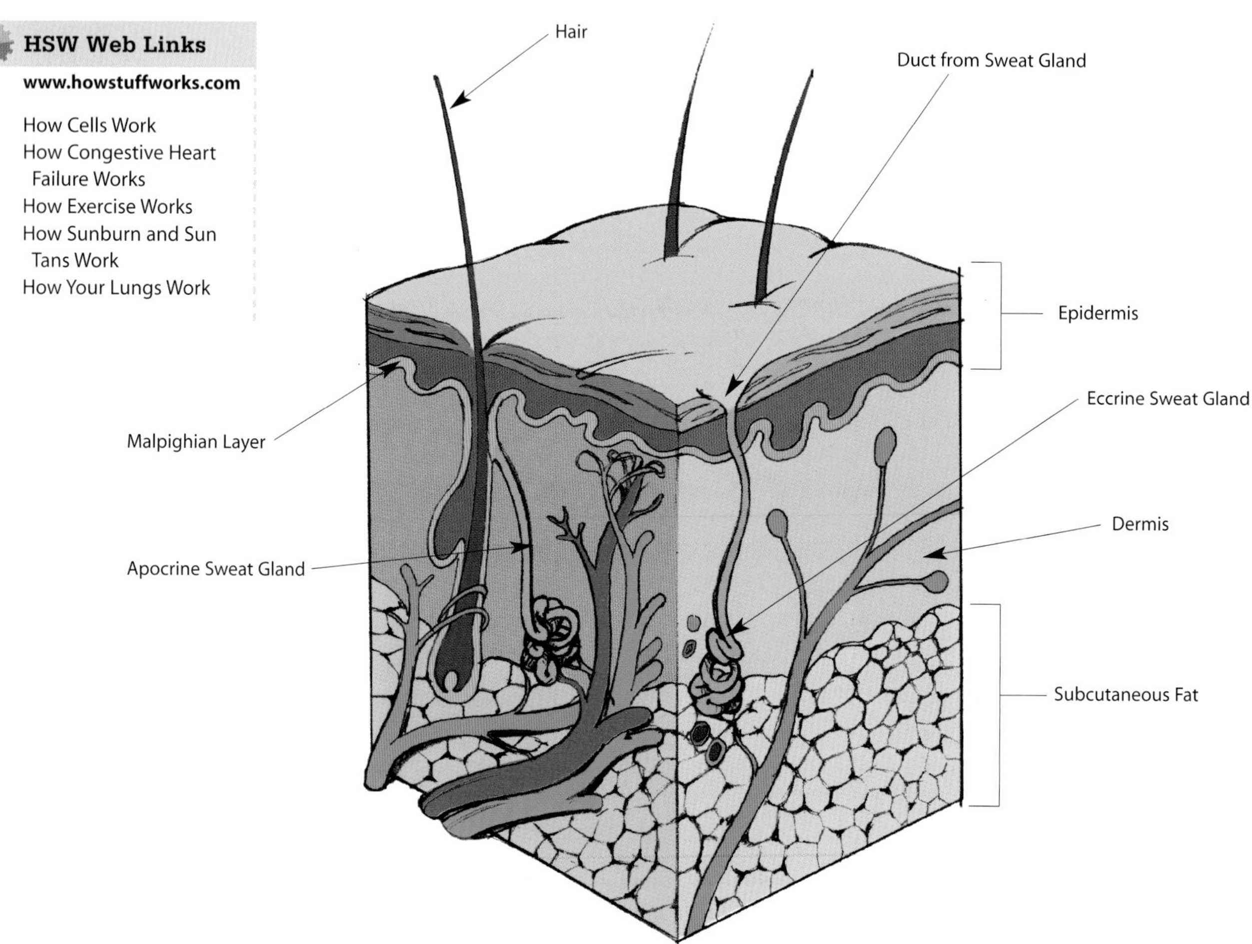

Sweat glands are distributed over the entire body (except for a few places, like the lips) in the dermis layer of the skin. Basically, a sweat gland is a long, coiled, hollow tube of cells. The coiled part of a gland makes the sweat, and the long portion connects the gland to the opening or pore on the skin's outer surface. Sympathetic nerve cells connect to the sweat glands. There are two types of sweat glands:

- **Apocrine**—Mostly confined to the armpits and the anal-genital area, these sweat glands typically end in hair follicles rather than pores. Apocrine glands become active only at puberty.
- **Eccrine**—These glands are all over the body, particularly on the palms of the hands, soles of the feet, and forehead. Compared to apocrine glands, they are smaller, are active from birth, and produce protein-free, fatty acid-free sweat. The eccrine glands are the ones that keep you cool.

Sweat Production

When a sweat gland is stimulated, the cells secrete a fluid that is mostly water. It is very similar in composition to blood plasma (which is the fluid part of blood that is left behind if you take out all the red blood cells). Sweat has a high concentration of sodium chloride and a low concentration of potassium. It also does not contain the proteins and fatty acids that are normally found in plasma.

In eccrine sweat glands, the fluid for sweat comes from the spaces between the cells. This fluid comes from the blood vessels (or capillaries) in the dermis. A sweat gland collects the fluid with the coiled-up portion of the gland. The fluid travels from the coiled portion up through the straight duct. What happens in the straight duct depends on the rate of sweat production or flow:

- **Low sweat production (during rest, at a cool temperature)**—Cells in the straight duct reabsorb most of the sodium chloride from the fluid because they have plenty of time. In addition, water is reabsorbed osmotically. So not much sweat reaches the outside.
- **High sweat production (during exercise, at a hot temperature)**—Cells in the straight portion do not have enough time to reabsorb all the sodium chloride or water. So a lot of sweat makes it to the surface. The sodium chloride concentration is about 50% less than that of plasma, and the potassium concentration is about 20% higher than that of plasma.

Sweat is produced in apocrine sweat glands in the same way as in the eccrine glands; however, it contains proteins and fatty acids, which make it thicker and give it a milky or yellowish color. This is why underarm stains in clothing appear yellowish. Although sweat itself has no odor, bacteria (for example, from skin and hair) metabolize the proteins and fatty acids to produce an unpleasant odor. This is why you need deodorants and antiperspirants on your underarms and not on your whole body.

The Cooling Effect of Sweat

When sweat evaporates from the surface of your skin, it removes excess heat and cools you. Converting water from a liquid to a vapor takes a certain amount of heat, called the heat of vaporization. As sweat evaporates, it carries away heat. Not all the sweat always evaporates—some may run off your skin.

A major factor that influences the rate of evaporation is the relative humidity of the air around you. If the air is humid, then it already has water vapor in it, probably near saturation, and it cannot take any more. Therefore, sweat does not evaporate and cool your body as efficiently as it does when the air is dry.

When the water in sweat evaporates, it leaves salts (sodium, chloride, and potassium) behind on your skin, which is why your skin tastes salty. The loss of excessive amounts of salt and water from your body can quickly dehydrate you, which can lead to circulatory problems, kidney failure, and heatstroke. This is why it is important to drink plenty of fluids when you exercise or are outside in high temperatures. Sports drinks contain some salts to replace those lost in the sweat.

Perspiration or sweat is your body's way of cooling itself, whether that extra heat comes from hardworking muscles or from over stimulated nerves. Humans have created some incredible machines. So, it isn't surprising that the human body is, itself, an equally fascinating machine of sorts.

And Another Thing...

Sweat glands often respond to your emotional state. So when you are nervous, anxious, or afraid, there is an increase in sympathetic nerve activity in your body as well as an increase in adrenaline. These substances act on your sweat glands, especially those on the palms of your hand and in your armpits, to make sweat that you feel as cold sweat.

Excessive Sweating

Excessive sweating—usually on the palms of the hand or the armpits—that is not caused by emotional or physical activity is called diaphoresis or hyperhidrosis. It is often an embarrassing condition. The cause is unknown, but it may be due to the following:

- Hormonal imbalances
- An overactive thyroid gland (the thyroid hormone increases body metabolism and heat production)
- Certain foods and medications (such as coffee, with its high amount of caffeine)
- Overactivity of the sympathetic nervous system

Excessive sweating can be treated with medications and surgical procedures.

How **SLEEP** Works

Sleep is one of those things a human being just has to have—your body does not give you a choice. People feel great after a good night's sleep, and they feel lousy after they sleep poorly.

HSW Web Links

www.howstuffworks.com

How Allergies Work
How Anesthesia Works
How Caffeine Works
How Your Heart Works
How Your Immune System Works

Cool Facts

- The average human sleeps away more than one-third or his or her life, yet approximately 100 million adults are considered to be clinically sleep deprived.
- Cows can sleep standing up but can dream only when lying down.
- A newborn baby may sleep for up to 20 hours per day. By the time that same child is an elderly adult, he or she may sleep only 4 to 6 hours per day.

When you see someone sleeping, you recognize a number of characteristics. If possible, the person is lying down. The person's eyes are closed, and the person doesn't hear anything except possibly loud noises. The person breathes in a slow, rhythmic fashion. The person's muscles are completely relaxed, and if the person is sitting up, he or she might fall out of his or her chair as sleep deepens. Occasionally during sleep, the person rolls over or reconfigures his or her body. This happens once or twice each hour, and it may be the body's way of making sure that no part of the body or skin has its circulation cut off for too long. In addition to these outward signs, things are happening inside, too. The heart slows down and the brain does some pretty unusual things.

A sleeping person is unconscious to most things happening in the environment. For any animal living in the wild, it just doesn't seem very smart to design in a mandatory 8-hour period of near-total unconsciousness every day. Yet that is exactly what nature has done.

Who Needs Sleep

Reptiles, birds, and mammals all sleep. They become unconscious to their surroundings for periods of time. Some fish and amphibians reduce their awareness but do not ever become unconscious like the higher vertebrates do. Insects do not appear to sleep, although they may become inactive in daylight or darkness.

By studying brainwaves, it is known that reptiles do not dream. Birds dream a little. Mammals all dream during sleep.

Different animals sleep in different ways. Some animals, like humans, prefer to sleep in one long session. Other animals (dogs, for example) like to sleep in many short bursts. Some animals sleep at night, and others sleep during the day. Some animals sleep standing, and many lie down.

How Much Sleep You Need

Most adult people seem to need 7 to 9 hours of sleep per night. This is an average, and it is also subjective. For example, you probably know how much sleep you need in an average night to feel your best. The amount of sleep you need decreases with age. A newborn baby might sleep 20 hours each day. By the age of 4, the average is 12 hours per day. By age 10, the average falls to 10 hours per day. Older adults can often get by with 6 or 7 hours per day.

Sleep and the Brain

If you attach electrodes to a person's scalp to monitor voltage changes, you find that the brain generates a variety of voltage patterns, depending on what the person is doing. These voltage patterns are caused by groups of neurons in the cortex working together. Patterns have been classified into several different groups:

- **Gamma waves**—Greater than 30 cycles per second (Hertz [Hz]); common during focused attention on a topic or a problem.
- **Beta waves**—Between 12 and 30 Hz; different states of wakefulness.
- **Alpha waves**—Between 8 and 12 Hz; relaxation.
- **Theta waves**—Between 3 and 8 Hz; light sleep.
- **Delta waves**—Less than 3 Hz; deep sleep.

During sleep, theta waves and delta waves take over. As a person falls asleep and sleep deepens, the patterns of brain waves slow down. The slower the brain wave patterns, the deeper the sleep.

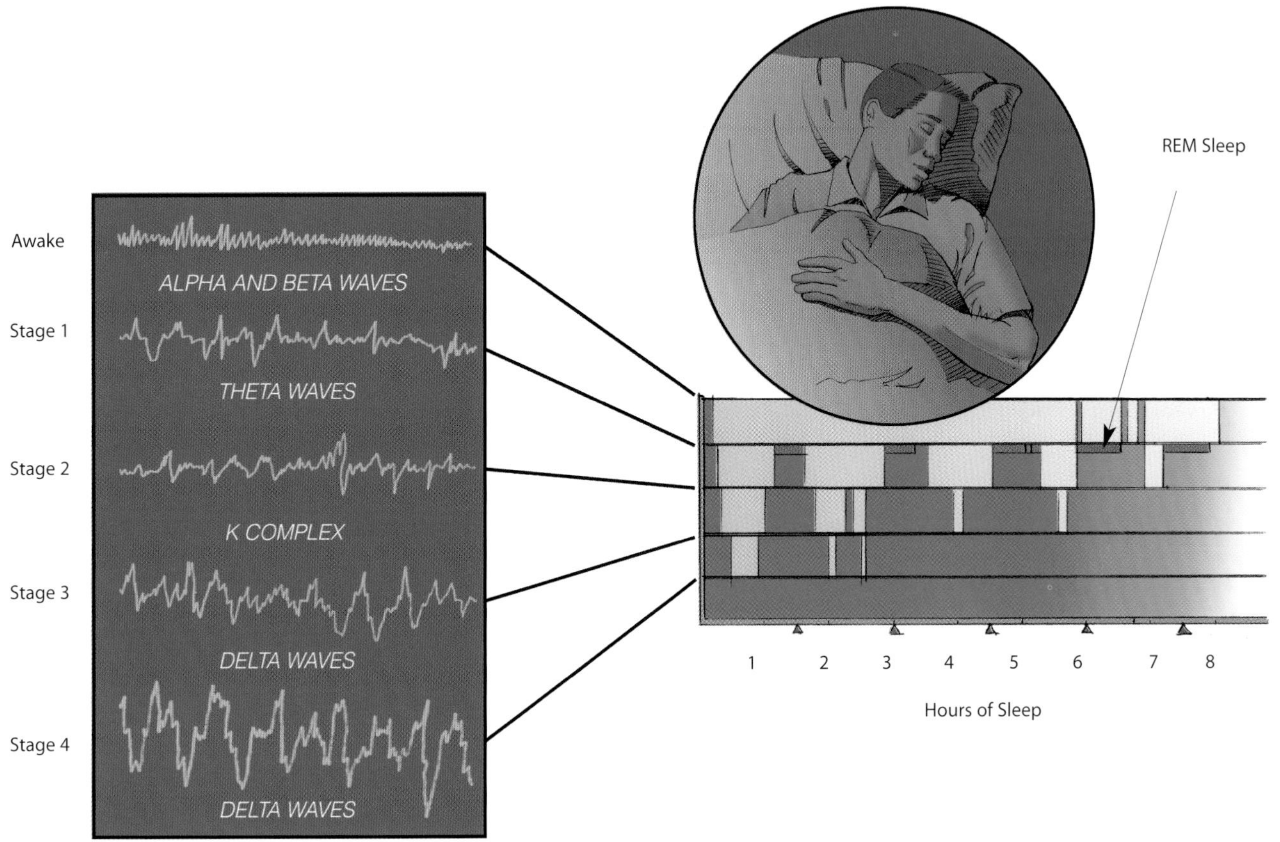

At several points during a long sleep, rapid eye movement (REM) sleep occurs. Most people experience three to five intervals of REM sleep per night, and brainwaves during this period speed up, sometimes to conscious levels. If you watch someone experiencing REM sleep, you see his or her eyes flicker back and forth rapidly. In many dogs and some people, arms, legs, and facial muscles twitch during REM sleep. Periods of sleep other than REM sleep are known as non-REM (NREM) sleep.

REM sleep is when you dream. If you wake a person during REM sleep, the person can vividly recall dreams. If you wake a person during NREM sleep, generally the person will not be dreaming. You must have both REM and NREM sleep to get a good night's sleep. A normal person will spend about 25% of the night in REM sleep and the rest in NREM. A REM session, also known as a dream, lasts 5 to 30 minutes.

Why We Sleep

One way to understand why we sleep is to look at what happens when we don't get enough:

- As you know if you have ever pulled an all-nighter, missing one night of sleep is not fatal. You will generally be irritable the next day and will either slow down and become tired easily or be totally wired because of your extra adrenaline.
- If you miss two nights of sleep, things get worse. Concentration is difficult, your attention span falls drastically, and you make extra mistakes.
- After three days without sleep, you start to hallucinate, and clear thinking is impossible. With continued wakefulness, you can lose your grasp on reality. Rats forced to stay awake continuously eventually die because of it, proving that sleep is essential. A person who gets just a few

hours of sleep per night can experience many of the same problems over time.

Two other things are known to happen during sleep. Growth hormone in children is secreted during sleep, and chemicals that are important to the immune system are secreted during sleep. You can become more prone to disease if you don't get enough sleep, and a child's growth can be stunted by sleep deprivation.

So why do we need to sleep? No one really knows. There are all kinds of theories, including these:

- Sleep gives the body a chance to repair muscles and other tissues, replace aging or dead cells, and so on.
- Sleep gives the brain a chance to organize and archive memories. Some think that dreams are part of the process.
- Sleep lowers energy consumption, so you need three meals a day rather than four or five. Since you can't do anything in the dark anyway, you might as well "turn off" and save the energy.
- Sleep may be a way of recharging the brain, using adenosine—the chemical that makes you sleepy—as a signal that the brain needs to rest. Adenosine levels in the brain rise during wakefulness and decline during sleep. Caffeine is a drug that blocks adenosine. (See "How Caffeine Works," page 264, for details.)

These are all possibilities, and the real reason humans need sleep may be a combination of them. We do know that with a good night's sleep, everything looks and feels better in the morning—both the brain and the body are refreshed and ready for a new day!

Snoring

Snoring is one of those funny and annoying things—like sweaty palms, and food allergies —that some people suffer with while other people never experience. Up to 20% of the population may experience problems with snoring. The rest of us have no problem with it, unless we're married to someone in the 20%.

Snoring is an anatomy problem involving the soft tissue at the back of the throat. This is the same tissue, by the way, that allows you to swallow, gargle, and talk like Donald Duck. When snoring, the problem is too much tissue. The tissue (including the soft palate, uvula, and tonsils) relaxes and vibrates against the back of the throat while breathing, creating quite a bit of noise. Think of the noise that a balloon makes when you let the air out of it. That noise is not unlike snoring, and it shows how soft structures can create noise when they flap against each other.

Improved Sleep

You can do a number of things to improve your sleep:

- Exercise regularly. It helps tire and relax your body.
- Avoid caffeine consumption after 3 p.m. or so.
- Avoid alcohol before bedtime. Although alcohol promotes relaxation, it disrupts the brain's normal patterns during sleep. Keep in mind that the proper balance between REM and NREM sleep is important, and alcohol, along with many other drugs, disrupts the balance.
- Stick to regular bedtime and waking times, even on weekends.

Dreams

Why do we have such crazy, kooky dreams? For that matter, why do we dream at all? It is known that, at certain times during the night, random signals pulsate through the brain. It is likely that the thinking parts of the brain try to make sense of these random signals in whatever way they can. Here are some of the things you may have noticed about your dreams:

- Dreams tell a story. They are like a TV show, with scenes, characters, and props.
- Dreams are egocentric. They almost always involve you.
- Dreams incorporate things that have happened to you recently. They can also address deep wishes and fears.
- A noise in the environment is often worked into a dream in some way, giving some credibility to the idea that dreams simply are the brain's response to random impulses. For example, if your phone rings while you're dreaming, a dreamed-up phone might ring and you might carry on a conversation with the person who called.
- You usually cannot control a dream. In fact, many dreams emphasize your lack of control by making it impossible to run or yell.

Dreaming is important. In sleep experiments where a person is woken up every time he or she enters REM sleep, the person becomes increasingly impatient and uncomfortable over time.

Not only is sleep a chance in your day to relax, but it's crucial to your ability to adapt and exist. Getting enough sleep prevents you from falling asleep at the wheel when driving, keeps you sharp, keeps your responses quick, and leaves you feeling rested and ready to take on the day.

How **VIRUSES** Work

When you catch a cold or the flu, you know that you are the lucky recipient of a virus. With the viral invasion, you get a bundle of symptoms that may include fever, congestion, coughing, sore throat, and muscle aches. It is not fun. Many other serious and sometimes deadly diseases—such as Acquired Immune Deficiency Syndrome (AIDS), Ebola fever, hepatitis, and herpes—are caused by viruses as well.

A virus is an amazing set of molecules that rides on a fine line that divides the living from the nonliving. A virus is odd because, on its own, it cannot reproduce. But if it can find an appropriate cell, it can reproduce by hijacking the internal chemical machinery of that cell. That is what makes viruses so fascinating and so dangerous.

A virus particle is a tiny package that has the following parts:

- A set of genetic instructions in the form of DNA or RNA
- A protein coating that surrounds and protects the genetic instructions
- In some viruses (for example, influenza), a lipid membrane that surrounds the protein; these viruses are called enveloped viruses as opposed to naked viruses, which have no membrane

A virus is about one-millionth of an inch (17 nm to 300 nm long), about a thousand times smaller than a bacterium. Viruses vary widely in shape and complexity. Some look like round popcorn balls, and others have a complicated shape like a spider.

Unlike human cells or bacteria, viruses do not contain the chemical machinery (enzymes) needed to carry out the chemical reactions for life. Instead, viruses carry only one or two enzymes that decode their genetic instructions. A virus must have a host cell—a bacterium, a plant, or an animal cell—in which to live and make more viruses. Outside a host cell, viruses cannot function. Most scientists agree that viruses are alive because of what happens when they infect a host cell.

HSW Web Links 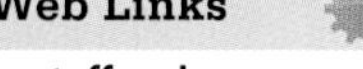

www.howstuffworks.com

How Cells Work
How Mad Cow Disease Works
How Your Immune System Works
How Your Lungs Work

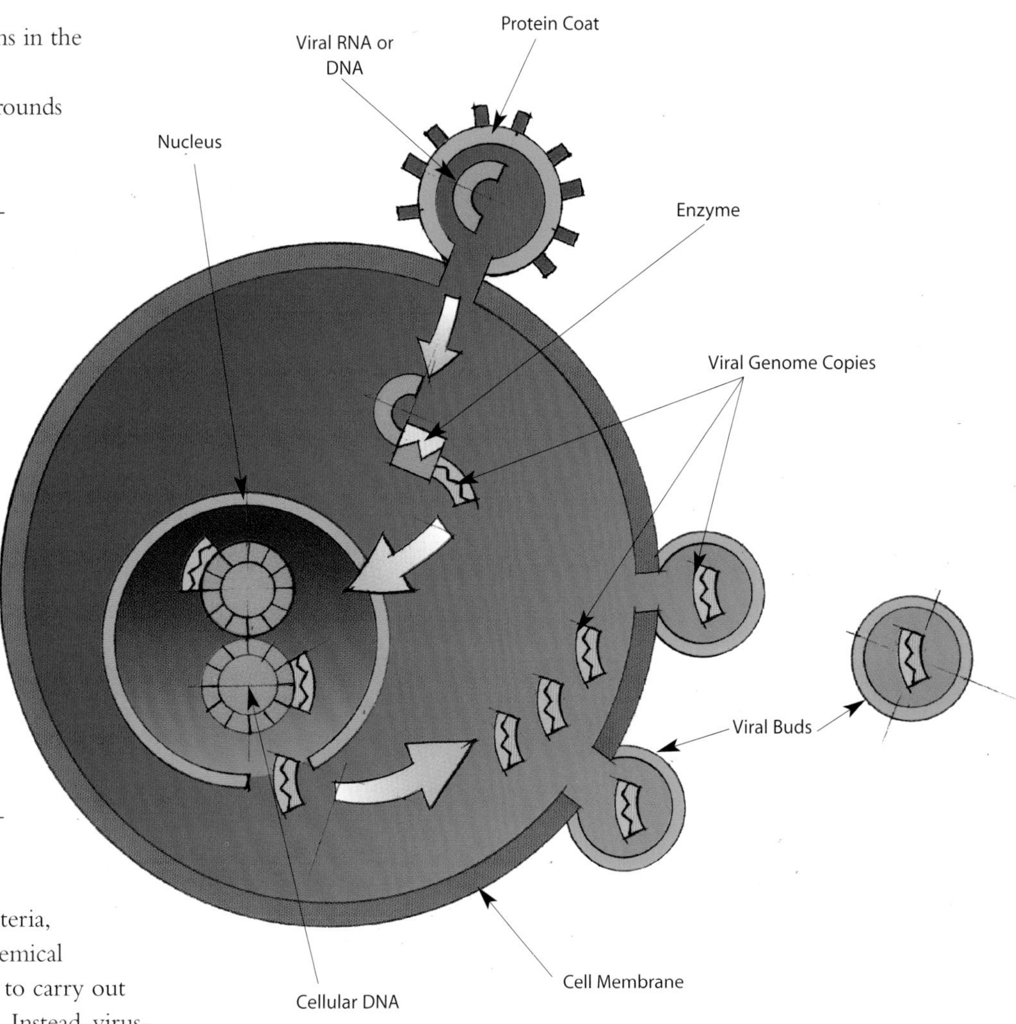

Did You Know?

A virus is about 10 times smaller than a bacterium. A bacterium is about 10 times smaller than a typical human cell. A typical human cell is about 10 times smaller than a human hair. Therefore, a virus is about 1,000 times smaller than a human hair.

Virus Attacks

Viruses can be harbored in other animals or cells or lie around the environment all the time, sometimes for years, just waiting for a host cell. Some viruses can enter a human through the nose, mouth, or breaks in the skin, such as the cold virus, while others, such as the human immunodeficiency virus (HIV) that causes AIDS, can be passed by exchanging body fluids (blood, semen, vaginal fluids). When a virus finds its way inside a person, it finds a host cell to infect. For example, cold and flu viruses attack cells that line the respiratory or digestive tracts, while HIV attacks the cells of the immune system. What happens next depends upon the type of virus. Viruses like HIV or herpes can lie dormant for several years in what is called the lysogenic cycle, while cold viruses act immediately in what is called the lytic cycle.

HIV follows these steps in the lysogenic cycle:

1) HIV attaches to the host cell by using a type of protein on the outside coat or envelope that feels or recognizes the proper host cell (immune helper T-cell). 2) HIV gets engulfed by the host cell.
3) Once inside, the viral genetic instructions (DNA) mix with the host's genetic instructions (DNA) in the cell's nucleus.
4) The virus lies dormant for several years, which is why a person who is infected with HIV can live without showing symptoms of AIDS for years, but he or she can still spread the virus to others.
5) At some point, the virus becomes reactivated, probably by some environmental factors.
6) The viral DNA takes over the host cell's enzymes and makes new viral DNA and viral proteins.
7) The new viral DNA and proteins get packaged.
8) The T-cells burst, releasing new virus particles and, thereby, killing the cell. The T-cell deaths render the person helpless against other types of infections, which eventually kill them.
9) The new viruses are free to infect other T-cells.

Cold viruses follow the same basic steps as the second half of the lysogenic cycle, called the lytic cycle:

1) A virus particle attaches to a host cell by using a type of protein on the outside coat or envelope that feels or recognizes the proper host cell(s). Some enveloped viruses dissolve right through the cell membrane, which is made of lipid, like the envelope is.
2) The virus injects its genetic instructions into the cell. Enveloped viruses simply release their contents when they get inside.
3) The injected genetic material recruits the host cell's enzymes.
4) The enzymes make parts (such as genetic instructions and proteins) for more new virus particles.
5) The new parts are packaged into new viruses.
6) The new viruses break free from the host cell. Either they break the host cell open (called lysis) and destroy the host cell, or they pinch out from the cell membrane and break away (called budding) with a piece of the cell membrane surrounding them. This does not destroy the host cell.

When it is free of the host, the new virus can attack other cells. Because one virus particle can reproduce thousands of new virus particles, viral infections can spread quickly throughout the body.

The sequence of events that occurs when you come down with the flu or a cold demonstrates how a virus works. First, an infected person sneezes near you. You inhale a virus particle, and it attaches to cells lining your nose and sinuses. Then the virus attacks those cells and rapidly reproduces new viruses. The host cells break and new viruses spread into your bloodstream and lungs. Because you have lost cells lining your sinuses, fluid can flow into your nasal passages and give you a runny nose. Viruses in the fluid that drips down your throat attack the cells lining your throat to give you a sore throat. Viruses in your bloodstream can attack muscle cells to give you muscle aches.

Your immune system can respond to the infection in several ways:

- It might produce chemicals called pyrogens that increase your body temperature. This is where the fever comes from in the flu.
- The fever slows down the rate of viral reproduction because most of your body's chemical reactions have an optimal temperature of 98.6°F (37°C).
- Antibodies and white blood cells attack the virus particles or cells that harbor new virus particles and begin eliminating them from your system. This immune response continues until the viruses are eliminated from your body.

The Spread of Viruses

Viruses can be spread through carrier organisms such as mosquitoes and fleas, the air (in the form of sneezes), direct transfer of body fluids (for example, saliva, sweat, nasal mucus, blood, semen, vaginal secretions) from one person to another, and surfaces (such as faucets, counters, and toys) on which body fluids (such as saliva and nasal mucus) have landed and even dried.

To reduce the risk of spreading or contracting viruses, here are some things you can do:

- Cover your mouth or nose when you sneeze or cough.
- Wash your hands frequently, especially after going to the bathroom or preparing food.
- Avoid contact with the bodily fluids of others.

Medicines That Help Stop Viruses

Contrary to popular belief, antibiotics have no effect on a virus because viruses do not carry out their own biochemical reactions. Immunizations work by pre-infecting the body so that it knows how to produce the right antibodies as soon as the virus starts reproducing. However, because viruses reproduce so quickly and so often, they can often change slightly. Therefore, one year's batch of vaccine might not be as effective against the same type of virus the next year. This is why new vaccines must be produced constantly to fight viral infections and prevent outbreaks.

Given the huge number of virus particles in the environment and the incredible number of different viruses, it is a miracle that we are not sick constantly. That shows how well the immune system works to fight off these tiny invaders.

Foot and Mouth Disease

Foot and mouth disease is caused by a virus that affects hoofed animals such as cows, pigs, sheep, and goats. Although the virus does not affect humans, it can be carried by humans. The virus causes blisters on the mouths and feet of infected animals as well as fever, lameness, lack of appetite, shivering, and reduced milk production. The virus can be spread in several ways:

- Direct contact between infected and uninfected animals within a herd (it spreads quickly within a herd)
- Virus-containing aerosols can travel several miles with the prevailing winds
- By traveling on the soles of shoes or the tires of vehicles

The virus can survive in frozen conditions (such as a meat freezer), but can be killed by heat, dryness, and disinfectants. Although the infected animals can recover from the disease within 2 to 3 weeks, officials agree that the best way to contain the disease is to destroy the infected animals. At present, there is no vaccine to prevent foot and mouth disease because the virus changes (mutates) rapidly.

How CAFFEINE Works

Around 90% of people in the United States consume caffeine every single day. More than half of all U.S. adults consume more than 300 mg of caffeine every day, making it by far the most popular drug in the United States. Caffeine is in foods and beverages such as coffee, tea, cola, and chocolate.

HSW Web Links

www.howstuffworks.com

How Chocolate Works
How Drip Coffee Makers Work
How Nicotine Works
How Performance-Enhancing Drugs Work
How Sleep Works

The chemical name for caffeine is trimethylxanthine ($C_8H_{10}N_4O_2$). In pure form, caffeine is a white, crystalline powder that tastes very bitter.

Doctors use caffeine medically as a cardiac stimulant and as a mild diuretic. People use caffeine recreationally to provide a boost of energy or a feeling of heightened alertness. Many people use it to help stay awake.

Caffeine is an addictive drug. Among its many actions, it operates using the same mechanisms that amphetamines, cocaine, and heroin use to stimulate the brain. Caffeine's effects are milder than those of amphetamines, cocaine, and heroin, but caffeine manipulates the same brain channels as those other drugs, and that is one of the things that gives caffeine its addictive qualities. If you feel like you cannot function without caffeine and must consume it every day, then you are addicted to caffeine.

Caffeine in the Diet

Caffeine occurs naturally in many plants, including coffee beans, tea leaves, and cacao beans. It is therefore found in all sorts of food products. It is added artificially to many other foods. Here are the most common sources of caffeine for people in the United States:

- **Typical drip-brewed coffee**—100 mg per 6-ounce cup
- **Typical brewed tea**—70 mg per 6-ounce cup
- **Typical cola**—50 mg per 12-ounce can
- **Extra-caffeine cola (such as Jolt)**—70 mg per 12-ounce can
- **Typical milk chocolate**—6 mg per ounce

Over-the-counter products:

- **Anacin**—32 mg per tablet
- **No-Doz**—100 mg per tablet
- **Vivarin and Dexatrim**—200 mg per tablet

Caffeine in Your System

Why do so many people consume so much caffeine? Why does caffeine wake you up? By understanding the drug's actions inside the body, you can see why people use it so much.

Adenosine is an important chemical in the brain—it is the chemical that makes you

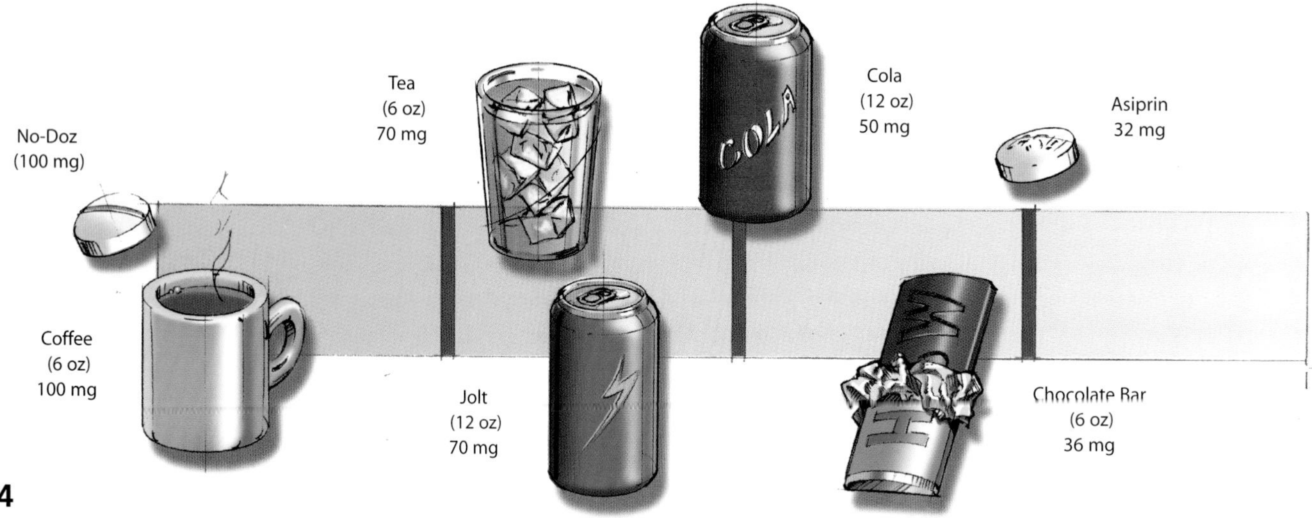

sleepy. Adenosine binds to adenosine receptors on nerve cells in the brain. The binding of adenosine causes you to get drowsy by slowing down nerve cell activity. In the brain, adenosine binding also causes blood vessels to dilate, probably to let more oxygen in during sleep.

To a nerve cell, caffeine looks just like adenosine. Caffeine therefore binds to the adenosine receptors. However, it doesn't slow down the cell's activity like adenosine would. The cell cannot detect adenosine anymore because caffeine is taking up all the receptors adenosine binds to. Instead of slowing down because of the adenosine attached to them, the nerve cells speed up.

You can see that caffeine also causes the brain's blood vessels to constrict, because it blocks adenosine's ability to open them up. This is why some headache medicines, such as Anacin, contain caffeine—if you have a vascular headache, the caffeine will close down the blood vessels and relieve it.

By blocking adenosine, caffeine increases neuron firing in the brain. The pituitary gland sees all the activity and thinks some sort of emergency must be occurring, so it releases hormones that tell the adrenal glands to produce adrenaline (epinephrine). Adrenaline is the "fight or flight" hormone, and it makes your body do a number of things:

- Your pupils dilate.
- Your breathing tubes open up (this is why people suffering from severe asthma attacks are sometimes injected with epinephrine).
- Your heart beats faster.
- Blood vessels on the surface constrict to slow blood flow from cuts and also to increase blood flow to muscles. Blood pressure rises.
- Blood flow to the stomach slows.
- The liver releases sugar into the bloodstream for extra energy.
- Muscles tighten up, ready for action.

This explains why, after consuming a big cup of coffee, your hands get cold, your muscles tense up, you feel excited, and you can feel your heartbeat increasing.

Caffeine also increases dopamine levels in the same way that amphetamines, heroin, and cocaine do: They slow down the rate of dopamine reuptake. Dopamine is a neurotransmitter that, in certain parts of the brain, activates the pleasure center. Obviously, caffeine's effect is not nearly as intense as heroin's, but it is the same mechanism. It is suspected that the dopamine connection contributes to caffeine addiction.

You can see why your body might like caffeine in the short term, especially if you are low on sleep and need to remain active. Caffeine blocks adenosine reception, so you feel alert. It injects adrenaline into your system to give you a boost. And it manipulates dopamine production to make you feel good.

The problem with caffeine is the longer-term effects, which tend to spiral. For example, when the adrenaline wears off, you face fatigue and depression. So what do you do? You take more caffeine to get the adrenaline going again. As you might imagine, having your body in a state of emergency all day long isn't very healthy, and it also makes you jumpy and irritable.

The most important long-term problem with caffeine is the effect it has on sleep. Adenosine reception is important to sleep—especially to deep sleep. The half-life of caffeine in your body is about 6 hours. That means that if you consume a big cup of coffee with 200 mg of caffeine in it at 3 p.m., then by 9 p.m., about 100 mg of that caffeine is still in your system. You may be able to fall asleep, but your body will probably miss out on the benefits of deep sleep. That deficit adds up fast. The next day you feel worse, so you need caffeine as soon as you get out of bed, and the cycle continues day after day.

When you get in the cycle of using caffeine, you have to keep taking the drug. Even worse, if you try to stop taking caffeine, you get very tired and depressed and you get a terrible, splitting headache as blood vessels in the brain dilate. These negative effects force you to run back to caffeine even if you want to stop.

Did You Know?

Have you ever heard that chocolate is bad for dogs? Guess what, it's true! It turns out that, for dogs, a chemical in chocolate called theobromine is the source of the problem. Theobromine is similar to caffeine. Reportedly, theobromine is toxic to a dog that ingests between 100 and 150 milligrams per kilogram of body weight. Different types of chocolate contain different amounts of theobromine, so it would take 20 ounces of milk chocolate to kill a 20-pound dog, but only two ounces of baker's chocolate or 6 ounces of semisweet chocolate. It is not that hard for a dog to get into something like an Easter basket full of chocolate eggs and bunnies and gobble up a pound or two of chocolate. If the dog is small, it could be deadly.

It turns out that chocolate poisoning is actually not as unusual as it sounds. For a human being, caffeine is toxic at levels of 150 milligrams per kilogram. That's exactly the same as dogs! Humans generally weigh a lot more than dogs, but small children can get into trouble with caffeine or chocolate if they consume too much. Infants are especially vulnerable because they don't eliminate caffeine from the bloodstream nearly as quickly as adults.

How **BREATH ALCOHOL TESTERS** Work

You hear and read about drivers having a blood alcohol content (BAC) that is over the legal limit. For example, a driver might have a BAC of 0.15, and the legal limit is 0.08. But what do those figures mean, and how do police officers measure a driver's BAC?

HSW Web Links

www.howstuffworks.com

How Alcohol Works
How Beer Works
How Blood Works
How Your Lungs Work

To measure a person's BAC, you could take a blood test, but that can be inconvenient in some cases, such as when a police officer has pulled over a driver he or she suspects of driving while intoxicated (DWI). In those instances, it's much easier to measure BAC by using a breath alcohol tester, such as a Breathalyzer.

When you drink alcohol, it is absorbed into your bloodstream. Alcohol is not digested or chemically changed in the bloodstream. As the alcohol-laden blood goes through the lungs, some of the alcohol moves across the membranes of the lung's air sacs (alveoli) into the air because alcohol evaporates easily from a solution. The concentration of the alcohol in the alveolar air is related to the level of alcohol in your blood. The ratio of breath to blood alcohol is 2,100 to 1, which means that 2,100 ml of alveolar air will contain the same amount of alcohol as 1 ml of blood. The legal standard for drunkenness in most states is 0.08, which means that there is 0.08 g of alcohol per 100 ml of blood.

The Chemistry of Alcohol

The alcohol found in alcoholic beverages is ethyl alcohol (ethanol). The molecular structure of ethanol looks like this:

$$\begin{array}{c} \quad\ \ H \\ H_3C-C-O-H \\ \quad\ \ H \end{array}$$

where C is carbon, H is hydrogen, O is oxygen, and the hyphen is the chemical bond between the atoms. The bonds of the three hydrogen atoms to the left carbon atom are not shown for clarity.

The OH (O-H) group on the molecule is what makes it an alcohol. So, there are four types of bonds in this molecule:

- carbon-carbon (C - C)
- carbon-hydrogen (C - H)
- carbon-oxygen (C - O)
- oxygen-hydrogen (O - H)

The chemical bonds between the atoms are shared pairs of electrons. Chemical bonds are much like springs. They can bend and stretch. These properties will be important in detecting ethanol in a sample by infrared (IR) spectroscopy.

Three major types of breath alcohol testers are in use today, and each is based on a different principle:

- A Breathalyzer uses a chemical reaction involving alcohol that produces a color change.
- An infrared (IR) device detects alcohol by IR spectroscopy.
- Fuel cell devices detect a chemical reaction of alcohol in a fuel cell.

Regardless of the type, each device has a mouthpiece or tube through which the suspect blows air into a sample chamber. What the rest of the device is like varies depending on the type.

An analyzer such as a Breathalyzer can be used as follows:

1) The person being tested exhales the alveolar air that contains the alcohol into the Breathalyzer.
2) The alcohol undergoes a chemical or an electrochemical reaction inside the Breathalyzer so that the device can determine the amount of alcohol in the alveolar air.
3) The Breathalyzer relates the amount of alveolar air to the BAC, and it displays the BAC level.

Using a breath alcohol tester, an officer can test a driver's breath on the spot to determine whether there is a reason to arrest the person.

Breathalyzers

A Breathalyzer device contains a system to sample the breath of the suspect, two glass vials containing the chemical reaction mixture, and a system of photocells connected to a meter to measure the color change during the chemical reaction. To measure alcohol, a suspect breathes into the device. The breath sample bubbles through a mixture of sulfuric acid, potassium dichromate, silver nitrate, and water. The sulfuric acid removes the alcohol from the air into a liquid solution, and the alcohol reacts with potassium dichromate to produce chromium sulfate, potassium sulfate, acetic acid, and water. The silver nitrate speeds up the reaction, and the sulfuric acid provides the acidic condition needed for this reaction.

During this reaction, the reddish-orange dichromate ion changes color and becomes a green chromium ion. The degree of the color change is directly related to the level of alcohol in the expired air. To determine the amount of alcohol, the Breathalyzer compares the reacted mixture to a vial of unreacted mixture with the photocell system, which produces an electric current that is related to the color difference between the two vials. The electric current causes the needle in the meter to move from its resting place. Then the operator rotates a knob to bring the needle back to the resting place and reads the level of alcohol from the knob.

IR Detectors

IR devices identify molecules based on the way they absorb IR light. Molecules are constantly vibrating. These vibrations change when they absorb IR light—that is, various bonds bend and stretch). Each type of bond within a molecule absorbs IR light at different wavelengths. So, to identify ethanol in a sample, you have to look at the wavelengths of the bonds in ethanol (C-O, O-H, C-H, C-C) and measure the absorption of IR light. The wavelength helps to identify it as ethanol, and the amount of IR light absorption tells you how much ethanol is there.

In the Intoxilyzer IR detector, a lamp generates a broadband IR beam that passes through the sample chamber and focuses on a spinning filter wheel. The filter wheel contains narrowband filters specific for the wavelengths of the bonds in ethanol. The photocell detects the light passing through each filter and converts it to an electrical pulse. The electrical pulse is relayed to the microprocessor, which interprets the pulses and calculates the BAC.

Fuel Cell Detectors

A fuel cell detector has two platinum electrodes with a porous acid electrolyte material sandwiched between them. As the exhaled air from the suspect flows past one side of the fuel cell, the platinum oxidizes any alcohol in the air, producing acetic acid, protons, and electrons. The electrons flow through a wire from the platinum electrode. The wire is connected to an electrical current meter and to the platinum electrode on the other side. The protons move through the lower portion of the fuel cell and combine with oxygen and the electrons on the other side to form water. The more alcohol that becomes oxidized, the greater the electrical current. A microprocessor measures the electrical current and calculates the BAC.

Operators of any breath alcohol tester must be trained to use and calibrate the devices, especially if the results are to be used as evidence in DWI trials. Law enforcement officers can carry portable breath testing devices that use the same principle as full-size devices. Court cases can turn on the perceived accuracy of a breath test, however, so prosecutors depend on the results from full-size devices.

The Oxidation of Alcohol

If you strip off hydrogens from the right carbon of ethanol in the presence of oxygen, you will get acetic acid, the main component in vinegar. The molecular structure of acetic acid looks like this:

$$\begin{array}{c} \quad\;\; O \\ \quad\;\; \| \\ H_3C-C-O-H \end{array}$$

Where C is carbon, H is hydrogen, O is oxygen, the hyphen is a single chemical bond between the atoms, and the || symbol is a double bond between the atoms. The bonds of the three hydrogen atoms to the left carbon atom are not shown for clarity. When ethanol is oxidized to acetic acid, two protons and two electrons are also produced.

How TATTOOS Work

It's impossible to walk through a mall for more than a couple minutes without spotting people of all ages with tattoos of all sizes and designs—from eagles to Celtic crosses to Betty Boop—on virtually every part of the body, including ankles, shoulders, arms, legs, chests, bellies, and bottoms.

HSW Web Links

www.howstuffworks.com

How Anesthesia Works
How Hair Replacement Works
How Tattoo Removal Works
How Your Immune System Works

A tattoo is a piece of body art that is created by injecting ink into the skin. The ink intermingles with cells in the dermis and shows through the epidermis, which is the layer of skin that you see and the skin that is constantly being replaced. The cells of the dermis are remarkably stable, so the tattoo's ink will last, with minor fading and dispersion, for your entire life.

Tattoo Application

The tattoo artist uses a hand-held tool that contains a fine needle. The tool moves the needle up and down at a rate of several hundred vibrations per minute and penetrates the skin by about 1 mm. Ink on the needle is pushed into the dermis in the process. Much of the tattoo application process focuses on safety, since any puncture wound—and that's what a tattoo machine is doing to your skin—holds the potential for infection and disease transfer.

Two main types of materials are needed to create a tattoo:

- **Reusable materials**—Reusable materials are things like the needle bar, vibrating mechanism, and tube, which must be completely clean.
- **Single-use items**—Most tattoo materials are used only once to eliminate the possibility of contamination. These items include tattoo application items (such as inks, ink cups, and needles), razors used to shave the skin (since hair clogs up the tubes and hinders application), latex gloves worn by the tattoo artist, and plastic barriers (or bags) used on spray bottles, tattoo machines, and clip cords to prevent cross contamination.

The tattoo artist washes his or her hands and inspects them for cuts or abrasions, disinfects the work area with a virucidal agent, dons fresh gloves, and follows this general procedure:

1) The artist shaves and disinfects the area to be tattooed, using an antiseptic soap.
2) The artist draws an outline (or uses a predrawn stencil) on the area to be tattooed.
3) The artist tattoos the outline of the design with a thin line to make it permanent. The artist starts at the bottom of the design, using one single-tipped needle and works up, so the stencil won't be lost when the artist cleans a permanent line. The artist uses thin black ink for the outline because thin ink can be easily wiped away from the skin without smearing. When the outline is complete, the artist cleans the area thoroughly with antiseptic soap and water.
4) The artist thickens the outline and adds shading by using a combination of needles.
5) The artist cleans the tattoo area again.
6) The artist applies color, overlapping each line of color to ensure solid, even hues with no "holidays" (that is, uneven areas where color has either lifted out during healing or where the tattoo artist simply missed a section of skin).
7) The artist sprays the tattoo with antiseptic, cleans it, and applies pressure, using a disposable towel to remove any blood and plasma excreted during the tattooing process.

The amount of time the process takes depends on the individual style of the artist and the intricacy of the tattoo design. A

Sterilization

The only acceptable method of sterilization for killing living microorganisms is to use an autoclave, a heat/steam/pressure unit (used in hospitals) that achieves and maintains 250°F (121°C) under 10 lb/in² for 30 minutes or up to 270°F (132°C) under 15 lb/in² for 15 minutes. Indicator strips on the packages change color when sterilization is complete.

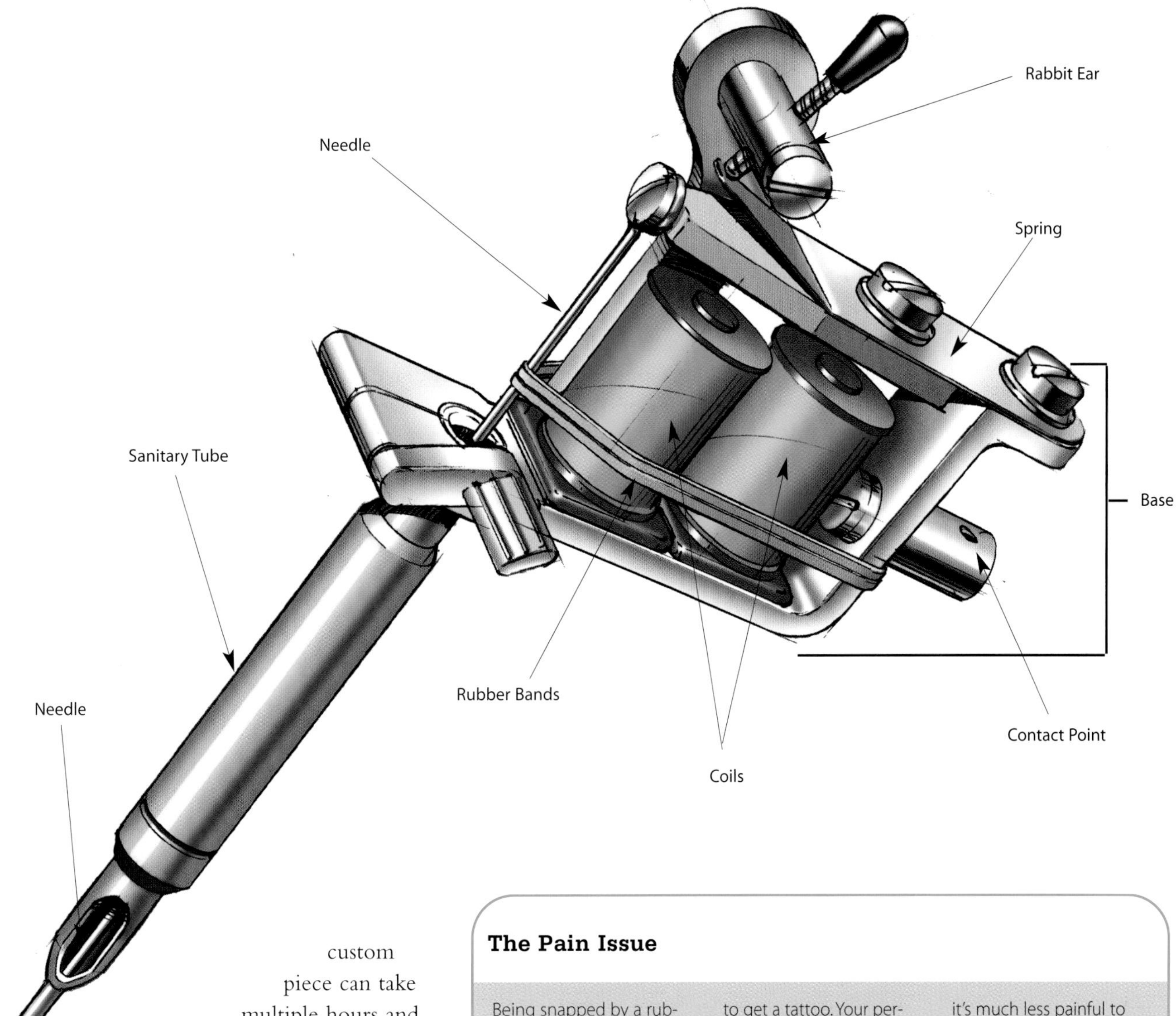

custom piece can take multiple hours and require a number of visits to the tattoo parlor.

Skin Healing

fter you leave the tattoo parlor, it takes several days for your skin to heal. During that time, you need to avoid disturbing the ink and prevent infection. You need to do the following:

- Avoid rubbing or scratching the tattoo.
- Wash the tattoo with soap and water to keep it clean—pat it dry, and do not rub it.
- Avoid swimming, baths, and pounding showers.
- Avoid the sun.
- Call your doctor if you see signs of infection.

The Pain Issue

Being snapped by a rubber band, a slight tickling, a bee sting, a sunburn, being pinched, "pins and needles" like when your foot's asleep, numb, pinpricks, tingling, like a drill going into your skin, uncomfortable—all these phrases have been used to describe what it feels like to get a tattoo. Your personal tolerance for pain, the size and type of your tattoo, and the skill of the artist help determine the amount of pain involved in getting a tattoo. Pain also depends on the location of your tattoo. The lower back and ankle are popular places for tattoos, but it's much less painful to get one on your chest or upper arm. This is because skin right above your bones tends to be more sensitive to needles, and the extra body mass in areas such as the upper arm and chest cushions the bones.

As you can see, if you or someone you know is considering a tattoo, there is a lot to think about. The next time you see an interesting piece of body art on someone's arm or ankle, you will truly be able to appreciate what was involved in the process of putting it there.

How **SUNBURNS & SUN TANS** Work

There is something mysterious about the sun and skin. If you go out on a bright summer day and spend an hour in the sun, your skin might burn. However, if you are more careful about it, you can expose yourself to the sun gradually and get a tan that prevents sunburn. Or you can use sunscreen and avoid burning and tanning altogether.

HSW Web Links

www.howstuffworks.com

How Cells Work
How Corrective Lenses Work
How Cosmetic Dentistry Works
How Sleep Works
How the Sun Works

What is the difference between a tan and a burn? Why can you spread a little blob of lotion on yourself and be protected, but if you forget you are miserable? This all makes sense if you understand how your skin and the sun interact.

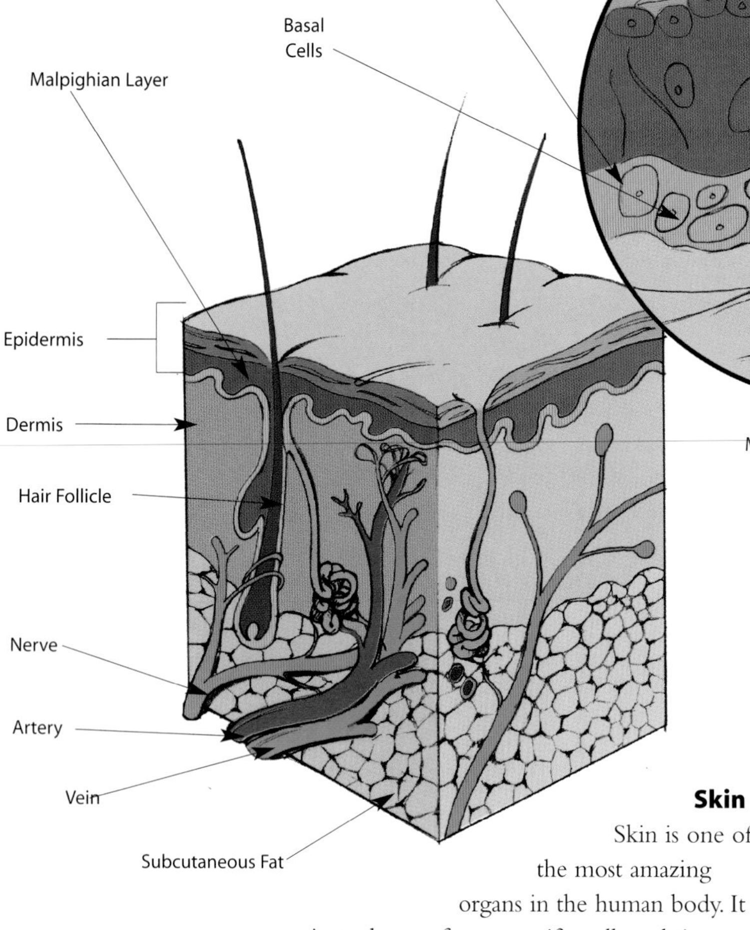

Skin

Skin is one of the most amazing organs in the human body. It is made up of very specific cells and tissues, and their collective purpose is to act as the boundary between you and the world. But skin does have to deal with the real world. Skin is loaded with sensors, and it has a very tough, layered design so that it can handle realities of the environment such as abrasion and sunlight.

If you take a look at a cross-section of typical skin (like the skin on your arm or leg), you find that it is made up of two main layers: the epidermis on the outside and the dermis on the inside. The epidermis is the barrier, and the dermis is the layer that contains all the "equipment"—things like nerve endings, sweat glands, sebaceous glands, and hair follicles (and a little erector muscle on each follicle that makes your hair stand on end).

In the subcutaneous layer (you may have heard of subcutaneous fat—this is where it lives), you find lots of blood vessels. These vessels branch into the dermis to supply the equipment there with blood. They also fan out into the capillary bed of the dermis, which satisfies the nutritional needs of the cells in the dermis. Capillaries also help the skin perform an important cooling function in humans.

The epidermis is your interface to the world, and it is quite interesting. It has two main layers: the inner, which is living, and the outer, which is dead. The dead skin cells of the outer layer are what you can actually see. They are constantly flaking off and being replaced by new cells that are pushed outward.

The living, inner layer of the epidermis is called the malpighian layer. This layer creates the dead cells that you can see. It is where the sun affects the skin during tanning.

One important part of the malpighian layer is the basal layer, which is made up of the basal cells. This is where basal cell carcinoma (cancer) starts.

Living among the basal cells is another type of cell, called a melanocyte. Melanocytes produce melanin, which is a pigment that creates the brown of a suntan. Not only do melanocytes produce a tan, but they are also responsible for the form of cancer called melanoma. Melanoma is caused by ultraviolet (UV) radiation damage to melanocytes. Repeated exposure to UV radiation can cause cancerous mutations.

Sunlight

Sunlight arrives on earth in three forms: infrared light (heat), visible light, and UV light.

UV light is classified into three categories:

- UVA , also known as black light, which causes tanning
- UVB, which causes damage in the form of sunburn
- UVC, which is filtered out by the atmosphere and never reaches earth

Ninety-nine percent of the sun's UV radiation at sea level is UVA. UVB causes most of the problems related to sun exposure—things like aging, wrinkles, and cancer—although research is increasingly implicating UVA as well.

Sun Tan

When you get a tan, your melanocytes are producing melanin pigment in reaction to the UV light in sunlight. In other words, UV light stimulates melanin production. This pigment has the effect of absorbing the UV radiation in sunlight, so it protects skin cells from UV damage. Melanin production takes a fair amount of time—that is why most people cannot get a tan in one day.

The previous paragraph applies to Caucasians. In people of a variety of other races, melanin production is continuous, so the skin is always pigmented to some degree. In these races, the incidence of skin cancer is much lower because cells are constantly protected from UV radiation by melanin.

Sunburn

When you get sunburn, you're getting cellular damage from UV radiation. The UV light kills skin cells near the surface. The body's response to cellular damage is to increase blood flow to the capillary bed of the dermis in order to bring in cells and building materials that can repair the damage. The extra blood in the capillaries causes the redness—if you press on sunburned skin it will turn white, and then return to red as the capillaries refill. The stinging pain is a natural reaction to cellular damage, too—the skin is wired so that it can warn your brain (with pain) whenever the skin is damaged. These nerve endings detect the damage caused by UV rays and send pain signals.

Sunscreen

Sunscreens contain chemicals that absorb UV radiation and turn it into heat instead of letting it reach skin cells. By applying a sunscreen, you keep UV rays from ever reaching or damaging your skin cells. With no UV radiation to trigger them, the melanocytes do not produce their pigment, so you don't get a tan either.

There is something about this whole discussion that is fascinating. On your body is an organ—the skin—and it responds in all of these interesting ways to sunlight. If you were to hear that your heart or lungs were suffering from cellular damage, you would probably find that pretty alarming. When you get sunburn, that's exactly what's happening to the largest organ of your body!

Sunless Tanning

There are several different kinds of sunless-tanning products available today. People have been able to pour on a tan since 1960, when Coppertone came out with the first sunless-tanning product—QT or Quick Tanning Lotion. If you are old enough to remember this, then you are probably thinking of the incredibly orange hue this lotion produced. Since then, there have been several advancements made on the sunless-tanning front. You can smooth, swipe, or spray on a light bronze glow or a deep, dark tan. Typically products take 45 minutes to one hour to start taking effect, and once you factor in drying time, you could be looking at about 3 hours spent achieving that sun-free tan.

According to the American Academy of Dermatology, the most effective products available are sunless- or self-tanning lotions that contain dihydroxyacetone (DHA) as the active ingredient. DHA is a colorless sugar that interacts with the dead cells located in the stratum corneum of the epidermis. As the sugar interacts with the dead skin cells, a color change occurs.

Every day, millions of dead skin cells are sloughed off or worn away from the surface of your skin. In fact, every 35 to 45 days, you have an entirely new epidermis. This is why tans from sunless- or self-tanning lotions will gradually fade—as the dead cells are worn away, so is your tan. For this reason, most of these products suggest that you reapply the sunless-or self-tanner about every 3 days to maintain your "tan."

How **SUNGLASSES** Work

If it's a bright, clear day outside, you may instinctively reach for your sunglasses when you head out the door. And you probably do it without much thought. The one time when you do think about sunglasses, however, is the day you go to select a new pair.

HSW Web Links

www.howstuffworks.com

How Corrective Lenses Work
How Light Works
How Refractive Vision Problems Work
How Sunburns and Sun Tans Work

A pair of sunglasses seems so simple—it's just two pieces of tinted glass or plastic in some sort of plastic or metal frame. But it turns out that there are many different things that you can do to those two pieces of glass or plastic, and these things can have big effect on you when you use the lenses.

A good pair of sunglasses should do four things for you:

- Provide protection from ultraviolet (UV) rays in sunlight
- Provide protection from intense light
- Provide protection from glare
- Eliminate specific frequencies of light

When you buy a random pair of cheap sunglasses, you often give up all these benefits and can even make some of these things worse. Buying the right pair of good sunglasses for the conditions in which you use them gives you maximum protection and performance.

The most important feature of a pair of sunglasses is the lenses, and the quality of the lenses comes from the materials used to make them. Optical-quality polycarbonate and glass lenses are free of distortions. If the lenses are not optical quality, they distort your vision—they bend the rays of light that pass through the lens so that things look wavy, blotchy, fuzzy, or out of place. Dis-tortion is extremely common in cheap sunglasses.

A variety of technologies are used in sunglasses lenses to eliminate the problems that sunlight can cause. The following sections discuss the different technologies currently in use:

- Tinting
- Polarization
- Photochromic lenses
- Mirroring
- Scratch-resistant coatings
- Antireflective (AR) coatings
- UV coatings

Cool Facts

Here's an easy way to tell if the lenses in a pair of sunglasses are of good quality. Find a surface with repeating lines, like the tiles on a floor. Hold the sunglasses a short distance away from your face and cover one eye. Look through one of the lenses at the lines while moving the sunglasses slowly from left to right then up and down. The lines should stay straight as you look at them. If they wiggle or waver in any way, then the lenses are not optical quality and will distort your vision. Distortion is extremely common in cheap sunglasses.

Tinting

The color of the tint in the lens determines the parts of the light spectrum that the lens absorbs. Manufacturers use different colors to produce specific results. Many manufacturers use a process called constant density to tint the lenses. It is the oldest method of creating sunglasses, and it produces a glass or polycarbonate mixture that has a uniform color throughout. Using this method, the tint is built right into the lenses when they are created.

Tinting can also be accomplished by applying a coat of light-absorbing molecules to the surface of clear polycarbonate. The most common method for tinting polycarbonate lenses is to immerse the lenses in a special liquid that contains the tinting material. The tint is slowly absorbed into the plastic. To make a darker tint, the lenses are simply left in the liquid longer.

Polarization

Lightwaves from the sun, or even from an artificial light source such as a lamp, vibrate and radiate outward in all directions. When its vibrations are aligned into one or more planes of direction, light is said to be polarized. For example, when light reflects off water, it becomes horizontally polarized. Polarized sunglasses take advantage of polarization to eliminate the light. Almost all the painfully bright reflections from water can be eliminated with polarized filters.

Polarized filters are most commonly made of a chemical film applied to a transparent plastic or glass surface. The chemical compound used is typically composed of molecules that naturally align in parallel to one another. When applied uniformly to the lens, the molecules create a microscopic filter that blocks any light that does not match their alignment.

The polarized lenses in sunglasses are fixed at an angle that allows only vertically polarized light to enter. You can see this for yourself by putting on a pair of polarized sunglasses and looking at a horizontal reflective surface, like a lake or the hood of a car. Slowly tilt your head to the right or left. You will notice that the glare off the surface brightens as you adjust the angle of your view.

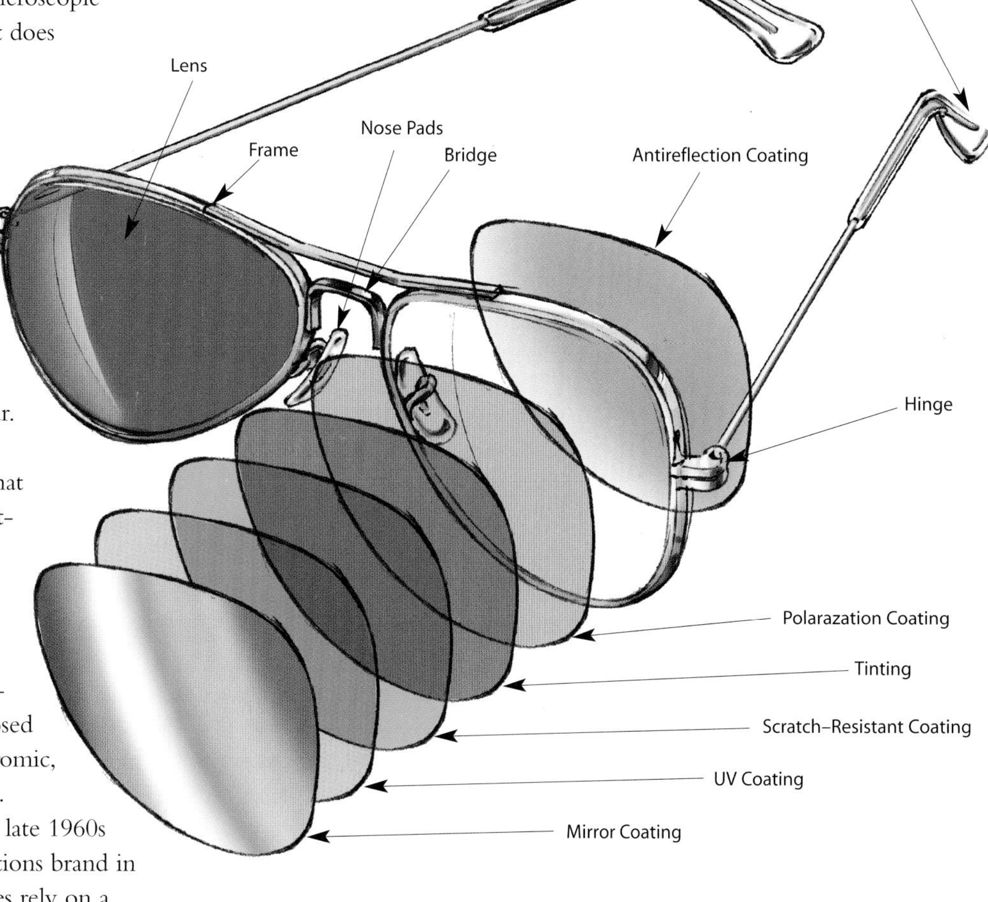

Photochromic Lenses

Sunglasses or prescription eyeglasses that darken when exposed to the sun are called photochromic, or sometimes photochromatic. Developed by Corning in the late 1960s and popularized by the Transitions brand in the 1990s, photochromic lenses rely on a specific chemical reaction to UV radiation.

Photochromic lenses have millions of molecules of substances, such as silver chloride or silver halide, embedded in them. The molecules are transparent to visible light in the absence of UV light, which is the normal makeup of artificial lighting. But when they are exposed to UV rays in sunlight, the molecules undergo a chemical change. The new molecular structure absorbs portions of the visible light, causing the lenses to darken. The number of molecules that change varies with the intensity of the UV rays.

One problem with photochromic sunglasses is that they don't work in a car. The car's windows block all the UV radiation so that the glasses do not activate.

Mirroring

The lenses in mirrored sunglasses have a reflective coating applied in a very thin, sparse layer—so thin that it's called a half-silvered surface. The name half-silvered comes from the fact that the reflective molecules coat the glass so sparsely that only about half the molecules needed to make the glass an opaque mirror are applied. At the molecular level, there are reflective molecules speckled all over the glass in an even film, but only half of the glass is covered. The half-silvered surface reflects about half the light that strikes its surface, and it lets the other half go straight through.

Often the mirror coating is applied as a gradient that gradually changes shades from top to bottom. This provides additional protection from light that comes from above while allowing more light to come in from below or straight ahead. That means that if you are driving, the sun's rays are blocked, but you can see the dashboard. Sometimes the coating is bigradient, shading from mirrored at the top and bottom to clear in the middle.

The key problem with reflective sunglasses is that the coating can easily be scratched. Apparently, sunglasses manufacturers have not been able to successfully apply a scratch-resistant layer on top of the reflective coating—only below it, on the lens itself.

Scratch-Resistant Coatings

Glass is naturally scratch resistant, but most plastics are not. To compensate, manufacturers have developed a variety of ways to apply optically clear hard films to a lens. Films are made of materials such as diamond-like carbon and polycrystalline diamond. Through a process of ionization, a thin but extremely durable film is created on the surface of a lens.

AR Coatings

A common problem with sunglasses is backglare. Light hits the back of the lenses and bounces into the eyes. The purpose of an AR coating is to reduce these reflections off the lenses. Similar to scratch-resistant coating, AR coating is made of a very hard, thin film that is layered on the lens. It is made of material that has an index of refraction that is somewhere between that of air and that of glass. This causes the intensity of the light reflected from the inner surface and the light reflected from the outer surface of the film to be nearly equal but phase-shifted, so they cancel each other out.

UV Coatings

Several very serious eye problems can be linked to UV light. UV light is often separated into two categories—UVA and UVB—based on the frequency and wavelength of the light. As a natural protection mechanism, the cornea of your eye absorbs all the UVB and most of the UVA light. But over time, this absorption can lead to cataracts. And the small amount of UVA that gets past your cornea can eventually lead to macular degeneration, which is the leading cause of blindness in people over age 65.

A good UV coating on your sunglasses can eliminate UV radiation, and you should check to make sure that your sunglasses filter out 100% of both types of UV rays. There should be a statement on the label telling you how much UV protection the sunglasses have. You want 100% protection against both UVA and UVB radiation.

Choosing the Perfect Pair

Here are some of the most important features to compare when you buy a pair of sunglasses:

- **Lens material**—There are several types of lens material. CR-39 is a plastic made from hard resin that meets optical quality standards. A synthetic plastic material, polycarbonate lenses tend to be lighter and are more impact-resistant. Glass lenses are heavier but are much more resistant to scratches.
- **Lens quality**—Optical-quality polycarbonate and glass lenses are free of distortions, such as blemishes or waves, and have evenly distributed color across each lens.
- **Lens darkness**—For most purposes, like going to the beach or driving, look for a tint that absorbs or blocks 70% to 90% of light. Tints that offer less than 60% blockage are mainly good for fashion since they offer only mild protection.
- **Frame and lens design**—Normal frames similar to prescription eyeglass frames filter the light coming through the lenses but offer no protection from ambient light, direct light, and glare from other angles. Wrap-around frames, larger lenses, and special sidereal attachments can keep this extra light from your eyes.
- **Frame material**—The material used to make the frames is often a huge factor in cost and durability. Most inexpensive sunglasses use simple plastic or wire frames, while many name brand sunglasses use high-strength, light-weight composite or metal frames. Also, the better sunglasses usually have features like tension springs connecting the arms to the face instead of just using screws.

Why Should I Wear Sunglasses?

In addition to preventing cataracts and macular degeneration, there are a number of other reasons to wear sunglasses. Ultraviolet (UV) light may be responsible for nonmalignant tumors of the eye, called pterygium. Temporary, but very painful damage of the cornea can result from exposure to UV light from tanning beds, and the reflection off snow, sand or pavement.

It has been reported that every year something like 35,000 Americans suffer eye injuries so serious that a trip to the emergency room is necessary. Some of these injuries, many of which occur outdoors, could have been prevented or, at least lessened, had the person been wearing sunglasses or other protective eye gear. Although UV-absorbing contact lenses are available from many companies, these are not a replacement for sunglasses. These contact lenses do not provide enough protection from UV radiation; the delicate skin around the eye and the eyelids need to be protected from the sun as well.

chapter eleven

JUST FOR FUN

How **LIQUID MOTION LAMPS** Work

Liquid motion lamps are one of those amazing inventions that have basically no practical application. But they are so cool that millions of people own them simply to gaze at them. Something about things that float is fascinating—maybe that explains the popularity of aquariums, too.

HSW Web Links

www.howstuffworks.com

How Helium Balloons Work
How Submarines Work
How Thermometers Work
How Water Heaters Work

Liquid motion lamps are actually fairly simple devices that are based on very basic scientific principles.

Parts of a Liquid Motion Lamp

A liquid motion lamp consists of only a few simple components:

- A compound that makes up the floating blobs
- A compound that the blobs float in
- A lamp that illuminates the display and provides the heat that is necessary to move the blobs

To create the floating blobs, the two compounds in a motion lamp must be immiscible, or mutually insoluble—which simply means that Liquid A does not dissolve in Liquid B—the two don't mix and you see two separate liquids, one floating on top of the other. The classic example of immiscible compounds is oil and water. If you fill a jar with common vegetable oil and water, you'll get a water layer with a layer of oil floating above it. This combination of water and oil in a jar looks similar to a commercial motion lamp with its light turned off.

The Right Combinations

The cool thing about motion lamps, of course, is that they produce distinct amorphous blobs that rise and fall inside the lamp on their own. To produce this effect, you need to choose the two insoluble compounds very carefully. In an oil-and-water jar, the water ends up on the bottom because it has a much higher density than oil.

To get blobs that float around, you need two substances that are very similar in density. Then you need to be able to change the density of one of the compounds so that sometimes it is lighter than the other compound (so that it floats to the top) and sometimes it is heavier (so that it sinks back to the bottom). By having compounds that have very close densities, you get a situation where the blobs can switch between rising and sinking.

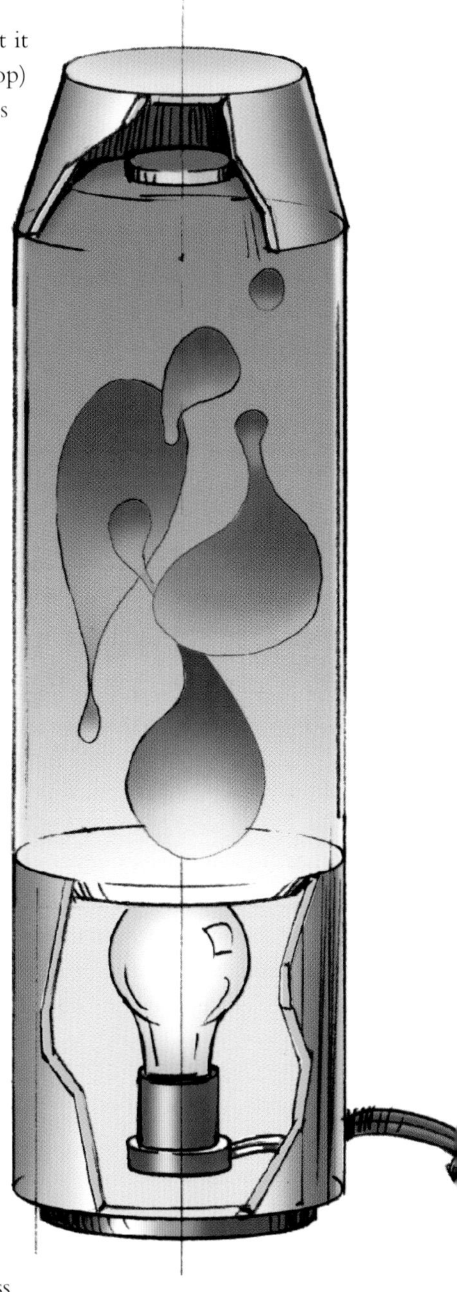

The most common way to change the density of compounds is to change their temperature. Heating a compound activates the molecules so that they spread apart, making the compound less dense.

If you look inside a motion lamp when it's turned off, you'll see a solid waxy compound on the bottom of the globe. This solid compound is only slightly denser than the surrounding liquid compound. When

Did You Know?

The invention of the liquid motion lamp is generally credited to a man named Edward Craven Walker, although there is a certain amount of controversy surrounding its actual origins. One widespread account is that Walker came up with the basic design in the 1950s while developing a complicated egg timer in England. Another version of the story is that Walker got the idea from a simpler liquid motion lamp he saw in a pub. In any case, Walker was definitely the man who molded this idea into its current form and started it on its way to becoming a culture icon.

you turn on the light at the base of the globe, here's what happens:

1) The solid quickly turns into a thick liquid and expands, giving it a lower density than the surrounding liquid.
2) This warm blob is now slightly less dense than the surrounding liquid, so it rises to the top of the globe.
3) Because it is farther away from the heat source, the blob cools slightly, becoming denser than the surrounding liquid (it does not cool down enough to change back into a solid, however).
4) The blob sinks to the bottom of the globe, where it heats up enough to rise again.

This all happens very slowly because heat absorption and dissipation are fairly slow processes, and the density changes we are discussing here are very slight. The blobs always have a density very close to that of the liquid, so they rise and sink slowly.

This is a pretty simple idea, but it's actually fairly complicated to balance all the elements—the compounds, the heat source, and the size of the globe—so that the blobs are constantly moving around. In fact, the major motion lamp manufacturers keep their ingredients top secret.

In recent years, all sorts of new twists on the motion lamp have surfaced in novelty stores. They all make use of the calming effects of liquids in motion, but the technologies being used to move the liquids are all different. To provide the motion, manufacturers now use heat, pumps, bubbles, and motorized plates or cylinders.

Homemade Liquid Motion Lamps

A number of web sites offer instructions on how to create your own liquid motion lamp. You won't find any plans with the exact same material found in a commercial lamp, however, because this information is still top secret.

Motion lamp recipes are so popular because coming up with exactly the right compound combination is an exciting chemistry puzzle. It's a real trick finding two compounds that will form the free-floating blobs you see in commercial lamps.

One of the simpler plans out there uses mineral oil as the blobs and a combination of 70% and 90% rubbing alcohol as the surrounding liquid. The trick with these ingredients is to balance the two different grades of rubbing alcohol until you get a good balance between the alcohol's density and the density of the oil. You can color the blobs by adding insoluble dye, such as the sort used in permanent markers. For best results, you should find a fairly tall glass container for the globe. For a safe and effective lamp, don't use anything hotter than a 40-watt bulb to heat the globe.

Other homemade motion lamp designs use more complicated combinations of materials to create better displays. In most cases, acquiring all the necessary ingredients for building a lamp yourself is actually more expensive than simply buying a commercial model. It is an interesting project, however, and it's an excellent way to fully explore how liquid motion lamps work.

If you are interested in making your own lamp, keep in mind that dealing with these sorts of materials and constructing any device with a heating unit is potentially hazardous. Don't undertake such a project unless you have a good knowledge of the materials involved or you are assisted by someone who does. Make sure you aren't handling any dangerous chemicals and, as with any chemistry experiment, wear eye protection and an apron. Most of the sites that feature motion lamp plans also include disclaimers disavowing any responsibility because they are fully aware of the potential dangers involved.

The effects that liquid motion lamps create range from the bizarre to the sublime. For example, you might see a lamp in which iridescent fluids create effects similar to the look of Jupiter's atmosphere, or large sparkles move in a think, opaque fluid, propelled by either heat or a pump. Or you might find a lamp in which bubbles float to the surface in a tube or through a set of obstacles. Basically, any liquid effect that looks cool can be harnessed to create a liquid motion lamp!

And Another Thing...

Walker worked on his motion lamp, which he called the Astro Light, for almost a decade before finally launching it in 1963. His U.K.-based company, Cresworth, achieved some success with the device, but the lamp design really took off when the U.S. company Lava Manufacturing Corp. formed in 1965. Two Chicago entrepreneurs, Adolph Wertheimer and Hy Spector, who discovered Walker's Astro Light at a trade show in Germany, founded the company. They acquired the U.S. patent rights and began producing a line of the motion lamps that soon caught on as a must-have decoration for the counterculture crowd. This company, now called Lava World International, still produces most all of the commercial models sold in U.S. stores. Their product is called a Lava Lite, not a Lava Lamp, but both terms are protected by copyright. Like Xerox and Kleenex, "Lava lamp" is commonly used as a generic term, but copyright law does not recognize it as such. Lava World International closely monitors how publishers print the word "lava" in connection with motion lamps.

How a **MAGNA DOODLE** Works

In 1974 four engineers from Pilot Pen Corporation invented a "dustless chalkboard." Their invention has become one of the most popular drawing toys ever created, with more than 40 million sold. This invention is the Magna Doodle, and it is a great example of how a very simple idea lies at the heart of a product that seems complex.

HSW Web Links

www.howstuffworks.com

How Ballpoint Pens Work
How Compasses Work
How Electromagnets Work
How Maglev Trains Will Work
How Magnetic Resonance Imaging Works

Etch-a-Sketch

An Etch-a-Sketch is actually a pretty amazing device! It is essentially a manually operated plotter with a built-in erasing system. Anyone who has used an Etch-a-Sketch before will recognize the familiar red plastic toy with the glass drawing-window. If you remove the outer case and the glass, what you find inside is a stylus mounted on a pair of orthogonal rails. These rails move when you turn the knobs. Also located inside is a mixture of extremely fine aluminum powder and beads. The beads help the powder to flow evenly. When you turn the Etch-a-Sketch upside down and shake, it coats the inside face of the glass with aluminum powder. The key to the Etch-a-Sketch mechanism is a tightly strung and very thin steel wire that connects the knobs to the horizontal and vertical bars through a pretty complicated pulley system. The wires connect both ends of each bar to its respective knob. When you turn a knob it moves its bar, and this moves the stylus. The stylus scratches off the aluminum dust coating the glass to create a line on the screen.

If you have ever drawn something on a Magna Doodle, you know that it is a very interesting drawing system. As you draw, a very nice, dark black line forms within the white screen.

This is weird and amazing to see in action—look at it carefully the next time you are at a toy store. The drawn line is very bold and dark on a very white background. However, you really don't have to apply much pressure on the pen to create the line—a light touch draws the same line as pressing hard. The drawing action is very different from that of a crayon or pencil, where pressure controls the density of the line.

Parts of a Magna Doodle

To make a drawing system that works this way, you need four basic parts:

- There are two sheets of plastic, one on top of the other. The top one is translucent and it is the surface that the pen touches when drawing.
- Between these two sheets is a honeycomb lattice that acts as a separator. You can see the lattice when you look at the screen. The lattice keeps the two sheets spaced out at a uniform thickness and also divides the interior between the sheets into cells.
- In the cells between the sheets is an opaque white liquid, about the consistency of cream.
- Very fine black magnetic particles of iron oxide are mixed in with the liquid.

Magna Doodle Writing and Erasing

When you apply a small magnet—the blunt metal tip of a Magna Doodle pen—to the front screen, it draws the black magnetic particles through the liquid to the surface, and they become visible. They don't sink back into the liquid because of the thickness of the liquid and because the particles and liquid are perfectly matched to have the same density.

To erase the screen, the slider moves a long magnet across the back of the screen. This magnet drags the magnetic particles toward the back screen, and the white liquid covers them.

Any magnet applied to the front screen will work to create a dark area on the screen. Some Magna Doodles come with magnetic stamps that take advantage of this property.

The best part of a Magna Doodle, especially if you have small children, is the fact that its pen contains no ink—it is a magnet. This makes it impossible for the pen to write on anything but the Magna Doodle screen, and it has probably saved parents in the United Staes several million gallons of paint since the Magna Doodle was invented!

How a **MECHANICAL SEE 'N SAY** Works

Just about everyone in the United States has owned a See 'N Say at some point. These toys play a selection of sounds when you pull the string (or, on modern mechanical ones, pull a lever). For example, if you point the dial at the cow and pull the string, the toy says, "The cow says 'moo'." Mechanical See 'N Says use no batteries and are nearly indestructible—many See 'N Says are submerged in bathtubs, thrown down the steps, dropped onto concrete patios, and so on, and they work fine afterward!

A See 'N Say is basically a wind-up gramophone record player that is nearly indestructible. Thomas Edison would be completely familiar with everything inside a See 'N Say because the mechanism is nearly identical to a mechanism he developed over a century ago.

Phonograph to Gramaphone to See 'N Say

Edison created the first device for recording and playing back sounds in 1877. He used a simple mechanism to store an analog sound wave. In Edison's original phonograph, a diaphragm directly controlled a needle, and the needle scratched an analog signal onto a tin cylinder. You spoke into Edison's device while rotating the cylinder, and the needle recorded what you said onto the tin.

This is an incredibly simple way to store sound. As the diaphragm vibrates, so does the needle, and those vibrations impress themselves onto the tin. To play the sound back, the needle moves over the groove scratched during recording. During playback, the vibrations pressed into the tin cause the needle to vibrate, which in turn causes the diaphragm to vibrate and play the sound.

Emil Berliner improved this system in 1887 by producing the gramophone, another purely mechanical device with a needle and diaphragm. The gramophone's major improvement over Edison's device was that it used flat records with spiral grooves, and that made mass production of the records easy. The See 'N Say uses this same gramophone technology more than 100 years later!

Parts of a See 'N Say

The heart of a See 'N Say is a plastic disc that is very similar to a gramophone record. A track containing the sound wave for each of the See 'N Say's sounds is embossed on the disc. A typical See 'N Say has between 10 and 26 sounds, and all these sounds are stored concentrically. On the edge of the disc, you can see the starting points for the different concentric tracks.

Sound reproduction is extremely simple. A metal needle rides in one of the grooves and transmits the vibrations embossed in the groove directly to a plastic cone speaker. The needle is embedded in a piece of plastic. The needle picks up the vibration pattern embedded in the disc and transmits it to the plastic, which transmits it to the tip of the cone. What you are hearing when a See 'N Say speaks is the purely mechanical vibration of that plastic cone.

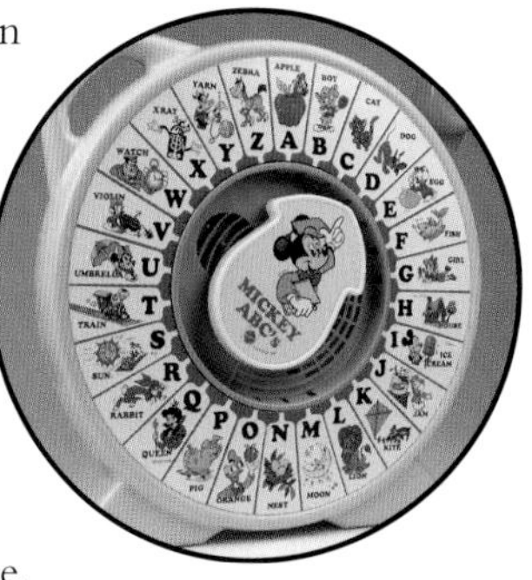

When you pull the See 'N Say's cord or pull its lever, the needle lifts and pulls back to the outer edge of the disc. Pulling the string or pulling the lever also winds a spring, and the energy in the spring turns the disc to play the sounds.

The only other part inside a See 'N Say is a simple but very effective governor that controls the speed of the disc when it rotates. The governor is made of two spring-loaded weights. When the disc spins too fast, the weights fly outward and apply pressure to the inside of the cylinder, slowing down the disc. At equilibrium, the governor keeps the disc spinning at exactly the right speed. This governor is a close cousin to the centrifugal clutch in a chainsaw.

Mechanical See 'N Says like the ones described here are quickly being replaced by electronic ones. The electronic See 'N Says, of course, need batteries so they would not survive a plunge in the bathtub.

HSW Web Links

www.howstuffworks.com

How Analog and Digital Recording Work
How Dancing Monsters Work
How Radio-Controlled Toys Work
How Singing Fish Work

Did You Know?

Almost all toys today, including the See 'N Say, talking dolls, and even the singing bass, have gone digital. They contain digitized music and voices stored on a ROM chip. A simple processor reads the data off the ROM chip and sends it to a digital-to-analog converter. The analog signal from the converter goes through a simple amplifier to the speaker. Companies can manufacture these digital sound circuits so cheaply that they can even be found in disposable greeting cards.

How **HELIUM BALLOONS** Work

Balloons are incredibly neat things, and helium balloons captivate people young and old. If you buy one at a circus or fair, you can hold its string, and it will ride along above you. But if you let go of the string, the balloon will fly away until you cannot see it anymore.

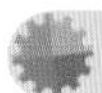

HSW Web Links

www.howstuffworks.com

How Airplanes Work
How Blimps Work
How CargoLifter's Airship Will Work
How Hot Air Balloons Work
The Helium Donald Duck Effect

Lifting Yourself

When you go to an amusement park and see someone selling helium balloons, the person is often holding a huge bouquet of balloons. How many balloons have to be in the bouquet to lift the person up?

Let's say one balloon is 1 foot in diameter. It holds 0.5 ft^3 of helium, so it can lift 14 grams. Let's say the balloon and the string weigh 4 grams, so the balloon can actually lift 10 grams besides what it is already lifting. If the person selling the balloons weighs 150 pounds, he or she weighs 67,200 grams, so the person would have to be holding more than 6,720 balloons to be lifted into the air.

In order to better understand how things float in air, it's good to know how water flotation works.

Water Flotation

Most of us feel comfortable with the idea of something floating in liquid. We see that happen every day. Cereal floats in milk, toys float in the bathtub, and people can float in water. So people have a way of directly experiencing water flotation. The reason things float in liquid applies to air as well, so let's start by understanding water flotation.

Try this little experiment:

1) Empty the soda from a plastic 1-l soda bottle.
2) Put the cap back on the bottle so that you have a sealed bottle full of air.
3) Turn the bottle upside-down and tie a string around the cap end, like you would with a balloon.
4) Dive down to the bottom of the deep end of a swimming pool with the bottle.

Because the bottle is full of air, it will want to rise to the surface. You can sit on the bottom of the pool, holding the bottle by the string, and it will float under the water just like a helium balloon does in air. If you let go of the string, the bottle will quickly rise to the water's surface.

The reason this soda bottle "balloon" wants to rise in the water is simple: Water is a fluid, and the 1-l bottle is displacing 1 l of that fluid. The bottle and the air in it weigh an ounce at most (1 l of air weighs about 1 g, and the bottle is very light as well). The 1 l of water it displaces, however, weighs about 1,000 g (about 2.2 lb). Because the weight of the bottle and the air inside it are less than the weight of water the bottle displaces, the bottle floats. This illustrates the law of buoyancy.

Helium Floatation

Helium balloons work by using exactly the same law of buoyancy as an empty soda bottle floating in water. When you stand under water at the bottom of a swimming pool, you are standing in a pool of water that is maybe 10 ft deep. When you stand in an open field, you are standing at the bottom of a pool of air that is many miles deep. So, a helium balloon that you hold by a string floats in a pool of air. The helium balloon displaces an amount of air (just like the empty bottle displaces an amount of water). If the helium plus the balloon is lighter than the air that the inflated balloon displaces, the balloon floats in the air.

Helium is a lot lighter than air. The difference is not as great as the difference between water and air (1 l of water weighs about 1,000 g, and 1 l of air weighs about 1 g), but it is significant. Helium weighs 0.1785 g/l, and nitrogen weighs 1.2506 g/l. And because nitrogen makes up about 80% of the air humans breathe, 1.25 g is a good approximation for the weight of 1 l of air. So, if you were to fill a 1-l soda bottle full of helium, the bottle would weigh about 1 g less than the same bottle filled with air. That doesn't sound like much—because the bottle weighs more than 1 g, the bottle will not float. However, in large volumes, the 1-gram-per-liter difference between air and helium can really add up.

Now you know what makes helium balloons so interesting. So, next time you see one, you'll know exactly how and why it floats and what happens when it floats away.

How **BOOMERANGS** Work

When you throw a boomerang correctly, it travels right back to you. A young child's first reaction to such a device is what the inventors of the boomerang might well have thought: This stick is obviously possessed with magical proprieties! The Australian Aborigines who discovered the returning boomerang had come upon an amazing application of some complex laws of physics.

If you throw a straight piece of wood it will simply keep going in one direction, turning end over end, until gravity pulls it to the ground. Why does changing the shape of that wood make it stay in the air longer and travel back to you?

The Shape of a Boomerang

A classic boomerang is simply two wings joined together in a single unit. The wings are at a slight tilt, and they have an airfoil design—they are rounded on one side and flat on the other, just like an airplane wing.

Basically, a boomerang is a propeller that isn't attached to anything. It would be reasonable to assume, then, that a boomerang would fly off in one direction as it spins. If you held it horizontally when you threw it, you would assume that the motion of the propeller would be up because that's the direction in which the axis is pointing. The boomerang should fly up into the sky like a helicopter taking off. If you held it vertically when you threw it, it seems it would simply fly off to the right or left. Obviously this isn't what happens because boomerangs come back to you when you throw them!

Why a Boomerang Comes Back

You throw a boomerang, so in addition to its spinning propeller motion, it also has the linear motion of the throw. The wing that is at the top of the spin at any moment ends up moving in the same direction as the forward motion of the throw, and the wing that is at the bottom of the spin is moving in the opposite direction of the throw. This means that, although the wing at the top is spinning at the same speed as the wing at the bottom, it is actually moving through the air at a higher rate of speed. And the wing that is moving faster generates more lift.

When you push something from the top, say a chair, you tip the thing over and it falls to the ground. Why doesn't this happen when you push on the top of a spinning boomerang? When you push on one point of a spinning object, such as a wheel, an airplane propeller, or a boomerang, the object doesn't react in the way you might expect. It reacts to the force as if you pushed it at a point 90 degrees from when where you actually pushed it. This is the same thing that is happening in a boomerang.

The uneven force caused by the difference in speed between the two wings applies a constant force at the top of the spinning boomerang, which is actually felt at the leading side of the spin. So the boomerang constantly turns to the left or right. If you throw it at the correct speed and angle, it travels in a circle and comes right back to you!

The five forces that act on a boomerang as it spins through the air are gravity, the force caused by the propeller motion, the force of your throw, the force caused by the uneven speed of the wings, and the force of any wind in the area. For a boomerang to actually travel in a circle and come back to its starting point, all five of these forces have to be balanced in just the right way. To accomplish this, you need a well-designed boomerang and a correct throw. Any boomerang enthusiast will tell you that the only way to consistently make good throws is to practice good technique.

HSW Web Links 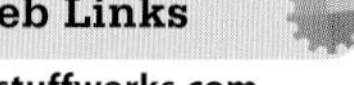

www.howstuffworks.com

How Airplanes Work
How Force, Power, Torque, and Energy Work
How Gyroscopes Work
How Helicopters Work
How the Physics of Football Works

Did You Know?

Boomerangs, also known as throwing sticks, were used as hunting weapons. Although the Australian aborigines are most commonly associated with this type of hunting device, several other groups of people around the world used throwing sticks. Anthropologists have found evidence that people in ancient Egypt, Eastern Europe, and the Hopi of North America used similar hunting weapons.

How GUITARS Work

One of the most popular musical instruments in use today is the guitar, and it is used in a huge range of musical styles: Rock, country, and flamenco music all use the same stringed instrument to create wildly different sounds. The guitar has been around since the 1500s, but it has undergone several transformations during its history. The development of the electric guitar is the most obvious recent mutation, and it had a huge effect on the popularity of the guitar.

HSW Web Links

www.howstuffworks.com

How Analog & Digital Recording Works
How CDs Work
How MP3 Works
How Singing Fish Work
How Tape Recorders Work

A guitar is a musical instrument with a distinctive shape and a distinctive sound. The best way to learn how a guitar produces its sound is to start by understanding all the different parts that make up the instrument.

Parts of a Guitar

A guitar has three main parts:

- The hollow body
- The neck, which holds the frets
- The head, which contains the tuning pegs

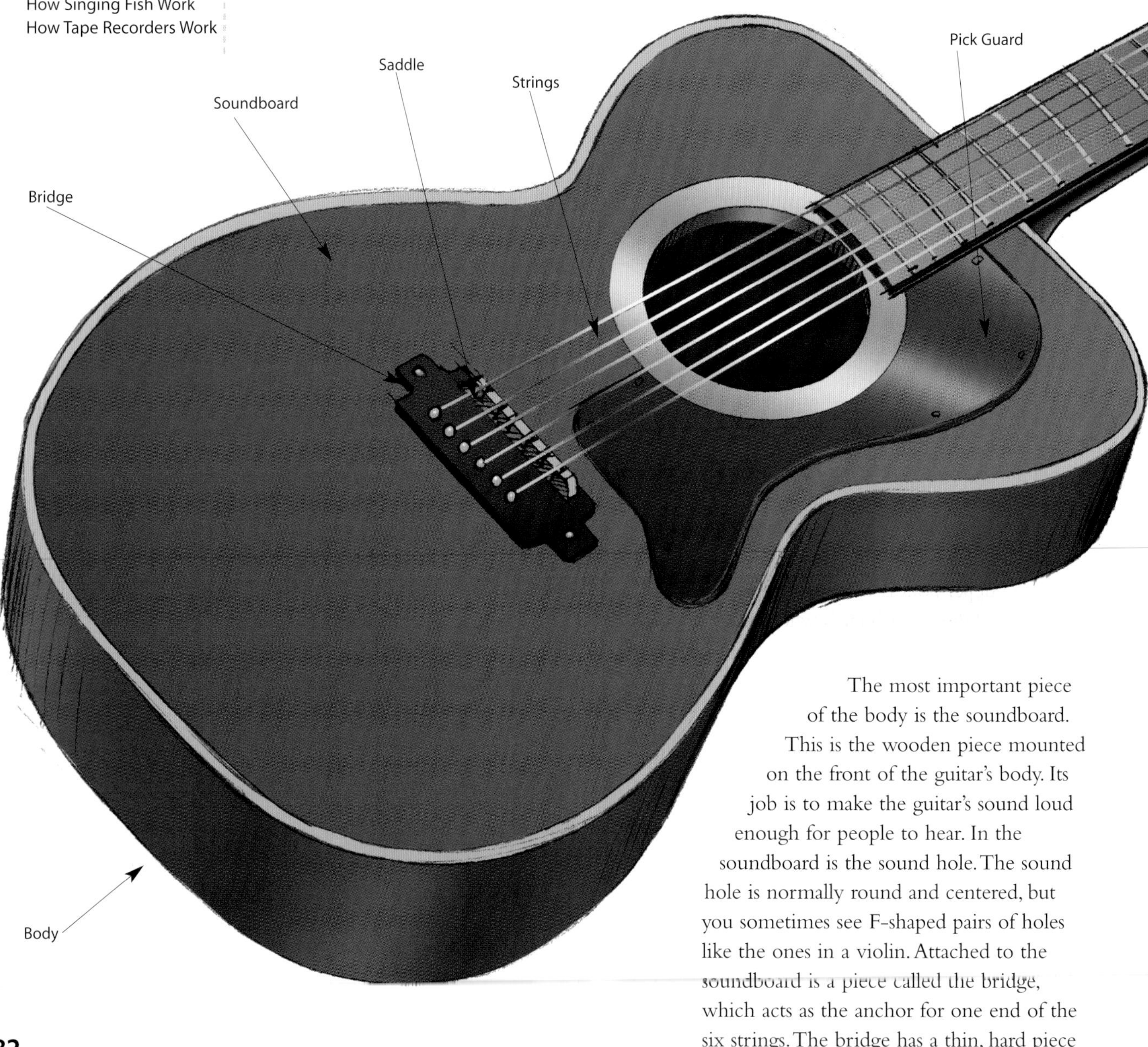

The most important piece of the body is the soundboard. This is the wooden piece mounted on the front of the guitar's body. Its job is to make the guitar's sound loud enough for people to hear. In the soundboard is the sound hole. The sound hole is normally round and centered, but you sometimes see F-shaped pairs of holes like the ones in a violin. Attached to the soundboard is a piece called the bridge, which acts as the anchor for one end of the six strings. The bridge has a thin, hard piece

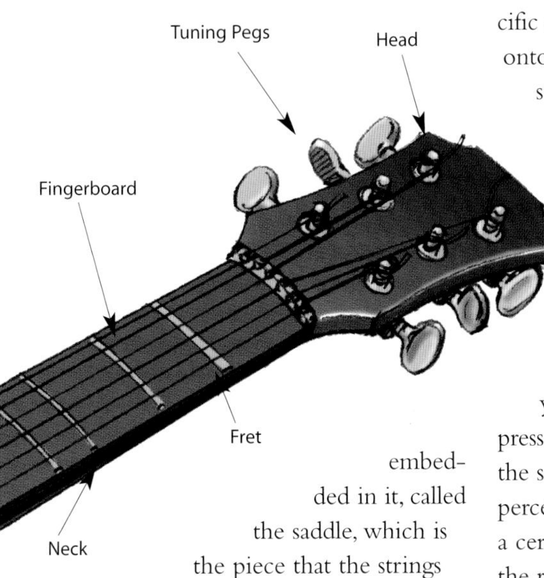

embedded in it, called the saddle, which is the piece that the strings rest against.

When the strings vibrate, the vibrations travel through the saddle to the bridge to the soundboard so that the entire soundboard vibrates. The body of the guitar forms a hollow sound box that amplifies the vibrations of the soundboard. If you touch a tuning fork to the bridge of a guitar, you will be able to prove that the vibrations of the soundboard are what produce the sound in an acoustic guitar. The soundboard greatly increases the amount of surface area that is vibrating, which greatly amplifies the sound.

The body of an acoustic guitar commonly has a "waist," or a narrowing. This narrowing makes it easy to rest the guitar on your knee. The two widenings are called bouts. The upper bout is where the neck connects, and the lower bout is where the bridge attaches.

The size and shape of the body and the bouts have a lot to do with the tone that a given guitar produces. Any two guitars that have different body shapes and sizes will sound subtly different. The two bouts affect the sound differently. If you drop a pick into the body of a guitar and rattle it back and forth in the lower bout and then the upper bout, you will be able to hear a difference. The lower bout accentuates lower tones, and the upper bout accentuates higher tones.

The face of the neck, which contains the frets, is called the fingerboard. The frets are metal pieces cut into the fingerboard at specific distances. By pressing a string down onto a fret, you change the length of the string and therefore the tone it produces when it vibrates.

Sound, Tones, and Notes

Sound is any change in air pressure that your ears are able to detect and process. For your ears to detect it, a change in pressure has to be strong enough to move your eardrums. The more strongly the pressure changes, the louder you perceive the sound to be. For your ears to be able to perceive a sound, the sound has to occur in a certain frequency range. For most people, the range of perceivable sounds falls between 20 Hertz (Hz), or oscillations per second, and 20,000 Hz. Most people cannot hear sounds below 20 Hz or above 20,000 Hz.

A tone is made up of one frequency or very small number of related frequencies. The alternative to a tone is a combination of hundreds or thousands of random frequencies—called noise. When you hear the sound of a river, or the sound of wind rustling through leaves, or the sound of paper tearing, you are hearing noise.

A musical note is a tone. However, a musical note is a tone that comes from a small collection of tones that are pleasing to the human brain when used together. For example, you might pick a set of tones at the following frequencies:

- 264 Hz
- 297 Hz
- 330 Hz
- 352 Hz
- 396 Hz
- 440 Hz
- 495 Hz
- 528 Hz

This particular collection of tones, which sound pleasant together, is known as the major scale. Notice that the two notes at either end are separated by exactly a factor of 2. That is, 264 is one-half of 528. This is the basis of octaves. Any note's frequency can be doubled to "go up an octave," and

Cool Facts

- The first guitar may have been played in the ancient Persian Gulf area of the Middle East, as early as 1900 BC.
- As the guitar made it from Rome to Egypt and then to Europe, its use became an art form. Many players were also composers, although it was difficult to publish music because of royal scrutiny.
- Ancient guitar players often decorated their guitars, sometimes very elaborately. Artifacts have been found with mother-of-pearl inlays, tortoise shell, and ivory.
- The Spanish conquistadors introduced the guitar to the New World.
- As the Industrial Revolution blossomed, so did factory production of guitars, revolutionizing the guitar manufacturing process and making guitars more affordable.

any note's frequency can be halved to "go down an octave."

Over time, most of the musical world came to agree on a scale called the tempered scale, with the A note set at 440 Hz and all the other notes tuned from that. In the tempered scale, all the notes are offset by the 12th root of 2 (roughly 1.0595). That is, if you take any note's frequency and multiply it by 1.0595, you get the frequency of the next note.

Strings and Frets

A guitar uses vibrating strings to generate tones. Any string under tension will vibrate at a specific frequency that is controlled by several factors:

- The length of the string
- The amount of tension on the string
- The weight of the string
- The "springiness" of the string's material (for example, a rubber band is a lot springier than kite string)

On a guitar you can see that the different strings have different weights. The first string is like a thread and the sixth string is much thicker and heavier. You control the tension on the strings with the tuning pegs. When you press down on a string at a fret, you change the length of the string and therefore its frequency when vibrating.

The frets are spaced out so that the proper frequencies are produced when the string is held down at each fret. The magic number to use in positioning frets is 17.817. Let's say that the scale length for a guitar is 26 inches (66 cm). The 1st fret should be located 1.46 inches (3.7 cm) down from the nut (26 in., or 17.817 cm), or 24.54 inches (62.33 cm) from the saddle; the 2nd fret should be 1.38 inches (3.5 cm) down from the first fret (24.54 in., or 17.817 cm), or 23.16 inches (58.82 cm) from the saddle; and so on. The 12th fret should be exactly halfway between the nut and the saddle.

The Guitar's Sound

Have you ever noticed that a piano, a harp, a mandolin, a banjo, and a guitar all play the same notes (frequencies) using strings, but they all sound very different from one another? If you hear the different instruments, you can easily recognize each one by its sound. For example, anyone can hear the difference between a piano and a banjo.

When the strings on an acoustic guitar vibrate, they transmit their vibrations to the saddle. The saddle, in turn, transmits its vibrations to the soundboard. The soundboard and body amplify the sound from the soundboard, and the sound comes out through the sound hole. The particular shape and material of the soundboard, along with the shape of the body and the fact that a guitar uses strings, give a guitar its sound.

There are a number of different ways to modify sounds to get the particular voice of the instrument. One modification in a guitar is to add harmonics to it. For example, when you pluck one string, the guitar plays the pure note, but the string also rings at harmonics such as two, three, and four times the pure tone. Other strings also pick up the vibrations from the saddle and add their own vibrations, so the sound you hear from a guitar for any given note is actually a blend of many related frequencies.

A guitar adds more to any note it plays: The note doesn't just start and stop abruptly—it builds and trails off. Over the course of the note, the loudness (amplitude) of the note changes.

Whether you are a musician or you simply enjoy listening to music, the guitar is a fascinating instrument. It is incredible the way the soundboard, strings, and the size and shape of the body all come together—much like the notes of an intricate piece of music do—to form something so creative.

Electric Guitars

If you have ever compared an electric guitar to an acoustic guitar, you know that they have several important things in common. Acoustic and electric guitars both have six strings, they both tune those strings with tuning pegs, and they both have frets on a long neck. The body end is where the major differences are found. Electric guitars have solid bodies, magnetic pickups, and several knobs instead of the hollow resonating cavity of an acoustic guitar. If you pluck a string on an electric guitar that is not plugged in, the sound is barely audible. Without a soundboard and the hollow body, there is nothing to amplify the string's vibrations.

To produce sound, an electric guitar senses the vibrations of the strings electronically and routes that electronic signal to an amplifier and a speaker. The sensing occurs in a magnetic pickup mounted under the strings on the guitar's body. The pickup consists of a bar magnet wrapped with perhaps thousands of turns of fine wire. Coils and magnets can turn electrical energy into motion, and they can turn motion into electrical energy. (See "How Electromagnets Work," page 90, for more information.) In the case of an electric guitar, the vibrating steel strings produce a corresponding vibration in the magnet's magnetic field and therefore a vibrating current in the coil. The amplifier amplifies the signal from the coil and plays it through a speaker.

How **QUARTZ WATCHES** Work

If you wear a watch, then there's a very good chance that it is a quartz watch. If it is, it keeps very good time. And it might have been very inexpensive—quartz watches can be cheap enough to appear in a box of cereal! The amazing thing is that a small piece of quartz—essentially a little piece of a rock you might find on the ground—can keep time with amazing precision

Every clock needs an oscillator that can tick precisely. In a pendulum clock, the pendulum swings precisely once every second or so. In a quartz watch, the oscillator is a sliver of quartz that resonates thousands of times per second.

Piezoelectric Materials

Quartz seems like nothing special. In lots of places you can walk out in your yard and find quartz rocks lying around on the ground. The thing that makes quartz useful as an oscillator is the fact that it is a piezoelectric material.

One of the most common places to find piezoelectric materials is in a gas grill that has a pushbutton lighter. Inside the lighter is a piezoelectric crystal. When you push the button on the lighter, a small hammer hits the crystal. On both sides of the crystal are metal electrodes. When a piezoelectric crystal is deformed—for example, by being hit—it generates a voltage. When you strike a piezoelectric crystal with a hammer, it can create thousands of volts. In a gas grill the voltage is high enough to jump a spark gap and light a fire.

The other thing about piezoelectric crystals is that, if you apply a voltage to them, they change shape very slightly. A piezoelectric speaker or beeper—typically what you hear when a watch with an alarm goes off—takes advantage of this effect. The crystal attaches to a diaphragm that creates the sound.

Inside the Watch

In a quartz watch, the quartz crystal takes advantage of both of the effects of piezoelectric materials. A sliver of quartz is ground to a very precise shape and plated with two electrodes. A voltage is applied so that the crystal changes shape, and then the voltage is released so that the crystal snaps back. In the process of snapping back, the crystal generates a voltage because it is changing shape. This generated voltage can be fed back to cause the crystal to change shape again, and so on. Through this process, the crystal resonates at a frequency that is controlled by its thickness. Essentially it rings like a bell, but at a very high frequency.

In a quartz watch, the circuit that makes the crystal oscillate generates a digital signal from the oscillations. For example, the crystal might oscillate very precisely a million times per second. A digital counter divides that signal by 1 million to create a once-a-second heartbeat for the watch. The heartbeat either drives a small motor that makes the second hand move once each second, or it drives a chain of digital counters that create a numeric display on a liquid crystal display.

Quartz oscillators are used in all sorts of places that need a precise frequency base. One common place is in radio transmitters, where the transmitter needs to send radio waves at a very precise frequency. You also find them in music synthesizers, where they control the frequency of the notes produced. And you find them in microprocessors, where they control the speed of the microprocessor's operation. It turns out that there are tiny slivers of quartz all around you!

HSW Web Links

www.howstuffworks.com

How Atomic Clocks Work
How LCDs Work
How Oscillators Work
How Pendulum Clocks Work

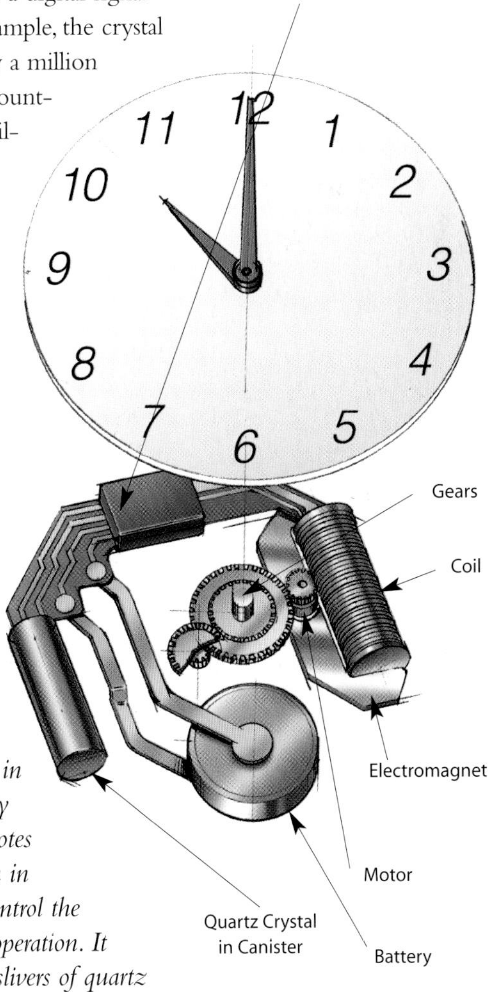

How TIME Works

Time is one of the most illusive things that you measure, but you measure it constantly, in almost everything you do. The day is an obvious starting point for time. A day consists of a period of sunlight followed by night. Humans are tuned in to this cycle through sleep, so each morning you wake up to a new day. No matter how primitive the culture, the concept of a day arises as an obvious and natural increment. People use clocks to divide the day into smaller increments and calendars to group days together into larger increments. Both of these systems have incredibly interesting origins!

HSW Web Links

www.howstuffworks.com

How Atomic Clocks Work
How Digital Clocks Work
How Pendulum Clocks Work
How Quartz Watches Work
Inside a Wind-up Alarm Clock

A day is the amount of time it takes for the earth to rotate once on its axis. But how long does it actually take the earth to rotate? That is where things become completely arbitrary. The world has decided to standardize on the following increments:

- A day consists of two 12-hour periods, for a total of 24 hours.
- An hour consists of 60 minutes.
- A minute consists of 60 seconds.
- Seconds are subdivided on a decimal scale into things like hundredths of a second and millionths of a second.

It's pretty bizarre to divide up a day by dividing it in half, then dividing the halves by 12ths, then dividing the 12ths into 60ths, then dividing by 60 again, and then switching to decimal for the smallest increments. It is not clear where the 12ths and 60ths come from, although Egypt and Babylon are the likely sources.

a.m., p.m., and Seconds

The terms a.m. and p.m. are of Roman derivation. These abbreviations stand for *ante meridiem*, before midday, and *post meridiem*, after midday. The Romans simply divided the day into two halves.

Today we base time on seconds. We divide the day into 86,400 seconds, and a second is officially defined as 9,192,631,770 oscillations of a Cesium-133 atom in an atomic clock.

Time Zones

Everyone on the planet wants the sun to be at its highest point in the sky (crossing the meridian) at noon. If there were just one time zone, this would be impossible because the earth rotates 15 degrees every hour. The idea behind multiple time zones is to divide the world into 24 15-degree slices and set the clocks accordingly in each zone. All the people in a given zone set their clocks the same way, and each zone is 1 hour different from the one next to it.

All time zones are measured from a starting point centered at England's Greenwich Observatory. This point is known as the Greenwich meridian, or the prime meridian. Time at the Greenwich meridian is known as Greenwich mean time (GMT), or universal time. The eastern time zone in the United States is designated as GMT minus 5 hours. When it is noon in the eastern time zone, it is 5:00 p.m. at the Greenwich Observatory. The international date line is located on the opposite side of the planet from the Greenwich Observatory.

Daylight Saving Time

During World War I, many countries started adjusting their clocks during part of the year. The idea was to try to adjust daylight hours in the summer to more closely match the hours that people are awake, in order to conserve fuel by lowering the need for artificial light. The United States and several other countries still use some variation on this system. By an act of Congress, the United States starts daylight saving time on the first Sunday in April by advancing the clocks one hour, and it ends daylight saving time on the last Sunday in October, when the clocks are moved back one hour. "Spring forward; fall back" is the phrase many people use to remember when the time changes and in which direction.

The Calendar

As mentioned earlier, the day is an obvious unit of time for people. But what about weeks, months, and years?

Years are fairly straightforward. Humans created the concept of a year because seasons repeat yearly. The ability to predict seasons is essential to life if you are planting crops or trying to prepare for winter. Most plants sprout and bear fruit on a yearly schedule, so a year is a natural increment.

A year is defined as the amount of time it takes for the earth to orbit the sun one time. It takes about 365 days to do that. If you measure the exact amount of time it takes for the earth to orbit the sun, the number is actually 365.242199 days. What happens with the extra part of a day each year? How do we account for that?

By adding one extra day to every 4th year—called a leap year—we get an average of 365.25 days per year, which is fairly close to the 365.242199 days it takes the earth to actually orbit the sun. This is why we have leap years that are 1 day longer than normal years. To get even closer to the actual number, every 100 years is not a leap year, but every 400 years is a leap year:

- If the current year is divisible by 4 or 400, then it is a leap year.
- If the current year is divisible by 100, then it is not a leap year.

This system brings the average length of a year to 365.2425 days, which is extremely close to the exact number, 365.242199 days. Adding leap seconds at the end of certain years adjusts for yet more precision.

The concept of a month comes from the cycle of the moon. Many ancient cultures used months whose lengths were 29 or 30 days (or something close to this) to chop up a year into increments. The main problem with this sort of system is that moon cycles at 29.5 days do not divide evenly into the 365.25 days of a year. When you look at the modern calendar, the months are extremely confusing. One has 28 or 29 days, some have 30 days, and the rest have 31 days. The names of the months and the number of days each are Roman inventions that have persisted today.

Days, months, and years all have a natural basis, but weeks do not. They come straight out of the Bible: "Remember the Sabbath day, to keep it holy. Six days shalt though labor, and do all thy work but the seventh day is the Sabbath of the Lord thy God" (Exodus 20:8). The tradition of the week persisted, and the Romans gave names to the days of the week based on the sun, the moon, and the names of the five planets that were known at the time. These names actually carried through to European languages fairly closely, and in English the names of Sunday, Monday, and Saturday made it straight through. The remaining four days were renamed in English for Anglo-Saxon gods.

Time is fairly elusive. We can't see it or sense it in any way. It just happens. Human beings have therefore come up with ways to measure time that are pretty arbitrary and also fairly interesting.

Time Measurements

The measurement of time covers an incredible range. Here are some common time spans, from the shortest to the longest:

- **1 picosecond (1 trillionth of a second)**—About the shortest period of time we can currently measure accurately
- **1 nanosecond (1 billionth of a second)**—2 to 4 nanoseconds is the length of time that a typical home computer spends executing one software instruction
- **1 microsecond (1 millionth of a second)**
- **1 millisecond (1 thousandth of a second)**—The typical fastest time for the exposure of film in a normal camera; a picture taken in 1/1,000 second will usually stop all human motion
- **1 centisecond (1 hundredth of a second)**—The length of time it takes for a stroke of lightning to strike
- **1 decisecond (1 tenth of a second)**—A blink of an eye
- **1 second**—The average time between human heartbeats
- **60 seconds**—One minute; a long TV commercial
- **2 minutes**—About as long as a person can hold his or her breath
- **60 minutes**—An hour; about as long as a person can sit in a classroom without glazing over
- **8 hours**—The typical workday in the United States, as well as the typical amount of sleep a person needs every night
- **24 hours**—A day; the amount of time for the earth to rotate one time on its axis
- **7 days**—A week
- **40 days**—About the longest a person can survive without food
- **365.24 days**—A year; the amount of time it takes for the earth to complete one orbit around the sun
- **10 years**—A decade
- **75 years**—The typical life span for a human being
- **5,000 years**—The span of recorded history
- **50,000 years**—The length of time *Homo sapiens* has existed as a species
- **65 million years**—The length of time dinosaurs have been extinct
- **200 million years**—The length of time mammals have existed
- **3.5 to 4 billion years**—The length of time that life has existed on earth
- **4.5 billion years**—The age of the earth
- **10 to 15 billion years**—The suspected age of the universe since the big bang

How PENDULUM CLOCKS Work

Have you ever looked inside a grandfather clock or a small mechanical alarm clock, seen all the gears and springs, and thought, "Wow—that's complicated!" Although clocks are normally fairly complicated, they do not have to be confusing or mysterious. In fact, as you learn how a clock works, you can see how clock designers faced and solved a number of interesting problems to create accurate timekeeping devices.

HSW Web Links

www.howstuffworks.com

How Digital Clocks Work
How Gears Work
How Quartz Watches Work
How Time Works

When you look at a pendulum clock from the outside, you see several parts that are important to all pendulum clocks:

- The face of the clock, with its hour and minute hands
- One or more weights (or a keyhole used to wind a spring inside the clock—we will stick with weight-driven clocks here)
- The pendulum itself

In many wall clocks that use a pendulum, the pendulum swings once per second. In small cuckoo clocks, the pendulum might swing twice per second. In larger clocks, the pendulum swings once every 2 seconds or more.

The Weight

The idea behind the weight is to act as an energy storage device so that the clock can run, unattended, for relatively long periods of time. It is like a battery. When you "wind" a weight-driven clock, you pull on a cord that lifts the weight. That gives the weight potential energy in the earth's gravitational field.

Say that you want to use a falling weight to create the simplest possible clock—a clock that has just a second hand on it.

Escapement

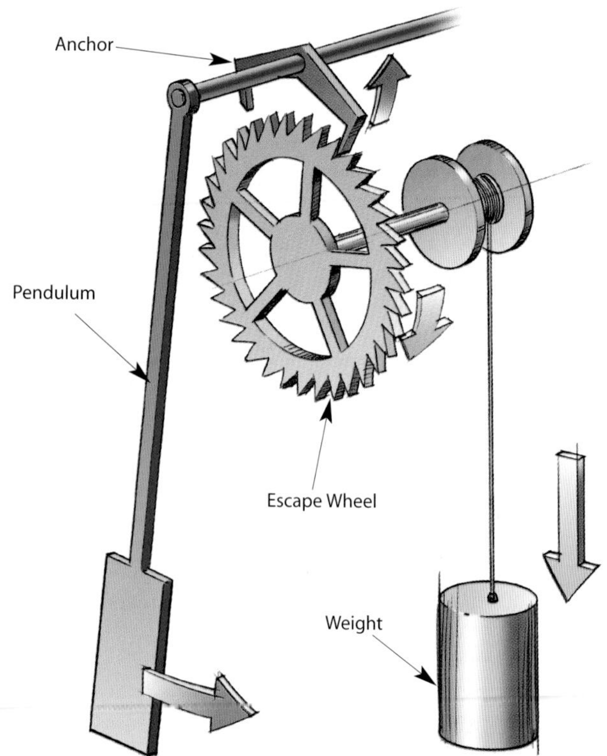

You want the second hand on this simple clock to work like a normal second hand on any clock, making one complete revolution every 60 seconds. One simple approach would be to attach the weight's cord to a drum and then attach a second hand to the drum as well. This, of course, would not work. Releasing the weight would cause it to fall as fast as it could, spinning the drum at about 1,000 revolutions per minute.

Still, it's headed in the right direction. Say you put some kind of friction device on the drum—some sort of brake that would slow down the drum. With this you might be able to work out some scheme to get the second hand to make approximately one revolution per minute. But as the temperature and the humidity in the air change, the friction in the device will change, so the second hand will not keep very good time.

The Pendulum

In the 1600s, people who wanted to create accurate clocks were trying to solve the problem of how to cause the second hand to make exactly one revolution per minute. The Dutch astronomer Christiaan Huygens first suggested using a pendulum. The period of a pendulum's swing (the amount of time it takes for a pendulum to go back and forth once) is related only to the length of the pendulum and the force of gravity. Because gravity is constant at any given spot on the planet, the only thing that affects the period of a pendulum that you can control is the length of the pendulum. The amount of weight does not matter, nor does the length of the arc that the pendulum swings through. Only the length of the pendulum matters.

When someone noticed this fact about pendulums, the phenomenon became the heart of an accurate clock. A pendulum clock needs a gear with teeth of some special shape and a device attached to the pendulum that can engage the teeth of the gear. The basic idea is that for each swing of the pendulum back and forth, one tooth of the gear is allowed to escape. Therefore, the device that is attached to the pendulum that engages the gear teeth is called an escapement.

For example, if the pendulum is swinging toward the left and passes through the center position, then the pendulum continues toward the left, and the lefthand stop attached to the pendulum will release its tooth. The gear will then advance one-half the width of a tooth and hit the righthand stop. In advancing forward and running into the stop, the gear will make a sound, such as "tick" or "tock." This is where the ticking sound of a clock or watch comes from.

The Gear and the Gear Ratio

One thing to keep in mind is that pendulums will not swing forever. Therefore, one additional job of the gear is to nudge the pendulum enough to overcome friction and allow it to keep swinging.

Although it would be accurate, this clock would have two problems that would make it less than useful:

- Most people typically expect a clock to have hour and minute hands as well as a second hand.
- You would have to wind the clock about every 20 minutes. Because the drum makes one revolution every minute, the weight would unwind to the floor very quickly.

The second problem is easy to solve. You can create a high-ratio gear train that causes the drum to make perhaps one turn every 6 to 12 hours. (See "How Gears Work," page 104, for more information on gear trains.) This would give you a clock that you only had to wind once a week or so.

If you create another gear train, you can have hour and minute hands as well. A 60:1 gear train can drive the minute hand, and another 12:1 gear train can connect the minute hand to the hour hand.

The hour, minute, and second hands use tubular shafts on the gears and then arrange the gear trains so that the gears driving the hour, minute, and second hands share the same axis. The tubular gear shafts are aligned one inside the other. Look closely at any clock face, and you can see this arrangement.

Parts of a Pendulum Clock

You can see that, even though all the gears in a clock make it look complicated, what a pendulum clock is doing is really pretty simple. There are five basic parts:

- **The weight or spring**—This provides the energy to turn the hands of the clock.
- **The weight gear train**—A high-ratio gear train gears the weight drum way up so that you don't have to rewind the clock very often.
- **The escapement**—Made up of the pendulum and the escapement gear, the escapement precisely regulates the speed at which the weight's energy is released.
- **The hand gear train**—This gear train gears things down so that the minute and hour hands turn at the right rates.
- **The setting mechanism**—The setting mechanism somehow disengages, slips, or ratchets the gear train so that the clock can be wound and set.

How RADIO-CONTROLLED TOYS Work

You have probably seen the ads on Saturday morning television proclaiming the amazing abilities of the Super Ultra Road-Rippin' Devastator or some other radio-controlled (RC) car. And you may have seen people at a park flying a model airplane or blimp, or controlling a miniature boat in a pond. All these RC toys use the same basic principles so that you can drive them from a distance.

HSW Web Links

www.howstuffworks.com

How Ham Radio Works
How Radio Scanners Work
How the Radio Spectrum Works
How Radio Works

If you were to take apart a basic radio-controlled car, you would find that it contains the following parts:

- An electric motor that makes the car go forward and backward
- Another motor that can steer the front wheels
- A battery to power both of these motors
- A receiver that accepts the radio commands that you send with the transmitter and that turns the two motors on and off

Signal Frequency

To operate an RC toy, you hold a transmitter that sends radio waves to the receiver to tell it what to do. The transmitter has a power source, usually a 9-volt battery, that provides the power to transmit the radio signal.

Most RC toys operate at either 27 megahertz (MHz) or 49 MHz. The FCC allocated this pair of frequencies for all sorts of basic radio-based consumer items, such as garage door openers, baby monitors, walkie-talkies, and RC toys. All these devices are low power and might transmit several hundred feet at the most. Advanced RC models, such as more sophisticated RC airplanes, use the 72-MHz or 75-MHz frequencies that are allocated just for this purpose.

Almost every RC toy has a label on the bottom that shows the frequency range in which it operates. Usually this label shows whether the toy is a 27-MHz or a 49-MHz model. Most RC toy manufacturers make versions of each model for both frequency ranges. That way, you can operate two of the same models simultaneously, for racing or playing together, without having to deal with interference between the two transmitters. Some manufacturers also provide more specific information about the exact portion of the frequency band that the toy operates in. For example, some manufacturers offer the option to race sets of up to six toys, with each model tuned to a different part of the 27-MHz frequency range.

Signal Transmission

An RC toy uses the simplest possible radio transmissions between the transmitter and the receiver. Pulse modulation is the system that RC toys use. (See "How Radio Works," page 220, for more information on pulse modulation.) In pulse modulation, the carrier frequency (for example, 27 MHz) simply turns on and off to encode information.

A typical RC signal transmission has two parts:

- **Synchronization pulses**—Four pulses are sent, and each one is 2.1 milliseconds long, separated by a 0.7-millisecond gap.
- **Burst sequence**—Information that tells the toy what to do. In the burst sequence, 0.7-millisecond pulses are separated by 0.7-millisecond gaps.

The number of pulses in the burst sequence tells the toy what to do. Here are the different burst sequences you might see in a typical toy:

- Forward: 16 pulses
- Reverse: 40 pulses
- Forward/left: 28 pulses
- Forward/right: 34 pulses
- Reverse/left: 52 pulses
- Reverse/right: 46 pulses

The transmitter sends bursts of radio waves that oscillate with a frequency of 27,900,000 cycles per second (that is, 27.9 MHz). The toy is constantly monitoring the assigned frequency, 27.9 MHz, for a signal. The pulse

Did You Know?

Radio controlled toys, which transmit at 27 Mhz and 49 Mhz, share their spectrum with a whole bunch of other radio-based devices. The most common devices on these frequencies are toy walkie-talkies, baby monitors (one-way walkie-talkies) and older analog cordless phones (glorified two-way walkie-talkies). Many garage door openers operate in the same frequencies, which makes sense since a garage door opener is a radio-controlled device. If you have radio-controlled lights or a wireless alarm system, it may also operate in the same frequency range. The peculiar encoding of radio controlled toy signals helps avoid confusion between all of the different devices sharing the spectrum.

sequence is sent to an integrated circuit (IC) in the toy's receiver, which decodes the sequence and starts the appropriate motor. For example, if the pulse sequence is 16 pulses (forward), then the IC sends positive current to the motor that is running the wheels. If the next pulse sequence is 40 pulses (reverse), the IC inverts the current to the same motor to make it spin in the opposite direction.

transistors to amplify the signal, and this signal is sent to the transmitter's antenna.

The receiver can use a simple tuner and diode to receive the pulses that the transmitter sends. An IC in the receiver counts the pulses and decides which motors to turn on and off. The IC typically uses a transistor amplifier or a small relay to turn the motors on and off.

The simplicity of the transmitter and receiver is the reason simple RC toys can be so inexpensive. Each toy requires just a handful of electronic components, and things like ICs and transistors are extremely inexpensive—just a few pennies each for the manufacturer. Besides the transmitter and receiver, an RC toy is no different from any other motorized toy.

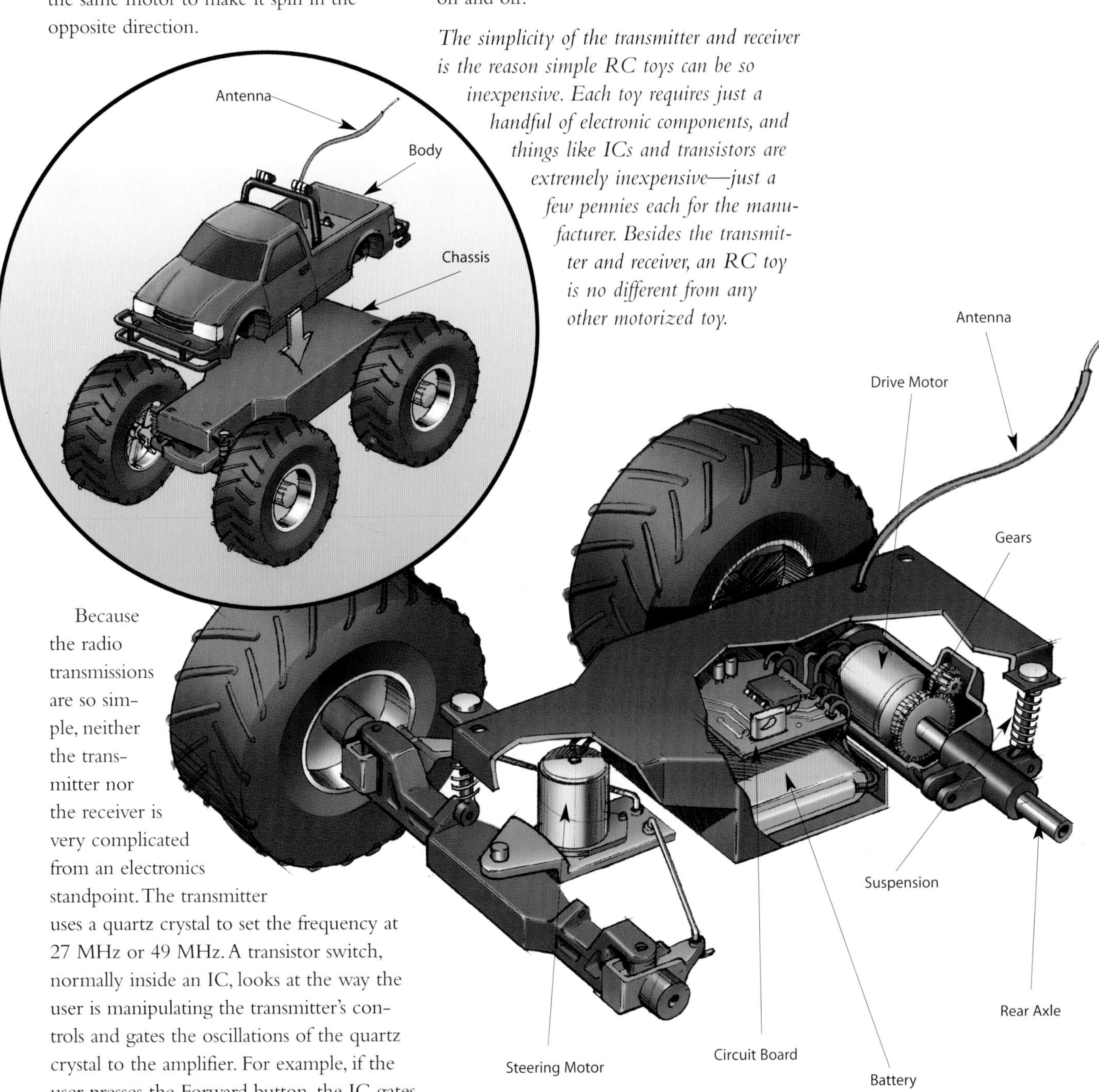

Because the radio transmissions are so simple, neither the transmitter nor the receiver is very complicated from an electronics standpoint. The transmitter uses a quartz crystal to set the frequency at 27 MHz or 49 MHz. A transistor switch, normally inside an IC, looks at the way the user is manipulating the transmitter's controls and gates the oscillations of the quartz crystal to the amplifier. For example, if the user presses the Forward button, the IC gates the pulses for the synchronization pulses and the burst sequence to the amplifier. The amplifier normally contains two or three

How ROLLER COASTERS Work

If you're studying physics, there are few more exhilarating classrooms than a roller coaster. Roller coasters are driven almost entirely by basic inertial, gravitational, and centripetal forces — they all come together for a great ride. Amusement parks keep upping the ante, building faster and more complex roller coasters, but the fundamental principles at work remain the same.

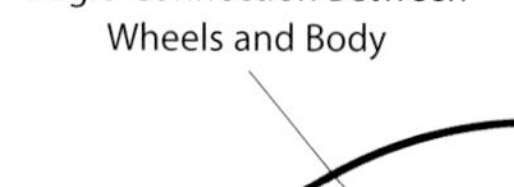

HSW Web Links

www.howstuffworks.com

How Force, Power, Torque and Energy Work
How Weightlessness Works
How Diesel Locomotives Work
How Maglev Trains Will Work

At first glance, a roller coaster is something like a passenger train. It consists of a series of connected cars that move on tracks. Unlike a passenger train, a roller coaster has no engine or power source of its own. For most of the ride, only the forces of inertia and gravity move a roller coaster. All of the energy for the ride comes at the very beginning of the ride—the coaster cars climb up a steep hill or they are launched by compressed air or linear induction motors.

What Goes Up Must Come Down

The purpose of the hill at the beginning of most rides is to build up a reservoir of potential energy. The concept of potential energy, often referred to as energy of position, is very simple: As the coaster gets higher in the air, there is a greater distance through which gravity can pull it down. You experience this phenomenon all the time—think about driving your car, riding your bike, or pulling your sled to the top of a tall hill. The potential energy you build going up the hill can be released as kinetic energy, energy of motion, as soon as you start coasting down the hill.

A roller coaster's energy is constantly changing between potential and kinetic energy. At the top of the first lift hill, there is maximum potential energy because the train is as high as it gets. As the train starts down the hill, this potential energy converts into kinetic energy—the train speeds up. At the bottom of the hill, there is maximum kinetic energy and little potential energy. The kinetic energy propels the train up the second hill, building up the potential-energy level. In this way, the course of the track is constantly converting energy from kinetic to potential and back again. This fluctuation in acceleration is one thing that makes roller coasters so much fun.

In almost any roller coaster, the hills decrease in height as you move along the track.

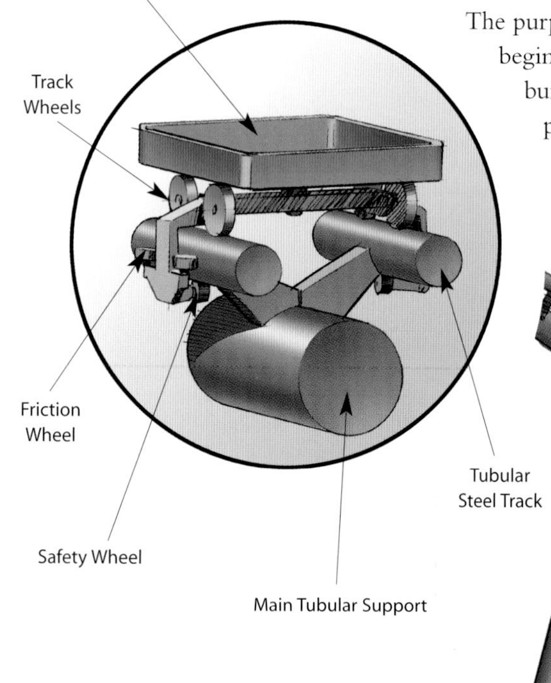

The total energy reservoir built up in the lift hill is gradually lost to friction between the train and the track, as well as between the train and the air. When the train coasts to the end of the track, the energy reservoir is almost completely empty. At this point, the train either comes to a stop or is sent up the lift hill for another ride.

Staying on Track

There are two major types of roller coasters, distinguished mainly by their track structure. The tracks of wooden roller coasters are something like traditional railroad tracks. The metal wheels of the coaster train roll on a flat metal strip, measuring 4- to 6-inches (10- to 15-cm) across. This strip is bolted to a sturdy running track made of laminated wood. In most coasters, the car wheels have the same flanged design as the wheels of a train—the inner part of the wheel has a wide lip that keeps the car from rolling off the side of the track. The car also has another set of wheels (or sometimes just a safety bar) that runs underneath the track. This keeps the cars from flying up into the air.

Wooden coaster tracks are braced by wooden crossties and diagonal support beams. The entire track structure rests on an intricate lattice of wooden or steel beams, just like the beam framework that supports a house or skyscraper. With these materials, designers can combine the hills, twists, and turns into an infinite variety of course layouts, and they can even flip the train upside down (though this is rare in modern wooden coasters). But, since the track and support structure are so cumbersome, a wooden track is fairly inflexible. This makes it difficult to construct complex twists and turns. In wooden coasters, the exhilarating motion is mainly up and down.

Tubular steel tracks consist of two long steel tubes. These tubes are supported by a sturdy, lightweight superstructure made out of slightly larger steel tubes or beams. In tubular steel coasters, the train wheels are typically made from polyurethane or nylon. In addition to the traditional wheels that sit right on top of the steel track, the cars have wheels that run along the bottom of the tube and wheels that run along the sides. This design keeps the car securely anchored to the track, which is absolutely essential when the train runs through the coaster's twists and turns. The train cars in tubular steel coasters may rest on top of the track, like in a traditional wooden coaster, or they may attach to the track at the top of the car, like in a ski-lift. In suspended coasters, the hanging trains swing from a pivoted joint, adding an additional side-to-side motion. In an inverted coaster, the hanging train is rigidly attached to the track, which gives the designer more precise control.

A tubular steel track is prefabricated in large, curved segments. The steel-manufacturing process allows for a smoothly curving track that tilts the coaster train in all directions. On a wooden roller coaster, the ride is punctuated by the rattling sensation of the coaster rolling over the joints that connect the pieces of the wooden track. In a tubular steel coaster, the track pieces are perfectly welded together, making for an incredibly smooth ride. As any coaster enthusiast will tell you, each sensation has its own distinctive charm.

Feeling Loopy

A roller coaster loop-the-loop is a sort of centrifuge. Your own inertia creates a sort of false gravity that points toward the bottom of the car even when you're upside down. You need a safety harness for security, but in most loop-the-loops, you would stay in the car whether you had a harness or not.

As you move around the loop, the net force acting on your body is constantly changing. At the very bottom of the loop, the acceleration force is pushing you down in the same direction as gravity. Since both forces push you in the same direction, you feel especially heavy at this point. As you move straight up the loop, gravity is pulling you into your seat while the acceleration force is pushing you into the floor.

At the top of the loop, when you're completely upside down, gravity is pulling you out of your seat, toward the ground, but the stronger acceleration force is pushing you into your seat, toward the sky. Since the two forces pushing you in opposite directions are nearly equal, your body feels very light. As you come out of the loop and level out, you become heavy again.

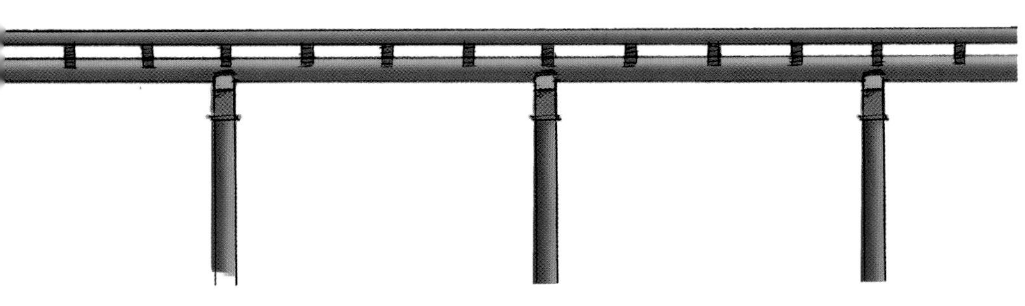

Starting and Stopping

The traditional lifting mechanism is a long length of chain (or chains) running up the hill under the track. The chain is fastened together in a loop, which is wound around a gear at the top of the hill and another one at the bottom of the hill. The gear at the bottom of the hill is turned by a simple motor. This turns the chain loop so that it continually moves up the hill like a long conveyer belt. The coaster cars grip onto the chain with several chain dogs. A chain dog is basically a hinged hook. When the train rolls to the bottom of the hill, the hook catches into the first chain link. Once the chain dog is hooked, the chain simply pulls the train to the top of the hill. At the summit, the chain dog is released and the train starts its descent down the hill.

In some newer coaster designs, the train is set in motion by a catapult launch. There are several sorts of catapult launches, but they all do basically the same thing. Instead of dragging the train up a hill to build up potential energy, these systems start the train off by building up a good amount of kinetic energy in a short amount of time. The most popular catapult systems are linear-induction motors. The main advantages of this system are its speed, efficiency, durability, precision, and controllability. Compressed air is another technique.

Like any train, a roller coaster needs a brake system so that it can stop precisely at the end of the ride or in an emergency. In roller coasters, the brakes aren't built into the train—they're built into the track. This system is very simple. A series of clamps are positioned at the end of the track and at a few other braking points. A central computer operates a hydraulic system that closes these clamps when the train needs to stop. The clamps close in on vertical metal fins running under the train, and this friction gradually slows the train down.

How GLOW-IN-THE-DARK STUFF Works

You see glow-in-the-dark stuff in all kinds of places, but it is most common in toys. There are glow-in-the-dark yo-yos, glow-in-the-dark balls, glow-in-the-dark mobiles, and even (if you can believe it) glow-in-the-dark pajamas!

If you have ever seen glow-in-the-dark products, you know that they all have to be "charged." You hold them up to a light and then take them to a dark place. In the dark they will glow for about 10 minutes. Usually it is a soft green light, and it is not very bright. You need to be in almost complete darkness to notice it.

Phosphors

All glow-in-the-dark products contain phosphors. A phosphor is a substance that radiates visible light after being energized. The two places where you most commonly see phosphors are in a TV screen and in fluorescent lights. In a TV screen, an electron beam strikes the phosphor to energize it. (See "How Television Works," page 190, for details.) In a fluorescent light, ultraviolet light energizes the phosphor. In both cases, you see visible light. A color TV screen actually contains thousands of tiny phosphor picture elements that emit three different colors: red, green, and blue. Fluorescent lights normally contain a mixture of phosphors that together create light that appears white.

Chemists have created thousands of chemical substances that behave like phosphor. Phosphors have three characteristics:

- The type of energy they require to be energized
- The color of the visible light they produce
- The length of time that they glow after being energized (known as the persistence of the phosphor)

To make a glow-in-the-dark toy, a phosphor that is energized by normal light and that has a very long persistence is mixed into plastic and molded. Two phosphors that have these properties are zinc sulfide and strontium aluminate. Strontium aluminate is newer—it's what you see in the "super" glow-in-the-dark toys.

Non-Light-Charged Phosphors

Occasionally you will see something glowing but it does not need to be charged. The most common place is on the hands of expensive watches. In these products, the phosphor is mixed with a radioactive element (such as radium), and the radioactive emissions energize the phosphor continuously. (See "How Nuclear Power Works," page 30, for more details.)

Another glowing object you may have noticed is the face of a certain type of wristwatch. The entire face glows and is called an electroluminescent panel. A very thin panel uses high voltage to energize the phosphor atoms that produce light. You start with a thin glass or plastic layer and coat it with a clear conductor. You coat that with a very thin layer of phosphor, coat the phosphor with a thin plastic, and then add another electrode. Essentially you have two conductors (a capacitor) with phosphor in between. When you apply 100 to 200 volts of alternating current to the conductors, the phosphor energizes and begins emitting photons.

Creating the high voltage can be a problem in a wristwatch. The watch has only a small 1.5-volt battery. To produce 100 to 200 volts, a 1:100 transformer is used. By charging the primary coil of the transformer with a transistor that is switching on and off rapidly, the secondary coil rises to 150 volts or so.

The next time you check the glow-in-the-dark face of your watch in a darkened movie theater or room, you'll understand exactly how you are able to tell the time without turning on the lights.

HSW Web Links 

www.howstuffworks.com

How Quartz Watches Work
How Digital Clocks Work
How Gears Work
How Quartz Watches Work
How Time Works

Electroluminesence

Electroluminescence is the conversion of electricity directly into light. This is not how an incandescent bulb works. In an incandescent bulb the electricity produces heat and the heat produces light. Electroluminescence is much more efficient because it converts the electricity directly to light.

The most common example of electroluninescence that we see on a regular basis is a neon light. In a neon light, high voltage energizes the electrons in neon atoms and, when the electrons de-energize themselves, they emit photons.

How **FIREWORKS** Work

If you have ever been to an aerial fireworks show, then you know that fireworks have a special and beautiful magic. This magic is created by a combination of chemical combustion and chemical explosion, along with an artistic flair for creating incredible patterns of light.

HSW Web Links

www.howstuffworks.com

How Light Works
How Liquid Motion Lamps Work
How Nightvision Works

Just about everyone has seen both sparklers and firecrackers. If you understand these two pyrotechnic devices, then you are well on your way to understanding aerial fireworks!

Bottle Rocket

Sparklers and Firecrackers

At its most basic, an aerial firework is just a bunch of sparklers packaged together with a firecracker and a fuse. The sparkler-like parts produce the bright, sparkling light from a firework and the firecracker-like part creates an explosion that shoots the sparkling lights through the sky.

Firecrackers have been around for hundreds of years. They consist of black powder (also known as gunpowder) in a tight paper tube with a fuse to light the powder.

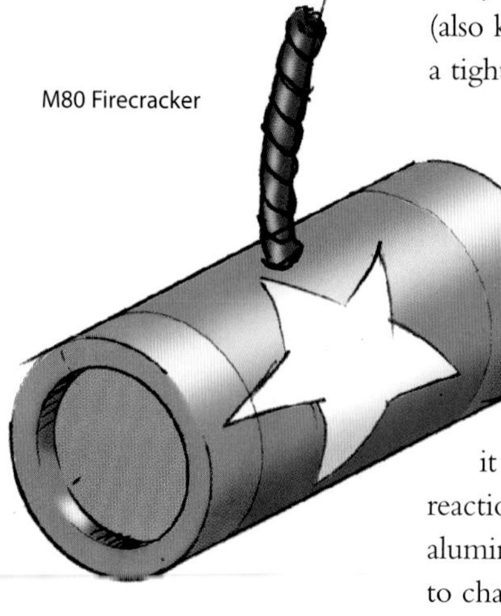

M80 Firecracker

Black powder contains charcoal, sulfur, and potassium nitrate. Charcoal and sulfur are the fuel, and potassium nitrate is the oxidizer—it provides oxygen to speed the reaction. The powder might include aluminum instead of or in addition to charcoal in order to brighten the explosion.

Sparklers are very different from firecrackers. A sparkler burns over a long period of time (up to a minute) and produces extremely bright and showery light. Sparklers are often referred to as snowball sparklers because of the ball of sparks that surrounds the burning portion of the sparkler. A sparkler consists of several compounds:

- A fuel
- An oxidizer
- Iron or steel powder
- A binder

The oxidizer is aluminum perchlorate or barium nitrate. The fuel is charcoal and sulfur, as in black powder. The binder can be sugar or starch. Mixed with water, these chemicals form a slurry that can be coated on a wire (by dipping) or poured into a tube. When the slurry dries, you have a sparkler. When you light it, the sparkler burns from one end to the other, like a cigarette. The fuel and oxidizer are proportioned with the other chemicals so that the sparkler burns slowly rather than exploding like a firecracker.

Aerial Fireworks

A aerial firework is normally formed as a shell that consists of four parts:

- **A container**—The container is usually pasted paper and string formed into a cylinder.
- **Stars**—Stars are spheres, cubes, or cylinders with a sparkler-like composition.
- **A bursting charge**—The bursting charge is a firecracker-like charge at the center of the shell.
- **A fuse**—The fuse provides a time delay so that the shell explodes at the right altitude.

The firework shell is launched from a mortar. The mortar might be a short steel pipe with a lifting charge of black powder that explodes in the pipe to launch the shell. When the bursting charge fires to launch the shell, it lights the shell's fuse. The shell's fuse burns while the shell rises to the correct altitude, and then the fuse ignites the contents of the shell so that it explodes.

A simple shell may consist of a paper tube filled with stars and black powder. Stars come in all shapes and sizes, but you can imagine a simple star as something like sparkler compound formed into a ball the size of a pea or a dime. It is very common for fireworks to also contain aluminum, iron, steel, zinc, or magnesium dust in order to create bright, shimmering sparks. The metal flakes heat up until they are incandescent and shine brightly or, at a high enough temperature, actually burn. A variety of chemicals can be added to create colors. The stars are poured into the tube and then have black powder sifted over them.

When the fuse burns into the shell, it ignites the black powder, causing the shell to explode. The explosion ignites the outside of the stars, which begin to burn with bright showers of sparks. Because the explosion throws the stars in all directions, you get the huge sphere of sparkling light that is familiar at fireworks displays.

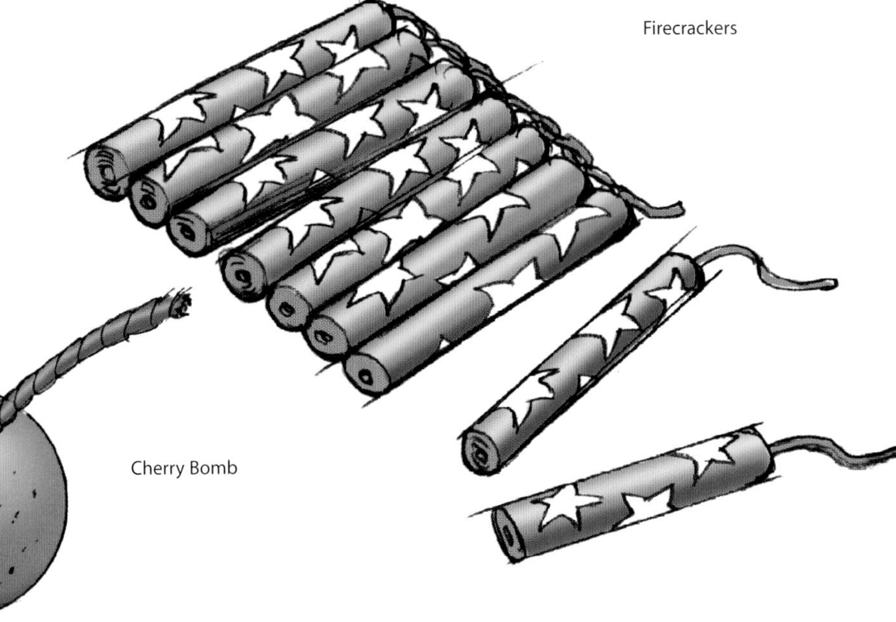

Firecrackers

Cherry Bomb

Sparkler

Aerial Firework Shells

There are several different types of aerial fireworks shells. The descriptions listed here will help you identify them the next time you witness a big fireworks show!

- **Palm**—Contains large comets, or charges in the shape of a solid cylinder, that travel outward, then curve downward after exploding, like the limbs of a palm tree.
- **Round shell**—Explodes in a spherical shape, usually of colored stars.
- **Ring shell**—Explodes to produce a symmetrical ring of stars.
- **Willow**—Contains stars whose high charcoal composition makes them long burning. The stars fall in the shape of willow branches and may even stay visible until they hit the ground.
- **Roundel**—Bursts into a circle of maroon shells that explode in sequence.
- **Chrysanthemum**—Bursts into a spherical pattern of stars that leave a visible trail, with an effect somewhat suggestive of the flower.
- **Pistil**—Similar to a chrysanthemum shell, but has a core in a different color from the outer stars.
- **Serpentine**—Bursts to send small tubes of incendiaries skittering outward in random paths, which may culminate in exploding stars.

Index

C

D

E

I

K

L

M

Q

R

S

W

X

Z